ATOMIC
COLLISIONS IN SOLIDS

Volume 1

ATOMIC COLLISIONS IN SOLIDS

Volume 1

Edited by

Sheldon Datz, B.R. Appleton, and C.D. Moak

Oak Ridge National Laboratory
Oak Ridge, Tennessee

SPRINGER SCIENCE+BUSINESS MEDIA, LLC

Library of Congress Cataloging in Publication Data

International Conference on Atomic Collisions in Solids, 5th, Gatlinburg, Tenn.,
  1973.
  Atomic collisions in solids.

  Includes bibliographical references and index.
  1. Solids—Congresses. 2. Collisions—(Nuclear physics)—Congresses. I. Datz, Sheldon,
ed. II. Appleton, B. R., 1937-   ed. III. Moak, C. D., 1922-   ed. IV.
United States. National Laboratory, Oak Ridge, Tenn. V. Title.
QC176.A1I53 1973   539.7'54   74-26825

ISBN 978-1-4684-3119-3   ISBN 978-1-4684-3117-9 (eBook)
DOI 10.1007/978-1-4684-3117-9

First half of the Proceedings of the Fifth International Conference on Atomic Collisions in Solids held in Gatlinburg, Tennessee, September 24-28, 1973, sponsored by
Oak Ridge National Laboratory

United Kingdom edition published by Plenum Press, London
A Division of Plenum Publishing Company, Ltd.
4a Lower John Street, London, W1R 3PD, England

Perhaps the most controversial aspect of this volume is the number (V) assigned to the conference in this series. Actually, the first conference to be held under the title "Atomic Collisions in Solids" was held at Sussex University in England in 1969 and the second at Gausdal, Norway in 1971, which would logically make the conference held at Gatlinburg, Tennessee, U.S.A. in 1973 the third (III). However, the appearance of the proceedings of the 1971 Gausdal Conference (published by Gordon and Breach) bore the number IV. The reasoning behind this was that, in fact, two previous conferences had been largely dedicated to the same subject area. The first of these was at Aarhus, Denmark in 1965 and the second in 1967 was held in Chalk River, Canada. Hence, the number V for the 1973 meeting.

Actually, the conference can easily be traced back to Paris, France in 1961 when it went under the colorful title of "Le Bombardement Ionique." In 1962 a small conference was held at Oak Ridge, Tennessee, U.S.A. at which the discovery of channeling was first formally annunciated. This was followed by conferences at Chalk River, Canada in 1963 and at Harwell, England in 1964. Moreover, immediately following the Chalk River conference in 1967 there was a conference on higher energy collisions at Brookhaven, New York, U.S.A. Thus, strictly speaking, the Gatlinburg meeting is the tenth (X) in the series.

The accent of the Gatlinburg conference was on primary atomic collision processes, states of ions penetrating solids and their relation to macroscopic effects. Applications of the results to various technologies such as ion implantation and analysis of solid structures have grown large enough to warrant separate conferences.

The effect of the crystal lattice on the motion of penetrating particles (channeling) was first mentioned by M. T. Robinson at the Paris meeting (1961) and the study of these effects grew to the point where the Gausdal 1971 meeting was subtitled "The Physics of Channeling." The phenomenon had been sufficiently well investigated to permit its application to a broad range of problems, principally the location of lattice impurities and defects. At Gatlinburg, papers on channeling were generally restricted to those aspects which related to scattering in solids. Even with this restriction four of the ten sessions were devoted exclusively to channeling papers and the remaining sessions contained many papers which utilized the technique.

The increasingly detailed knowledge of the consequences of inelastic atomic collisions and of charge exchange processes in solids are accented in three sessions. All of these observations contribute to a new level of understanding which permits the treatment of such macroscopic effects as stopping power and radiation damage on a more fundamental basis.

CONTENTS OF VOLUME 1

SECTION V: X RAYS

CONTENTS OF VOLUME 2

SECTION IX: CHANNELING

SECTION X: DECHANNELING

SECTION I
ELECTRONIC STOPPING

STOPPING POWER OF FAST CHANNELED PROTONS IN THE IMPACT PARAMETER TREATMENT OF ATOMIC COLLISIONS

K. DETTMANN

Institut für Festkörperforschung der KFA Jülich
517 Jülich, W. Germany

The energy loss of protons, passing through crystals in a random direction with an energy $E \geq 1$ MeV, is due to electronic excitation and can be explained quantitatively by the Bethe-Bloch formula [1] for the average energy loss per unit path length or the electronic stopping power $S_r = - dE/dx$. In a channeling situation, however, small impact parameters are avoided and the Bethe-Bloch theory no longer applies.

To account for the channel stopping power S_c, Lindhard [2] referred to the equipartition rule for a free electron gas. According to this rule small momentum transfers or plasmon excitation and large momentum transfers or single particle excitation supply equal ammounts to S_r. Assuming single particle excitation to be proportional to the local electron density $n(r) \simeq 0$ in the channel, Lindhard concludes a reduction of S_c by 50%. This was confirmed by experimental data [3] for S_i, but subsequent measurements [4] for Ge supplied reductions by 75%.

We therefore suggest an approach from a tight binding approximation, where the energy loss is calculated for individual inelastic collisions of the protons with single crystal atoms. This is done in the impact parameter treatment of atomic collisions [4], describing each electron by a scaled hydrogenlike groundstate. We calculate the average energy loss $\Delta E(b_m,E)$, due to a collision with the crystal atom m (Fig. 1) in first Born approximation. The average over the channel cross section amounts to an integral over all b and yields the random Bethe-Bloch formula. For fast protons $\Delta E(b)$ decreases only slowly ($\sim 1/b^2$) for b much larger than the dimensions of the electron distribution a_0, and many atoms in a layer perpendicular to the channeled proton trajectory contribute substantially to S_c. Then it is obvious that for very fast protons there is little reduction of S_c as compared with S_r. Therefore equipartition and channeling are not related, though also in the tight binding approximation equipartition holds for small and large momentum transfers. For $b \leq a_0$ the energy loss $\Delta E(b)$ is slowly varying with a finite limiting value $\Delta E(b=0) \sim 1/E$.

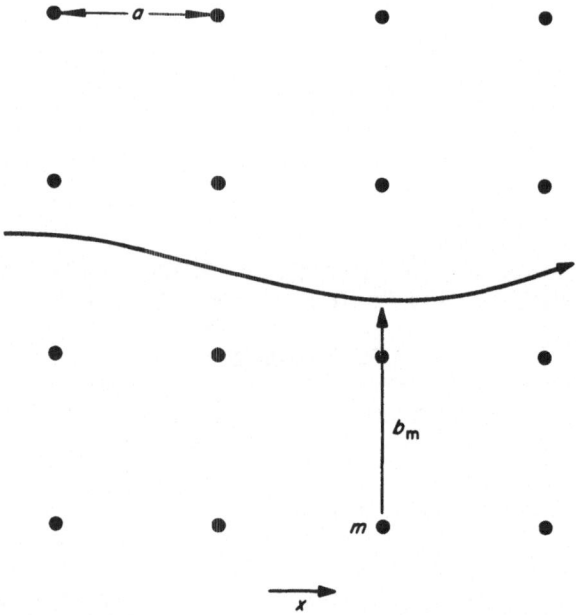

Fig. 1. Channeled proton trajectory (schematic for axial channeling in a simple cubic lattice with lattice constant a).

     We apply our results to the best channeled 4 MeV protons for
axial channeling in Si and Ge. The valence electrons supply the
full random value because the main contribution is due to impact
parameters much larger than the channel radius. For the core electrons to be excited the best channeled protons in <110> direction
are not fast enough. Therefore the different ratios of $S_c^{<110>}/S_r$
for Si and Ge are due to the different electron numbers in the L
and M shell of Si and Ge, which contribute to S_r but not to $S_c^{<110>}$
for the best channeled protons. The results for $S_c^{<110>}/S_r$ and for
$S_c^{<100>}$, $S_c^{<111>}$ are in good agreement with the experimental data
by Clark [5] et al.

Theory

     In the following theory we assume all electrons i of the
crystal to be localized, which amounts to a tight binding approximation. This is well justified for the core electrons. For the
loosely bound valence electrons only the binding energy will enter
into the channel stopping power for MeV protons, so that the
approximation of localized wave functions is not reflected by S_c.
In this model the stopping power of the channeled proton with
energy E in Fig. 1 is

$$S_c = -\frac{dE}{dx} = \frac{1}{a} \sum_m \Delta E(b_m, v) \qquad (1)$$

where $\Delta E(b_m, v)$ is the average energy loss due to the collision with atom m, when the proton of velocity $\underline{v}$ passes with impact parameter $\underline{b}_m$ (Fig. 1). The sum extends over all lattice sites m in a layer perpendicular to $\underline{v}$, and a is the layer distance. The random stopping power S_r results from (1) by averaging S_c over the channel cross section F:

$$S_r = \frac{1}{F} \int_F df S_c = \frac{1}{Fa} \sum_m \int_F d^2\underline{b}_m \, \Delta E(\underline{b}_m, v)$$

$$= \rho \int_\infty d^2\underline{b} \Delta E(b, v) \qquad (2)$$

The average energy loss ΔE can be written

$$\Delta E(b, v) = \hbar \sum_{n>o} \omega_n P_n(\underline{b}, v) \qquad (3)$$

where $P_n(b, v)$ is the probability to excite an atom into the state $|n>$ with energy $\hbar\omega_n$ above the ground state $|0>$. With (3) Eq. (2) can be cast into the common form

$$S_r = \rho \sum_{n>o} \hbar\omega_n \sigma_n(v) \qquad (2a)$$

where $\sigma_n = \int P_n(\underline{b}, v) d^2\underline{b}$ is the cross section to excite a crystal atom into $|n>$.

We now calculate $P_n(\underline{b}, v)$ in the impact parameter treatment for inelastic atomic collisions (Fig. 2). The proton p moves with impact parameter $\underline{b}$ and velocity $\underline{v}$ on a straight line trajectory $\underline{R}(t) = \underline{b} + \underline{v}t$. The electron at $\underline{r}$ with mass m, binding energy ε_o and spatial extension a_o can be either a valence electron (centered half way between two lattice sites in Si and Ge) or a core electron. The concept of a classical trajectory works for $\lambda \ll a_o$, where λ is the de Broglie wavelength of the proton. This is equivalent with $E \gg \frac{m}{M}\varepsilon_o$, where M is the proton mass so that $E \gtrsim 1$ keV would be sufficient. But we also assume a straight line trajectory, which means $\theta \simeq V_{pn}(a_o)/E \simeq \varepsilon_o/E \ll 1$. Here θ is the scattering angle for the typical distance a_o. For the K shell of Si or the M shell of Ge, for instance, $\varepsilon_o \simeq 1-2$ keV, and the scattering angle is small for $E \gtrsim 10$ keV. So we see, that even for the low energy region $E \lesssim 100$ keV the impact parameter treatment holds quantitatively.

The Schrödinger equation for the wave function $\psi(\underline{r}, t)$ of the system in Fig. 2 is

$$H\psi = i\hbar\dot{\psi}, \quad H = H_o + V(\underline{r} - \underline{R}(t)) \qquad (4)$$

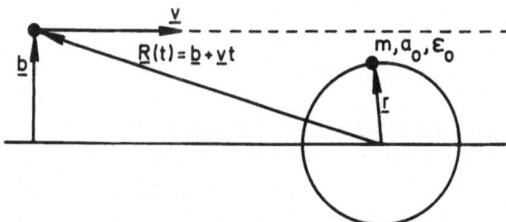

Fig. 2. Semiclassical collision of the proton with a single crystal atom.

with the unperturbed Hamiltonian H_0 for the atom and the Coulomb interaction $V = -e^2/|\underline{r}-\underline{b}-\underline{v}t|$ between the proton and the electron under consideration. The initial condition is $\psi(t \to -\infty) = |0>$. After some operator algebra [6] one obtains

$$P_n(\underline{b},v) = |fn(\underline{b},v)|^2$$

$$f_n = -\frac{i}{\hbar} \int_{-\infty}^{\infty} dt\; e^{i\omega_n t} <n|V(t)|\psi(t)> \tag{5}$$

For fast protons we have $v \gg v_0$, with the orbit velocity v_0 of the electron, and we can in first Born approximation replace $|\psi>$ by the unperturbed ground state $|0>$:

$$f_n \simeq -\frac{i}{\hbar} <n|\tilde{V}(\omega_n,\underline{b},v)|0> \tag{6}$$

with

$$\tilde{V}(\omega_n) = \int dt\; e^{i\omega_n t} V(\underline{r}-\underline{R}(t))$$

We now treat each electron with $v_0 \ll v$ in a scaled hydrogen model and neglect the innermost shells with $v_0 \gtrsim v$, since they contain fewer and more tightly bound electrons. The scaling is achieved by

$$\varepsilon_0 = \frac{1}{2} mv_0^2 = \frac{\hbar^2}{2ma_0^2} \;,\quad <\underline{r}|0> = \frac{1}{\sqrt{\pi a_0^3}} e^{-r/a_0} \tag{7}$$

where ε_0 for each subshell is determined by the ionization potential or the energy levels from x-ray data.

With (2a), (5) and (6) we obtain for the random case

$$S_r = \rho \sum_{n>o} \hbar\omega_n \int d^2\underline{b} \, \frac{1}{\hbar^2} \, |<n|\tilde{V}(\omega_n)|0>|^2 \qquad (8)$$

Fourier transforming $\tilde{V}(\omega_n)$ with respect to $\underline{r}$ the sum over n in (8) for $v>>v_o$ can be calculated [7] conveniently by the closure relation. The result is

$$S_r(v>>v_o) = \frac{4\pi e^4}{mv^2} \, \rho \, \ell n \, \frac{2mv^2}{\varepsilon_o} \qquad (9)$$

This is the well known Bethe-Bloch formula [1], rederived here in the semiclassical impact parameter treatment.

In the Fourier transformed form the contributions $\hbar\omega_n \stackrel{>}{<} \varepsilon_o$ in (8) correspond to momentum transfers $k \stackrel{>}{<} 1/a_0$ and can be calculated separately. It turns out that large momentum transfers ($k > 1/a_0$) or ionization ($\hbar\omega_n > \varepsilon_o$) and small momentum transfers ($k < 1/a_0$) or excitation ($\hbar\omega_n < \varepsilon_o$) both contribute equal amounts to S_r. This equipartition rule also holds for the energy loss of fast protons in a free electron gas [8], and corresponds there to single particle excitation and plasmon excitation.

Along the same lines as S_r in (8) one can also calculate the average excitation energy

$$\hbar\bar{\omega}(v>>v_o) = \frac{\sum\limits_{n>o} \hbar\omega_n \sigma_n}{\sum\limits_{n>o} \sigma_n} = 2\varepsilon_o \qquad (10)$$

Eq. (10) shows that fast protons, exciting bound electrons, supply on the average twice the binding energy.

For the channel stopping power in (1) we calculate $\Delta E(b,v)$ with (3), (5) and (6):

$$\Delta E(b,v) = \frac{1}{\hbar} \sum_{n>o} \omega_n |<n|\tilde{V}(\omega_n)|0>|^2 \qquad (11)$$

The sum in (11) can be calculated by the closure relation, if we replace ω_n approximately by $\bar{\omega}$ of (10). The result is

$$\Delta E(b,v) \simeq \Delta E_{\bar{\omega}} = \frac{\bar{\omega}}{\hbar} \{<0||\tilde{V}(\bar{\omega})|^2|0> - |<0|\tilde{V}(\bar{\omega})|0>|^2\} \qquad (12)$$

Integrating $E_{\bar{\omega}}$ over all b according to (2) the exact random stopping power S_r in (9) results for $v>>v_o$. This shows that the $\bar{\omega}$ approximation is self-consistent in the average over all b. The matrix elements in (12) can be calculated [7] analytically for $v \gg v_0$ and $b \gg a_0$, $b \ll a_0$. We obtain

$$\Delta E(b, v \gg v_0) \simeq \begin{cases} 8\epsilon_0 (\frac{v_0}{v})^4 (\frac{v_B}{v_0})^2 [K_1^2(\frac{b}{b_c}) + K_0^2(\frac{b}{b_c})]; & b \gg a_0 \\[3mm] 2\epsilon_0 (\frac{v_B}{v})^2 (\frac{\pi}{3} - 1)^3 & b \ll a_0 \end{cases}$$

$$(13)$$

with $v_B = e^2/\hbar$, $b_c = \frac{v}{v_0} a_0 = \frac{v}{2\omega_0}$.

For $a_0 \ll b \ll b_c$ Eq. (13) yields $\Delta E \sim 1/b^2$, which is a slow decrease of $\Delta E(b)$ for large impact parameters up to b_c. In fact the main contribution to S_r in (2) is supplied by $a_0 \ll b \lesssim b_c$ because of the weight $2\pi b$ in the b-integration leading to the log-term in (9). The limiting impact parameter b_c is an average adiabacity criterion, which is only half the value proposed by Erginsoy [4] from qualitative arguments. Our b_c has the intuitive meaning, that protons passing an atom with this impact parameter have the same angular velocity as the electron in the orbit. The interesting feature is the long range of $\Delta E(b)$ for fast protons ($v \gg v_0 \to b_c \gg a_0$), which is due to the long range Coulomb interaction inducing a dipole moment into the atom. From this we see that for sufficiently fast protons b_c becomes much larger than any channel radius even for the core electrons. Then in (1) these large impact parameters supply a large contribution and there is little reduction of S_c compared with S_r leading to the conclusion*

$$R(v \gg v_0) = S_c/S_r(v \gg v_0) \simeq 1 . \qquad (14)$$

The large b_c values in (13) show that equipartition in momentum transfer $k \gtrsim 1/a_0$ is not related to a fixed b in the order of a channel radius. Instead the b-value corresponding to $k = 1/a_0$ is energy-dependent and can become larger than the channel radius.

For $b < a_0$ Eq. (13) shows the finite limiting value $\Delta E(b=0, v \gg v_0)$ decreasing as $1/v^2$, while Figs. 3, 4 show the complete b-dependence computed by numerical integration of the matrix elements (12) [7]. One clearly recognizes in Fig. 3 the increasing range of $\Delta E(b)$ with velocity. Fig. 4 shows that for fixed b a small and a large velocity provide the same ΔE.

---

*The long range of $\Delta E(b)$ is enhanced by the $\bar{\omega}$-approximation in (12), so that a more accurate calculation might reduce $R(v \gg v_0)$ in (14).

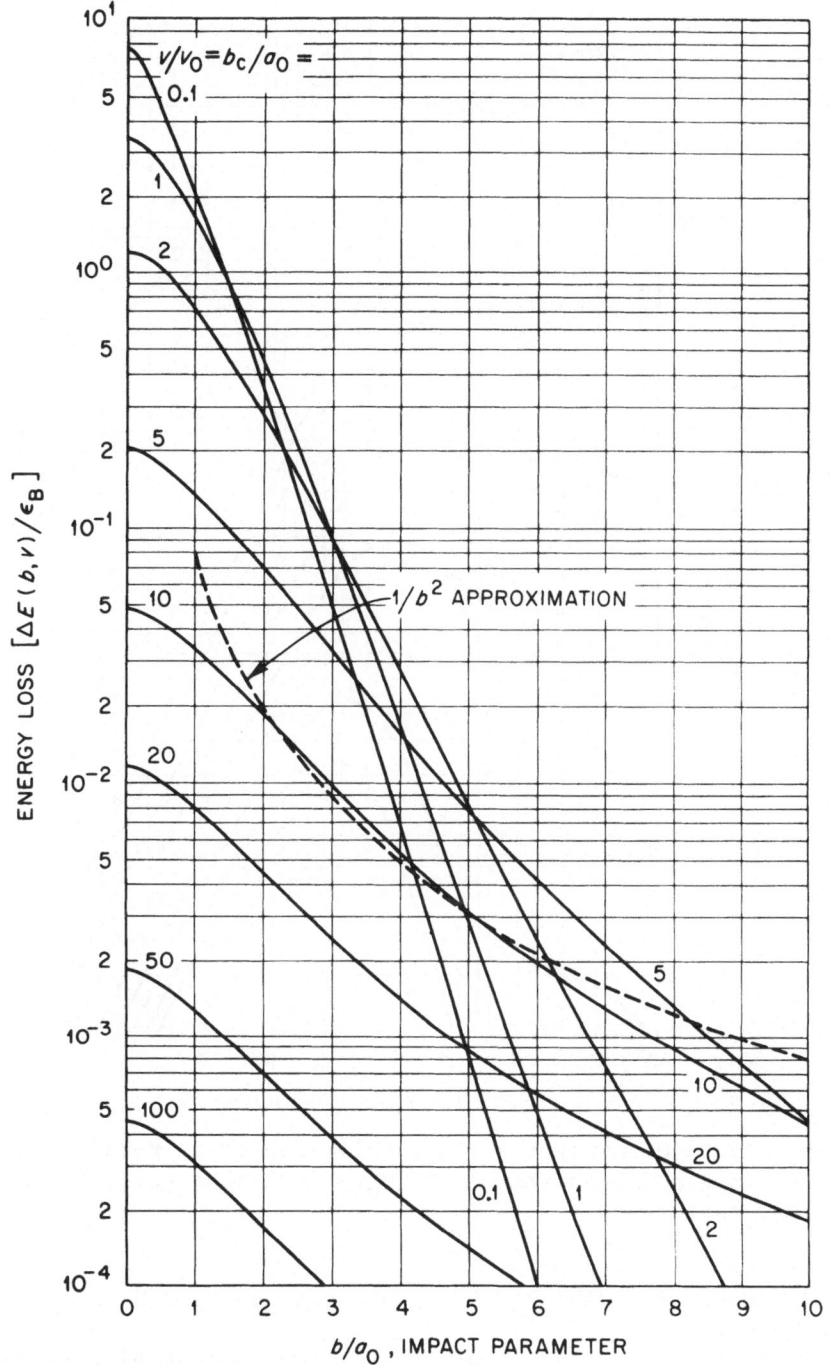

Fig. 3. Average energy loss in a collision of a proton with an electron in a hydrogenlike orbit ($v_0 = v_B$ = Bohr velocity, $a_0 = a_B$ = Bohr radius, ε_0 = 13.6 eV). The dashed curve shows the asymptotic $1/b^2$ dependence for the case $v/v_0 = 10$.

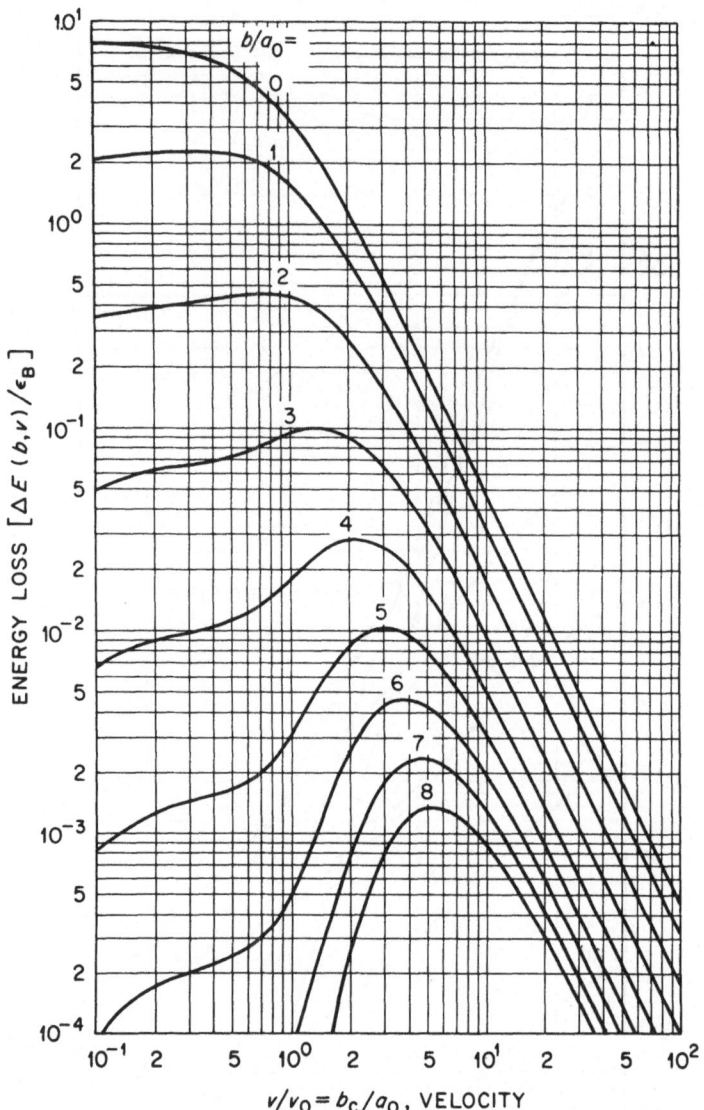

Fig. 4. Average energy loss in a collision of a proton with an electron in a hydrogenlike orbit ($v_0 = v_B$ = Bohr velocity, $a_0 = a_B$ = Bohr radius, ε_0 = 13.6 eV).

Application to Stopping Power Measurements in Si and Ge

In this section the theory of the previous section is applied to the stopping power measurements of Clark et al. [5] for 4 MeV protons in different axial channels of Si and Ge (Fig. 5). In either case the lattice type is diamond structure with nearly the

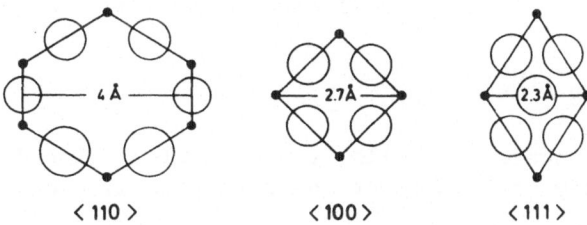

⟨110⟩   ⟨100⟩   ⟨111⟩

Fig. 5.   Cross sections of axial channels in Si and Ge.

same lattice constant of about 5.5 Å, and with four outer valence electrons supplied by each lattice atom. We confine ourselves to the best channeled particles, which move along the channel center and can be selected experimentally by the leading edge of the channeled energy loss spectra [5]. In a first approximation we neglect oscillations of the protons around the channel axis, so that the impact parameter for the atomic collisions does not change with penetration depth.

Valence electrons

The valence electrons are described by hydrogenic ground states centered between two next neighbour lattice sites with a binding energy $\varepsilon_v \simeq 8$ eV corresponding to the half width of the valence band plus the band gap. With (7) and (13) we obtain for 4 MeV

$$a_v \simeq 0.7 \text{ Å}, \frac{v}{v_v} \simeq 16.4, \, b_v \simeq 11.3 \text{ Å} .$$

Since the maximum channel radius is $r_{<110>} \simeq 2$ Å $<< b_v$, many terms in (1) contribute to S_c^v, so that the sum can be replaced by an integral leading to the random value S_r^v according to (2) for any channel. With the atomic density $\rho \simeq 5 \cdot 10^{-2}(4.4 \cdot 10^{-2})/\text{Å}^3$ for Si(Ge) (9) yields with 4 valence electrons and $\varepsilon_v \simeq 8$ eV

$$4S_r^v = 0.84(0.74) \text{eV/Å}$$

Core electrons

The next inner shell contains 8L electrons in Si and 18M electrons in Ge with a binding energy of 50-150 eV. For 100 eV we obtain from (7) and (13) for 4 MeV

$$a_{L(M)} \simeq 0.2 \text{ Å}, \frac{v}{v_{L(M)}} \simeq 4.7, \, b_{L(M)} \simeq 0.9 \text{ Å} .$$

The random value for the core electrons results from (9) with $\varepsilon_{L(M)} \simeq 100$ eV

$$8(18)S_r^{L(M)} = 1(2.1) \text{ eV/Å} .$$

We neglect the innermost shells, the K shell in Si and the L shell in Ge, since they contain fewer and more tightly bound electrons.

<110> Channel

The best channeled particles in <110>-direction have too large impact parameters, $b \simeq r_{<110>} \simeq 2$Å, to excite the core electrons with $b_{L(M)} \simeq 0.9$ Å for 4 MeV. Therefore, the energy loss is due to the valence electrons alone supplying the random value

$$S_c^{<110>} \simeq 4S_r^V = 0.84(0.74)\text{eV/Å} \tag{15}$$

in reasonable agreement with the experimental value 0.68 (0.70) eV/Å for Si (Ge). The ratio R is

$$R_{<110>}(4\,\text{MeV}) = \frac{S_c^{<110>}}{S_r} = \frac{4S_r^V}{4S_r^V + 8(18)S_r^{L(M)}} = 0.46(0.26) \tag{16}$$

in excellent agreement with the experimental values 0.43(0.28). The difference in $R_{<110>}$ for Si and Ge is due to the different electron numbers in the L shell of Si and the M shell of Ge and has nothing to do with the equipartition rule.

<100> and <111> channels

For the best channeled particles in <100> direction with $b \simeq r_{<100>} \simeq 1.35$ Å (Fig. 5) the core contribution $(\Delta S)_{<100>}$, additional to $4S_r^V$, can be calculated taking into account only the four rows of atoms bordering the <100> channel. This is due to the small $b_{L(M)} \simeq 0.9$ Å. We obtain with 0.73 atoms/Å and 8L electrons for Si with (13)

$$(\Delta S)_{<100>} = 0.73 \cdot 8 \cdot \Delta E(1.35 \text{ Å}, 4 \text{ MeV}) = 0.16 \text{ eV/Å} \tag{17}$$

while the experimental value [5] is 0.17 eV/Å.

In <111> direction it is sufficient to take into account only the two nearest rows in the smaller channel dimension, so that $b = 1.1$ Å (Fig. 5). Eq. (13) yields with 0.42 atoms/Å for Si

$$(\Delta S)_{<111>} = 0.42 \cdot 8 \cdot \Delta E(1.1 \text{ Å}, 4 \text{ MeV}) = 0.2 \text{ eV/Å} \tag{18}$$

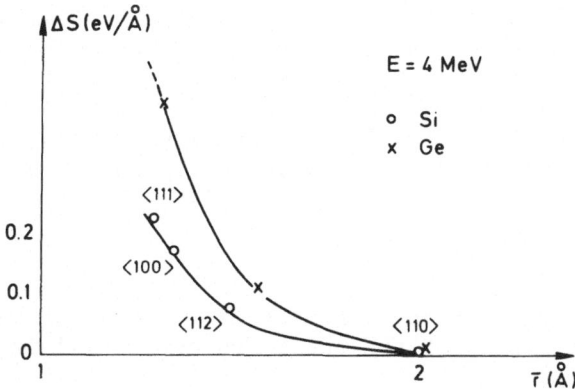

Fig. 6. Stopping power of 4 MeV channeled protons along the major axes in Si and Ge; ──── theory, o, x experimental data [5].

again in very good agreement with the experimental value 0.22 eV/Å. Fig. 6 shows the good agreement of the calculated and experimental ΔS values. The Ge-data scale according to $(\Delta S)^{Ge} = 18/8 (\Delta S)^{Si}$ corresponding to the 18M electrons in Ge with a similar binding energy as the 8L electrons in Si.

Conclusions

     The theory described above, based on the impact parameter treatment of single inelastic atomic collisions, yields excellent agreement for the core contribution to S_c (16), (17), (18) and reasonable agreement for the valence electrons (15). The localized, single electron model is a tight binding approximation, which can account only qualitatively for the nonlocalized crystal wave functions of the valence electrons. However, the valence electrons enter into the channel stopping power for MeV protons only by their random value, which is rather insensitive with respect to a tight binding approximation. Treating the valence electrons in a free electron approximation [8] the binding energy $\varepsilon_o \simeq 8$ eV in (9) is replaced by the plasmon energy $\hbar\omega_p \simeq 16$ eV, which is a small change for fast protons with $2mv^2 >> \varepsilon_o, \hbar\omega_p$. Either approximation supplies equipartition for small and large momentum transfer, which corresponds to excitation of plasmons and single particles in the free electron approximation and to excitation and ionization in the tight binding approximation.
     The impact parameter treatment of the two preceding sections which is necessary to describe a channeling situation shows that there is no energy independent correspondence for equipartition in momentum transfer and impact parameter. Instead, the impact parameter $b_c'(v)$ with equal energy loss for $b \gtrless b_c'(v)$ increases with

proton velocity v and can become larger than the channel radius at least for the valence electrons. A promising experiment would be the channeling of 160 MeV protons in Si, where the L-shell energy is $\varepsilon_L \simeq 100$ eV and the corresponding critical impact parameter $b_L \simeq 5$ Å. This is 4.3 $r_{<100>}$, so that the L shell of the Si atoms should be excited for <100> channeling with an efficiency comparable to random conditions, which would show the trend $R \to 1$.

Acknowledgments

I should like to thank G. Dearnaley for many discussions concerning the application of the theory to the Si and Ge data.

[1] H. Bethe, Ann. Phys. 5, 325 (1930).
[2] J. Lindhard, Kgl. Danske Videnskab. Selskab., Mat.-Fys. Medd. 34, No. 14 (1965).
[3] C. Erginsoy, H. E. Wegner, and W. M. Gibson, Phys. Rev. Letters 13, 530 (1964).
[4] A. R. Sattler and G. Dearnaley, Phys. Rev. Letters 15, 59 (1965); B. R. Appleton, C. Erginsoy, and M. W. Gibson, Phys. Rev. 161, 330 (1967).
[5] G. J. Clark, D. V. Morgan and J. M. Poate in "Atomic Collision Phenomena in Solids, Eds. D. W. Palmer, M. W. Thompson and P. D. Townsend (North Holland, Amsterdam, 1970), p. 388.
[6] K. Dettmann, Springer Tracts in Modern Physics, 58, 119 (1971).
[7] K. Dettmann and M. T. Robinson, Phys. Rev., to be published.
[8] J. Lindhard and A. Winther, Kgl. Danske Videnskab. Selskab., Mat. Fys. Medd. 34, No. 4 (1964).

THE Z_1 OSCILLATIONS IN ELECTRONIC STOPPING

J. S. BRIGGS and A. P. PATHAK
Theoretical Physics Division
A.E.R.E. Harwell, Didcot, Berkshire
England

ABSTRACT

An expression for the energy loss cross-section of a particle of mass m_1 and velocity $\underline{v}_1$ colliding with particles of mass m_2 having a distribution of velocities $\underline{v}_2$ is obtained. The energy loss cross-section is proportional to the momentum transfer cross-section which depends upon the elastic scattering intensity in the target-projectile center-of-mass system. When applied to the scattering of electrons by incident atoms moving at velocities $v_1 \sim v_2$ the momentum transfer cross-section shows strong oscillations with Z_1 in the same regions as those observed in the electronic stopping power. These oscillations are connected with the shell structure of the periodic table by an analysis of the phase shift induced in the electron wavefunctions of different angular momentum. By considering the Coulomb scattering from a bare incident particle, the connection with other theories of stopping, when $v_1 \ll v_2$ and $v_1 \gg v_2$ respectively, is established.

Introduction

The electronic contribution to the stopping power of solids for low-velocity heavy ions exhibits a periodic dependence on the charge Z_1 of the incident ion [1]. This variation of stopping power is enhanced under channeling conditions. In previous calculations [2], the Z_1 oscillations have been reproduced most successfully by modifying the stopping power theory of Firsov [3]. In this theory, momentum is transferred between colliding atoms by an electron flux which is taken to be a function of the electron density in a plane mid-way between the atoms. This density is dependent upon the shell structure and hence the Z_1 oscillations appear principally as an atomic size effect. Each electron which takes part in the interaction is considered to transfer a momentum mv where v is the velocity of the incident nucleus relative to the target and m is the electron mass. The major shortcoming of this theory is an under-estimate of the magnitude of the oscillations.

Lindhard and Finneman have suggested an interpretation which considers that the oscillations in stopping power arise from the variation in scattering cross-section of target electrons near the Fermi level from the field of the incident ion. Quantum-mechanically, the variation in scattering cross-section arises from a variation in the phase shift induced in the electron wavefunction by the presence of the field of the moving ion. In this model, the stop-ping power has been shown to be proportional to the electron momen-tum transfer cross-section [4]. The variation of this cross-section for low relative velocities of electron and ion does indeed exhibit a periodicity with Z_1 which is very similar to that shown by the stopping power. In particular, deep minima occur, i.e., the inci-dent ion appears "transparent" to the target electrons. In this paper, the model will be developed further from a semi-classical approach. That is, the collision between the heavy ion and a tar-get electron will be treated classically but the quantum-mechanical expression will be used to calculate the elastic scattering ampli-tude as a function of scattering angle. In the section, Energy-loss Due to Collision with Electrons, the expression for stopping power is derived and in the section, Calculations of Q_d, the relation of the variation of stopping power to the shell structure of the atom is exposed. The section, Connection with Other Theory, contains discussion of the relationship of the theory given here to other theories of heavy-ion stopping.

Energy-loss Due to Collision with Electrons

In this section we will consider the energy lost when an atom, initially moving with a velocity $\underline{v}_1$ is in collision with a free electron, initially moving with a velocity $\underline{v}_2$. The electron-atom collision will be considered to be elastic. Gerjuoy [5] has con-sidered the collision of two moving particles of masses m_1 and m_2 and shown that the energy gain by particle 2 as seen in the labora-tory system is

$$T_2 = \mu \underline{V} \cdot (\underline{v} - \underline{v}'), \tag{1}$$

where $\mu = m_1 m_2 / (m_1 + m_2)$ is the reduced mass, $\underline{V} = \dfrac{(m_1 \underline{v}_1 + m_2 \underline{v}_2)}{m_1 + m_2}$

is the center-of-mass velocity in the laboratory fram and $\underline{v}$, $\underline{v}'$ are the relative velocities before and after the collision respectively. Taking $\bar{\theta}$ and $\bar{\theta}'$ as the angles between the direction of $\underline{V}$ and the relative velocity vectors $\underline{v}$, $\underline{v}'$ respectively we have

$$T_2 = \mu v V (\cos\bar{\theta} - \cos\bar{\theta}') \tag{2}$$

since $v = |\underline{v}| = |\underline{v}'|$.

For given initial velocities $\underline{v}_1, \underline{v}_2$ the vectors $\underline{V}$, $\underline{v}$ are fixed and the scattering in the center-of mass system is described by a scattering intensity $I_0(\chi; \underline{v}_1, \underline{v}_2)$ where χ is the angle between $\underline{v}$ and $\underline{v}'$. The total cross-section for energy transfer Q_T is obtained by integrating over all final directions of $\underline{v}'$

$$Q_T(\underline{v}_1, \underline{v}_2) = \mu v V \int d\Omega' \ (\cos\bar{\theta} - \cos\bar{\theta}') \ I_0(\chi; \underline{v}_1, \underline{v}_2), \tag{3}$$

where $\bar{\Omega}'$ is an abbreviation for the angles $\bar{\theta}$, $\bar{\phi}'$. The elastic scattering amplitude $f(\chi, \underline{v}_1, \underline{v}_2)$, is expanded in Legendre polynomials of angular momentum ℓ

$$f(\chi) = \sum_\ell A_\ell P_\ell \ (\cos\chi) \tag{4}$$

where

$$A_\ell = \frac{2\ell + 1}{2 \ i \ k} \ [\exp \ (2i\eta_\ell) - 1 \ ] \tag{5}$$

with $k = \mu v$ and η_ℓ the phase shift of the ℓth partial wave. The intensity I_0 is equal to $|f|^2$. The cross-section Q_T of eq. (3) now becomes

$$Q_T(\underline{v}_1, \underline{v}_2) = \mu v V \sum_{\ell\ell'} A_\ell A_{\ell'}^* \int d \ \bar{\Omega}' (\cos\bar{\theta} - \cos\bar{\theta}') P_\ell (\cos\chi) P_{\ell'} (\cos\chi)$$

$$\tag{6}$$

Since χ is the direction between the angles $\bar{\Omega}$ and $\bar{\Omega}'$, the Legendre polynomials may be expanded

$$P_\ell (\cos\chi) = \frac{4\pi}{(2\ell+1)} \sum_m Y_{\ell m}^* \ (\bar{\Omega}) \ Y_{\ell m}(\bar{\Omega}') \tag{7}$$

and the integral over $\bar{\phi}'$ and the sum over m, m' performed to give

$$Q_T(\underline{v}_1, \underline{v}_2) = \mu v V \cos\bar{\theta} \{ \sum_\ell \frac{|A_\ell|^2}{(2\ell+1)}$$

$$- \sum_{\ell\ell'} A_\ell A_{\ell'}^* \int_{-1}^{1} P_1(x) P_\ell(x) P_{\ell'}(x) dx \} \qquad (8)$$

The factor in curly brackets can be shown, with the use of (5) to reduce to

$$\frac{4\pi}{k^2} \sum_\ell (\ell+1) \sin^2(\eta_\ell - \eta_{\ell+1}) \qquad (9)$$

which is known as the momentum transfer cross-section $Q_d(\underline{v}_1, \underline{v}_2)$. Hence,

$$Q_T(\underline{v}_1, \underline{v}_2) = \mu v V \cos\bar{\theta} \; Q_d \; (\underline{v}_1, \underline{v}_2), \qquad (10)$$

where the scalar product $\underline{v} \cdot \underline{V}$ is

$$v V \cos\bar{\theta} = (\underline{v}_1 - \underline{v}_2) \cdot (m_1\underline{v}_1 + m_2\underline{v}_2)/(m_1 + m_2), \qquad (11)$$

Equation (10) is exact for given initial velocities v_1 and v_2. Considerable simplification ensues when we take advantage of the conditions pertinent to our system of a target electron being scattered by a moving ion of several hundred keV. Here $m_1 \gg m_2$ and $v_1 \approx v_2$ so that (eq. 10) can be approximated as

$$Q_T(\underline{v}_1, \underline{v}_2) = m_2 v_1 \; (v_1 - v_2 \cos\theta) \; Q_d(\underline{v}_1 \cdot \underline{v}_2) \qquad (12)$$

with electron momentum

$$k \approx m_2 [v_1^2 + v_2^2 - 2v_1 v_2 \cos\theta]^{1/2}. \qquad (13)$$

The angle θ is the angle between $\underline{v}_1$ and $\underline{v}_2$. In a channeling experiment, $\underline{v}_1$ can be taken to be fixed along the channel direction and an average taken over the projection $v_2\cos\theta$ of the target electron velocity along this direction. Clearly, if $v_2 < v_1$ then all values of θ are possible, but if $v_2 > v_1$ there is no collision for values of θ such that $\cos^{-1}(v_1/v_2) > \theta$. For $v_2 < v_1$ both $(v_1 - v_2\cos\theta)$ and v span the region $(v_1 - v_2)$ to $(v_1 + v_2)$ but for $v_2 > v_1$, $(v_1 - v_2\cos\theta)$ has a lower limit of zero whilst v has a lower limit of $(v_2^2 - v_1^2)^{1/2}$.

The average energy transferred per collision is Q_T/Q_o where Q_o is the total elastic scattering cross-section and since there are nQ_o collisions per unit length when there is a density of n target electrons per unit volume, the average stopping power is given by

$$-\frac{dE}{dx} = m <\underline{v}_1 \quad (\underline{v}_1 - \underline{v}_2) \; n(\underline{v}_2) \; Q_d(\underline{v}_1,\underline{v}_2)> \; , \qquad (14)$$

where m is the electron mass and the brackets denote an average over the directions of $\underline{v}_2$. The expression (14) is the generalization of that given previously [4], where the target electrons were considered to be stationary. However, eq. (14) is still only strictly applicable to a gas of free electrons with a fixed speed v_2.

For small atoms or ions moving through wide channels in metals, eq. (14) applied to electrons with the Fermi velocity, will account for most of the electronic stopping. However, in insulators and semiconductors, electrons which are essentially bound to a target atom (in narrow bands) will make a major contribution.

Calculation of Q_d

The momentum transfer cross-section Q_d will be calculated in the approximation that the scattering intensity $I_o(\chi)$ is decided by the phase shift induced by the static potential of the incident particle, which will be taken as neutral for this purpose. Then the phase shift η_ℓ of the ℓth partial wave of an electron of momentum $\underline{k}$ is obtained from the asymptotic solution [6]

$$G(r) \sim \sin (kr - \frac{1}{2}\ell\pi + \eta_\ell), \; r \to \infty$$

of the radial equation

$$\frac{d^2G}{dr^2} + [k^2 + U(r) - \frac{\ell(\ell + 1)}{r^2} \;] \; G = 0 \qquad (15)$$

and U(r) is the atomic potential. Phase shifts for $\ell = 0,1,2,3$ for a range of k values $0 \leq k \leq 2.0$ and for all $Z_1 \leq 54$ have been calculated by a direct numerical solution of eq. (15) and the phase shifts evaluated by the procedure of Burgess [7]. The Thomas-Fermi potential was used for U(r) and the momentum transfer cross-section as a function of Z_1 is shown in Fig. 1 for k = 0.75. This cross-section shows striking oscillations as a function of Z_1, the extrema occurring very close to the positions of extrema observed in the electronic stopping power. Similar oscillations in the total elastic scattering cross-section from a simple representation of the atomic potential have been demonstrated long ago by Morse [8]. These oscillations depend upon the increase of certain phase shifts by π, which increase can be connected with the shell structure of the periodic table.

For a given ℓ value, the effective atomic potential is $-U(r) + \ell(\ell+1)/r^2$. If this quantity is positive for all r then the atom cannot support a bound state. For fixed ℓ, as Z increases, a value of Z is reached where the total potential becomes negative and capable of supporting a negative energy state. As Landau and Lifshitz [9] have shown, even a simple potential such as our Thomas-Fermi potential correctly predicts the Z regions where s, p, d, etc. states first appear in the periodic table. Moreover, from Levinson's theorem, the establishment of a bound state of given angular momentum ℓ is accompanied by an increase of π in the phase η_ℓ of the wavefunction of zero-velocity unbound electrons. Indeed, solution of eq. (15) for k = 0 confirmed increases in the phase shifts of π in the $\ell = 1, 2, 3$ wavefunctions at approximately the correct Z values. Vestiges of these sudden increases by π persist at the low velocities considered here. As Manson [10] has shown, the step function rise at a given Z is smeared by the finite velocity and our calculation indicates that the increase in π takes place roughly over the Z region where the appropriate subshell is filled. This provides the connection between shell structure and the oscillatory behaviour of Q_d.

Fig. 1 shows the separate contributions of the partial cross-sections of Eq. (9). The s and p phase shifts generally rise together and hence the $\ell = 0$ partial cross-section is small. The $\ell = 1$ partial cross-section is responsible for the first peak in Q_d, since the p-wave phase shift goes through π correspondingly to filling of the 2p shell whilst the d-wave phase shift is zero. Similarly, the second and third peaks arise from the $\ell = 2$ partial cross-section as the d-wave phase shift successively increases by π as the 3d and 4d shells fill whilst the f-wave phase shift is zero. The f-wave phase shift begins to rise around $Z_1 = 50$ and presumably another peak occurs before $Z_1 = 71$ corresponding to the filling of the 4f shell.

The smallness of Q_d i.e. the "transparency" of the atomic core to incident low-energy electrons occurs in the regions $Z_1 = 11, 29$ and 47 and a similar phenomenon is the reason for the high electrical conductivity of these elements. This formulation of stopping power is strictly valid for a gas of free electrons and hence is most applicable to the stopping by metals. However, for heavy ions it is clear that outer bound electrons of the target will also contribute to the stopping, since there will be overlap between them and the incident ion. The inclusion of a correct electron density $n(\underline{v}_2)$ is the major source of uncertainty in the application of eq. (14). The problem of stopping by more tightly bound electrons does not arise in the case of well-channeled particles where close collisions do not occur.

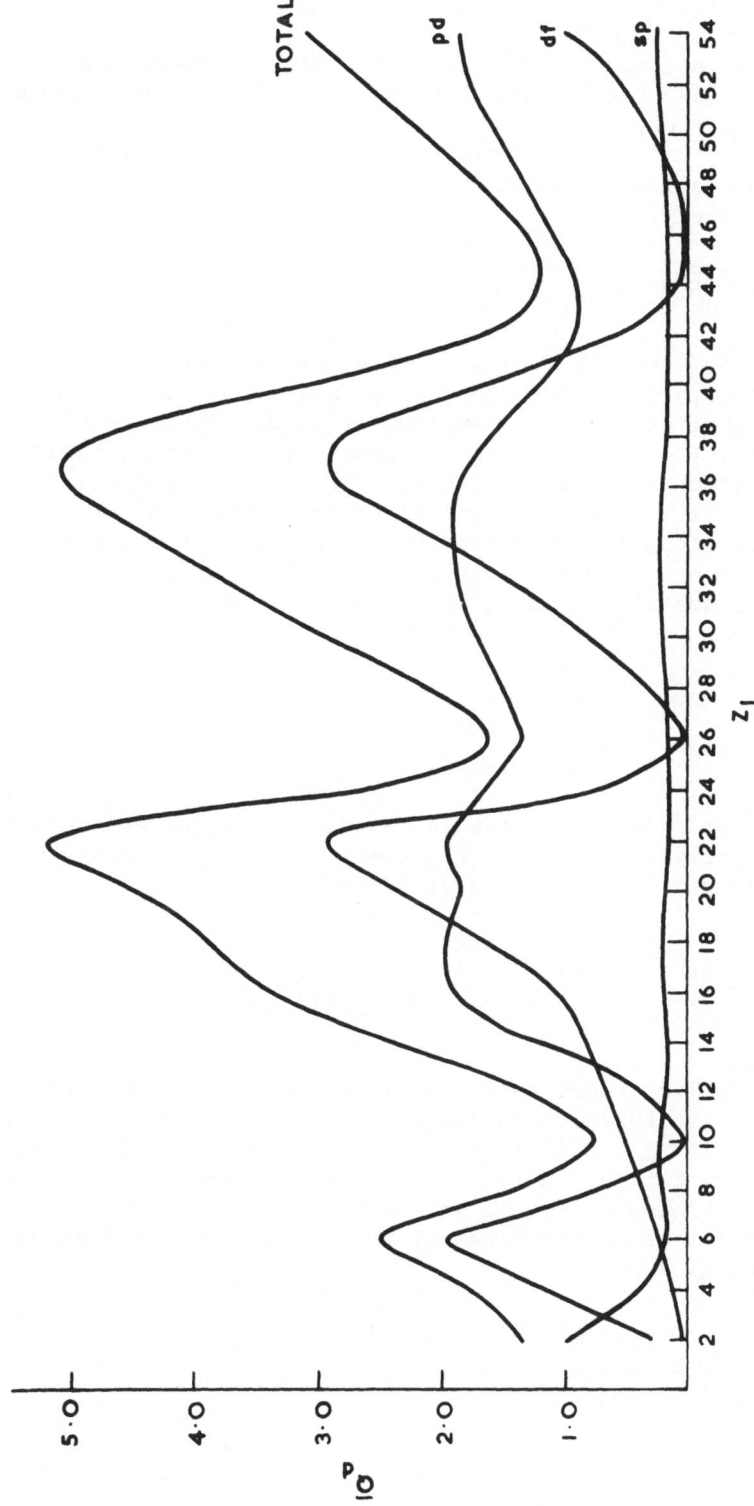

Fig. 1. Calculated Z_1 dependence of momentum transfer cross-section.

Connection With Other Theory

In this section it will be shown that the general stopping formula eq. (14) can be approximated to yield simple well-known stopping power expressions.

First of all, it is clear that whenever $v_1 \ll v_2$ the relative velocity may simply be replaced by v_2 so that the stopping power is approximately.

$$- \frac{dE}{dx} = mnv_1 v_2 \, Q_d(v_2) \tag{16}$$

and increases linearly with v_1. This approximation is the origin of the velocity dependent stopping of all low-velocity theories, including that of Firsov. However, it is not strictly valid in the region ($v_1 \sim 1.0$ a.u.) of its usual application although it should be recognized that a linear v_1 dependence could also arise if the product $\underline{v} \, Q_d \, (\underline{v})$ happens to be independent of v_1.

Simple stopping power expressions consider the scattering of electrons from a bare particle of charge Z_1. Under these conditions the elastic scattering is described by the Rutherford differential cross-section

$$I_o(\chi) = \frac{z_1^2 \, e^4}{m^2 v^4} \; \frac{1}{(1-\cos\chi)^2} \; . \tag{17}$$

Consider first the situation studied by Fermi and Teller, where a massive particle of charge Z is moving with a velocity $v_1 \ll v_2$. Then the linear dependence on v_1 holds and

$$- \frac{dE}{dx} = 2\pi \; mn \; v_1 v_2 \left(\frac{z_1^2 \, e^4}{m^2 v_2^4} \right) \int_0^\pi \frac{\sin d\chi}{(1-\cos\chi)} \tag{18}$$

Fermi and Teller argued that the divergence at $\chi=0$ does not arise because of electronic screening and the effective lower limit of integration is $\chi_{min} = (v_o/v_f)^{1/2} \simeq \lambda/a_o$, where λ is the electron gas screening length. Here, $v_2 = v_f$ the electron Fermi velocity and v_o is the atomic unit of velocity ($=e^2/\hbar$).

Substitution of the electron gas density and performing the integration gives

$$\frac{-dE}{dx} = \frac{2m^2 z_1^2 e^4}{3\pi\hbar^2} \; v_1 \; \ln\left(\frac{4v_f}{v_o}\right) \tag{19}$$

where the additional approximation $v_f > v_o$ has been made. To within a numerical factor of the order of unity this agrees with the results of Fermi and Teller [11].

Secondly, we can consider the situation where a fast incident particle collides with an atomic electron which we consider to scatter as if it were free. Here, $v_1 \gg v_2$ and substitution of eq. (17) into eq. (14) leads to the classical form of the stopping power

$$-\frac{dE}{dx} = \frac{2\pi Z_1^2 e^4}{mv_1^2} \; n \; \ln\left(\frac{2mv_1^2}{I}\right) \; ,$$

where $\chi = \pi$ gives $T_2 = 2mv_1^2$ and the minimum χ is decided by $T_2 = I$.

The foregoing is simply meant to demonstrate that the general form eq. (14) leads to the two different v_1 dependences of the prelogarithmic factor according as $v_1 \ll v_2$ or $v_1 \gg v_2$. In the first place the Coulomb scattering is justified since a bare particle scatters from quasi-free electrons. In the second case it is justified when v_1 is so large that the atomic binding is not dominant. In the energy region considered here, $v_1 \sim v_2$, it seems necessary to retain the full form of eq. (14) and since we are dealing with passage of a slow heavy-ion through the solid, to include the appropriate description of the scattering of electrons from the potential field of the atomic core.

References

[1] J. H. Ormrod, J. R. MacDonald and H. E. Duckworth, Can. J. Phys. 43, 275 (1965); B. Fastrup, P. Hvelplund and C. A. Sautter, Mat. Fys. Medd. Dansk. Vid. Selsk. 35, No. 10 (1966); L. Eriksson, J. A. Davies and P. Jespersgaard, Phys. Rev. 161, 219 (1967); F. H. Eisen, Can. J. Phys. 46, 561 (1968).

[2] I. M. Cheshire, G. Dearnaley and J. M. Poate, Phys. Letters 27A, 304 (1968); Proc. Roy. Soc. A311, 47 (1969) and Atomic Collision Phenomena in Solids (North Holland 1970) p. 351.

[3] O. B. Firsov, Soviet Physics J.E.T.P. 9, 1076 (1959).

[4] J. S. Briggs and A. P. Pathak, J. Phys. C. (Solid State Physics) 6, L153 (1973).

[5] E. Gerjuoy, Phys. Rev. 148, 54 (1966).

[6] E. Geltman, Topics in Atomic Collision Theory, p. 20 (Academic Press, New York) 1969.

[7] A. Burgess, Proc. Phys. Soc. 81, 442 (1963).

[8] P. M. Morse, Rev. Mod. Phys. 4, 577 (1932).

[9] L. D. Landau and E. M. Lifshitz, Quantum Mechanics (Nonrelativistic Theory), 1965 (Pergamon, Oxford) p. 257.

[10] S. T. Manson, Phys. Rev. 182, 97 (1969).

[11] E. Fermi and E. Teller, Phys. Rev. 72, 399 (1947).

Z_1^3-DEPENDENT STOPPING POWER AND RANGE CONTRIBUTIONS*

J. C. ASHLEY and R. H. RITCHIE
Health Physics Division, Oak Ridge National Laboratory
Oak Ridge, Tennessee 37830

and

W. BRANDT
Physics Department, New York University
New York, New York 10003

Z_1^3-Dependent Stopping Power Contributions

The stopping power of a target, composed of atoms with atomic number Z_2, for a projectile of charge $Z_1 e$ and velocity v_1, depends in first Born approximation on the projectile charge as $(Z_1 e)^2$. Recently [1], we extended this theory to include the $(Z_1 e)^3$ dependence in a classical treatment which is equivalent to a second Born approximation. The Z_1^3 contribution was calculated for the statistical model of the target atom in the Lenz-Jensen approximation. Expressing the projectile energy, E_1, in terms of the reduced parameter $x \equiv v_1^2/v_o^2 Z_2 = 40.2\ E_1$ (MeV)$/M_1 Z_2$, where v_o is the Bohr velocity and M_1 the projectile mass in amu, we write the stopping power, $S(x)$, in terms of the stopping power in first Born approximation, $S_o(x)$, in the form

$$S(x) = S_o(x)\ (1 + \frac{Z_1}{Z_2^{1/2}}\ \frac{\kappa(b,x)}{x})\quad. \qquad (1)$$

Since $S_o(x) \propto Z_1^2$, the second term in the parentheses represents the Z_1^3 contribution. The function $\kappa(b,x)/x$, where b is a parameter related to the choice of the lower impact-parameter cut-off in the classical description, is shown in Figure 1 as $\kappa(b,x)/x$ vs x and compares it, according to Eq. (1), with the data [2] on Al ($Z_2 = 13$) and Ta ($Z_2 = 73$) in the form $Z_2^{1/2}\ \Delta S(x)/Z_1 S_o(x) \equiv Z_2^{1/2}(S-S_o)/Z_1 S_o$. This comparison yields the present "best" value for b of $b = 1.8 \pm 0.2$ [3].

The Z_1^3 contribution accounts for the shorter ranges of positive particles (e.g., π^+) compared to the ranges of their

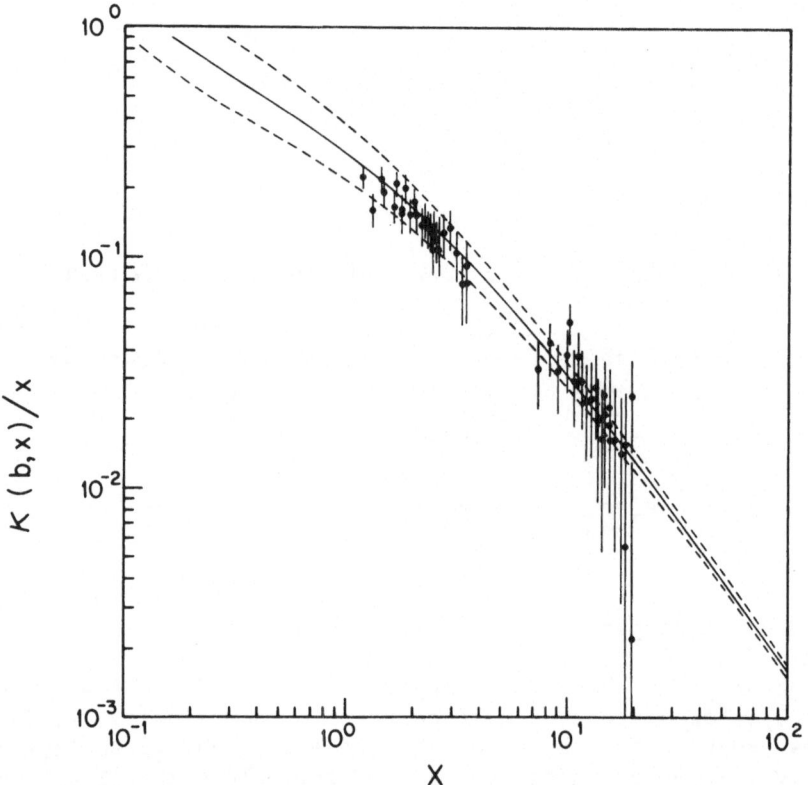

Fig. 1. Comparison of $\kappa(b,x)/x$, in Eq. (1) for b = 1.8 (solid
curve) with experiments by Andersen, Simonsen, and Sørensen [2].
The upper and lower dashed curves correspond to 10% changes in b,
viz., b = 1.6 and b = 2.0. The group of data near x = 2 comes
from measurements on a Ta (Z_2 = 73) target, the group near x = 10
from measurements on an Al (Z_2 = 13) target.

antiparticles (π^-) [1]. Most of these observations are made in
nuclear emulsions. We derive the $Z_1{}^3$ contribution for a compound
target under the assumption that Bragg's additivity rule of stopping power applies. For a target consisting of atomic constituents
Z_{2i}, present in atomic concentrations n_i, the stopping power becomes
$S_o = \Sigma\, n_i S_{oi}(x_i)/\Sigma\, n_i$, where $x_i \equiv v_1{}^2/v_o{}^2 Z_{2i}$. Noting that
$S_{oi} \propto L(x_i)/x_i$, the $Z_1{}^3$ contribution takes the form

$$\frac{S - S_o}{S_o} \equiv \frac{\Delta S}{S_o} = Z_1 \left(\frac{v_o}{v_1}\right)^2 \frac{\Sigma n_i Z_{2i}{}^{3/2} L(x_i)\, \kappa(b,x_i)}{\Sigma n_i Z_{2i} L(x_i)} \qquad (2)$$

where the function $L(x_1)$, derived and displayed in Ref. 1, is the stopping number per target electron.

For illustration we have evaluated Eq. (2) for the important case of standard emulsion [4] for which S_0 is tabulated [5]. The experimental points are derived from the work of Heckman and Lindstrom [6] on $\pi^\pm$ ranges in the emulsion and show reasonable agreement with the theory (see Fig. 2).

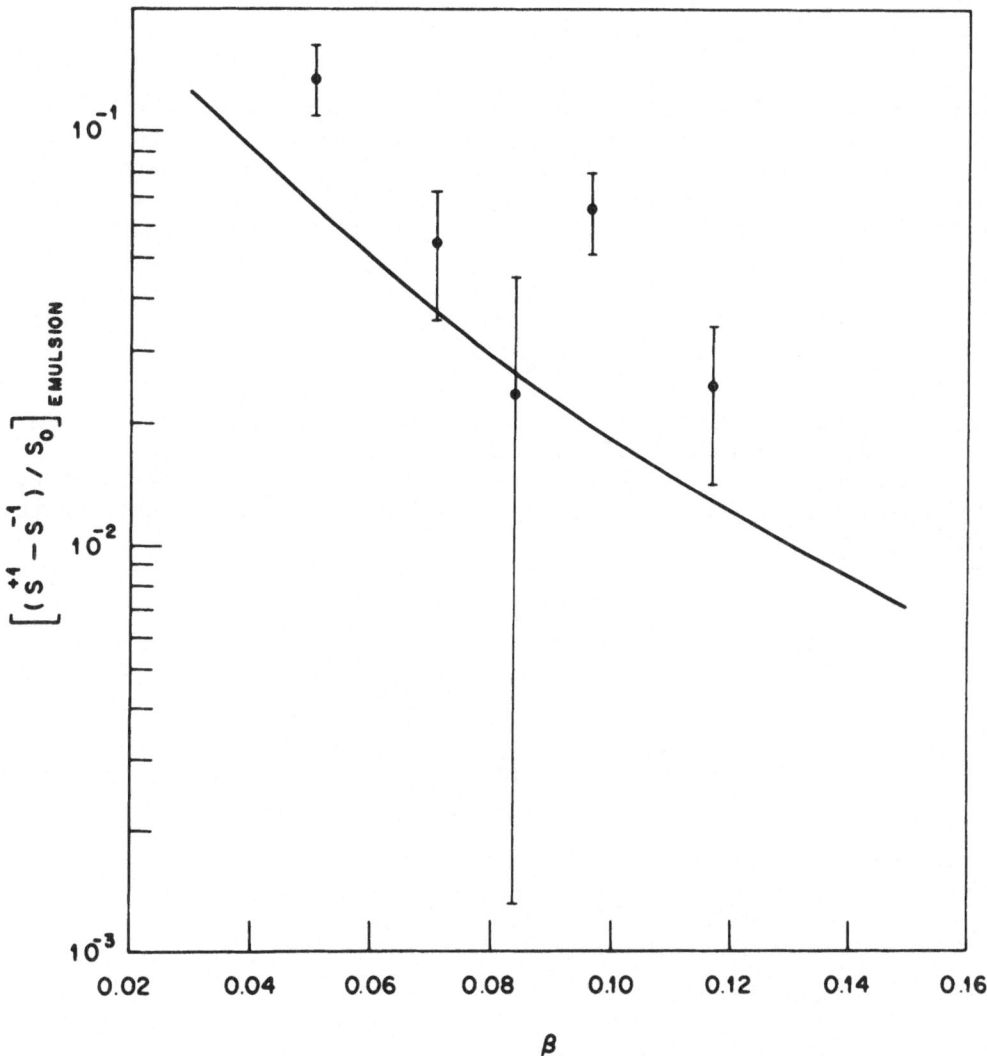

Fig. 2. The fractional difference in stopping powers for $Z_1 = \pm 1$ particles in emulsion as a function of β. The experimental results are derived from the work of Heckman and Lindstrom [6] with $\pi^\pm$.

$Z_1{}^3$–Dependent Range Contributions

With Eq. (1), the range $R(x)$, including the $Z_1{}^3$ contribution, becomes

$$R(x) = \frac{M_1 R Z_2}{m} \int^x \frac{dx'}{S(x')} \simeq R_o(x)$$

$$\cdot [1 - \frac{Z_1}{Z_2^{1/2} R_o} \frac{M_1 R Z_2}{m} \int^x \frac{\kappa(b,x')dx'}{x' S_o(x')}] \tag{3}$$

where $R = 13.6$ eV is the Rydberg energy. In the region where the $Z_1{}^3$ contribution is important the function $\kappa(b,x)$ may be replaced by a constant value $\kappa_o \simeq 0.32$ and $S_o(x) \propto x^{-1} \ln (x/q)$, where $q = 0.18$ gives good agreement with the statistical model when $x > 1$. With these approximations, Eq. (3) reduces to the simple form

$$R(E_1) = R_o(E_1) [1 - \frac{Z_1 \kappa_o}{Z_2^{1/2} q} \frac{R_o(\sqrt{QE_1})}{R_o(E_1)}] \ , \tag{4}$$

in terms only of known theoretical ranges R_o. Here E_1 is given in MeV and $Q(\text{MeV}) = 4.57 \times 10^{-4} M_1 \text{ (amu) } I_2 \text{ (eV)}$.

A range formula for compounds, including the $Z_1{}^3$ contribution, follows directly from Eq. (4) by averaging over atomic constituents. The result is

$$R(E_1) = R_o(E_1) \left\{ 1 - \frac{Z_1 Z_c^{1/2}}{\bar{Z}_c} \frac{\kappa_o}{q} [\frac{R_o(\sqrt{QE_1})}{R_o(E_1)} (1 + \ln \frac{\bar{Z}_c}{\bar{Z}_c'}) \right.$$

$$\left. - \frac{Q \ln(\bar{Z}_c/\bar{Z}_c')}{S_o(E_1) R_o(E_1)}] \right\} \tag{5}$$

with the abbreviations

$$Z_c^{1/2} = \Sigma n_i Z_{2i}^{3/2} / \Sigma n_i Z_{2i} \ ,$$

$$\ln\bar{Z}_c = \Sigma n_i Z_{2i} \ln Z_{2i} / \Sigma n_i Z_{2i} \ ,$$

$$\ln\bar{Z}_c' = \Sigma n_i Z_{2i}^{3/2} \ln Z_{2i} / \Sigma n_i Z_{2i}^{3/2} \ .$$

With the aid of tables of theoretical values of $S_0(E_1)$ and $R_0(E_1)$ Eq. (5) is a convenient and accurate formula for calculating $R(E_1)$, or for constructing curves of $(R-R_0)$ vs R_0 by using E_1 from tables of $R_0(E_1)$ as the connecting parameter, or for pairs of antiparticles $\Delta R = (R_- - R_+) = 2|R-R_0|$ vs R_0.

We have performed calculations of ΔR for π^- vs π^+ as a function of R_0 in nuclear emulsion from Eq. (5), using Tables 2 and 3 of Ref. 5. The result is shown graphically in Fig. 3 as the solid line.

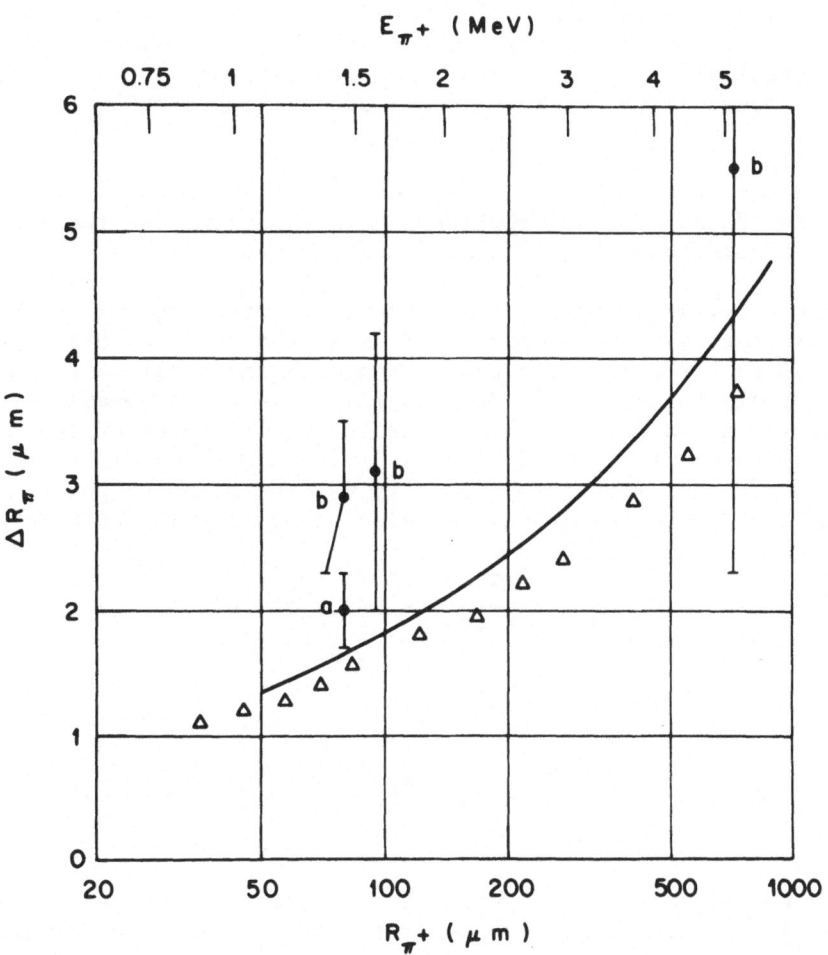

Fig. 3. The range difference for π^+ and π^- mesons in nuclear emulsion, $\Delta R_\pi \equiv R_{\pi^-} - R_{\pi^+}$, as a function of R_{π^+} and E_{π^+}, calculated according to Eq. (5) with range-energy tables given in Ref. 5. The triangles are estimates of the range difference derived from theoretical results presented in Ref. 7. The experimental points are from the work of Tovee et al. [9] (point labeled "a") and Barkas et al. [8] (points labeled "b").

The constants for standard nuclear emulsion are given by $Z_c^{1/2} = 5.62$, $\bar{Z}_c = 27.06$, $\bar{Z}_c^! = 35.47$ and $Q = 0.1234$ M_1 (amu). The results calculated with the analytical formula, Eq. (5), agree quite well with the results found in a similar calculation by Jackson and McCarthy [7] by numerical integration. The values estimated from Fig. 4 in Ref. 7 are shown as triangles in Fig. 3. The difference in the results at the higher energies is due to relativistic corrections which are not included in our analytical formulae.

The theoretical ΔR_π values from Eq. (5) are ca. 40% smaller than the range differences measured by Barkas et al. [8] at the π^+ ranges 80 and 90 μm. The theoretical value $\Delta R_\pi = 1.69$ at $R_{\pi+} = 80$ μm (see Fig. 3) agrees satisfactorily with the recent measurement $\Delta R_\pi = (2.0 \pm 0.3)$ μm by Tovee et al. [9]. At $R_{\pi+} = 725$ μm, the theoretical value $\Delta R_\pi = 4.35$ μm agrees, within the uncertainties, with the measured value $\Delta R_\pi = (5.5 \pm 3.2)$ μm reported by Barkas et al. [8].

Influence of the Z_1^3 Contribution to Stopping Power on the Evaluation of Mean Excitation Potentials and Shell Corrections

Recently, very careful measurements of stopping powers for 5–12 MeV protons and deuterons [10] have been used to obtain very exact values for mean excitation potentials and shell corrections for the elements $Z_2 = 20$ to $Z_2 = 30$ [11]. We will reexamine the data presented in Ref. 11 to illustrate the influence of the Z_1^3 effect in these measurements.

For a material of atomic density n_2, the stopping power, including the Z_1^3 contribution, may also be written in the form [1]

$$\frac{mc^2\beta^2}{4\pi(Z_1 e^2)^2 n_2 Z_2} \left(-\frac{dE_1}{dz}\right) = L(\beta, Z_2) + D(\beta, Z_1, Z_2) \qquad (6)$$

where

$$D(\beta, Z_1, Z_2) \equiv \frac{Z_1}{Z_2^{1/2}} \frac{F(1.8/x^{1/2})}{x^{3/2}} = \frac{Z_1}{Z_2^{1/2}} \frac{\kappa(1.8, x)L(x)}{x} \qquad (7)$$

and $x = \beta^2/\alpha^2 Z_2$ with $\alpha = e^2/\hbar c$. The functions F and L are defined and plotted in Ref. 1 and are tabulated in Ref. 12. The dimensionless function $L(\beta, Z_2)$ is given by the Bethe expression

$$L(\beta, Z_2) = f(\beta) - \ln I_2 - \frac{C}{Z_2} \qquad (8)$$

where I_2 is the mean excitation potential of the target, C/Z_2 represents the shell corrections and $f(\beta) \equiv \ln(2mc^2\beta^2/1 - \beta^2) - \beta^2$. If the experimental stopping power measurements [10,11] are written in the reduced form $L'(\beta, Z_1, Z_2)$ defined by

$$L'(\beta, Z_1, Z_2) \equiv f(\beta) - \ln I_2' - \left(\frac{C}{Z_2}\right)' , \tag{9}$$

then agreement between theory and experiment requires

$$L'(\beta, Z_1, Z_2) = L(\beta, Z_2) + D(\beta, Z_1, Z_2) . \tag{10}$$

If we also require that the measured shell corrections, $(C/Z_2)'$, be the same as the shell corrections that appear in the Bethe expression, C/Z_2, then the mean excitation potential measured on the basis of Eq. (9), I_2', is related to the I_2 in Eq. (8) by

$$\ln I_2' = \ln I_2 - D(\beta, Z_1, Z_2). \tag{11}$$

In the "asymptotic fitting" procedure used in Ref. 11, $\ln I_2'$ from Eq. (9) is plotted as a function of energy using two different theoretical forms for the shell corrections. Since the theoretical shell corrections approach the same value at high energies, a value of I_2' is extracted using the asymptotic value approached by the two $\ln I_2'$ curves at high energy. We have calculated I_2 from the data presented in Ref. 11 by taking values of $\ln I_2'$ from Figs. 1 and 2 of Ref. 11, calculating $\ln I_2'$ by Eq. (11), and redoing the asymptotic fitting procedure for Ca and Cu. From this very rough reanalysis we find values of I_2 for Ca and Cu to be within the error limits stated in Ref. 11 [13].

Given the mean excitation potentials calculated with the $Z_1{}^3$ contribution accounted for (the I_2's above), the shell corrections may be evaluated from experimental data. From the stopping powers tabulated in Ref. 10 we find $L'(\beta, Z_1, Z_2)$. The shell corrections are then given by Eqs. (8) and (10) as

$$\frac{C}{Z_2} = f(\beta) - L'(\beta, Z_1, Z_2) - \ln I_2 + D(\beta, Z_1, Z_2) . \tag{12}$$

For the mean excitation potentials we used the I_{asy}'s from Table I of Ref. 11 for this Z_2-range since only small changes were found in the I_2 values. The results for the shell corrections with the $Z_1{}^3$ contribution to stopping power accounted for are compared in Fig. 4 (solid curves) with the shell corrections reported in Ref. 11 (dashed curves) for Ca ($Z_2 = 20$) and Cu ($Z_2 = 29$). A significant change in the shape of the curves has been introduced by including the $Z_1{}^3$ effect. Note that any errors involved in the evaluation of

I_{asy} will not alter the shape of these curves but will simply shift
them up or down as we can see from Eq. (12). At the lowest energies
considered here, the $Z_1{}^3$ contribution to the stopping power is only
∿2%, but this produces changes of 20-30% in the vlaues derived for
 the shell corrections.

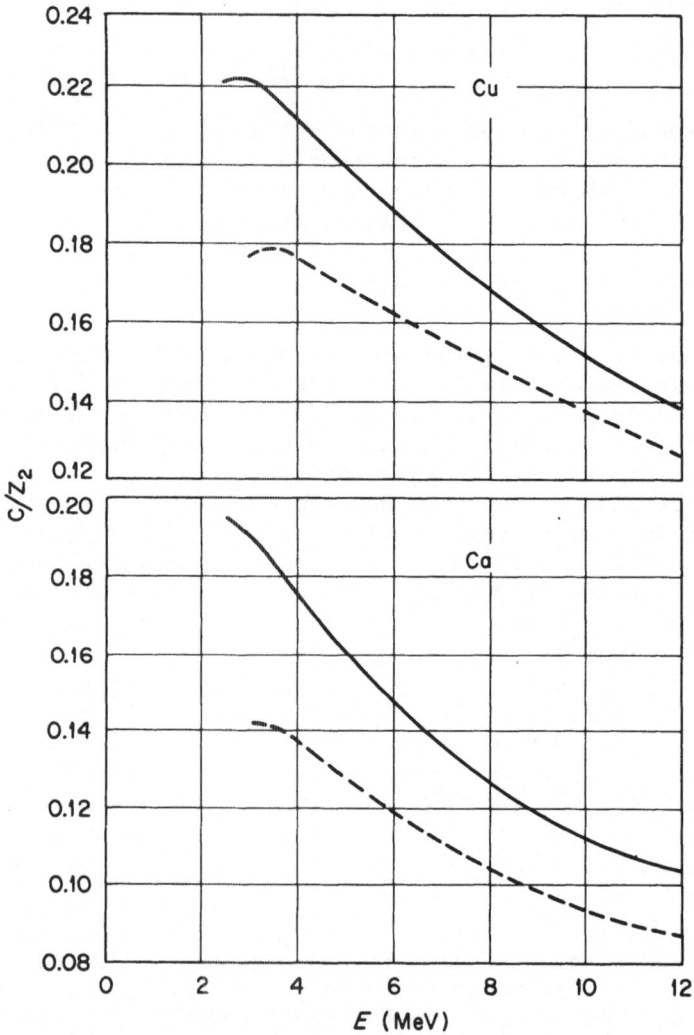

Fig. 4. Comparison of the shell corrections derived from experimental
stopping powers with the $Z_1{}^3$ contribution accounted for (solid curves)
with the results of Ref. 11 (dashed curves) for Ca (Z_2 = 20, I_{asy} =
193.6 eV) and Cu (Z_2 = 29, I_{asy} = 320.8) for protons of energy
2.50 - 12.00 MeV.

References

*Research sponsored by the U.S. Atomic Energy Commission under contract with Union Carbide Corporation.

[1] J. C. Ashley, R. H. Ritchie, and W. Brandt, Phys. Rev. B$\underline{5}$, 2393 (1972).

[2] H. H. Andersen, H. Simonsen, and H. Sørensen, Nucl. Phys. $\underline{A125}$, 171 (1969).

[3] Tables of the function $\kappa(b,x)$ over a wide range of x for b = 1.8 ± 0.2, as well as an extended discussion of the work described in this paper will soon be available in: J. C. Ashley, R. H. Ritchie, and W. Brandt, Phys. Rev. (to be published).

[4] W. H. Barkas, Nuovo Cimento $\underline{8}$, 201 (1958).

[5] See, e.g., M. M. Shapiro, "Nuclear Emulsions," in Handbuch der Physik, edited by S. Flügge (Springer-Verlag, Heidelberg, 1958) Vol. 45, p. 361 ff.

[6] H. H. Heckman and P. J. Lindstrom, Phys. Rev. Letters $\underline{22}$, 871 (1969).

[7] J. D. Jackson and R. L. McCarthy, Phys. Rev. B$\underline{6}$, 4131 (1972), have repeated our treatment for a somewhat different small impact-parameter cutoff and have included relativistic corrections.

[8] W. H. Barkas et al., Phys. Rev. $\underline{101}$, 778 (1956); Phys. Rev. Letters $\underline{11}$, 26 (1963); CERN Report 65-4 (unpublished).

[9] D. N. Tovee et al., Nucl. Phys. $\underline{B33}$, 493 (1971) (and private communication).

[10] H. H. Andersen et al., Phys. Rev. $\underline{175}$, 389 (1968).

[11] H. H. Andersen, H. Sørensen, and P. Vadja, Phys. Rev. $\underline{180}$, 373 (1969).

[12] J. C. Ashley, V. E. Anderson, R. H. Ritchie, and W. Brandt, "$Z_1{}^3$-Effect in the Stopping Power of Matter for Charged Particles: Tables of Functions," NAPS Document No. 02195, to be ordered from ASIS NAPS, c/o Microfiche Publications, 305 E. 46th St., New York, N. Y. 10017, remitting $1.50 for microfiche or $5.00 for photocopy up to 30 pages and $0.15 per each additional page over 30. Estimate 17 pages.

[13] A more definitive treatment of mean excitation potentials is planned based on the more recent stopping power data of H. Sørensen and H. H. Andersen (Phys. Rev., to be published). This later data extends the work of Ref. 10 to higher energies and includes several heavier elements.

DEPTH DISTRIBUTION OF DAMAGE DUE TO IONIZATION

K. B. WINTERBON
Atomic Energy of Canada Limited
Chalk River Nuclear Laboratories
Chalk River, Ontario, Canada

ABSTRACT

The depth distribution of the damage due to ionization caused by heavy ion irradiation is calculated by the transport equation method previously used for the calculation of range and damage distributions. The calculation differs from Brice's chiefly in that it allows for the transport of energy by recoils away from the primary projectile path. Like the damage distribution, it should be more accurate than Brice's at low energies.

It is found that a power-law limit of the ionization distribution can be constructed; thus one may take advantage of the simple scaling properties of power-law distributions.

The importance of energy transport by recoils is examined for both ionization and damage distributions.

A qualitative distinction between the ionization distribution and range or damage distributions is that the ionization distribution is discontinuous at the target surface. (This refers to the infinite target of the calculations, where the "surface" is merely a plane at which ion motion starts.) The differential-equation method of obtaining distributions from moments is generalized to allow inclusion of this discontinuity.

Introduction

The energy an implanted ion carries into a target material is spent, during the slowing-down time of the ion and its associated

cascade of recoiling target atoms, in creating several types of
excitation of the target material. It is customary to dichotomize
the distribution of the projectile energy into energy deposited in
motion of target atoms and energy deposited in their excitation.
The atom-motion portion of the energy, at the end of the slowing-
down time, becomes e.g. energy of dislocations and other defects,
of strain, or if sufficiently concentrated, of amorphization; it
is customary to call this the "deposited energy" and to more or
less identify it with "damage". The electronic excitation portion
of the energy causes, e.g., ionization, x rays, and the electron-
hole pairs which are observed in semiconductor radiation detectors.
The spatial distribution of this part of the energy will be called
the "ionization" distribution.

At low energies, and in metal or semiconductor targets, the
deposited energy has the more obvious consequences and there is
now a reasonably well-developed theory of the spatial distribution
of that energy; for a recent review see Ref. 1. However, the
energy going into atomic excitation has also been shown [2,3] to
cause observable changes in target materials, and it seems of
interest to discuss the spatial distribution of this portion of the
projectile energy. In a previous paper [4] I have pointed out how
this calculation could be done, as a simple modification of the
calculation of damage distributions. Also Brice [1,5] has calculated ionization distributions using methods he had previously
developed [6] for the calculation of damage (deposited energy)
distributions. The two calculations are complementary; the present
one is in principle exact, and is manageable at low energies, while
Brice's uses approximations appropriate for high energies.

All calculations presented here use the continuous, path-
independent, approximation for electronic stopping. Furthermore,
the stopping was assumed to be proportional to particle speed.

The equation for the moments of the ionization distribution
has already been presented in Ref. 4 and the calculations presented
here were done by the method used there. Thus no great amount of
algebraic manipulation is presented here.

In the section, Moment Calculations, some calculated first
and second moments are presented, from which one can estimate mean
ionization depths, and the spread about these depths, for most
projectile-target combinations. To this end, power cross section
calculations show the mass ratio dependence of these quantities
and Thomas-Fermi calculations for 3 mass ratios show the energy
dependence.

Brice has published several of his calculations of ionization
[1,5] (and of damage [6]). In the section, Energy Transport by
Recoils, the effect of his principal approximation, neglect of
transport of energy by recoils, is discussed. The dependence of
the importance of this approximation on mean depth and straggling[†]
is shown as a function of mass ratio and energy, using a similar

combination of power law and Thomas-Fermi calculations to that in
the section, Moment Calculations. We take the opportunity of looking at the effects of the approximation on damage distributions as
well. Results are threefold. First, the effect of the approximation on the first two moments is perhaps less than one might have
expected. Secondly, the results allow one to estimate corrections
to Brice's tables. Finally, Brice's distributions overestimate
the amount of damage near the surface.

The ionization distribution, unlike the range and damge distributions, has a step discontinuity at the target surface. (This is
discussed in the section, Energy Transport by Recoils). In the
section, Construction of Distribution from Moments, the differential-
equation method [4] of deriving the distribution from the moments
is modified to allow a known step discontinuity.

Moment Calculations

The integral equation for the moments of the ionization distribution has already been discussed [4]. To obtain a power cross
section distribution with the scaling properties that make the power
cross section so attractive, it is necessary not just to simplify
the cross section to the power form, but to go to a low energy
limit in which particle ranges are not affected by the presence of
electronic stopping. (Remember that in power-cross-section range
and damage distributions electronic stopping is not included.
Since we are here looking at the distribution of energy lost through
electronic stopping, we are obliged to include it in some form.)
Simplifying the cross section reduces the double power series
solution in Ref. 4, to a single power series; taking the low
energy limit reduces it to the first non-zero term. In this limit
the normalization of the ionization distribution is proportional
to the electronic stopping but its dimensions are independent of
electronic stopping. Hence apart from the normalization, the ionization distribution has the same scaling properties as the range
and damage distributions: all lengths are proportional to E^{2m},
where m is the "power" parameter of the power cross section, and
the shape of the distribution is independent of energy, depending
only on the projectile-target mass ratio (and on m). Mean ionization

---

†It has become customary to use the term "straggling" for the
second moment about the mean, or its square root, in analogy with
the terminology for range equations. It should be remembered that
the "straggling" for damage and ionization distributions contains
both the breadth of the damaged region due to one implanted ion,
and the "straggle" in this, (that is to say, the variation from
one ion to another).

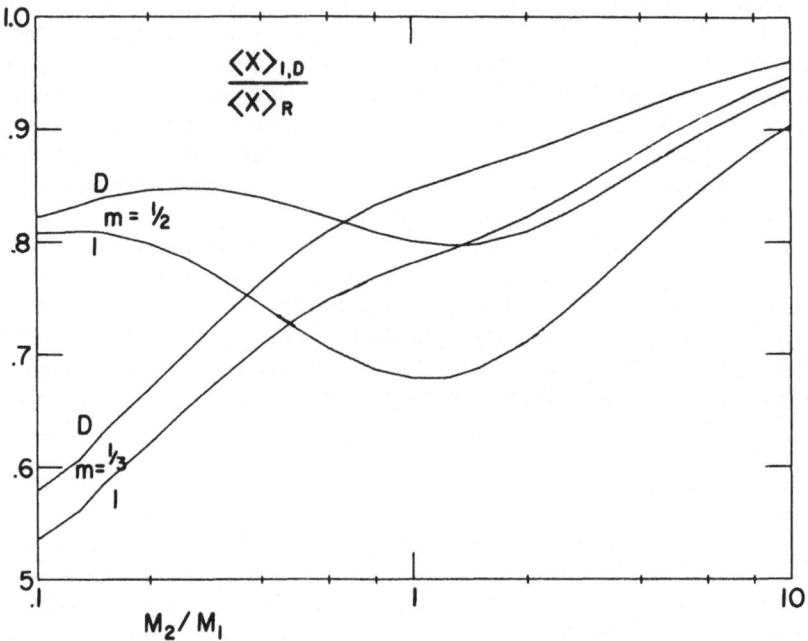

Fig. 1. Mass ratio dependence of mean ionization and damage depths relative to mean range. Power cross section values m=1/3 and m=1/2. Mass ratio M_2/M_1 = target atom mass/projectile mass.

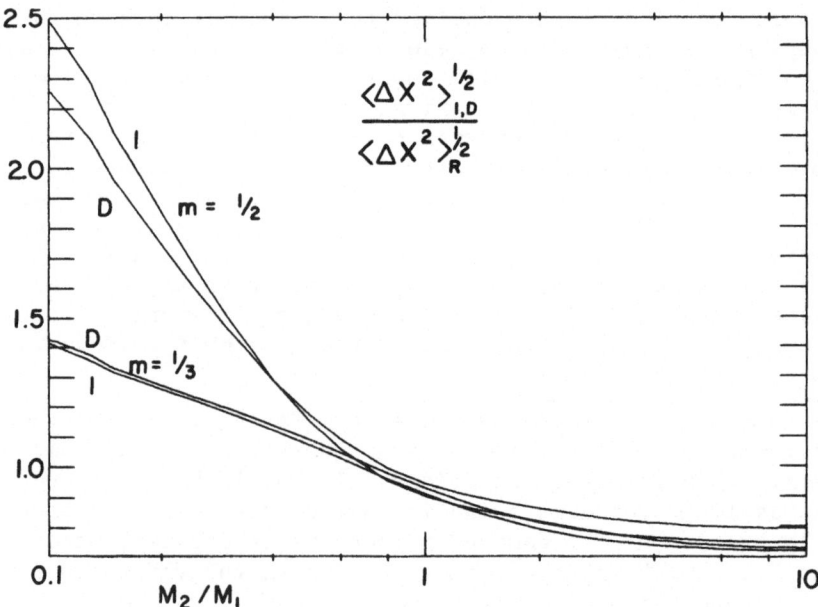

Fig. 2. Mass ratio dependence of ionization and damage stragglings relative to range straggling. See Fig. 1.

depths and straggling are compared with both mean damage depth and straggling and mean range and straggling. The ratios of mean ionization depth to mean range, and mean damage depth to mean range are shown in Fig. 1 for both m=1/3 and m=1/2 power cross sections, as functions of M_2/M_1 = target atom mass/projectile mass. The corresponding ratios of stragglings are shown in Fig. 2. The m=1/3 cross section should be a reasonable approximation if Lindhard's dimensionless energy ε is less than ~0.2, and m=1/2 for 0.08 $\lesssim$ $\varepsilon \lesssim$ 2. (However, see the discussion beginning on p. 31 of Ref. 7.)

The energy dependence of these ratios (for ionization only) is shown in Figs. 3 and 4 for mass ratios M_2/M_1 = 0.5, 1, and 2. The calculations presented are for ^{56}Fe, ^{28}Si, and ^{14}N slowing down in ^{28}Si; they may be applied to other projectile-target combinations at corresponding ε values and using the appropriate value of electronic stopping. In each case the calculation has been done for three values of the electronic stopping constant k, namely k/k_L = 0.8 (short dashes), 1.0 (solid line) and 1.2 (long dashes) where k_L is the standard Lindhard-Scharf value for that

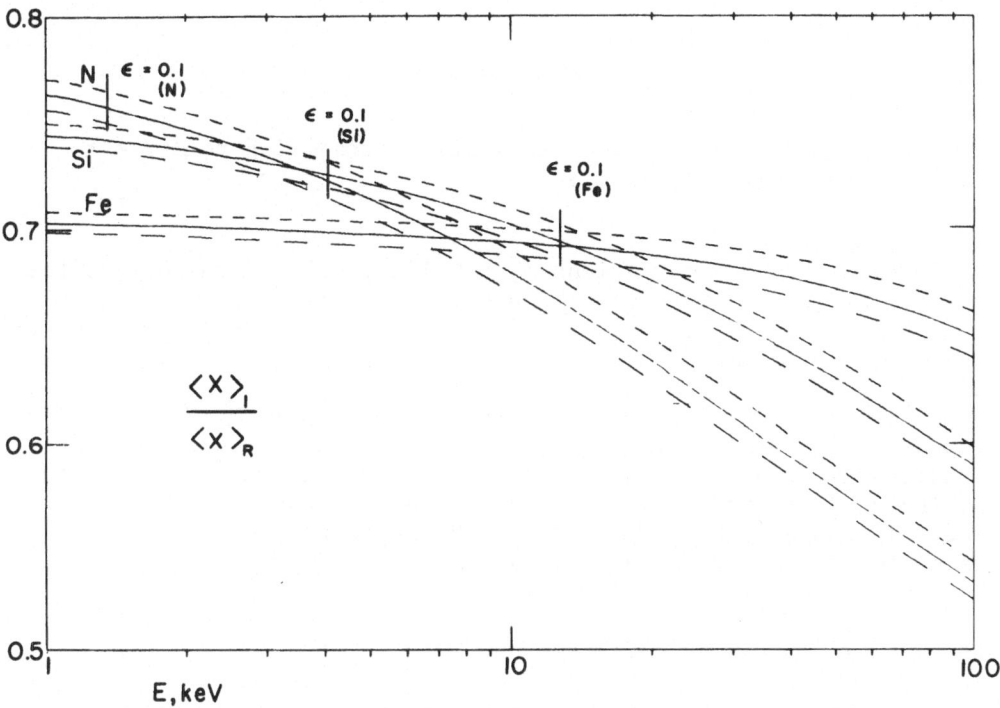

Fig. 3. Energy dependence of mean ionization depth relative to mean range. ^{56}Fe, ^{28}Si, and ^{14}N ions incident on ^{28}Si. Calculated for electronic stopping constant k/k_L = 0.8 (short dashes), 1.0 (solid line) and 1.2 (long dashes), where k_L is the standard Lindhard-Scharff value. The energies corresponding to ε = 0.1 are shown for each case; ε is Lindhard's dimensionless energy variable.

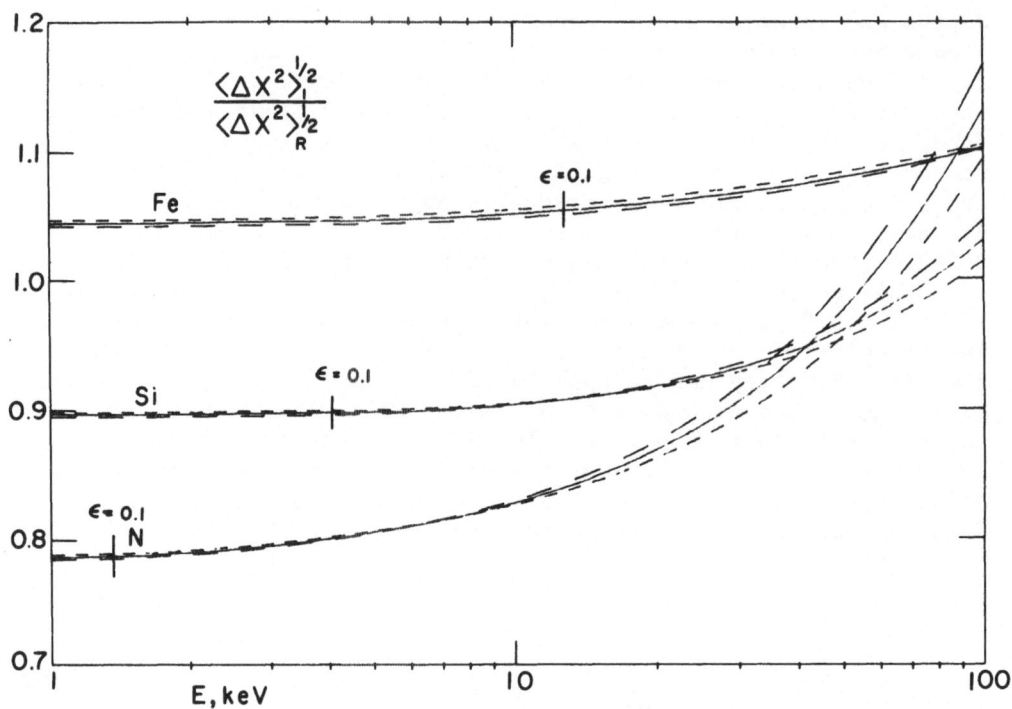

Fig. 4. Energy dependence of ionization straggling relative to range straggling. See Fig. 3.

projectile-target combinations (k_L= 0.1211, 0.1464, and 0.1975 for the Fe, Si, and N ions respectively).

From Fig. 1 it follows that one should most easily be able to distinguish ionization and damage distributions in the equal mass case.

Compare Figs. 1 and 3. For the Fe projectile M_2/M_1 = 0.5, the m=1/3 and m=1/2 values of ionization depth to range are nearly equal, suggesting that this ratio is nearly independent of energy, as indeed it is in Fig. 3. For both the Si and N projectile the ratio is decreasing with energy, in both figures, crossing the value for Fe, as suggested by Fig. 1.

Energy Transport by Recoils

In Brice's calculation he first constructs the depth distribution of the implanted ions, as a function of the energy to which they have slowed down. (He approximates this distribution by a gaussian; the effect of such an approximation is probably small, and is not considered further here.) From this depth distribution he calculates the ionization distribution due both to the original particle and to the recoils, with the ionization due to the recoils

assumed to occur at the point where the recoil was created. A
similar approximation is used in his damage calculations.

In the present calculation the moments of the ionization distribution are calculated exactly, including the transport of energy
by all generations of recoils but neglecting energy transport by
photons or ejected electrons. The distribution is then constructed
from the moments by an inversion procedure. However, the moments
are constructed as double power series in energy variables, and at
high energies it becomes difficult to sum the series. The difficulty
increases with the order of the moments, so it is impractical to
calculate many moments; to calculate the distribution from only a
few moments, one must have some preconceptions of the shape of the
distribution. In summary, at low energies transport of energy by
recoils is important, and the power series are readily summable so
the present method is superior. At high energies, transport by recoils is relatively less important, and the power series are hard
to sum, so the practical advantages of Brice's calculation make it
more useful than the exact-in-principle method used here.

Qualitative statements like those in the preceding paragraph
have been made before. It is possible to make some of these statements more quantitative. In particular, consider the effect of
neglect of energy transport by recoils on moments of the ionization
(or damage) distribution. This can be examined within the calculational framework used here by assuming all recoil ranges to be zero.
The ratios of approximate to exact mean depths, both for ionization
and for damage, are shown for the m=1/2 power cross section in
Fig. 5, to show the mass-ratio dependence; the corresponding ratios
of straggling are shown in Fig. 6. Figs. 7 and 8 show the energy
dependence of the ratios for mass ratios 0.5, 1, and 2. Provided
the projectile is not much heavier than the target atoms, the
approximation is not much heavier than the target atoms, the approximation is seen to be surprisingly good.

Another aspect of the distribution is the behaviour near the
surface. In both calculations, the target material is infinite,
and the "surface" is merely a reference plane at which ion motion
starts. Taulbjerg (private communication) has pointed out that
there must be a discontinuity of some derivative in the damage
distribution[†] and that if one is concerned with the distribution
of some quantity created with a finite cross section, such as
ionization, there will be a step discontinuity in that distribution
at the surface [8]. The discontinuity in the ionization distribution is the value of electronic stopping for the incoming projectile.
If transport by recoils is neglected, then there is an additional
step discontinuity in both distributions, which is the integrated
product of the cross section for producing recoils with the damage

---

[†]It may be shown (Winterbon, unpublished) that the discontinuity
in the damage distribution is proportional to x_+^P, $P = (1-m)/2m$,
where $x_+ = x$ for $x > 0$ and $x_+ = 0$ for $x < 0$; m is the power parameter
in the small-energy-transfer limit of the cross-section.

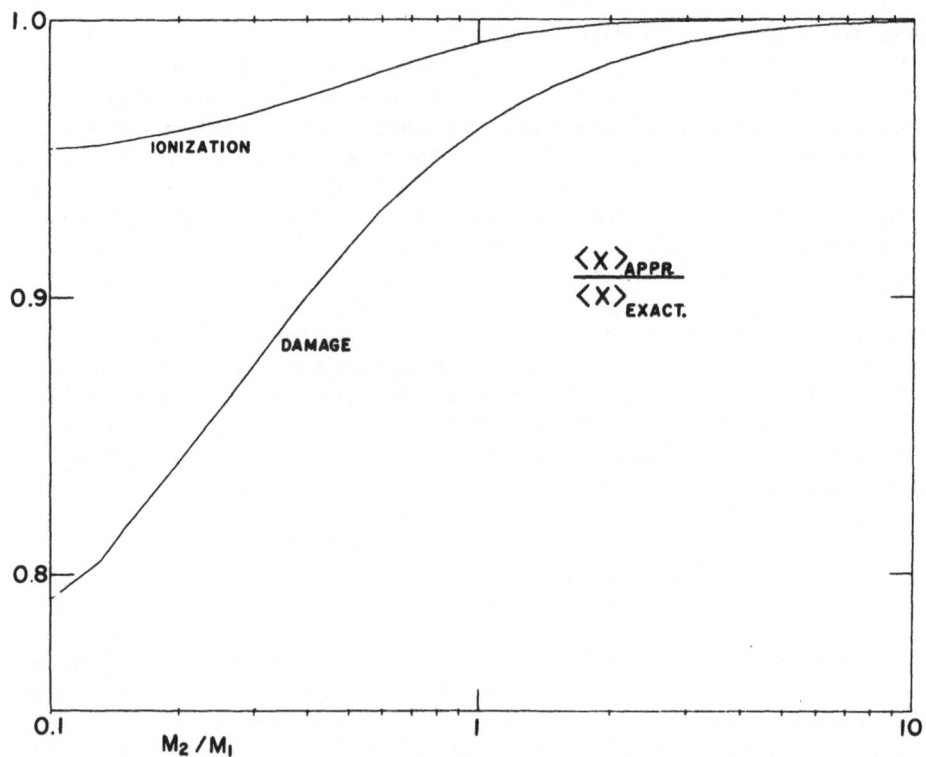

Fig. 5. Effect of energy transport by recoils. Mass ratio dependence of mean depths. Ratio of mean ionization and damage depth
neglecting energy transport by recoils to that including energy
transport by recoils. Results for m=1/2 power cross section.

or ionization the recoils will produce. This causes an overestimate
of the distribution at the surface, and in the case of the ionization distribution leads one to predict distributions decreasing
monotonically with depth, even at low energies.

Construction of Distribution from Moments

The differential equation method has been used [4] for constructing range and damage distributions from their moments. These
distributions are smooth or have only weak singularities that have
been ignored. The ionization distribution has a step discontinuity
at the origin; it is possible to modify the differential-equation
method to incorporate such a discontinuity. Assume that the distribution f(x) is smooth, apart from a step discontinuity of amplitude A at x = $-\xi$. Then one can write

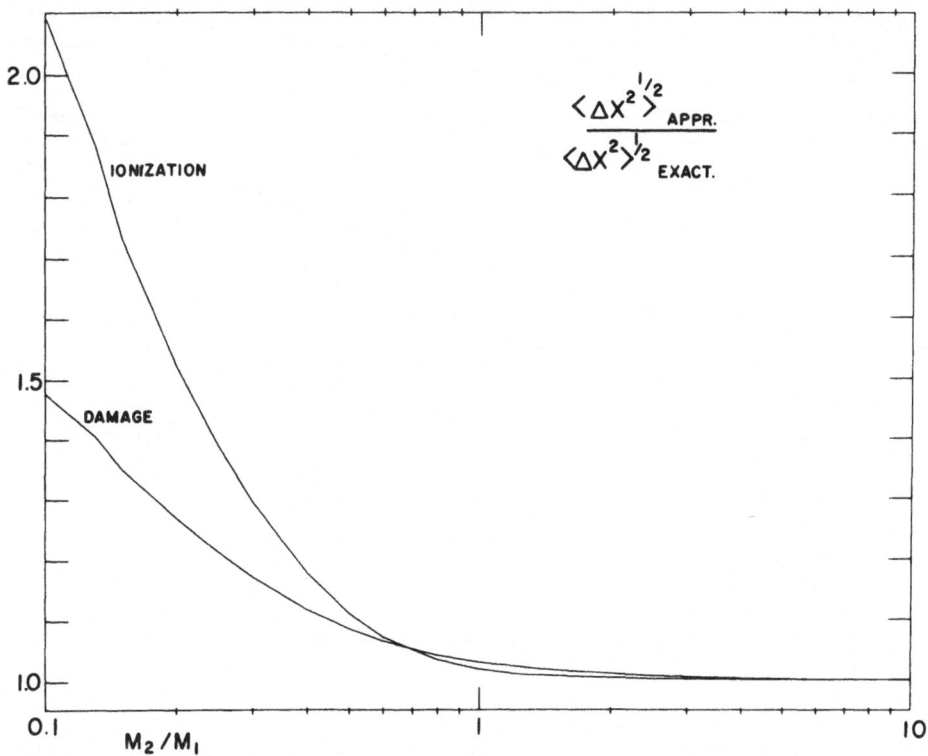

Fig. 6. Effect of energy transport by recoils. Mass ratio dependence of stragglings. See Fig. 5.

$$\frac{\partial f(x)}{\partial x} = \phi(x)f(x) + A\ \delta(x+\xi) \tag{1a}$$

The function $\phi(x)$ is approximated by a quotient of polynomials,

$$\phi(x) = P(x)/Q(x). \tag{1b}$$

Equation (1) reduces to the earlier [4] construction on putting A=0. From (1) it follows that

$$\int dx\ x^n\ [-Q(x)\ \frac{\partial f}{\partial x} + P(x)f(x)+AQ(x)\delta(x+\xi)] \equiv 0 \tag{2}$$

and from this equation one can derive recurrence relations for the moments, or, alternatively, a set of linear equations for the coefficients of P and Q. In the lowest (non-gaussian) approximation,

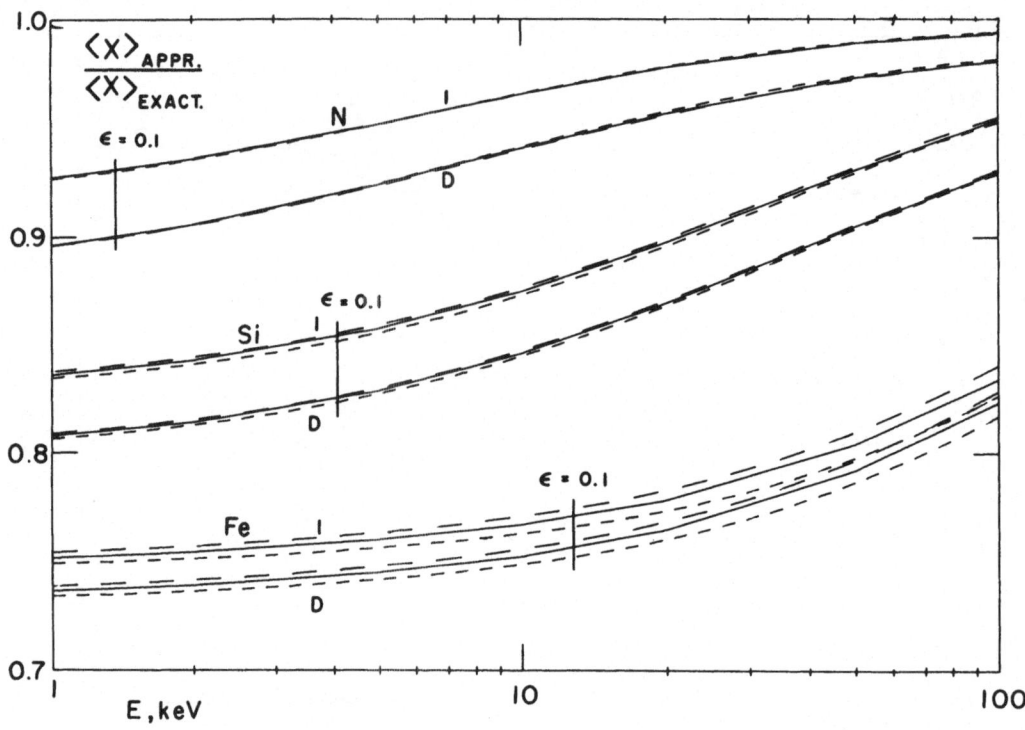

Fig. 7. Effect of energy transport by recoils. Energy dependence of mean depths. See Fig. 5. Calculated for same cases as Fig. 3.

$$P(x) = \alpha x^2 + \beta' x + y' \tag{3a}$$

and

$$Q(x) = x+b.$$

Let $u(x)$ be a solution of (1) when $A=0$. Then

$$u(x) = (x+b)^\gamma \exp(\alpha x^2/2 + \beta x), \tag{4a}$$

and

$$f(x) = (C+A\theta(x+\xi)/u(-\xi))u(x) \tag{4b}$$

where C is chosen to make $\int_{-\infty}^{\infty} dx\ f(x)=1$ and θ is the unit step function.

The primed quantities in (3a) are given in terms of the variables of (4a) by

$$\beta' = \beta+\alpha b, \quad \gamma' = \gamma + \beta b. \tag{5}$$

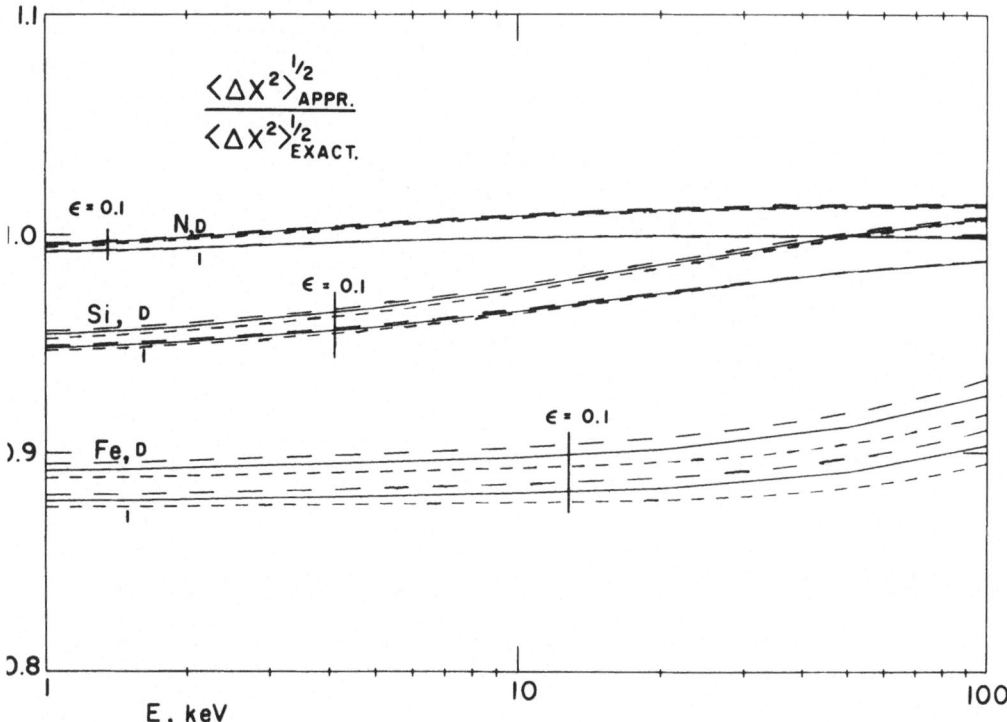

Fig. 8. Effects of energy transport by recoils. Energy dependence of straggling. See Fig. 5. Calculated for same cases as Fig. 3.

They are given in terms of the moments as follows. Renormalize the distribution, and transform coordinates to obtain zero mean and unit variance, i.e. $\langle x^0 \rangle = \langle x^2 \rangle = 1$ and $\langle x^1 \rangle = 0$. Write the skewness and kurtosis as $\langle x^3 \rangle = s$, $\langle x^4 \rangle = k$. The discontinuity A of this transformed distribution is the discontinuity in the original distribution, the electronic stopping, multiplied by the original variance, because of the coordinate transformation, and divided by the original normalization. Then

$$\alpha = (bs-2+A(\xi-b)(\xi^2-1+s\xi))/(k-1-s^2)$$

$$\beta = -b-\alpha(b+s)+A\xi(b-\xi)$$

$$\gamma = -1-\alpha-\beta b-A(b-\xi). \qquad (6)$$

The quantity b is left free; calculations should be made for several values of b. In the calculations shown here (Fig. 9) b has been taken to be -10 (E = 1 and 10 keV) or $+10$ (E = 100 keV), i.e. 10 standard deviations away from the mean. In each case values of the other sign resulted in positive values of α.

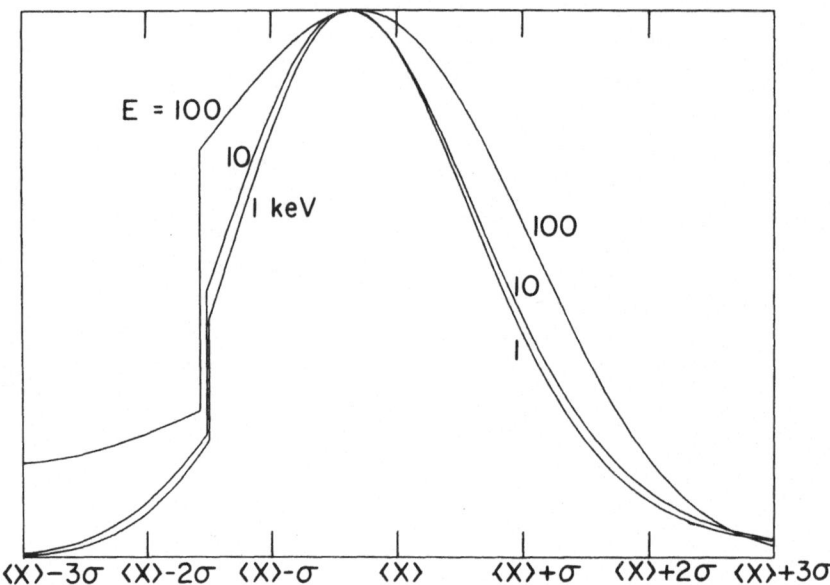

Fig. 9. Ionization distributions for Si in Si, calculated at 1, 10 and 100 keV, plotted with the same mean depth and straggling.

Figure 9 shows the calculated ionization distributions for 1, 10 and 100 keV Si in Si, all plotted with unit peak height and with the same mean depth and variance. The apparent increased breadth of the 100 keV distribution is due to a decrease in kurtosis to a value (2.55) near that for a rectangular distribution (1.8), which would extend out to ±1.732... standard deviations from the mean. The lower-energy distributions have kurtosis greater than 3, the gaussian value, so these distributions are narrower than gaussians near the peak and have larger tails.

References

[1] D. K. Brice, in Proc. III International Conference on Ion
    Implantation ..., Yorktown Heights, New York, December 1972.
[2] G. W. Arnold and F. L. Vook, Rad. Effects 14, 157 (1972).
[3] Other references are given in refs. 1 and 5.
[4] K. B. Winterbon, Rad. Effects 13, 215 (1972).
[5] D. K. Brice, Sandia Report SLA-73-0230, March 1973.
[6] D. K. Brice, Rad. Effects 11, 227 (1971).
[7] K. B. Winterbon, P. Sigmund, and J. B. Sanders, Mat. Fys.
    Medd. Dan. Vid. Selsk. 37, No. 14 (1970).
[8] I thank Knud Taulbjerg for insisting on this point sufficiently
    often.

INVESTIGATION ON ELECTRONIC STOPPING POWER IN ALKALI HALIDES BY MEANS OF COLOR CENTER PROFILES

A. PEREZ, P. THEVENARD and C. H. S. DUPUY
Département de Physique des Matériaux
Université Claude Bernard, Lyon I
43, Bd du 11 novembre 1918
69621 - Villeurbanne, France

ABSTRACT

F-center profiles created in LiF with 56 MeV α-particles and 28 MeV deuterons are obtained at room temperature by an optical method using a micro-spectrophotometer. These profiles are in a good agreement with the calculated electronic stopping power but a departure from linearity above 10^{14} MeV/cm^3 is observed. A saturation effect related with the dose is also observed and the F-center production is slightly lower, in the case of samples irradiated in the (110) direction, than in the (100) one. Taking into account the interaction tubes with F-centers, F-aggregate centers and cationic defects, we can give an interpretation of our results.

Introduction

Many studies have been done on the coloration of alkali halides by ionizing radiations and ion bombardments. A great number of the defects responsible for this coloration are well known and can be easily studied by optical absorption measurements. Ion bombardment, as a coloration mode, leads to high concentrations of defects located in a thin layer and three kinds of defects are present:
- the defects created by ionization processes,
- the defects created by direct displacement processes,
- the defects due to implanted ions themselves.

In the case of LiF, the anionic defects such as F-centers (electron trapped in a fluor vacancy) and F-aggregate centers can be connected with the electronic energy loss process [1]. The

cationic defects such as I-centers [2] (lithium in interstitial
position) can be related to the nuclear energy loss process [3].
On the other hand, metallic aggregates of implanted ions have been
found in the case of LiF bombarded with Li^+, Na^+ and K^+ ions [4].
Using LiF samples bombarded with high energy ions (56 MeV α-particles and 28 MeV deuterons) the penetration depths are sufficient
enough to obtain the distribution profiles of the defect by means
of a microspectrophotometer. These profiles can be compared with
the calculated energy loss profiles. Then we can obtain some
information on the interaction processes of the incident particles
with the target.

Experimental Procedure

The synchrocyclotron accelerator of the "Institut de Physique
Nucléaire de Lyon" was used to produce beams of 56 MeV α-particles
and 28 MeV deuterons. The beam current was measured and integrated
in a Faraday cup arrangement. The bombarded target area was 1 cm^2
and the dose rate $8.1 \cdot 10^8$ ions cm^{-2} sec^{-1}.
LiF samples were cleaved from a "Quartz et Silice" single
block prehardened by exposure to ^{60}Co γ-rays (10^6R) and then
annealed for several hours at 500°C to remove the induced damage.
The dimensions of the crystals irradiated in the (100) direction
were 15 x 5 x 0.1 mm^3. For irradiation in the (110) direction
samples were first mechanically polished to obtain a (110) face and
subsequently cleaved into plates of the same dimensions as the
first one. In order to compare the effects of the bombardment in
the (100) and (110) directions, the two kinds of samples were
mounted in the same holder and then irradiated in the same conditions.
After irradiation the F-center concentration was analysed in the
OX direction of incident ion penetration (Figure 1) and consequently
the light beam of the microspectrophotometer was perpendicular to a
(100) face.
The optical assembly used is shown in Figure 1. The image of the
entrance slit of the monochromator was focussed on the (100) face of
the sample using two "spectrosil" lenses. The light beam was diaphragmed
to use only the paraxial rays in order to obtain a clear image. This
image was 10 μm wide and the samples were thin enough (0.1 mm) to give
only a small variation of the image dimensions in the sample. The crystal holder was motor driven in the OX direction (0.25 mm/min). For
focussing of the image on the (100) face of the sample we used a microscope which could be mounted in place of the photomultiplier.

Results

Electronic stopping power for α-particles and deuterons in LiF

The electronic stopping power was computed using the Benton's
method [5] for particles in the 0.1 - 20 MeV/nucleon energy range.

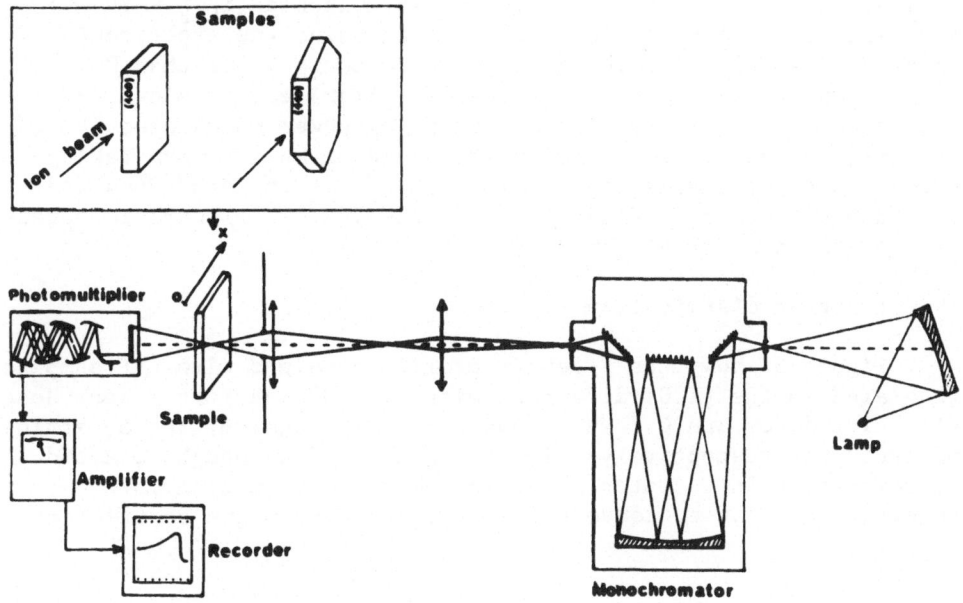

Fig. 1. Microspectrophotometric set-up.

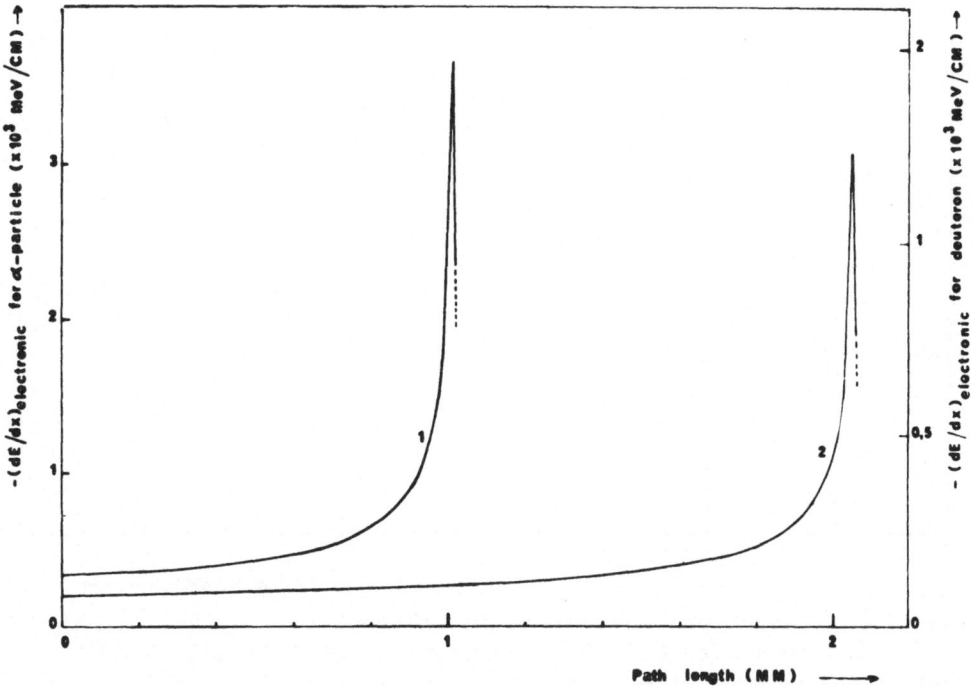

Fig. 2. Corrected electronic stopping powers for 56 MeV α-particles
(1) and 28 MeV deuterons (2) versus path length in LiF.

The mean ionization potential of the LiF molecule ($I_{LiF} = 75$ eV) which enters the calculation was obtained using the experimental values of Evans [6] for I_{Li} and I_F. For energy lower than 0.1 MeV/nucleon we have used the values of (dE/dx) electronic given by Bader et al. [7]. The electronic stopping power calculated for one particle must be corrected in the case of an ion beam taking into account the energetic distribution of the incident ions and the range straggling. Figure 2 shows the (dE/dx) electronic versus path length theoretical curves in LiF.

F-center distribution

Figure 3 shows the F-center profile obtained with LiF samples irradiated in the (100) direction with $8.1 \ 10^8$ ions/cm^2 · sec. dose rate. The doses are $6.7 \ 10^{10}$ and $2.8 \ 10^{11}$ ions/cm^2 for α-particles and deuterons respectively. These doses are low enough to avoid the saturation in F-center creation and the aggregation into M-centers. In order to convert the measured absorption coefficient

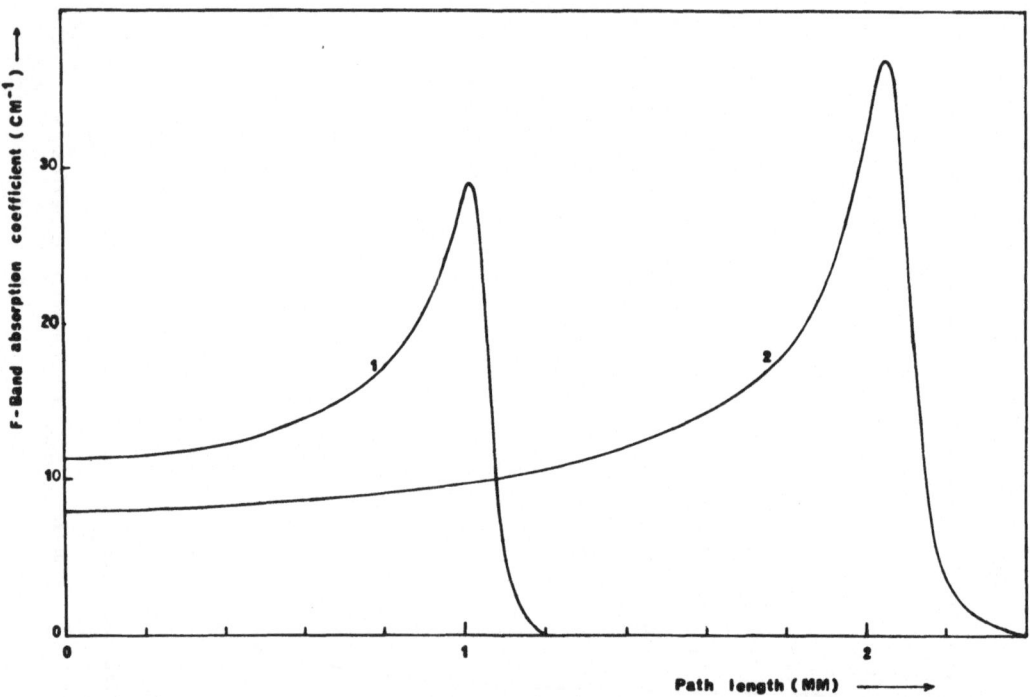

Fig. 3. F-center profile curves in LiF bombarded at R.T. in the (100) direction with $6.7 \cdot 10^{10}$ α-particles/cm^2 (1) and $2.8 \cdot 10^{11}$ deuterons/cm^2 (2). The F-center production is slightly lower for samples irradiated in the (110) direction.

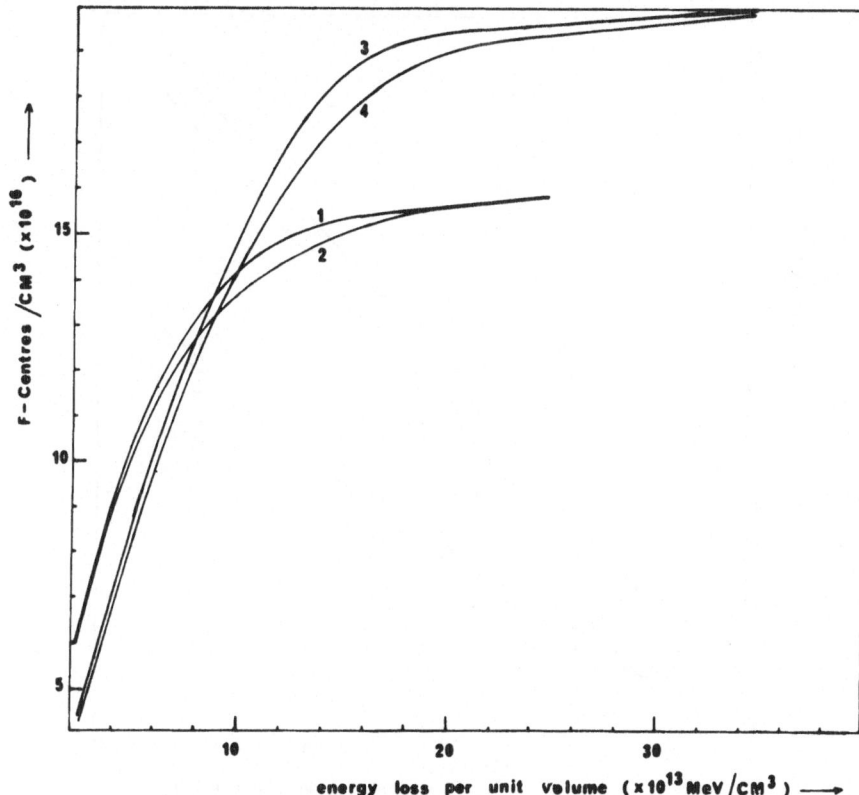

Fig. 4. F-center concentration versus energy lost into electronic
processes per cm^3 for LiF bombarded at R.T. with 6.7·10^{10}
α-particles/cm^2 in the (100) direction (1), (110) direction (2),
and 2.8·10^{11} deuterons/cm^2 in the (100) direction (3), (110)
direction (4).

of the F-band into F-centers/cm^3 (N_F), Smakula's equation [8,9]
was employed using an oscillator strength of unity and a F-band
hafl-width of 0.70 eV at room temperature. By comparing the concentration profile curves of the F-centers with the energy lost
into electronic processes per cm^3 (dE/dx per ion in MeV/cm times
incident ion dose in ions/cm^2) we obtain the F-center creation
efficiency for α-particles and deuterons (Fig. 4). The integration
of the F-center concentration profiles gives the total number of
F-centers per cm^2 of surface perpendicular to the penetration direction of the incident ions. This is shown in Figure 5. At
last, we have studied the F-center creation as a function of the
doses. Figure 6 shows the profiles obtained for doses between
8.1·10^9 and 5·10^{11} α-particles/cm and Figure 7 shows the profiles
obtained for doses between 1·10^{11} and 3·10^{12} deuterons/cm^2. Using

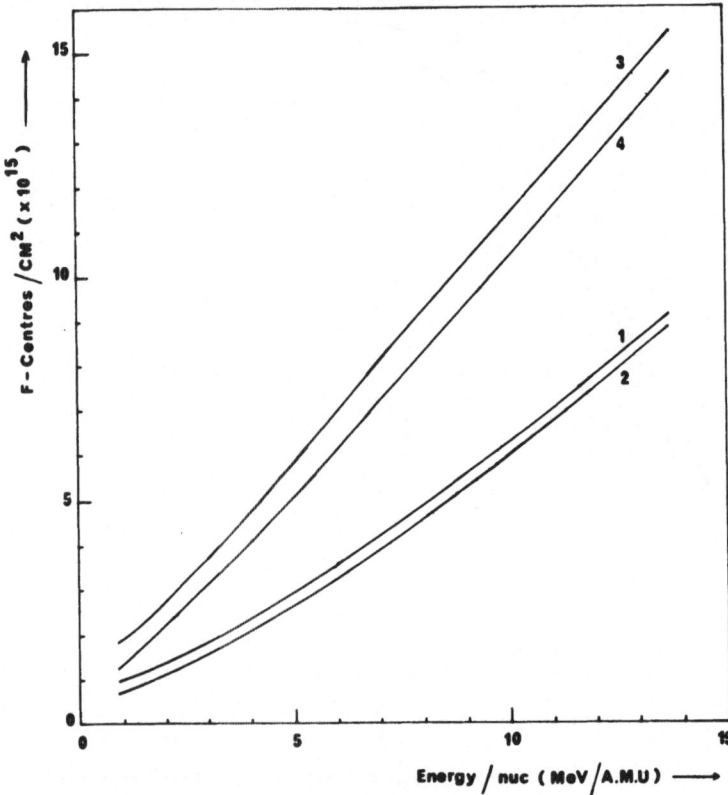

Fig. 5. Total number of F-centers per cm^2 versus energy of the incident α-particles and deuterons in LiF bombarded at R.T. in the (100) direction (1,3) and (110) direction (2,4).

these results we can plot the F-center growth curve at a definite energy loss (Fig. 8).

Discussion

The previous results show a good agreement between the F-center profiles and the electronic energy loss profiles. However, near the end of the incident particle path, at the ionization maximum, we observe a saturation effect for any irradiation dose. This saturation effect occurs above 10^{14} MeV/cm^3 for α-particles and $1.8 \cdot 10^{14}$ MeV/cm^3 for deuterons. Analogous phenomenon had been observed by Arnold et al. [10] with radiophotoluminescent glass. On the other hand, the F-center production in the case of samples irradiated in the (110) direction is slightly lower than

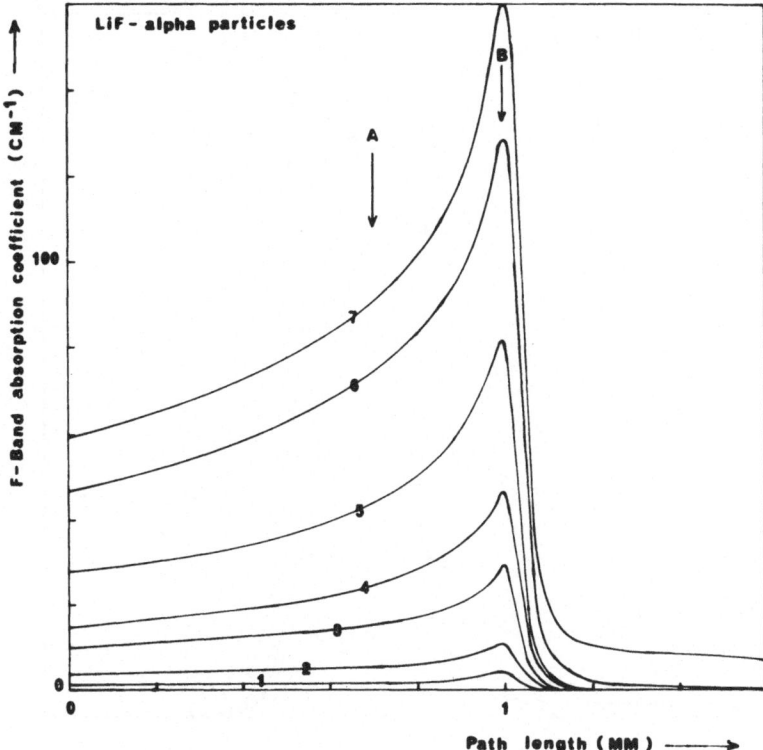

Fig. 6. F-center profile curves in LiF bombarded at R.T. with $8.1 \cdot 10^9$ α-particles/cm^2 (1), $3 \cdot 10^{10}$ (2), $6.5 \cdot 10^{10}$ (3), 10^{11} (4), $2 \cdot 10^{11}$ (5), $4.2 \cdot 10^{11}$ (6) and $5 \cdot 10^{11}$ (7).

in the (100) one. This can be explained if we take into account the cationic defect production by direct displacement processes which is more efficient in the (110) than in the (100) direction. These cationic defects can trap the electrons and then the F-center creation becomes lower. An evidence of cationic defects production is found with the I-centers (absorption band near 550 nm) which are present in LiF samples bombarded with high doses ($\sim 10^{15}$ ions/cm^2). The first experiments performed on the I-center profiles show that these profiles could be connected to the nuclear energy loss curve.

The F-center growth curves give a linear dependence of the F-center production with the dose, in a log-log plot system ($\sim t^{0.9}$). A departure from linearity above $5 \cdot 10^{11}$ α/cm^2 and 10^{12} deuterons/cm^2 is observed. However, the optical measurements become difficult in this case because of an important optical density. The saturation effects related to the dose and those

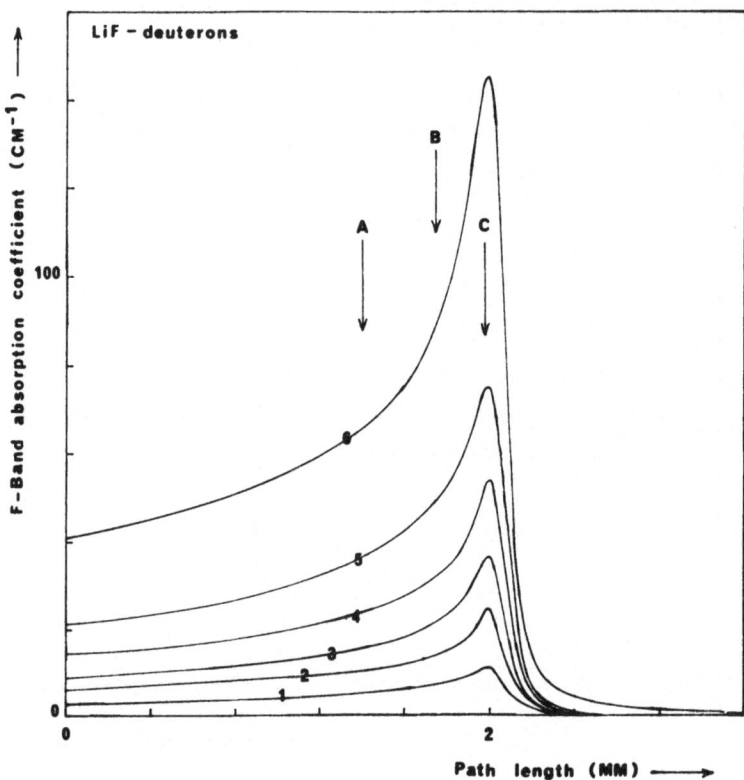

Fig. 7. F-center profile curves in LiF bombarded at R.T. with
10^{11} deuterons/cm^2 (1), $2 \cdot 10^{11}$ (2), $4.2 \cdot 10^{11}$ (3), $6 \cdot 10^{11}$ (4),
10^{12} (5) and $2.9 \cdot 10^{12}$ (6).

related to the electronic energy loss (near the ionization maximum) let us assume that the interaction of the particles with the
target occurs in a tube. In this case, the great local F-center
concentration and the presence of cationic defects can explain
some phenomena which are observed in our crystals: for instance,
the $F_2^+ \rightarrow F_3^+$ transformation [11] instead of $F_2^+ + e \rightarrow F_2$ and
the formation of colloidal centers in high dose irradiated samples.
An estimation of about 100 Å for the diameter of the interaction
tubes can be given in regard of the previous results.

Conclusion

     The study of the defects created in alkali halides by ener-
getic ions allows investigation of the different interaction
mechanisms of the particle with the target. In α-particle and

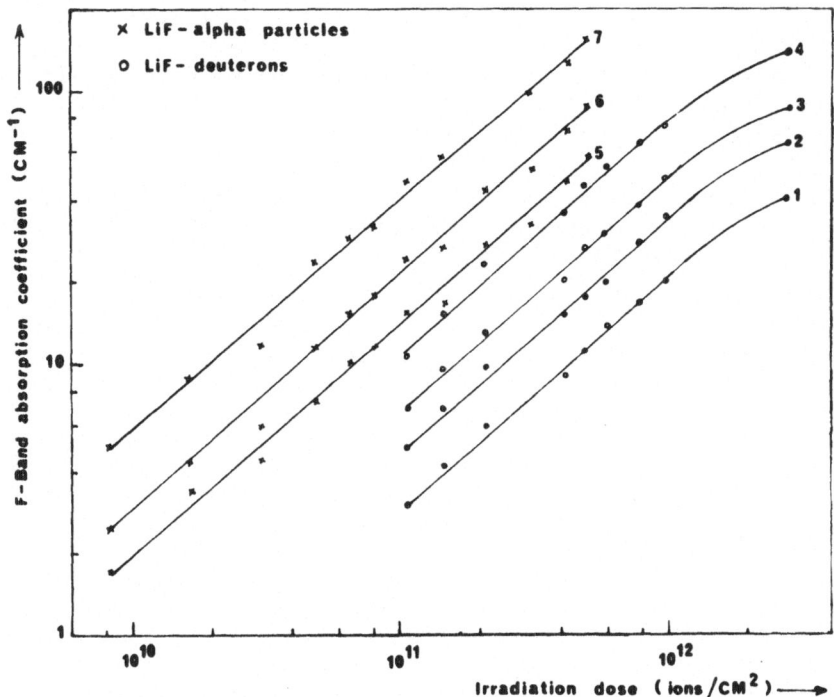

Fig. 8. F-center growth curves deduced from Figures 6 and 7 near the surface of the crystal (5,1) and at depths corresponding to the points A (6,2), B (7,3) and C (4).

deuteron bombarded LiF we observe an F-center production in a good agreement with the electronic stopping power. The experimental results induce us to consider the interaction tubes in which the cationic defects are present in addition to the F and F-aggregate centers.

References

[1] A. Perez, P. Thevenard, J. Davenas and C. H. S. Dupuy, Phys. Stat. Sol. (a) 18, 189 (1973).
[2] Y. Farge, Thèse. Paris (1967).
[3] D. Pooley, Advanced Study Institute on Radiation Damage Processes in Materials, Corsica 27 August - 9 September 1973.
[4] J. Davenas, A. Perez, P. Thevenard and C. H. S. Dupuy, to be published in Phys. State. Sol. (a), 19,(1973).
[5] E. V. Benton and R. P. Henke, USNRDL-TR-1102 (1966).

[6] R. D. Evans, The Atomic Nucleus, McGraw–Hill Publ. Co. (1955), p. 645.

[7] M. Bader, R. E. Pixley, F. S. Mozer, and W. Whaling, Phys. Rev. 103, 32 (1956).

[8] A. Smakula, Z. Phys. 59, 603 (1930).

[9] D. L. Dexter, Phys. Rev. 101, 48 (1956).

[10] G. W. Arnold and F. L. Wook, Radiat. Eff. 14, 157 (1972).

[11] P. Thevenard, A. Perez, J. Davenas, and C. H. S. Dupuy, Phys. Stat. Sol. (a) 10, 67 (1972).

VELOCITY DEPENDENCE OF THE STOPPING POWER
OF CHANNELED IODINE IONS*

C. D. MOAK, B. R. APPLETON, J. A. BIGGERSTAFF,
S. DATZ and T. S. NOGGLE
Oak Ridge National Laboratory
Oak Ridge, Tennessee 37830, U. S. A.

Recent measurements of some uranium ion stopping powers, in the energy range 30-90 MeV [1], in polycrystalline targets, together with earlier data for Br ions and I ions have shown that, in the energy region where the theories of Lindhard, Scharff and Shiøtt [2] and Firsov [3] predict that electronic stopping should obey the relation $S_e = k^{1/2}$, the data show that $S_e = a + bE^{1/2}$. Fig. 1 is illustrative of the fact that heavier ions show a larger value of a than that for light ions. Presumably this effect would be un-important for ions as light as oxygen. In particular, stopping power data for uranium ions has been compared with the theoretical predictions of the theories of Lindhard and Firsov for carbon, nickel and gold respectively, as shown in Figs. 2-4. The important point is that the data cannot be reconciled with theory by a simple change of slope.

It was expected that ions moving in crystal channels and having no close collisions with atoms might show the $S_e = kE^{1/2}$ behavior. Some data taken by Eriksson, Davies and Jespersgaard [4] with very low energy Xe ions in W crystals do appear to follow this relationship. At higher energies, 21.6 to 32.5 MeV, iodine ion stopping powers have been measured for particles hyperchanneled [5] in the [001] axis of a crystal of Ag. The results are shown in Fig. 5. Polycrystalline stopping powers measured by Moak and Brown [6] are shown as measured and, slightly below, adjusted to remove the estimated contribution of nuclear stopping [7]. Channeling data have not been adjusted for nuclear stopping since there is strong evidence for the conclusion that nuclear stopping is neg-ligible for channeled ions [8]. The energy loss pattern for iodine ions hyperchanneled in the [001] axis in Ag is shown schematically in the figure [9]. Beginning with particles showing the least energy loss (and running nearest the center of the channel) and including particles which fall in the class of ordinary axially channeled particles and finally including particles which run in

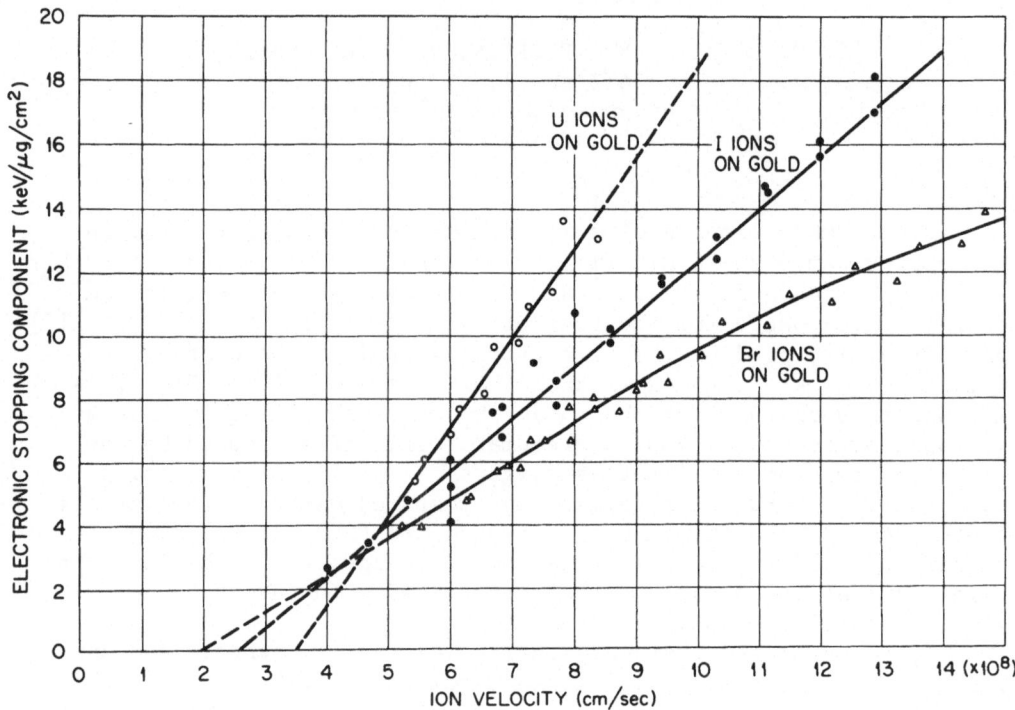

Fig. 1. Electronic stopping component for U, I and Br ions in Au.

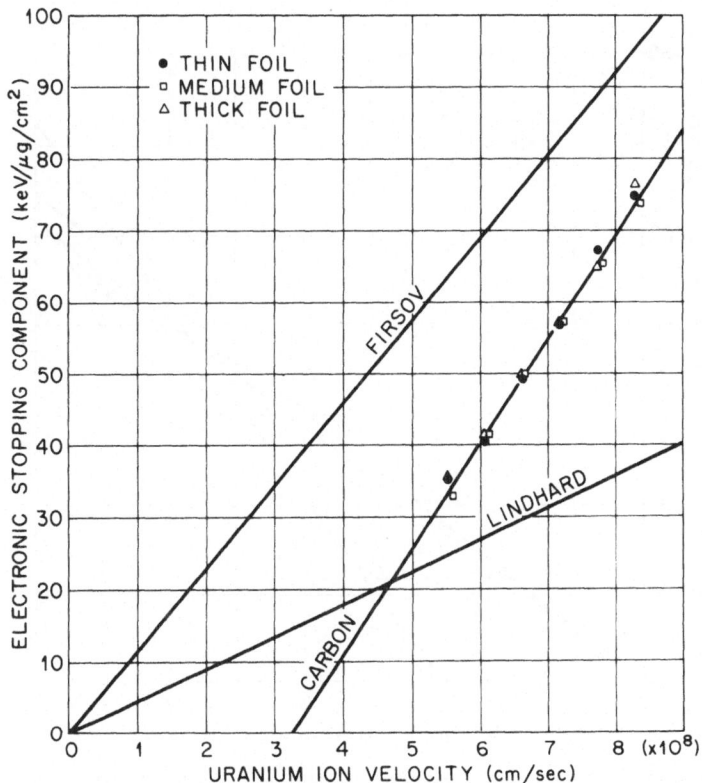

Fig. 2. Comparison of U electronic stopping component in C
compared with the Firsov [3] and Lindhard et al. [2] predictions.

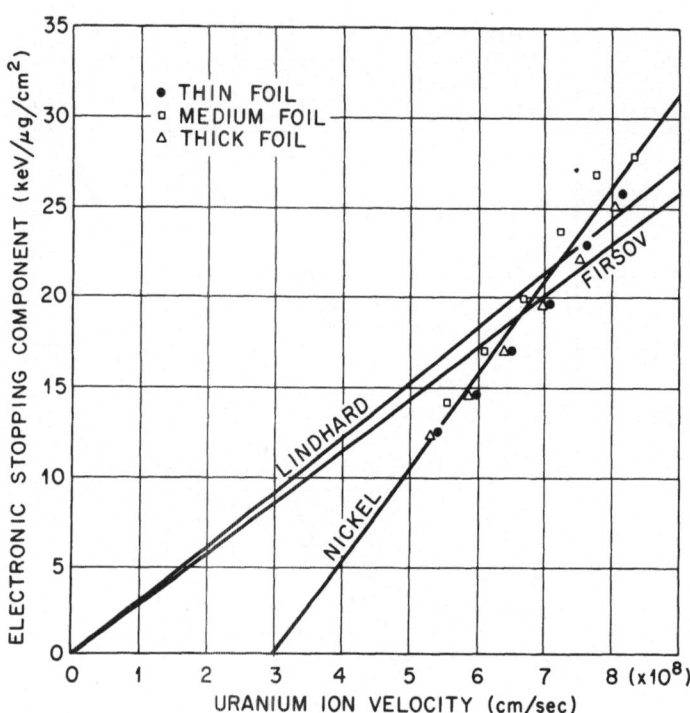

Fig. 3. Comparison of U electron stopping component in Ni compared with the Firsov [3] and Lindhard et al. [2] predictions.

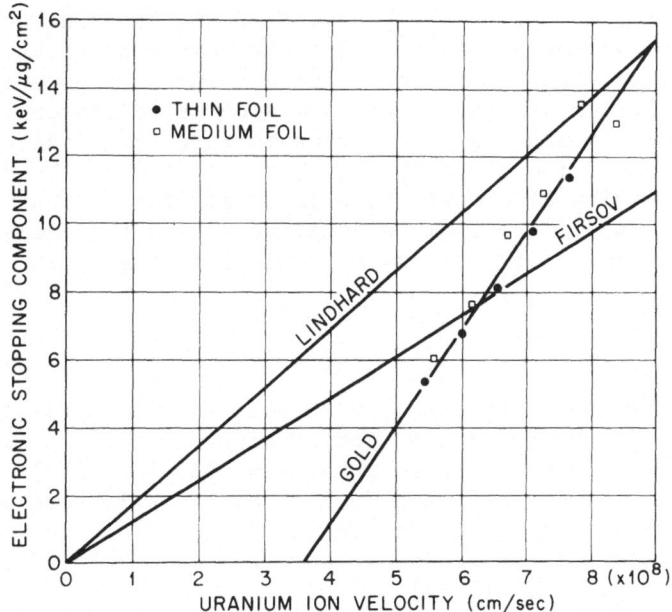

Fig. 4. Comparison of U electronic stopping component in Au compared with the Firsov [3] and Lindhard et al. [2] predictions.

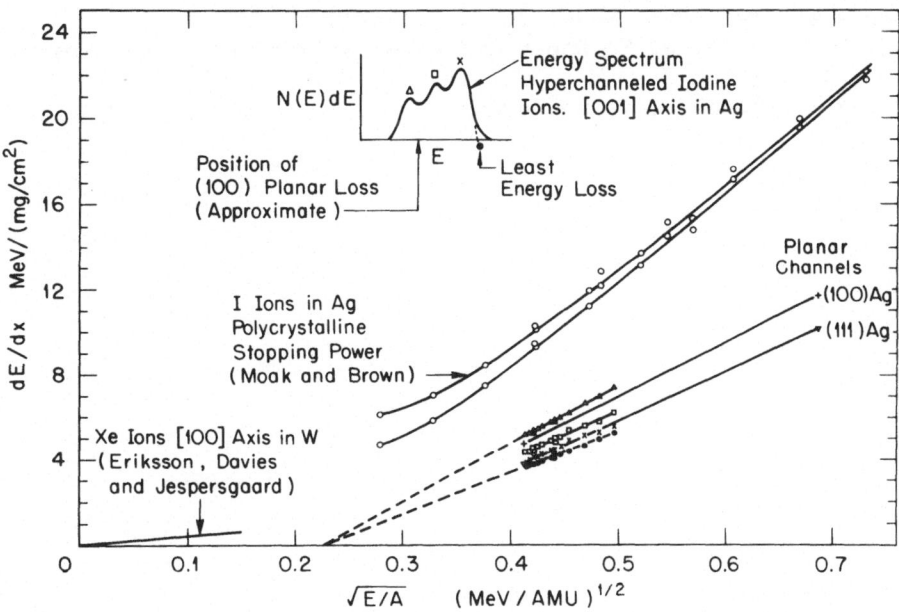

Fig. 5. Stopping powers for I ions in Ag in random and planar channeling directions.

the (100) planar channels, the behavior of the stopping power does not obey the relation $S = kE^{1/2}$. The data given by Eriksson, Davies and Jespersgaard have been included in the figure, even though both the ion and the stopping medium are different. At higher energies, where the iodine measurements were made, the stopping powers follow the relation $S_e = a + bE^{1/2}$. The data suggest that, below 21.6 MeV, there is a velocity region where the stopping power curve is more complicated and that some additions must be made to the theory for this velocity range.

References

*Research sponsored by the U. S. Atomic Energy Commission under contract with Union Carbide Corporation.

[1] M. D. Brown and C. D. Moak, Phys. Rev. B6, 90 (1972).

[2] J. Lindhard, M. Scharff and H. E. Schiøtt, Kgl. Danske Videnskab. Selskab, Mat.-Fys. Medd. 33, 14 (1963).

[3] O. B. Firsov, Zh. Eksperim. i Teor. Fiz. 36, 1517 (1959), [Sov. Phys. JETP 9, 1076 (1959)].

[4] L. Eriksson, J. A. Davies and P. Jespersgaard, Phys. Rev. 161, 219 (1967).

[5] B. R. Appleton, C. D. Moak, T. S. Noggle, and J. H. Barrett, Phys. Rev. Lett. 28, 1307 (1972).

[6] C. D. Moak and M. D. Brown, Phys. Rev. 149, 224 (1966).

[7] J. Lindhard, V. Nielsen and M. Scharff, Kgl. Danske Videnskab Selskab, Mat.-Fys. Medd. 36, 10 (1968).

[8] C. D. Moak, J. W. T. Dabbs and W. W. Walker, Rev. Sci, Instr. 37, 1131 (1966).

[9] B. R. Appleton, this conference.

CHARGE STATE DEPENDENCE OF STOPPING POWER FOR OXYGEN IONS CHANNELED IN SILVER*

S. DATZ, B. R. APPLETON, J. A. BIGGERSTAFF, M. D. BROWN[†]
H. F. KRAUSE, C. D. MOAK and T. S. NOGGLE
Oak Ridge National Laboratory
Oak Ridge, Tennessee 37830
U. S. A.

ABSTRACT

In previous work [1] it was shown that electron capture and loss cross sections for 30-40 MeV oxygen ions in Au channels were small enough ($\sim 10^{-19}$ cm^2) to preclude charge state equilibrium in crystals 1 μm thick. In fact, a large fraction of charges 8+ and 7+ passed through the crystal without any charge exchange. In this work we show that similar conditions obtain in Ag and we have measured the charge state dependence of the stopping power for 27.8 and 40 MeV O ions in the [110] axial channel. We find that for those ions which do not undergo charge exchange $S = kq^2$ for charges 6+, 7+ and 8+. Other features observed in the energy spectra of the emergent ions are attributable to charge exchange in the crystal. Screening by Fermi electrons is expected to be the same for channeled and randomly penetrating ions. The results indicate that this screening is not effective in reducing the energy loss of the ion.

Introduction

The effect of ionic charge on the stopping power of the ion in a solid has been a subject for conjecture since the observation by Lassen [2] that charge states of heavy ions emerging from solids were much higher than those found upon emergence from gaseous targets (e.g., the most probable charge state of 60 MeV I emerging from C is 22+ while only 12+ is attained in an N$_2$ target [3]). Yet the electronic stopping powers in the solid and gaseous media are quite closely the same. Two explanations have been offered to explain this apparent charge state independence.

    Betz and Grodzins [4] proposed in 1970 that the ion charge in
the solid is actually the same as in the gas but that a high degree
of steady state excitation exists which is relaxed by multiple
Auger electron ejection from the ion after it leaves the solid.
No direct experimental evidence has yet been found to support this
conjecture [5].

    Bohr and Lindhard [6] proposed in 1954 that any difference in
ion charge in the solid as against the gas was made up by a difference in screening by the electrons of the medium. In 1973 Brandt
et al.[7] elaborated on the notion of screening of fast heavy ions
and introduced the idea of dynamic screening to account for their
observed differences in Al K x-ray yields caused by differing incident charge states of the 12 – 68 MeV oxygen ions used to excite
the target. They noted the charge state dependence for x-ray
excitation cross-sections recently reported for single collisions
in gases [8] and found with their experiment using thin solid Al
targets (1–20 $\mu g/cm^2$) a charge state dependence which decreased
with increasing target thickness. They account for the observed
data in terms of the time required for the Fermi electrons in the
solid to respond to the charge imbalance created by the entering
projectile. This dynamic response rate λ is related to the plasma
frequency ω_p; $\lambda = (\beta/2\pi)\omega_p$, where $\beta \simeq 0.1$. From this they obtain
a dynamic screening equilibrium distance $D \simeq 40(v_1/v_0)$Å (e.g.,
400 Å for 40 MeV O ions). Further they conclude that the effective
charge states in a solid can be described in terms of the dynamic
screening of a moving particle by the target electron gas and that
the distinction between screening by bound states undergoing electron
capture and loss processes and by dynamic polarization is unimportant.

    In the present work we have measured the stopping power of 28
and 40 MeV O ions channeled in Ag <110> as a function of the number
of bound electrons. These measurements are possible because of the
low cross sections for electron capture and loss for these channeled
ions. Because of this we are able to measure the stopping powers for
ions of charge 6+, 7+ and 8+ which have not changed their charge
during transit through the solid.

Experimental

    The arrangement is shown in Fig. 1. Oxygen ions of 27.8 MeV
and 40 MeV were accelerated by the Oak Ridge Tandem Accelerator,
passed through a thin carbon foil and various beam charge states
were selected by a 30° analyzer magnet and then passed through a
single crystal of Ag. The crystal thickness was 0.56 μm and, when
oriented to the <110> axial channel at 45°, the effective crystal
path length was 0.8 μm. A set of collimators selected the central
core of the emergent channeling pattern and these particles were
charge-analyzed in an electrostatic analyzer. Particle energies
were measured with a surface barrier detector having an energy
resolution of 120 keV. Charge state distributions were measured

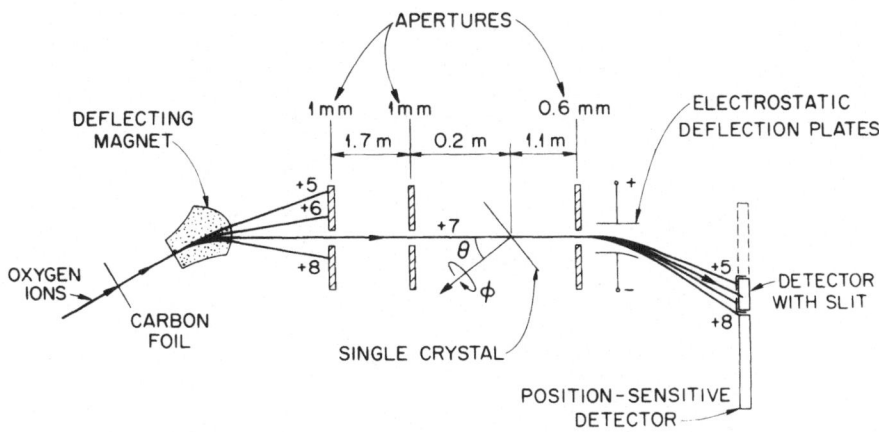

Fig. 1. Experimental arrangement.

with a position-sensitive detector for each case. Thus it was
possible to select a given emergent charge state from the center
portion of the channeling pattern, and to measure energy losses
of these particles.

Results and Discussion

     The exit charge-state distributions (CSDs) obtained from in-
jection of 40-MeV oxygen ions in initial charge states ranging from
6+ to 8+ and 27.8 MeV oxygen ions in initial charge states ranging
from 5+ to 8+ into a <110> axial-channeling direction in a 0.8 μm
thick Ag crystal are shown in Fig. 2, together with the CSDs obtained from injection in a "random" direction. The random distributions (dashed curves) are representative of charge-state equilibrium since they are independent of input charge state (6+ or 8+).
The channeled-ion CSDs are clearly non-equilibrium cases. Since
the 8+ fraction is enhanced for all initial charge states, it is
immediately obvious that the equilibrium CSD for channeled ions
will contain a higher fraction of higher charge states than the
random case. Moreover, it is clear that some of the charge-changing
cross sections for the channeled ions in attaining equilibrium
must be quite small, i.e., if we assume that equilibrium is attained
in <u>ca</u> 5 mean-free paths for charge exchange, then charge-changing
cross sections of $<<10^{-18}cm^2$ must be involved if equilibrium is
not achieved in 1 μm with target densities of $\sim 5 \times 10^{22}$ atoms/cm^3.
     These results are exactly analogous to our previously reported
results with O channeled in Au [1]. In that work we determined
charge-changing cross sections from measurement of CSDs as a function
of pathlength. For the <110> channel in Au the cross sections were

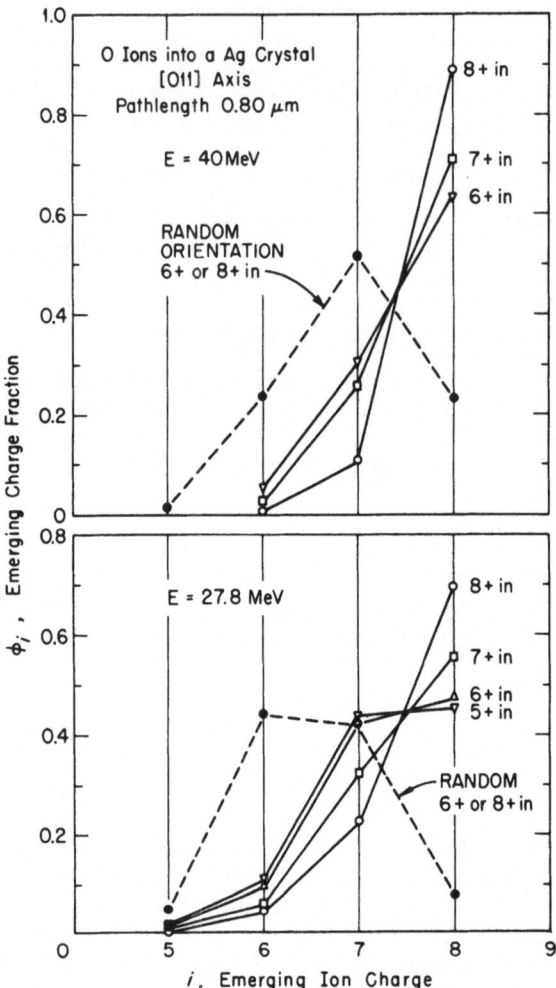

Fig. 2. Emergent charge state distributions of 27.8 and 40 MeV O ions channeled in Ag [011] for various input charges. For random directions the exit distribution is independent of input charge.

$\sigma_{8,7} = 1.5 \times 10^{-19}$ for electron capture by O^{8+} and $\sigma_{7,8} = 3.5 \times 10^{-19}$ for electron loss by O^{7+} at 40 MeV. $\sigma_{7,6}$ is comparable to $\sigma_{8,7}$ and $\sigma_{6,7} > \sigma_{7,8}$. A comparison with the present data indicates that the charge-changing cross sections in Ag are comparable and perhaps somewhat smaller. With these small cross sections ($\sim 10^{-19} cm^2$) it is possible for a large fraction of the entering ions to pass entirely through the crystal without ever undergoing any charge-changing collisions.

The energy loss spectra observed for 40 MeV oxygen 8+, 7+ and 6+ ions entrant and 8+, 7+, and 6+ ions emergent respectively are

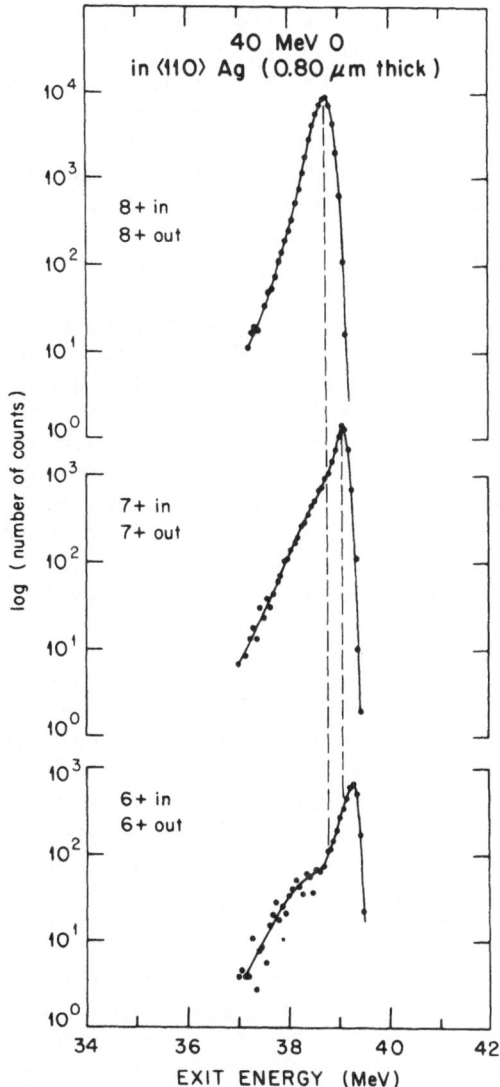

Fig. 3. Emergent energy spectra for 40 MeV O ions channeled in Ag [011], 0.8 μm pathlength, Δq = 0.

shown in Fig. 3. The peaks of these spectra are quite clearly displaced from each other. At the energy loss represented by the peak of the 6+ in 6+ out spectrum the 8+ in 8+ out spectrum is down 3 orders of magnitude from its peak. This reflects the extremely small cross section for sequential capture from 8 to 7 and from 7 to 6 (cf. Fig. 2). The slight shoulder on 7+ in 7+ out spectrum at the position of the 8+ peak may represent ions that have traveled most of the way through the crystal as 8+ ions; i.e., they have lost an electron close to the entrance point and picked one up again just prior to emerging from the crystal.

The trailing off to lower energy (higher energy loss) most evident in the 6+ spectrum is due to the increasing contribution of particles with higher transverse energy in the channel. As we demonstrated and discussed in our previous work on Au [1] particles with higher transverse energy, due either to smaller channel dimension or higher amplitude in planar channels, give a charge-state distribution closer to the random CSD. Thus we would expect a larger relative contribution at lower charge state (cf. Fig. 2). From other evidence where we measured the best channeled energy losses for 6, 7 and 8+ in the (111) planar channel and found systematically higher energy losses for the A_0 group we can assert that the peaks of energy spectra shown in Fig. 3 are those due to hyperchanneled [9] particles which have remained at their input charge state throughout the crystal. (For randomly penetrating ions where the charge-changing cross sections are ∿100 times higher no difference in energy loss with input charge is observed.)

The data taken for an identical experiment but at lower energy (27.8 MeV) are shown in Fig. 4. Here again the peak positions vary with charge state and the remaining features are similar. However, at the lower energy all charge changing cross sections are higher and most significantly $\sigma_{7,6}$ capture cross section is larger relative to the $\sigma_{6,7}$ loss cross section (cf. Fig. 2). These effects cause the double peak seen in the 6+ in 6+ out spectrum. The peak appearing near the same energy loss as the 7+ in 7+ out spectrum comes from ions which lost an electron close to their entrance point and picked on up again close to their exit point (i.e., their total energy loss is close to the characteristic 7+ energy loss).

We have measured spectra for all permutations of entrance and exit charge states out and all conform to the picture outlined above. As an example see Fig. 5 which shows the spectra obtained with 6+ in and 8+, 7+ and 6+ out for 28 MeV O ions. The dashed vertical lines indicate the energy loss for the $\Delta q = 0$ cases shown in Fig. 4. Since $\sigma_{6,7}$ and $\sigma_{7,8}$ are appreciable the ions emerging at 7+ have lost an electron rather early on in their path and have losses almost characteristic of the 7+ loss throughout the crystal. To reach 8+, however, takes a larger distance so that the peak in the 8+ out spectrum lies at a higher energy than the pure 8+ loss (i.e., a small part of its path has been as 6+ and a larger part of its path has been at 7+ prior to its losing its lost electron to form 8+).

The most probable energy loss for all permutations of input and output charge are plotted in Figs. 6 and 7 versus the square of the input charge state. Considering the cases for which no charge exchange has occurred ($\Delta q = 0$) the energy loss is accurately proportional to the square of the charge state (i.e., $s = kq^2$ where $k = .024$ MeV/μm at 40 MeV and $.032$ MeV/μm at 27.8 MeV. These k values correspond to the stopping power anticipated for a hypothetical singly charged channeled oxygen ion.

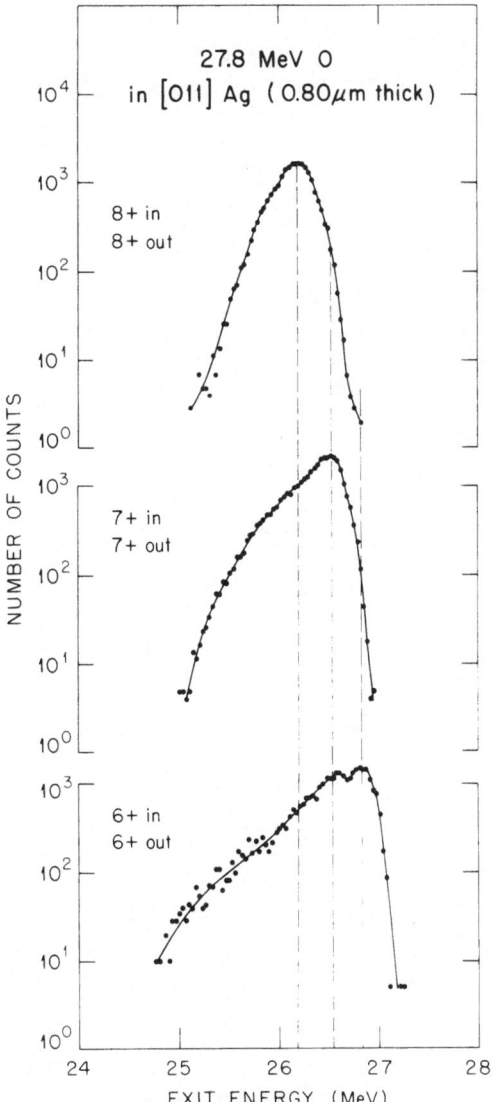

Fig. 4. Emergent energy spectra for 27.8 MeV O ions channeled in Ag [011], 0.8 μm pathlength, Δq = 0.

An interesting comparison may be made with the stopping powers of protons at the same velocities (MeV/nucleon). For protons at 2.5 and 1.74 MeV (corresponding to 40 and 27.8 MeV O) the stopping powers for protons in random Ag are listed as .061 and .077 MeV/μm respectively [10]. The k values for O are 40% of these and are

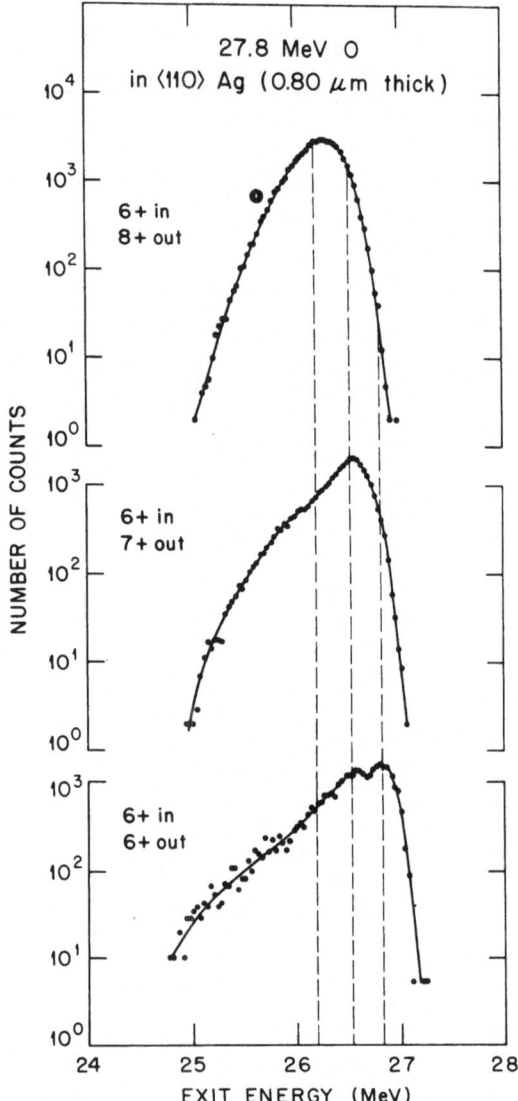

Fig. 5. Emergent energy spectra for 27.8 MeV O ions. Here the input charge is 6+ and the emergent charges are 8+, 7+ and 6+.

therefore in accord with the stopping powers anticipated for protons channeled in Ag.

From the above it is clear that the bare oxygen nucleus behaves very much like a point mass of charge 8+ and that the addition of bound electrons to form 7+ and 6+ ions reduce the

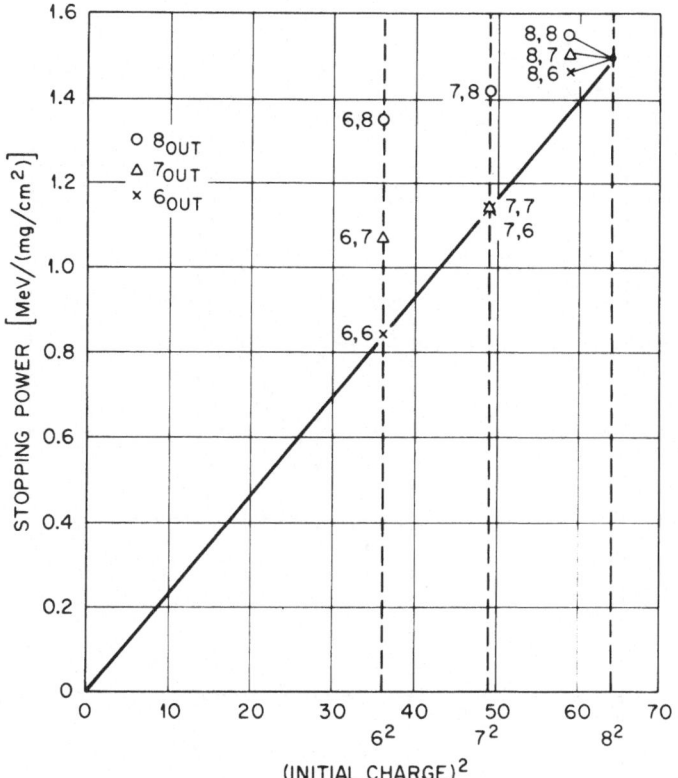

Fig. 6. Stopping powers of 40 MeV O ions in the [011] axis of Ag for various charge combinations q_{in} and q_{out}.

effective charge for stopping by one whole unit each. Moreover, it appears that the dynamic screening by polarized valence electrons is unimportant in decreasing the particles' effective charge. This in spite of the fact that the pathlength is 8000 Å compared to the ∿400 Å expected for the establishment of dynamic screening equilibrium [7].

To understand this let us consider in some detail the processes which contribute to the stopping of these ions. The initial interaction with conduction electrons occurs in the manner described by Brandt et al., i.e., the electron plasma responds to the intrusion of the positive ion in a time ≃$1/\omega_p$. However, for ions with velocities $v_1 \gg v_0$ the response of the plasma lags behind the ion's position in space. A calculation by Neelavathi and Ritchie [10] shows that the ion is followed by a wake in which there are oscillations in the electron density both above and below the unperturbed densities. Immediately following the ion there is an enhanced density which falls to zero at a distance $\pi v/\omega_p$. The number of electrons contained in this wake is equal to the charge on the

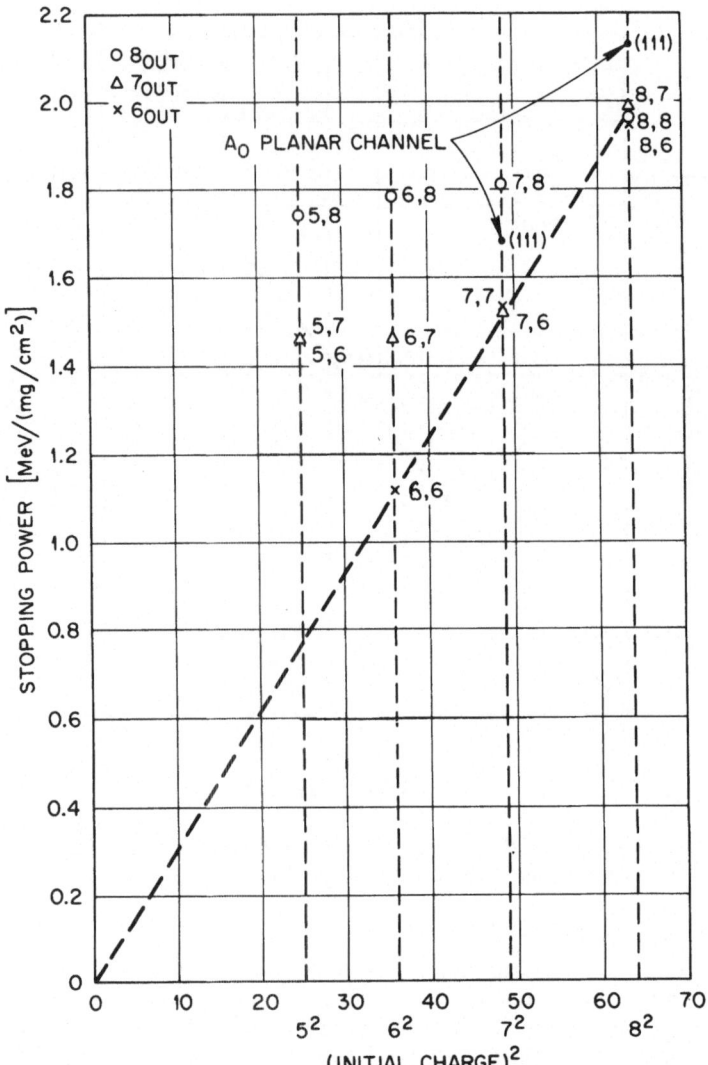

Fig. 7. Stopping powers of 27.8 MeV O ions in the [011] axis of Ag for various charge combinations q_{in} and q_{out}. (The points labeled (111) are the A_O group losses in planar (111) channels).

moving ion. For 40 MeV O ions the first enhanced density lobe is ∿60 Å long, it is approximately symmetric with a peak at a distance of about 30 Å from the nucleus, and the enhanced density at the nucleus is very small. The effect of adding a bound electron to the nucleus would be to totally screen one unit of the nuclear charge at a distance of about 0.1 Å (i.e., somewhat larger than $a_0/8$.

The Coulombic impulse due to the extended wake would be minimal when compared with the compact nucleus-bound electron system. The bulk of the stopping power arises from interactions with the 4f electrons which lie ∿0.5 Å from the best channeled ion's path. Hence the bound electrons would screen the nuclear charge by one unit each whereas the trailing electron wake would be ineffective in screening.

References

*Research sponsored by the U. S. Atomic Energy Commission under contract with the Union Carbide Corporation.
†Department of Physics, Kansas State University.

[1] S. Datz, F. W. Martin, C. D. Moak, B. R. Appleton and L. B. Bridwell, Radiation Effects 12, 163 (1972).

[2] N. O. Lassen, Kgl. Danske Videnskab. Selskab, Mat.-Fys. Medd. 26, No. 5 and 12.

[3] C. D. Moak, H. O. Lutz, L. B. Bridwell, L. C. Northcliffe and S. Datz, Phys. Rev. 176, 427 (1968).

[4] H. D. Betz and L. Grodzins, Phys. Rev. Letters 25, 211(1970).

[5] S. Datz, B. R. Appleton, J. A. Biggerstaff, M. G. Menendez and C. D. Moak, Bull. Am. Phys. Soc. 18, 662 (1973).

[6] N. Bohr and J. Linhard, Kgl. Danske Videnskab, Mat.-Fys. Medd. 28, No. 7 (1954).

[7] W. Brandt, R. Laubert, M. Morino and A. Schwartzchild, Phys. Rev. Letters 30, 358 (1973).

[8] J. R. Mowat, I. A. Sellin, D. J. Pegg, R. S. Peterson, M. D. Brown and J. R. MacDonald, Phys. Rev. Letters 30, 1289 (1973).

[9] J. H. Barrett, B. R. Appleton, T. S. Noggle, C. D. Moak, J. A. Biggerstaff and S. Datz, this volume.

[10] V. N. Neelavathi and R. H. Ritchie, this volume.

TRANSMISSION ENERGY LOSS OF PROTONS CHANNELED IN THIN SILICON SINGLE CRYSTALS AT MEDIUM ENERGY

G. DELLA MEA, A. V. DRIGO, S. LO RUSSO, P. MAZZOLDI
Unita GNSM-CNR, University of Padova - Italy

and

G. G. BENTINI
*Laboratorio Tecnologia dei Materiali per l'Elettronica-CNR
Bologna - Italy*

ABSTRACT

Measurements of stopping power of protons
in silicon in random and channeling conditions
(<111> direction) in the energy-range between
300 to 50 keV have been performed. The trend of
α-ratio has been reported.

Previous measurements [1] of transmission energy loss in channeling conditions have been extended at low energy from 300 keV
down to about 50 keV. Proton beams have been obtained from the 400
keV Cokroft of CNR-LAMEL in Bologna. The crystals used were Si<111>
3500 to 8000 Å thick. The energy of transmitted ions emerging within a small angle about the incident direction was measured using a
$\overline{Ex}\overline{B}$ velocity filter for protons with energy below 200 keV and both
filter and silicon surface barrier detector for higher energies.
Since in this energy region the energy loss is not negligible compared to incident energy, we used an iterative procedure to obtain
the stopping power from the energy loss. It is found that the
ratio α between channeling and random stopping power is a function
of incident energy and reaches values near the unity at about 80
keV energy, the same energy at which the maximum in the random
stopping power is observed (Figure 1).
The energy value at which the random stopping power is maximum
and the corresponding value of the stopping power are correlated,
according to the universal function given by Brandt [2]. In the
region where random stopping power is maximum only valence electrons
contribute to stopping power [3]. Since the latter is the major
contributer to stopping power under channeling conditions, the

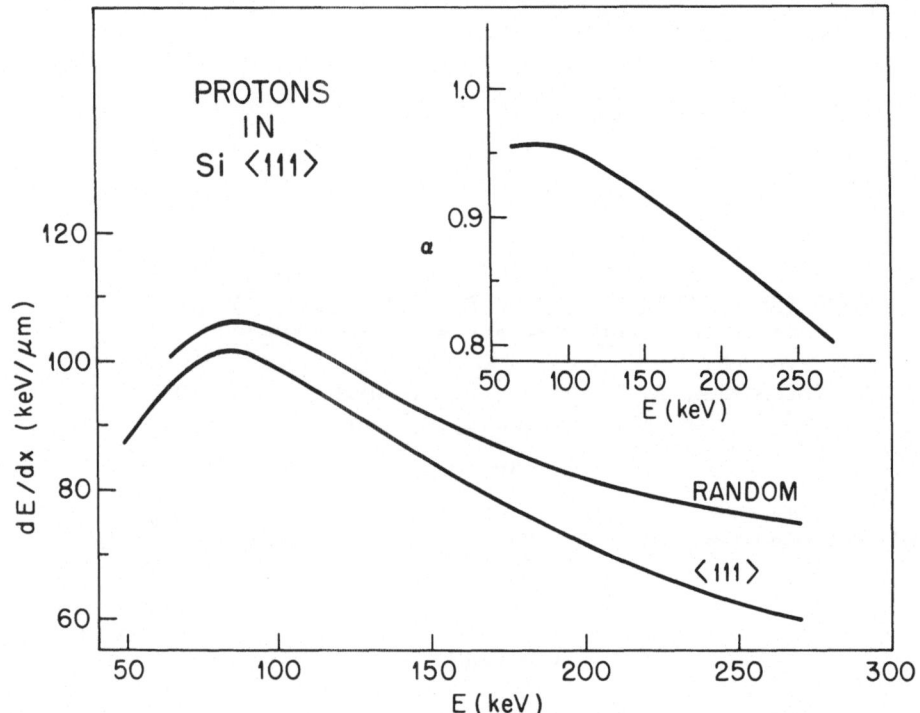

Fig. 1. Stopping power trend vs energy. The ratio α between the channeling and random stopping power is shown top right as a function of energy.

observed results concerning the α-ratio can be accounted for very easily. The deviation from unity may be due to deviations from a constant electron density of the valence electron gas because of the interaction with the ion core pseudopotential, producing different average electron densities in the two cases [4].

Acknowledgments

We are indebted to Prof. W. Brandt for clarifying some important points and to Dr. A. Desalvo and R. Rosa for useful discussions.

References

[1] G. Della Mea, A. V. Drigo, S. Lo Russo, P. Mazzoldi and G. G. Bentini, Phys. Rev. Lett. 27, 1194 (1971) and Rad. Effects 13, 115 (1972).
[2] W. Brandt, this conference.
[3] W. Brandt and J. Reinheimer, Phys. Rev. B2, 3104 (1970).
[4] W. Brandt, private communication.

A NEW METHOD TO DETERMINE THE ENERGY LOSS
OF HEAVY IONS IN SOLIDS

HORST SCHMIDT-BÖCKING, GERD RÜHLE and KLAUS BETHGE
II. Physikal. Institut der Univ. Heidelberg,
69 Heidelberg, Germany

When today's physicist needs data about the differential energy loss of heavy ions, he refers to the
tables by Northcliffe and Schilling (Nuclear Data Tables
A7) [1]. However, these data are obtained for heavy ions
in a semiempirical way and hardly examined experimentally
for ions heavier than oxygen in solids. We have thus
developed a method by which quick and reliable energy
loss data for heavy ions in solids can be determined.
If the elastic scattering cross section as a function
of the energy is known, then it is possible to determine
a larger energy range of the differential energy loss
from one single measurement.

Description of the Method

The basic idea of this method is the following: A beam of
heavy ions strikes an infinitely thick target with a very smooth
surface (<0.3 μ). The heavy ions penetrate the target to various
depths and in part are scattered out of the target. In a surface
barrier detector the heavy ions, that are scattered under the angle
ϕ (Fig. 1), are counted. If $\sigma_{elast.\ scatt.} = \sigma_{Ruth}$ the particle
spectrum, as a function of the energy, has the form shown in Fig. 2.
Disregarding all straggling processes in the target, we can attribute
to each energy interval ΔE_n (Fig. 3) a layer of thickness Δx_n in the
target, in which all particles of this energy interval were scattered.
The layer thickness Δx_n can be determined from the number of
particles ΔN_n registered in an energy interval ΔE_n with knowledge
of the scattering cross section $\sigma(E_n')$. Once all the Δx_n have been
calculated, the differential energy loss $dE(E_n)/dx \equiv f(E_n)$ can be
determined from the total energy loss of the ions ($E_0 - E_n$). Via
a recursive method, beginning with energy E_0, all $f(E_n)$ can be successively calculated, since E_1' equals E_0.

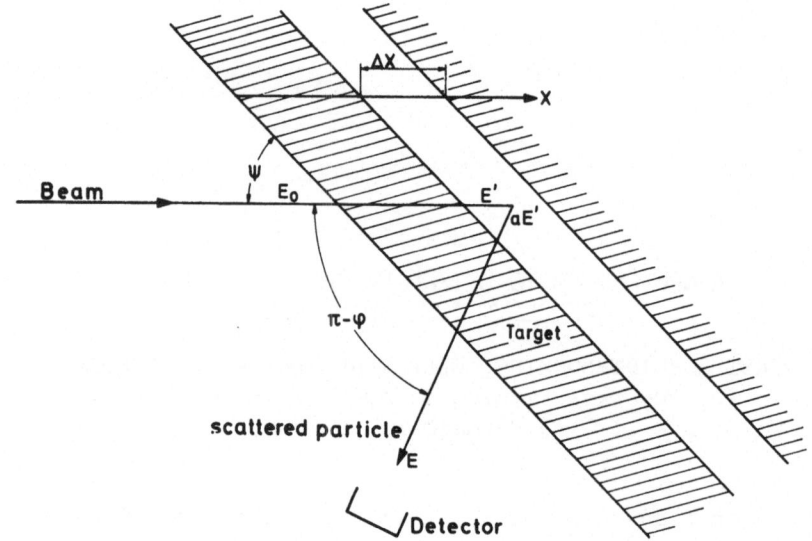

Fig. 1. Definition of the incident and scattering angles.

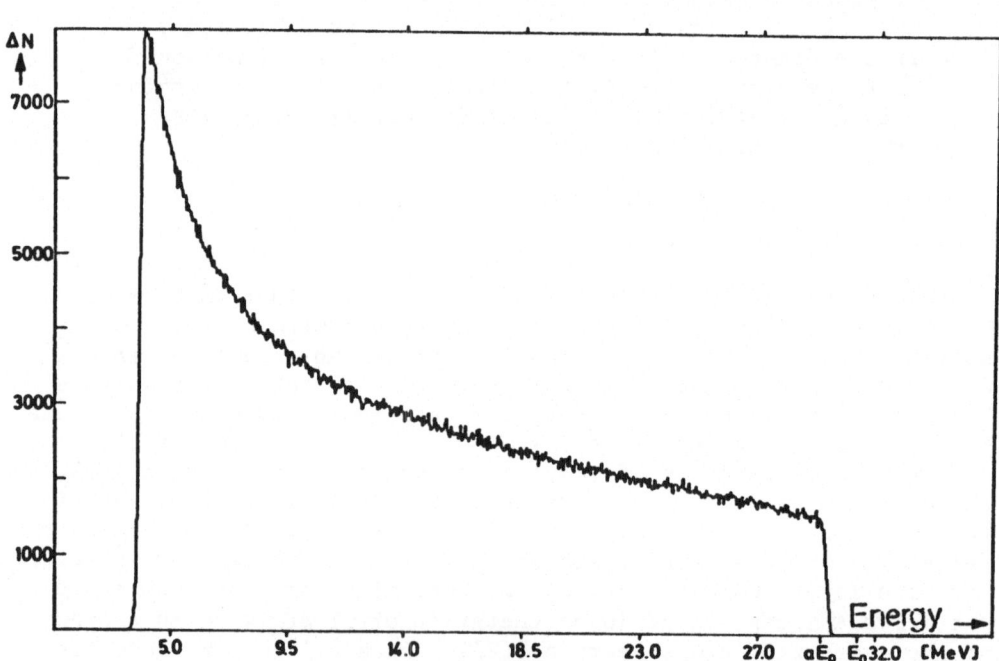

Fig. 2. Typical spectrum of elastically scattered heavy ions on an infinitely thick target (elastic scattering of ^{16}O on a thick gold-target E = 31.5 MeV, ϕ = 50°, ψ = 25°).

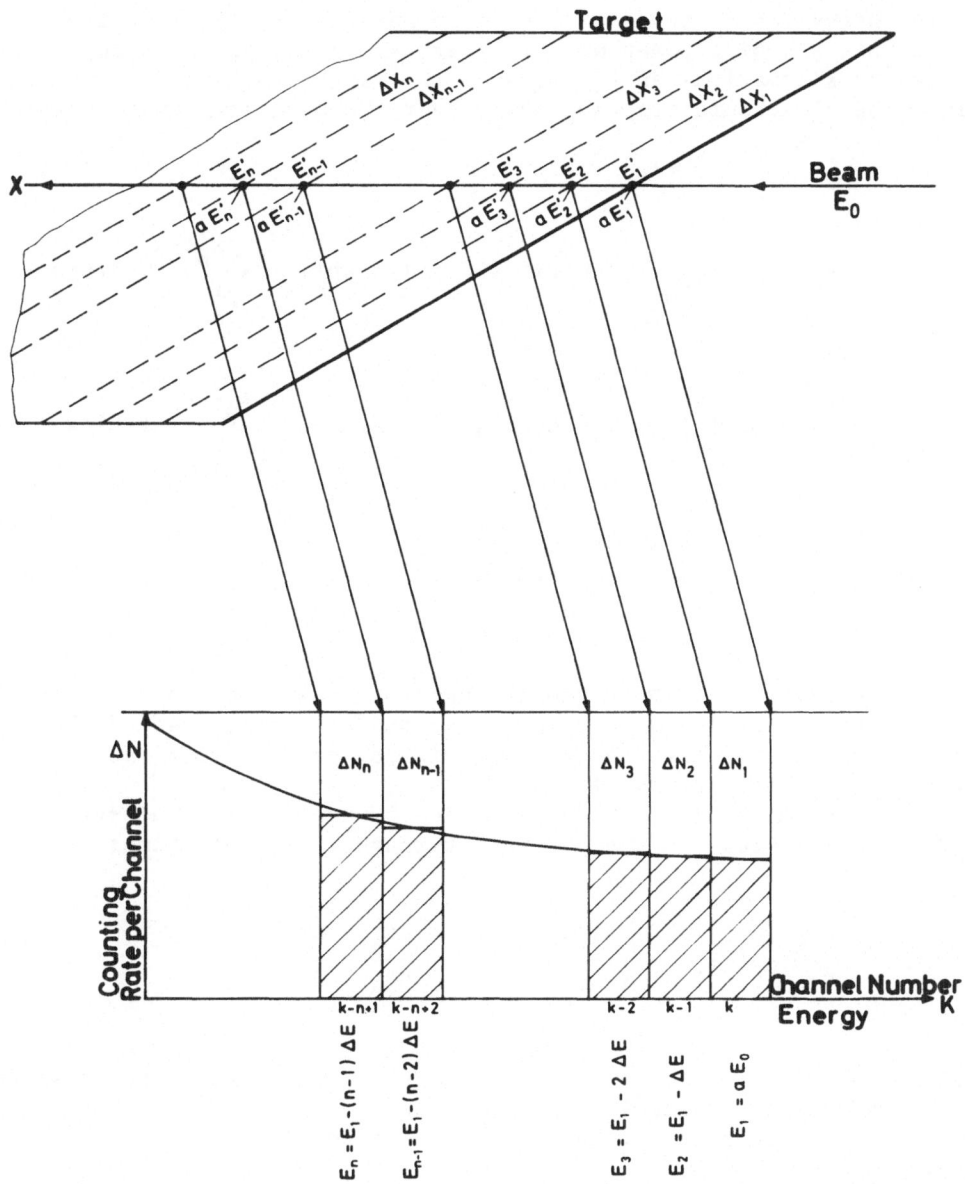

Fig. 3. Correlations between target layers and energy intervals in the multichannel analyzer [E_0 = bombarding energy, E_n' = energy of the particles just before scattering, aE_n' = energy of the particles just after they are scattered, a = kinematical factor (energy indepndnet), E_n = energy after leaving the target, x_n = length of the path of the ion in the target before scattering, λx_n = length of the path of the ion in the target after scattering].

The most important mathematical steps are as follows: The energy
of the heavy ion E_n is equal to the bombarding energy minus the
electronic stopping power over the path x_n to the place of scattering
minus the kinematical energy loss to the target nucleus ($E_n' - aE_n'$)
minus the electronic stopping power over the path $\lambda \cdot x_n$ to the place
of scattering.

$$E_n = E_0 + \int_0^{x_n} f(E(x))dx - (E_n' - aE_n') + \int_{x_n'}^{x_n' + \lambda x_n} f(E)dx$$

$$= E_0 + \int_0^{x_n} f(E)dx + \int_{x_n}^{x_n'} f(E)dx + \int_{x_n'}^{x_n' + \lambda x_n} f(E)dx$$

$$= E_0 + \int_{0'}^{x_n' + \lambda x_n} f(E)dx \tag{1}$$

Since the energy must be a smooth function of x, and E at the place
of scattering makes a jump through the sudden energy loss of the
heavy ions to the target nucleus, a virtual path x_n to x_n' must be
attributed to this energy loss.

From Equation (1), the association between energy interval ΔE_n
and layer thickness Δx_n is obtained through differentiating:

$$\frac{dE_n}{dx_n} = f(E_n) \cdot \left(\frac{af(E_n')}{f(aE_n')} + \lambda\right) \approx \frac{\Delta E_n}{\Delta x_n} = \frac{\Delta E}{\Delta x_n} \tag{2}$$

All ΔE_n are given through the experimental arrangement (energy width
of the kicksorter channels ΔE), $\Delta E_n \equiv \Delta E$.

For the energy E_n' at the place of scattering equation (1)
yields:

$$E_n' = E_{n-1}' - \frac{\Delta E/f(E_{n-1})}{a/f(aE_{n-1}') + \lambda/f(E_{n-1}')} \cdot \tag{3}$$

Between ΔN_n and Δx_n we have the following relation:

$$\Delta N_n = N_0 \cdot \rho(E_n') \cdot \Delta\Omega \cdot \Delta x_n \, , \tag{4}$$

where ρ = atoms/cm^3.
Combining Equations (2), (3) and (4) all Δx_n and x_n can be eliminated. The result is Equation (5):

$$f(E_n) = \frac{\Delta E}{\Delta N_n} \cdot \frac{c}{[af(E_n')/f(aE_n') + \lambda]}$$

$$x \frac{1}{\left[E_{n-1}' - \dfrac{\Delta E/f(E_{n-1})}{a/f(aE_{n-1}') + \lambda/f(E_{n-1}')}\right]^2} \tag{5}$$

Normally, $f(E_n)$ is a function of several $f(E_i)$ with $E_n < E_i < E_0$. c is a constant that can be calculated. Now $f(E_n)$ can be calculated by iteration. In the approximation of order zero the factor a is replaced by 1. By recursion all the $f_0(E_n)$ can now be calculated in zero order approximation. In the approximation of order k all the $f_{k-1}(E_n)$ obtained from the approximation of order $(k - 1)$ will be inserted in Equation (5). For energies E_i between E_1 and E_0 the $f(E_i)$ are obtained by extrapolation. The above iteration process converges very strongly. The approximation of second order deviates from the converged value over the whole energy range by less than 1%.

Experimental Arrangement

The heavy ion beam of the Heidelberg EN Tandem Van-de-Graff accelerator is well collimated (0.8·6 mm^2) and directed on a target of 2 mm thickness. The surface of the target is extremely smooth (<0.3 μ). Through the collimating system the direction of the beam is defined within an error of roughly 0.08°. The target and the detectors are mounted on a single segment which can be rotated. Their relative angle is adjusted with a tolerance of 0.05°. With the scattering chamber used (diameter: 960 mm) the resulting scattering angle was defined within <0.1°.
For the calculation of the absolute differential energy loss the exact number N_0 of incoming particles must be counted. To achieve this the whole scattering chamber was isolated and used as Faraday cup. Thus the error in measuring the incoming ion current could be made smaller than 3 pA. Since the vacuum in the beam pipe was better than $3 \cdot 10^{-6}$ Torr, charge exchange with the residual gas could be neglected. The scattered ions were detected in Ortec surface barrier detectors (depth: 100 μ, resolution for ^{16}O ions of 40 MeV: 180 keV).

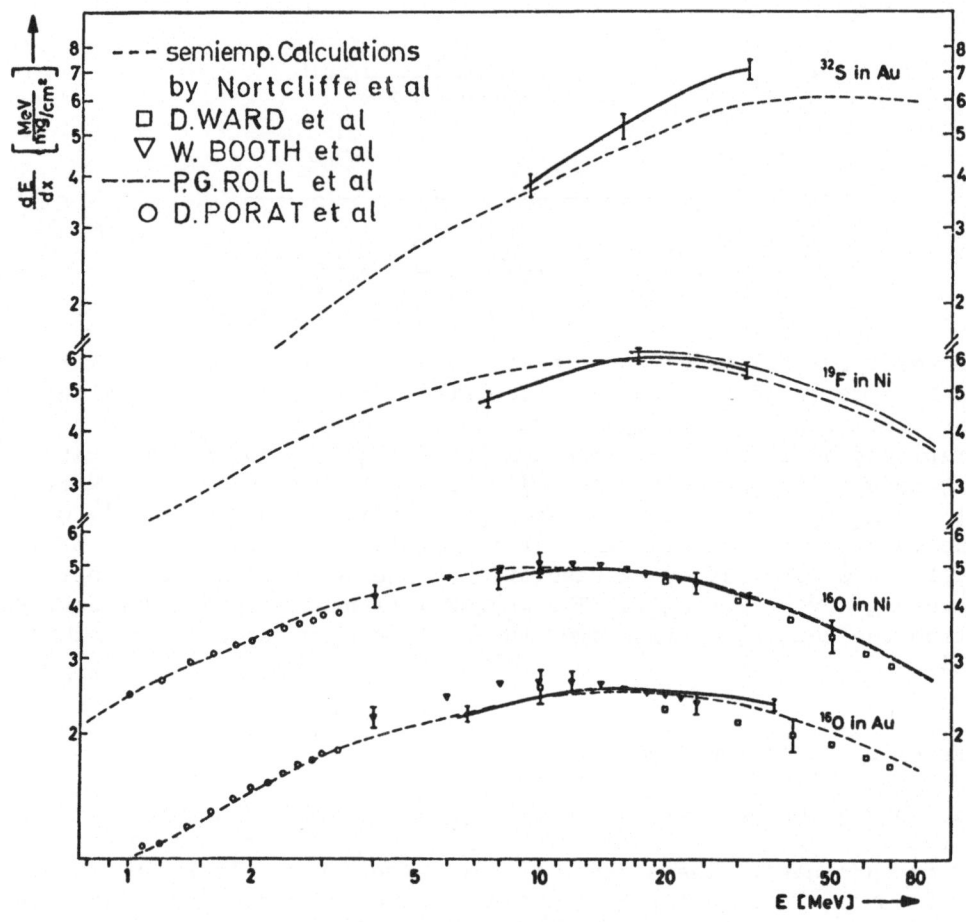

Fig. 4. Differential energy loss of ^{16}O, ^{19}F and ^{32}S. (This work: solid line.)

Results

With the arrangement described above the differential energy loss
of ^{16}O in nickel and gold, of ^{19}F in nickel and of ^{32}S in gold was
measured in the energy range from 8 to 40 MeV. Spectra were recorded
for various bombarding energies E_0 and for four different angles.
The energy distributions were processed by a computer program, and
an average was calculated from the individual measurements. Through
the averaging, the error in the absolute value of the energy loss
of ^{16}O in nickel and gold could be made as low as 2.5 to 4.5%,
according to the energy. The results of the measurements are given
in Fig. 4. The results for ^{16}O agree excellently with those given
by Booth et al. [2]. Over the whole energy range the semiempirical
calculations by Northcliffe and Schilling [1] deviate from our
measurements only within the error limits, partly by much less.
    In the energy range investigated by us, for ^{19}F in nickel and
for ^{32}S in gold no other experimental data have been published yet.
The values tabulated by Northcliffe and Schilling [1] are up to 5%
higher than our results in the energy range from 7 to 13 MeV, and
up to 3% lower in the range from 13 to 33 MeV.
    For ^{32}S our data lie up to 15% above those published by North-
cliffe et al. [1].
    In calculating the energy loss according to the method given
above, energy and angular straggling in the target were neglected.
From measurements we did ourselves [4] we found that for large energy
losses the energy straggling is less than 2 to 3% of the total energy
loss. Therefore its influence on our energy loss data lies within
the error limits given above.

References

[1] L. C. Northcliffe and R. F. Schilling, Nucl. Data Tables A7,
     233 (1970).
[2] W. Booth, I. S. Grant, Nucl. Phys. 63, 481 (1965).
[3] D. Ward et al., Progress Report 1971 (Aug.), Chalk River.
[4] H. Schmidt-Böcking et al., Jahresbericht 1972, MPI für Kernphysik
     Heidelberg.
[5] P. G. Roll et al., Nucl. Phys. 17, 54 (1960).
[6] P. G. Roll et al., Phys. Rev. 120 Nr. 2, 470 (1960).
[7] D. I. Porat et al., Proc. Phys. Soc. (London) 78, 1135 (1961).

SECTION II
RADIATION DAMAGE

THRESHOLD ENERGY FOR ATOMIC DISPLACEMENT IN RADIATION DAMAGE

PETER JUNG

Institut für Festkörperforschung
Kernforschungsanlage Jülich, Jülich, Germany

ABSTRACT

Threshold energies for atomic displacement
as obtained from electron irradiation of solids
are reviewed and compared to other properties
of these materials. Experimental problems as,
e.g., influence of impurities are discussed.
Results on the dependence of threshold energy
on irradiation direction and irradiation temperature are given.

Introduction

The threshold energy for atomic displacement is an important
quantity in radiation damage experiments and a fundamental quantity
in the theory of atomic displacement processes. It is defined as
the minimum energy that an irradiation particle must transfer to
a lattice atom to produce a stable defect in the lattice.

From the atomic structure of a solid it must be expected that
threshold energy depends on lattice direction. While in polycrystalline material only the minimum threshold energy is measurable,
single crystals experiments are necessary to determine the directional dependence of threshold energy.

The first part of this paper concerns the minimum threshold
energy as obtained in polycrystals, while in the second part results
on single crystals are reported, from which a profile of the directional dependence of threshold energy can be derived.

In principle, determination of threshold energy should be
possible by simply monitoring the onset of damage, when the energy
of irradiation electrons is increased.

As an example Fig. 1 shows damage rates of polycrystalline
platinum as a function of energy. The abscissa gives the maximum
transferred energy and the electron energy, respectively. The

damage rate on the ordinate is determined from the change of residual resistivity per electron dose.

In the enlarged scale we see a sharp rise of the damage rate
at a maximum transferred energy of about 34 eV. Such a determination of minimum threshold energy from resistivity measurements is
reproducible with an error of typically ± 2 eV, which is caused by
limited experimental sensitivity and by errors in the energy calibration.

Impurities

On the other hand some materials, for example copper and germanium, show a pronounced "tailing off" of damage rate at low
energies. This is shown in Figure 2 by two measurements on copper

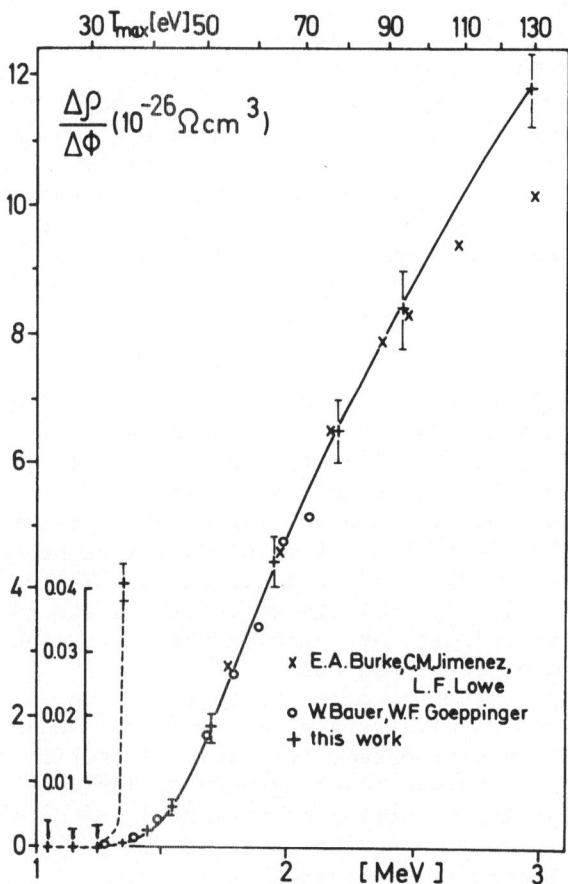

Fig. 1. Energy dependence of damage rate in polycrystalline
platinum from different authors as indicated.

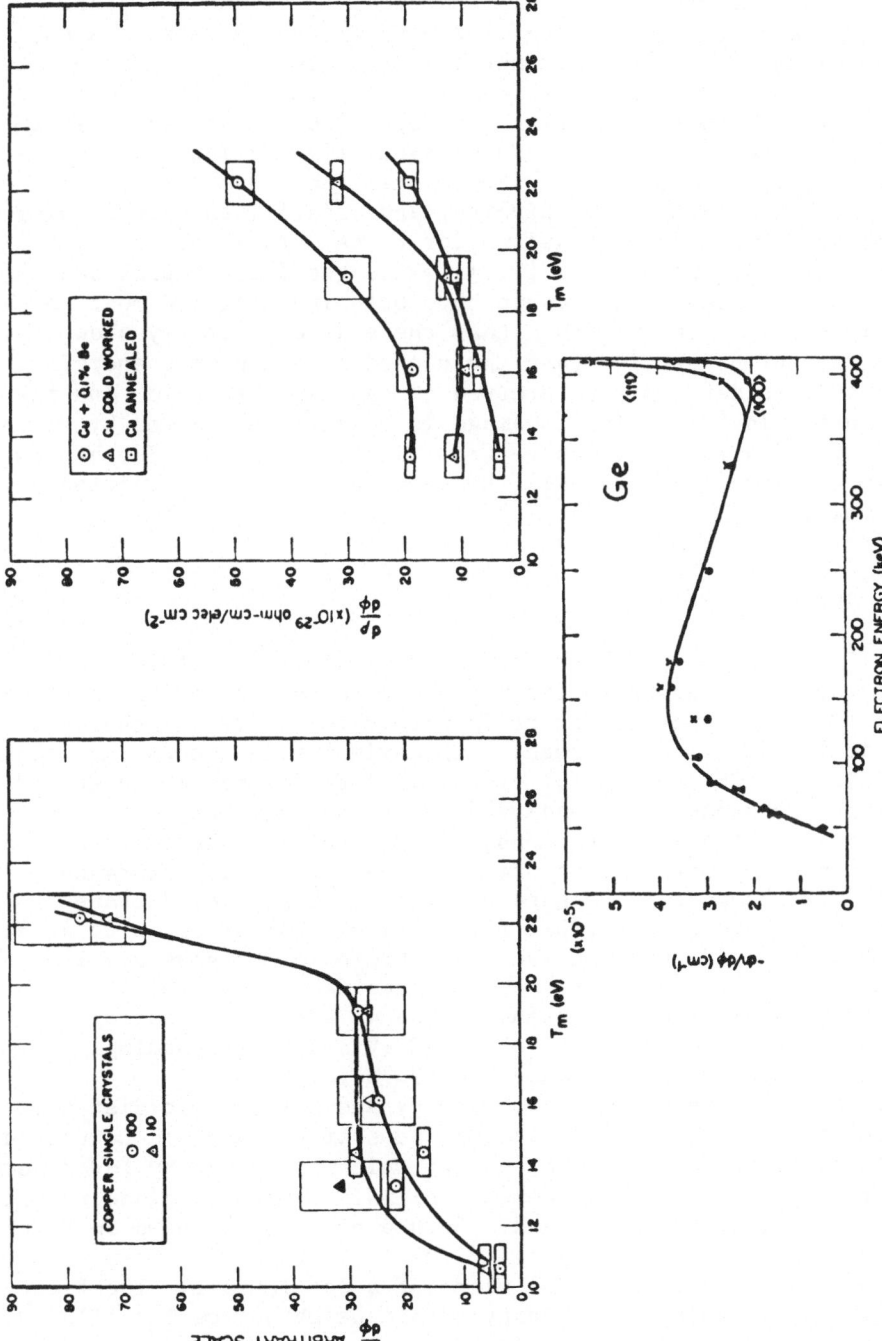

Fig. 2. So called subthreshold damage in copper crystals of different orientation and impurity content, and in polycrystalline germanium.

by Bauer and Sosin [1] for single crystal specimen of different
orientation and for samples of different impurity content and by
measurements on germanium by Chen and MacKay [2]. As a typical
feauture a tail of nearly energy independent damage rate is seen,
which tends to zero only at rather low energies.

From these and other experiments we see that "tailing off" is
independent of lattice direction but depends on impurities. This
"tailing off" behaviour which can be partially ascribed to impurities is especially seen using experimental techniques of high
sensitivity like dislocation pinning, and is still an open problem
in threshold energy determinations for those materials.

In germanium and silicon [2,3], where the defects produced in
the tail of the damage rate curve have been investigated most in-
tensively, it was further found that these defects really show
quite different properties than those produced at higher energies.
Therefore these defects are referred to as "subthreshold" and the
"real" threshold of intrinsic damage is usually determined by extrapolation of the damage rate curve from higher energies. Further on
the term "threshold energy" will be used for these extrapolated
values only.

Minimum Threshold Energy

In Figure 3 minimum threshold energy values as obtained from
the literature are shown in the upper figure as a function of atomic
number. Typical fluctuations of T_d within the different shells of
the periodic system can be seen, with maximum values occuring around
the middle of each shell; for example, 23 eV for nickel in the 3d-,
37 eV for molybdenum in the 4d- and 40 eV for tungsten in the 5d-
shell. Besides these fluctuations, we see that threshold energies are
slowly increasing with atomic number. The interatomic distance
r_{nn} which is shown in the figure below is a simple solid state
quantity which shows comparable fluctuations, but of opposite direction. From these figures we see that threshold energies are low in
materials with large interatomic distances independent of lattice
type. For example from tungsten to lead we see a decrease of
threshold energy from 40 eV to about 12 eV and correspondingly an
increase of interatomic distance from 2.74 to 3.5 Å.

The correlation of threshold energy and interatomic distance
was expected from theoretical calculations of Lehmann and Leibfried
[4]. In these calculations the energy which is necessary for defect
production by so-called replacement collisions along a close packet
lattice direction was determined. Such a process is shown in Figure
4.

The figure gives the [100] plane of a fcc-lattice. The first
knock-on atom transfers its energy to its neighbour on the <110>
close-packed lattice direction and occupies its site. Such replacements are repeated until energy is too low for further replacements

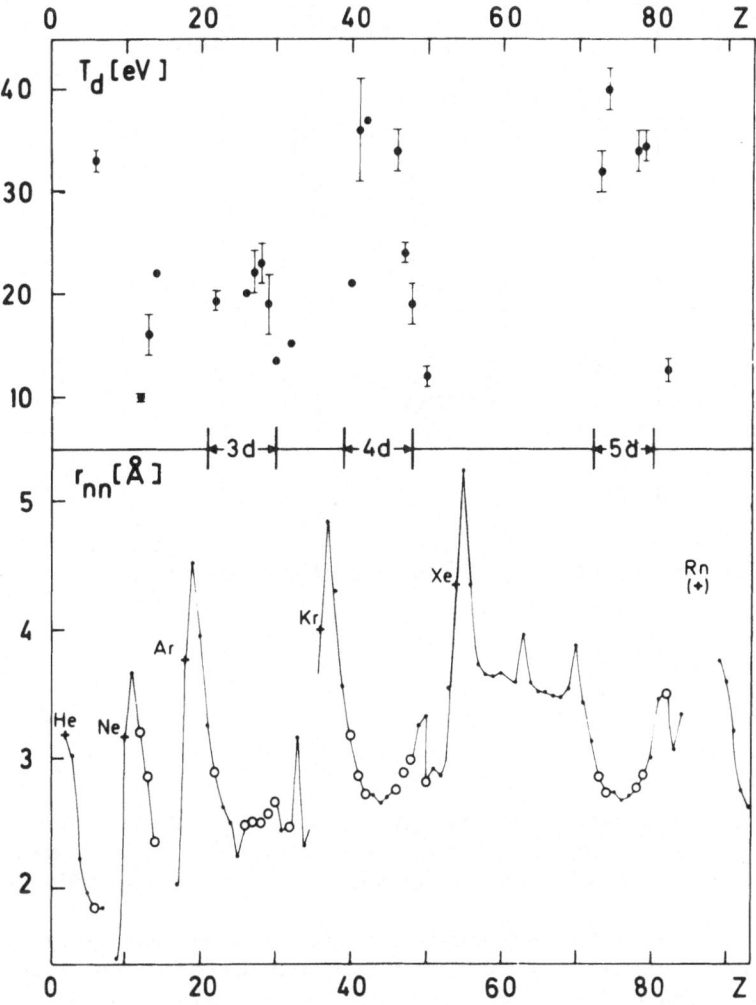

Fig. 3. Minimum threshold energy T_d of polycrystalline materials and interatomic distance r_{nn} as a function of atomic number z.

due to unavoidable losses to the surrounding lattice. After this process the lattice contains a stable defect consisting of a vacancy and an interstitial at a certain distance. Since the lattice is most open for atomic movements along the most close-packed direction, this process is expected to possess minimum threshold energy. The calculations give the threshold energy for this process approximately as

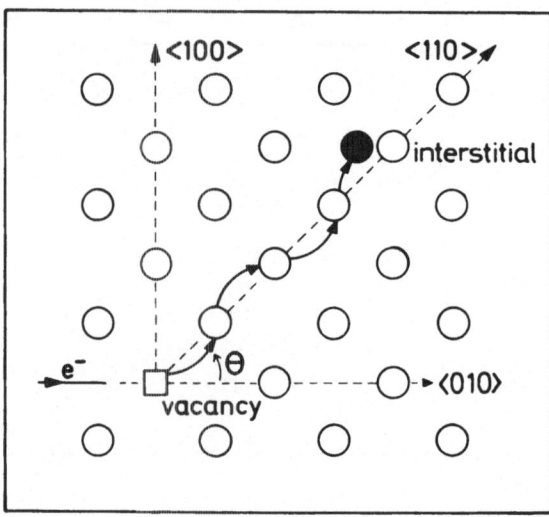

Fig. 4. Schematic view of a replacement collision chain along the
<110> direction in the [100] plane of a fcc-lattice.

$$T_d = \frac{A}{2} e^{-r_{nn}/2R}$$

A and R are parameters of the interatomic potential which was
assumed to be of the Born-Mayer type. As A and R vary slowly with
atomic number, this equation suggests an inverse dependence of T_d
on r_{nn}.

   As the potential parameters A and mainly R are only insuffici-
ently known for metals, this equation can help in deriving better
potential parameters from threshold energy values, compare Lucasson
and Lucasson [5] and Andersen and Sigmund [6].

   This interpretation of threshold energy in terms of collision
sequences strongly differs from the early estimate of Seitz [7] that
the threshold energy should amount to about 4 times the sublimation
energy.

   Deciding between the two models is not simple since most
elements with small interatomic distance have high sublimation
energies and vice versa. But the failure of the correlation between
threshold energy and sublimation energy is seen most clearly by
comparing the data of refractory metals with noble metals of the
same shell.

   The following table compares sublimation energies S, r_{nn} values
and the measured T_d values of Ti and Cu, Zr and Ag, Ta and Au, and
also of Pt and Au:

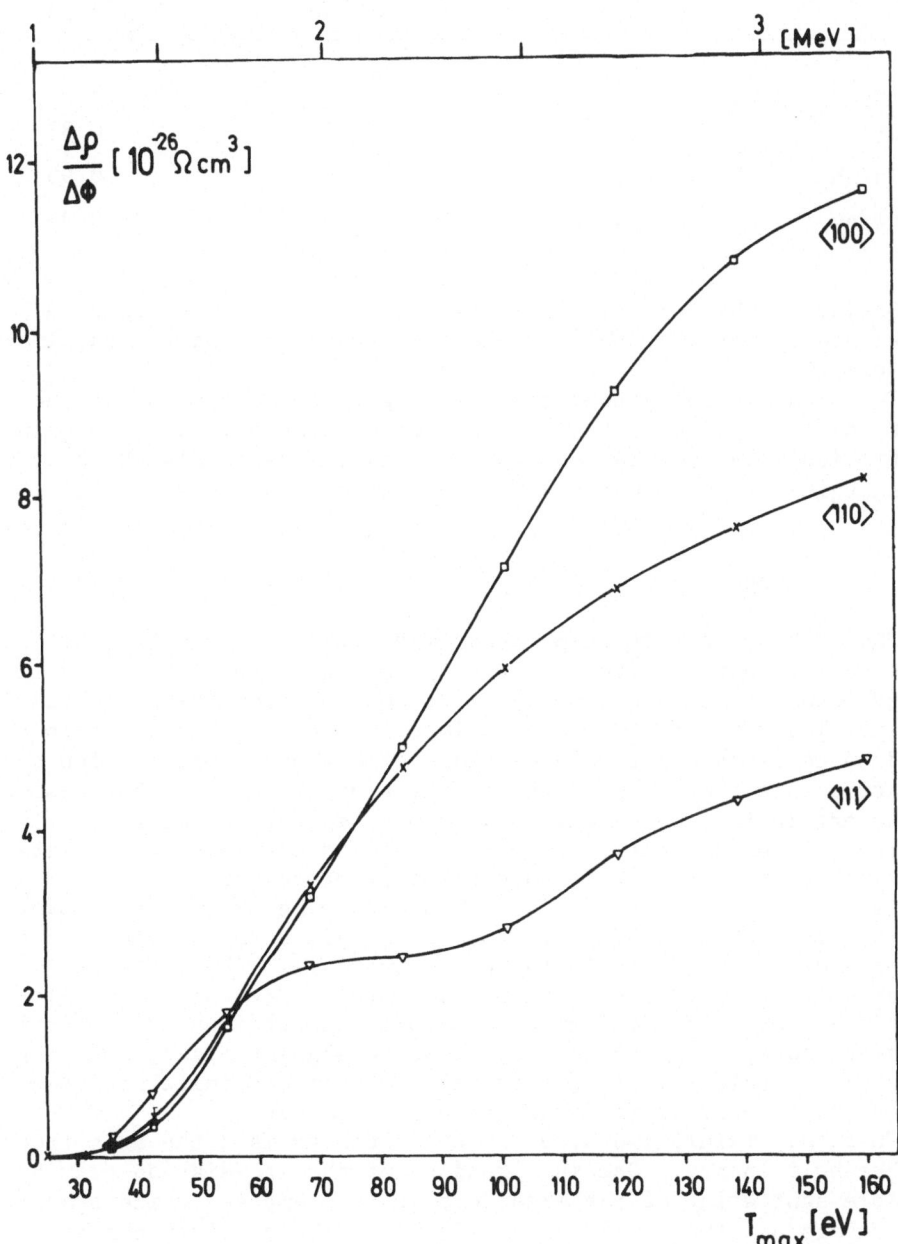

Fig. 5. Energy dependence of damage rate in tantalum single crystals of different orientations.

Table 1

Metals	S/S	r_{nn}/r_{nn}	T_d/T_d
Ti/Cu	1.45	1.14	1.01
Zr/Ag	1.95	1.10	0.87
Ta/Au	2.20	0.99	0.93
Pt/Au	1.49	0.96	0.98

Obviously the inverse correlation between the r_{nn}-values and T_d is much better than the correlation between T_d and the sublimation energy.

This results imply that the breaking of the bonds is not the determining factor of defect production but the necessity to gain a sufficient distance between vacancy and interstitials which at low energies is only possible by replacement sequences.

Directional Dependence

The problem of whether a replacement process along the most close-packed lattice direction really has the lowest threshold energy leads to the question of directional dependence of threshold energy. The straightforward experiment to answer this question would be to irradiate single crystals of different orientation and to extrapolate the damage rates to their onset energy. But besides the uncertainties, which have been mentioned for extrapolation in polycrystal measurements, further problems occur in single crystal experiments which frustrate such a simple procedure.

As an example Figure 5 shows damage rate measurements on single crystals of tantalum. The abscissa again gives the maximum transferred energy and the ordinate the damage rate. The different curves correspond to irradiations along the three main crystallographic directions as indicated. Despite the pronounced differences in damage rates at higher energies, which suggest highly anistropic threshold energies, it is impossible to obtain definite differences of the onset energies of damage by extrapolation.

There are mainly two effects which tend to mask the anisotropy of threshold energy. One is spreading of the irradiation beam by multiple scattering of the electrons in the samples, which practically smears out the irradiation direction. This effect can be reduced to some degree by using thin foils.

The second and more severe effect is that the Coulomb interaction between irradiation electrons and lattice atoms has a differential cross section $\frac{d\sigma}{d\Omega}$ (E,θ) which strongly favors impacts off the

irradiation direction. E describes the irradiation energy and
θ the angle between the irradiation direction Ω_0 and the impact
direction Ω of the knock-on atom. This means that the impact direction of the knock-on atoms does not predominantly coincide with
the irradiation direction.

The consequences are shown schematically in Figure 6. A very
simple dependence of threshold energy on lattice directions is
assumed with a constant value for all directions except small dips
along the direction A. The displacement cross section for such a
profile is calculated by summing the differential cross section
over all that angles where the transferred energy T is higher than
the threshold energy T_d:

$$\sigma_v(E,\Omega_0) = \int\limits_{T(E,\theta)\geq T_d(\Omega)} \frac{d\sigma}{d\Omega}(E,\theta)d\Omega \ , \ \theta = \sphericalangle(\Omega,\Omega_0)$$

The transferred energy is given by $T = T_{max}(E)\cos^2\theta$ where T_{max}
is only a function of irradiation energy. T is described for different T_{max}-values by the broken lines on the left hand side of
the figure. We see that for irradiation along A, damage production
starts at an energy T_0 with displacements in the dip. Then the
displacement cross section decreases like $1/T$ until T_1 is reached
where the transferred energy becomes higher than T_d outside the
dip also.

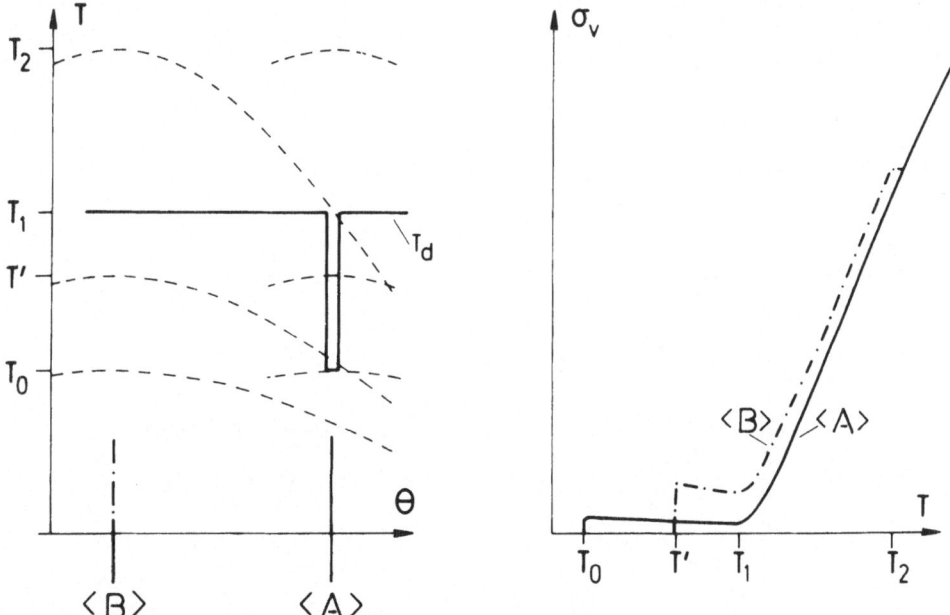

Fig. 6. Schematic view of damage rate curves (right) caused by
an anisotropic profile of threshold energy T_d (left).

For irradiation along a direction outside the dip, for example B, the energy transferred along any direction stays below T_d until T' is reached, where at first damage production in the dip also becomes possible. But because the scattering angle to the direction of the dip for irradiation along B is greater than for the A direction, the total cross section for B rises above the value of A. The further behaviour is then qualitatively the same for both directions.

From this figure we see that the onset energy of damage production for irradiation along A coincides with the threshold energy along this direction, while for irradiation along B the onset of damage does not occur at the real threshold energy T_1 of this direction but much earlier at an intermediate energy T'.

As real threshold energy profiles are expected to be much more complicated than in this figure, the only way to obtain them is to compute damage rate curves from an assumed profile and to compare these curves to experimental data. This fitting procedure needs high-speed computers. Further on many data points are desirable to increase the sensitivity of the fit.

Therefore, in a recent work on silicon by a British group [3] and in our works on tantalum [8] and on copper and platinum [9] foils of different orientation were not only irradiated normal to their surface, but by rotating these foils, irradiations along many other lattice directions were achieved.

Figure 7 shows, for example, how the <100> direction is obtained by rotating a <111>-foil around a <110> axis. The figure on the right hand side shows that the <100> direction is also attainable by rotating a <110> foil around another <100> direction. Similarly the other main crystallographic directions as well as the directions

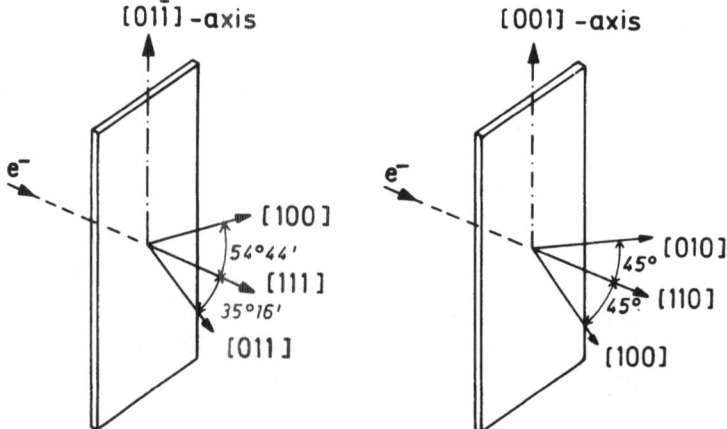

Fig. 7. Schematic view of changing the irradiation direction in single-crystal foils by rotation around different axis.

in between are attainable by rotating samples of different orientations.

By this method one obtains firstly damage rates for a large number of lattice directions, and secondly the relative precision of these measurements is much higher than in the case when damage rates from different samples must be compared.

As an example, in Figure 8 the damage rates for lattice directions between the main crystallographic directions <110>, <100> and <111> are shown for the case of platinum. Different curves correspond to different irradiation energies between 2.95 MeV and 1.45 MeV which is close to the minimum threshold energy of platinum. The data are normalized to the value of the <110>-direction. The plots show a pronounced anisotropy of damage rate far outside the measuring accuracy given by the error bars. For example, at 1.45 MeV the damage rates along <100> and <111> are reduced by more than

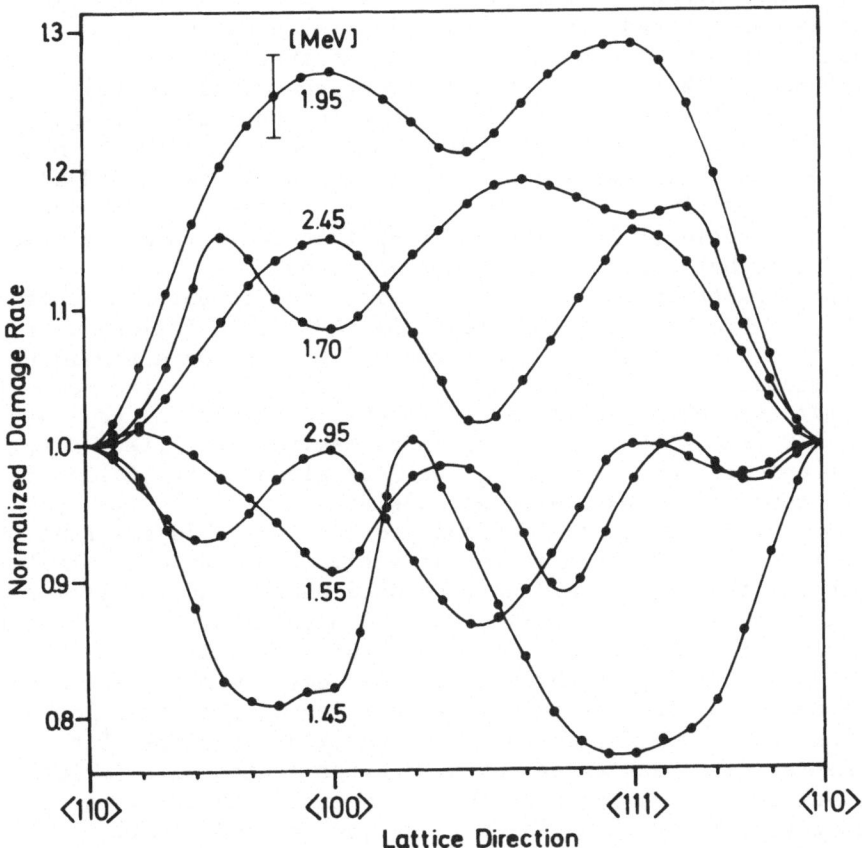

Fig. 8. Dependence of damage rate in platinum on lattice direction and electron energy. The damage rates are normalized to the <110> value.

20% with respect to <110> while at 1.95 MeV these directions already have damage rates nearly 30% higher than that of the <110>
direction. As already mentioned there is not first glance correlation between this anisotropy of damage rate and the directional
dependence of threshold energy profile, but the complicated connection between damage rates and threshold energies which was
described above makes a fitting program necessary. By this program damage rates for several energies and lattice directions were
calculated from an assumed threshold-profile. Then the program
tried to increase the overall agreement between the calculated and
the measured damage rates by systematically changing the profile.

The resulting directional dependence of threshold energy in
fcc-metals is given in Figure 9. The figures show the fundamental
triangle of the cubic system which is bounded by the three main
symmetry axes and which contains all lattice directions. The
position of the triangle in the elementary cube is shown in the
corner. For calculation this triangle was subdivided in about
40 equal-sized areas and the threshold energy value within each
element was determined. From these pictures areas with similar
threshold values were grouped together and average values are given
by the numbers. Both platinum and copper show qualitatively rather
similar threshold energy profiles. In both cases the threshold
is changing smoothly by about 100% within the profile. The minimum
energy regions which are shaded lie around the close packed directions <110> and <100> while there is a pronounced maximum along
the <111>-direction. It is interesting that the very minimum is
attained somewhat outside the <110>-direction.

The third triangle shows the results from computer calculations of the Brookhaven group [10]. As their calculations were
restricted to directions on the border of the triangle the inner
part is tentatively completed by smooth curves for better comparison. This theoretical profile also shows low-threshold regions
around <110> and <100> and a pronounced maximum at <111>. More
complete calculations on fcc-iron will be given in the following
paper in this volume by Beeler and Beeler.

Concerning the accuracy of our profiles I should say that it
is impossible to give an exact error estimation, but our calculations showed that the lower the threshold energy and the larger
the area corresponding to this value the more sensitive the fit is
on changing this value. On the other hand in small regions the
threshold profile may be changed to some degree without influencing
the fit markedly. For example, the position of the small low energy
region off the <100> direction in copper gives the best fit to
experimental data, but agreement with experimental data is still
sufficient if this region of values between 19 and 22 eV is ex-
changed with the region of 20 to 22 eV just around <100>.

Figure 10 shows corresponding results for bcc-materials. The
first triangle gives our results on tantalum, while the other triangles are threshold profiles, determined for bcc-iron. The one
is derived from experiments by Lomer and Pepper [11]

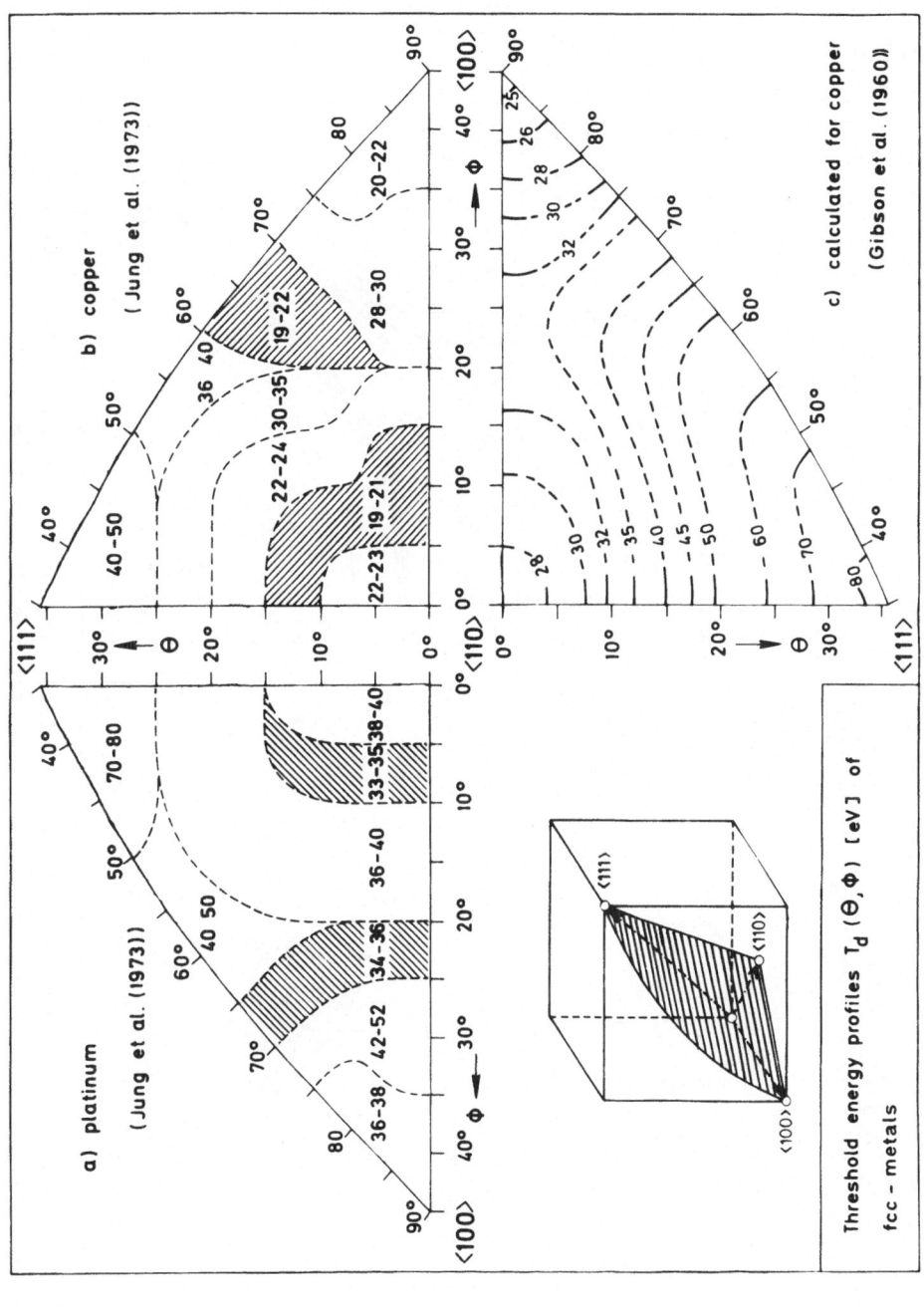

Fig. 9. Threshold energy profiles $T_d(\theta, \phi)$ [eV] of fcc-metals. The upper profiles are both derived from measurements, the lower one from computer simulations. The minimum threshold energy regions are shaded.

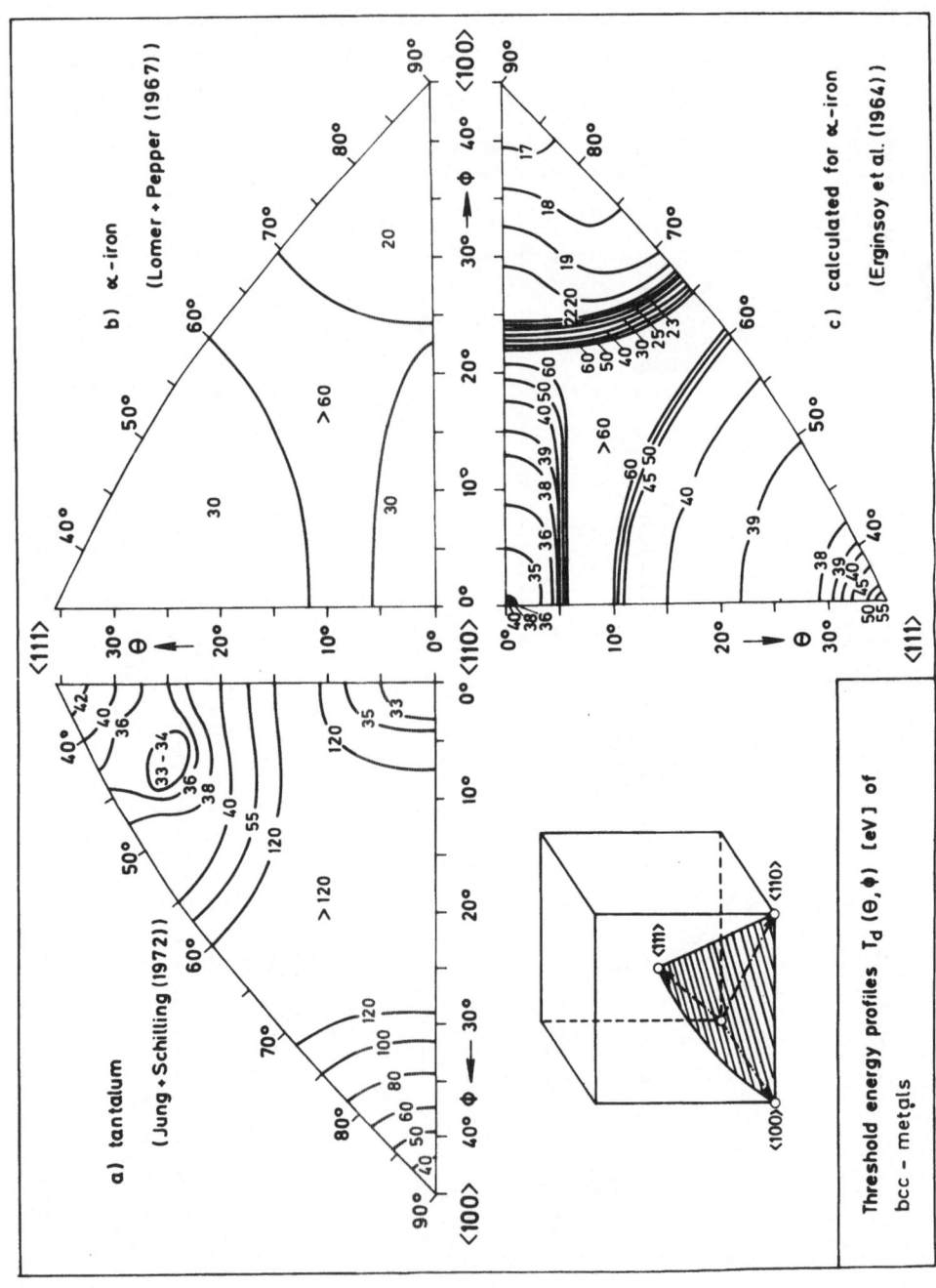

Fig. 10. Threshold energy profiles T_d (θ,ϕ) [eV] of bcc-metals. The upper profiles are both derived from measurements, the lower one from computer simulations.

and the other from computer calculations by the Brookhaven group
[12]. Obviously all these profiles show threshold minima around
all three main crystallographic axes and a steep maximum in the
middle of the triangle. On the other hand there are differences
concerning the relative values of the threshold minima. While in
iron the very minimum was found near <100>, we found for tantalum
the lowest threshold values near <111>.

While the experimental profile of iron was derived from only
a few data points and therefore should be viewed as somewhat preliminary, there remains a discrepancy between our tantalum results
and the calculated profile for ion. On the other hand, there is
of course no a priori argument that the relation of <111> to <100>-
threshold energy must be unique for all bcc-metals.

From Figure 9 we see that the experimental results in fcc-
metals confirm earlier estimates which have been made for aluminium
[13] and gold [14] that in fcc-lattices both the <110> and the
<100> directions have about the same threshold energy and that the
threshold energy profile is smoothly varying. We therefore can
say that experimental results for different fcc-metals as well as
the theoretical calculations of threshold energy profiles agree
quite well.

On the other hand, fewer experiments have been done on bcc-
materials and correspondingly some problems are still unsolved.
For example, it is impossible to say at the moment whether there is
one lattice direction in all bcc-metals which has the lowest threshold energy. But measurements and calculations both confirm that in
bcc-metals there is much more anisotropy in damage rate and in
threshold energy than in fcc-metals.

In the diamond-type elements silicon [3] and germanium [15]
little anisotropy of threshold energy has been found. The results
point to a minimum threshold energy along the most close-packed
direction <111>, while the threshold-values of other directions are
only slightly higher.

On hcp-materials only one damage rate measurement at low temperature was done by a French group on cobalt [16]. This work
shows a strong anisotropy of damage rate similar to the bcc-metals
and a minimum threshold energy occurs along the most close-packed
<1012>-direction. This is in contradiction to experiments at room
temperature on hexagonal materials where the minimum threshold energy
was found along the hexagonal c-axis.

Thus, we find that independent of lattice type, minimum threshold energy occurs along the most close-packed lattice direction.
But in hexagonal as in bcc-metals further results are necessary before safe conclusions are possible.

Temperature Dependence

In the experiments reported till now, electrical resistivity
was the measured quantity. But during the last five years there

is an increasing number of experiments where minimum threshold
energies and threshold energy profiles are determined by using an
electron microscope. Thereby the damage rate is derived from
observations of loop formation, probably consisting of interstitials.

A very important difference between these observations in the
electron microscope and the experiments just described is that
investigations in the elecgron microscope are usually done at room
temperature rather than helium temperature. This implies that close
Frenkel pairs which are predominant in low temperature irradiations
are not stable and therefore cannot contribute to damage rate. In
the electron microscope only defects can be seen which are produced
after free migration and eventually clustering of interstitials.

Also, until now most experimental work in the electron microscope concerning threshold energies was done on fcc-metals. Thereby about the same minimum threshold energies as in the low-tempera-
ture experiments were found within experimental error despite the
different types of defects involved. But when comparing threshold
energies of different lattice directions an important difference
to the results at helium temperature was found [17,18,19].

While the minimum threshold energy was again obtained along
<110> the onset energy for the <100> direction was appreciably
higher. This would imply that by low-energy-transfers along <100>,
only close pairs can be produced, while free interstitials are
obtained only at higher energies. On the contrary along the close-
packed direction <110> the energy needed to produce free interstitials is approximately the same as the threshold energy for the
production of close pairs.

This result would confirm theoretical estimations [20] that
collision processes along the most close-packed lattice directions
show much lower energy dissipation than collision chains along
other directions. (Compare also the paper of Holmes and Robinson
in this volume.)

Conclusions

Finally I will give a short summary of the results which are
obtained from activities on the field of threshold energy:
1. Reliable minimum threshold energy values are now present
   for a great number of polycrystal materials and they show
   a dependence on atomic number which can be interpreted in
   terms of collision sequences.
2. In fcc and diamond-type lattices it seems that the most
   close-packed lattice direction, <110> and <111> respectively, has the lowest threshold energy, and this correlation seems also to hold for bcc and hexagonal materials.
3. It is an experimental fact that damaging processes are
   much more anisotropic in bcc- and hexagonal materials
   than in fcc- and diamond-type lattices. This different
   behaviour cannot be understood by simple considerations
   concerning the lattice structure.

4. In fcc-metals, equal threshold energies along <110> and <100> are only obtained in low temperature experiments. At higher irradiation temperatures where only freely migrating defects contribute to damage rate, the collision processes along <100> need higher initial energies than those along <110> to gain sufficient separation between interstitials and vacancies.

Concluding I should say that damage rate measurements of course cannot give as precise threshold energy profiles and as much details of the atomic collision processes as computer simulations. But the experiments are necessary to guide the theoretical calculations which still suffer from uncertainties about the interatomic forces.

Therefore only the combination of computer calculations, damage rate measurements and annealing experiments can elucidate the production mechanism and to some degree the configuration of Frenkel pairs.

Acknowledgment

The author gratefully acknowledges that Ms. Maury communicated the results of the Orsay group on cobalt before publication.

References

[1] W. Bauer and A. Sosin, J. Appl. Phys. 35, 703 (1964).
[2] Y. Chen and J. W. MacKay, Phys. Rev. 167, 745 (1968).
[3] P. L. F. Hemment and P. R. C. Stevens, J. Appl. Phys. 40, 4893 (1969).
[4] C. Lehmann and G. Leibfried, Z. Physik 162, 203 (1961).
[5] P. Lucasson and A. Lucasson, J. de Phys. 24, 503 (1963).
[6] H. H. Andersen and P. Sigmund, Danish Atomic Energy Commission, Risö, Report No. 103 (1965).
[7] F. Seitz, Disc. Farad. Soc. 5, 271 (1949).
[8] P. Jung and W. Schilling, Phys. Rev. B 5, 2046 (1972).
[9] P. Jung, R. L. Chaplin, H. J. Fenzl, K. Reichelt and P. Wombacher, Phys. Rev. B 8, 553 (1973).
[10] J. B. Gibson, A. N. Goland, M. Milgram and G. H. Vineyard, Phys. Rev. 120, 1229 (1960).
[11] J. N. Lomer and M. Pepper, Phil. Mag. 16, 1119 (1967).
[12] C. Erginsoy, G. H. Vineyard and E. Englert, Phys. Rev. 133A, 595 (1964).
[13] L. R. Kirkland, Jr., Thesis, Clemson University, U.S.A., (1971).
[14] W. Bauer, A. I. Anderman and A. Sosin, Phys. Rev. 185, 870 (1969).
[15] W. L. Brown and W. M. Augustinyak, J. Appl. Phys. 30, 1300 (1959).
[16] F. Maury, P. Vajda, A. Lucasson and P. Lucasson, Phil. Mag. 22, 1265 (1970), further work to be published.

[17] A. Bourret, Phys. Stat. Sol. (a) $\underline{4}$, 813 (1971).
[18] L. M. Howe, Phil. Mag. $\underline{22}$, 965 (1970).
[19] M. J. Makin, Phil. Mag. $\underline{18}$, 637 (1968).
[20] M. W. Thompson, Defects and Radiation Damage in Metals (Cambridge, U.P., London) 1969.

DIRECTIONAL DEPENDENCE OF THE DISPLACEMENT ENERGY THRESHOLD FOR A FCC METAL

J. R. BEELER, JR. and M. F. BEELER
North Carolina State University
Raleigh, N.C. 27607 USA

ABSTRACT

Dynamical method computer experiments were used to map out the displacement energy threshold in a fcc metal as a function of the direction in which a primary knock-on atom is ejected from a lattice site. The displacement energy threshold for the [100] and [110] directions was about 22 eV and that for the [111] direction was 57 eV. A region of high displacement energy threshold was found in the interior of the stereographic triangle near the line between the [111] and [110] poles. The largest threshold energy in this region was 70-75 eV.

Introduction

Interest in computer experiments on the directional dependence of the displacement energy threshold has been stimulated by the recent work of Jung et al.[1,2] and by the use of the high voltage electron microscope to observe the accumulation of electron irradiation defects as they are produced by the microscope beam [3]. We report the results of computer experiments on the anisotropy of the displacement energy threshold (DET) in fcc iron. This work was done to provide an estimate of displacement energy anisotropy in austenitic (fcc) stainless steels.

Computational Method

Production of an isolated stable Frenkel pair in fcc iron was simulated using a dynamical method computer experiment program (DYNAM). DYNAM essentially is a copy of the GRAPE Program [4]

developed at Brookhaven by Gibson et al.[5]. DYNAM differs from
GRAPE in the aspect of having a faster routine for computing static
equilibrium configurations, but dynamically they are identical. The
interatomic potential used in the present work is a composite of
the Erginsoy-Vineyard potential [6] and Johnson's central force
potential [7] for alpha-iron. These two potentials were joined at
$r_j = 1.816$ Å (Table I). The composite uses the Erginsoy-Vineyard
potential for $r < r_j$ and Johnson's potential for $r > r_j$. Here r
is the interatomic distance. The fcc iron lattice constant was
assigned the value 3.64 Å.

Table I. Composite interatomic potential which melds the Erginsoy-
Vineyard and Johnson potentials for alpha-iron. This potential is
defined by the polynomial fit $V(r) = Ar^3 + Br^2 + Cr + D$. $V(r)$ is
in eV and r is in Angstroms, r is the interatomic distance. The
polynomial coefficients are listed below for six interatomic distance ranges.

r, Å	A	B	C	D
0 - 1.223	-2370.788	8678.215	-10728.93	4509.742
1.223 - 1.716	-182.5370	935.7322	-1614.107	940.9742
1.716 - 2.400	-11.56701	82.15697	-194.9803	154.4306
2.400 - 3.000	-.6392298	5.975194	-18.13979	17.75493
3.000 - 3.440	-1.115032	10.25744	-30.98654	30.60164
r 3.440	0	0	0	0

An estimate of the DET anisotropy was made in the following
way: An atom was ejected from a normal site with a given energy E
and along a specified direction with direction cosines (uvw). All
other atoms were initially at rest at normal sites. The evolution
of the lattice excitations induced by this perturbation were followed in a fully dynamical simulation until it was clear that a
stable Frenkel pair either had been produced or had not been produced. A lower bound E(-) and upper bound E(+) for the displacement
energy E_d(uvw) were estimated by increasing E in steps until a stable
Frenkel pair was produced, or by decreasing E in steps until a stable
Frenkel pair was not produced. The operational meanings of E(-) and
E(+), therefore, are as follows: E(-) is the largest energy assigned
for which no stable Frenkel pair was produced and E(+) is the smallest
energy assigned for which a stable Frenkel pair was produced. The
directional displacement energy threshold, E_d(uvw), lies between

E(−) and E(+). In most instances, the energy spread, [E(+) − E(−)] is 5 eV, however, a few directions remain for which a 10 eV spread exists. These larger spreads will be narrowed in future computations. It is important to emphasize that they occur in statistically insignificant instances.

Twenty directions were considered. These directions are defined in Table II. The polar angle, θ, is measured from the z-axis and the azimuthal angle, φ, is measured from the x-axis. Hence, the direction cosines of the ejection direction are, $u = \sin\theta \cos\phi$, $v = \sin\theta \sin\phi$, and $w = \cos\theta$.

Table II

Ejection Angles, Displacement Energy (E_d) and Interstitial Type Produced. FCC Iron.

Point	Angles (θ, ϕ)	E(−)	E(+)	Int. Type	Sep. Dis. in hlc
1	(90,0)	25 eV	25 eV	(110)	4.00
2	(80,10)	20	25	(211)	3.16
3	(72,20)	40	50	(111)	2.45
4	(64,30)	(45)	50	(111)	2.00
5	(55,45)	(55)	60	(211)	4.47
6	(59,41)	45	50	(100)	2.45
7	(89,15)	30	35	(100)	4.00
8	(89,30)	30	35	(100)	5.66
9	(90,45)	20	25	(100)	−
10	(88,43)	30	25	(100)	7.07
11	(72.5,45)	60	70	(111)	2.45
12	(84,15)	30	35	(211)	3.16
13	(79,30)	50	60	(111)	3.74
14	(72.5,42)	71	75	(110)	3.16
15	(89,44)	20	25	(100)	6.48
16	(80,20)	30	35	(100)	3.16
17	(89,22.5)	35	40	(100)	3.16
18	(86,11)	(25)	30	(111)	3.16
19	(70,32.5)	55	57	(100)	3.16
20	(80,40)	(50)	60	(211)	4.90

E_d is greater than E(−) and less than or equal to E(+).

hlc denotes one-half of the lattice constant.

Point Defect Types Observed

The vacancy observed was conventional. Six types of interstitials were observed in the dynamical simulations of Frenkel
pair production. Table III lists these defects along with their
configuration energies. The positions of the two central atoms in
the core configurations of the split interstitial types are given
in Table IV.

The lowest energy interstitial configuration was the [100]-
split configuration. The [211]-split, cited in Table II, more
accurately, is a [411]-split interstitial.

Table III.

FCC Iron		
Defect	Configuration Energy, eV	Run NC-
Vacancy	3.00	2226
Interstitials:		
Tetrahedral	3.12	2270
Octahedral	2.94	2269
(110)-Split	2.92	2600
(111)-Split	2.90	2376
(211)-Split	2.75	2555
(100)-Split	2.63	2229

Table IV.

Coordinates of the Two Central Core Atoms in a Split Interstitial.
FCC Iron.   Interstitial Center at (0,0,0)

Int. Type	Atom 1, hlc	Atom 2 hlc	Separation, hlc
(100)	(0.60,0,0)	(-.60,0,0)	1.20
(110)	(0.39,0.39,0.18)	(-.39,-.39,-.18)	1.16
(111)	(0.34,0.34,0.34)	(-.34,-.34,-.34)	1.18
(211)	(0.78,0.22,0.22)	(-.35,-.08,-.08)	1.21

Results

The results for E_d(uvw) are summarized in Table II, Fig. 1, and
Fig. 2. Recall that E_d(uvw) and $E_d(\theta,\phi)$ translate into one another
by means of the relations given in Sect. 3. Figure 1 indicates that
the [100] and [110] directions are the easy directions for atom displacement in fcc iron. In these instances, the threshold energy is
about 22 eV. The threshold displacement energy for the [111] direction is about 55 eV. This ordering of the DET for the low index
directions in a fcc crystal is in agreement of the results of Jung
et al.[2]. However, a part of our results is different from the
experimental data of Jung et al. and extrapolations [8,9] of the
computer experiment results of Gibson et al. for a fcc metal. We
find a 'difficult region' between the [1̄10] and [111] poles of the
stereographic triangle. This 'difficult region' extends into the
interior of the stereographic triangle to the point (θ = 80°,
ϕ = 25°). Figure 2 describes threshold energy zones on the stereographic triangle. The hatched regions, A and B, are easy displacement regions. Regions E and F are difficult regions and regions
C and D are intermediate difficulty regions.

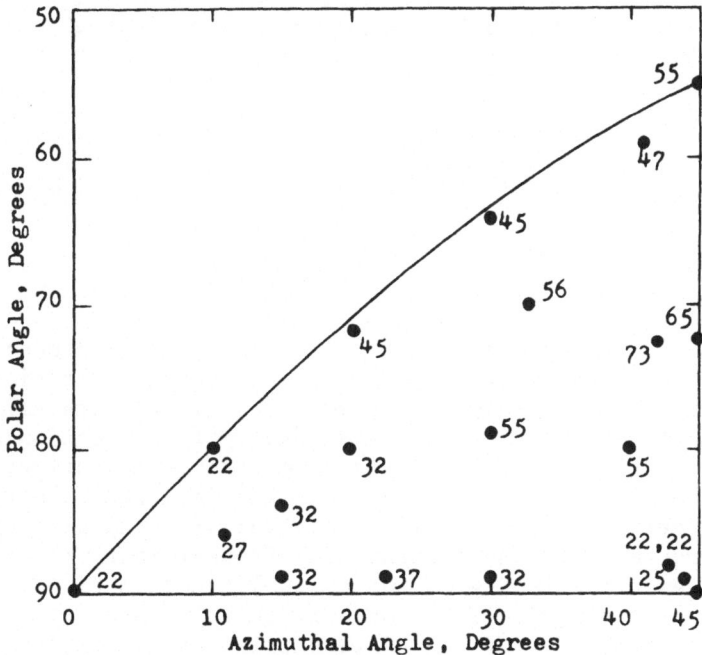

Fig. 1. Displacement energy threshold (DET) values for each of
20 directions in fcc iron. The DET is cited in eV. This set of
data should be applicable to austenitic stainless steel.

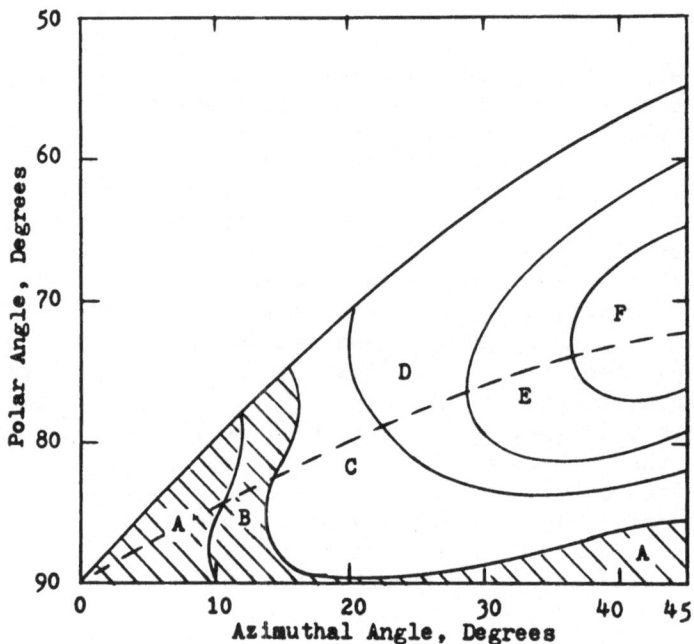

Fig. 2. DET zones in fcc iron. This set of data should be applicable to austenitic stainless steel. A = 22-25 eV, B = 25-30 eV, C = 30-45 eV, D = 45-55 eV, E = 55-60 eV, F = 60-75 eV.

Discussion

The difficult regions, E and F in Fig. 2, correspond to excitations which produced multiple vacancies and interstitials. The easier displacement regions mainly were associated with simple, one-vacancy, one-interstitial types of Frenkel pair production events. The multiple defects just mentioned were transient configurations which eventually relaxed into a single, stable Frenkel pair.

The geometry and time scale associated with the production of a simple, one-vacancy, one interstitial Frenkel pair production process is described in Fig. 3 for a [100] ejection direction event (Region A energy range). The time unit is 10^{-14} sec. In this particular displacement event, the interstitial configuration center oscillated between Sites 1195 and 1212 during the t = 5.0 to t = 24.8 time interval. This oscillatory behavior at the terminal position is characteristic of [100]-split interstitial formation from either a [100] ejection or a [110] ejection perturbation.

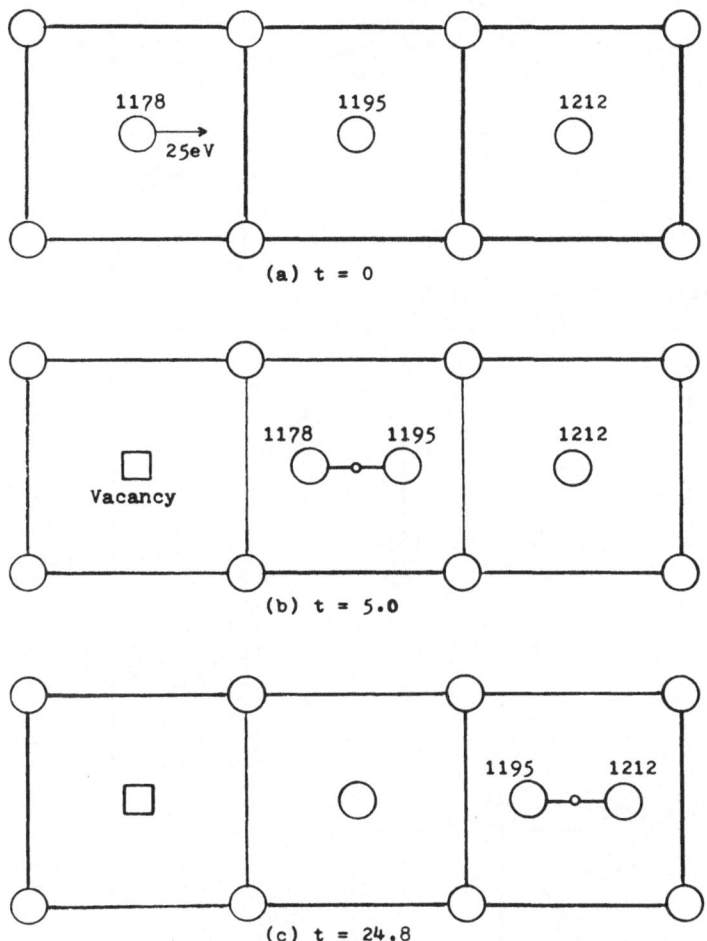

Fig. 3. A simple Frenkel pair production process in fcc iron. This process is a [100] replacement collision chain.

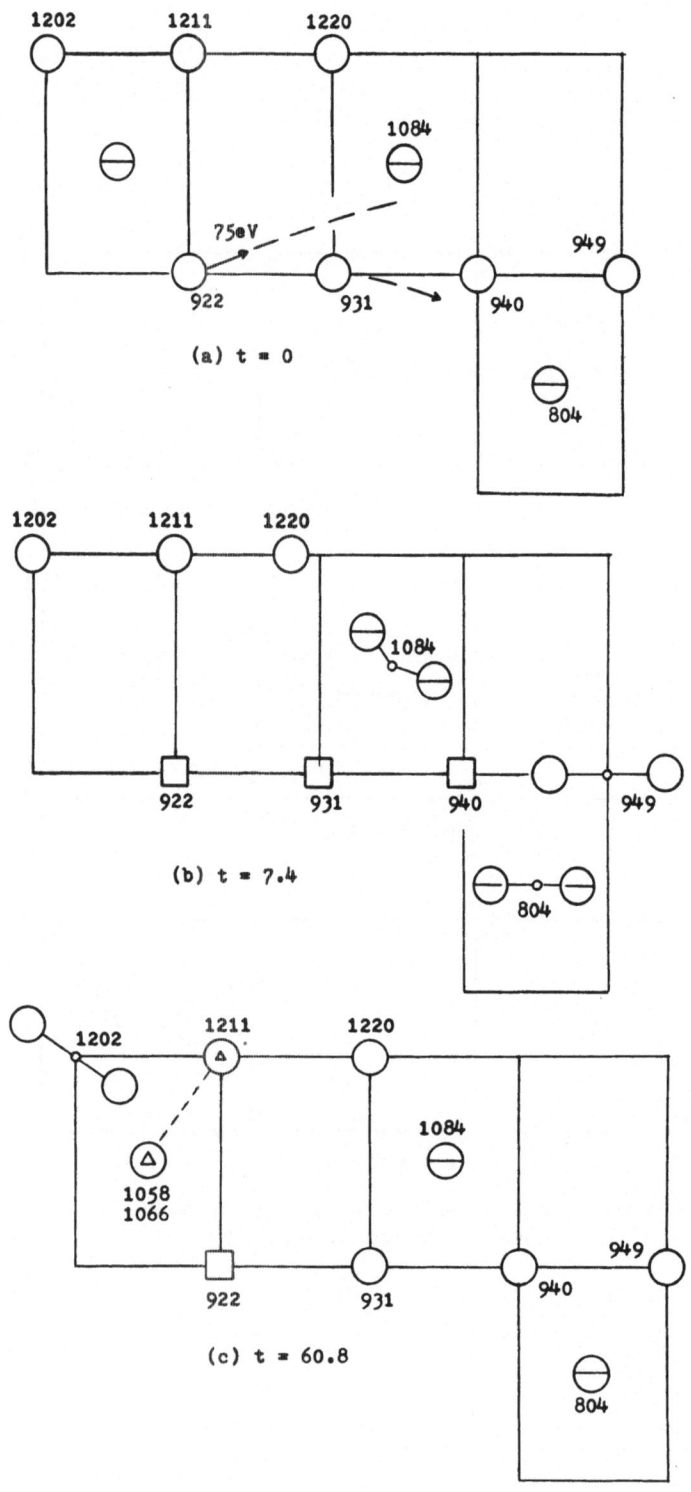

(a) t = 0

(b) t = 7.4

(c) t = 60.8

The 75 eV shot, associated with point 14 in Table II, produced three vacancies and three interstitials in the transient stage. Two of these vacancy-interstitial pairs were annihilated subsequently. The consequences and behavior of this shot are typical of those for difficult directions and atom energies near the DET. Initially, the close-packed line segment running from Site 922 to Site 949 was excited. Also excited were the atoms at Sites 804 and 1084 in the (110) plane behind the open circle atom plane, in Fig. 4, which contains Site 922. The initial excitations above Site 922 were caused by the upward deflection of the ejected atom from Site 922. The initial excitations below Site 922 were caused by the deflection of atom 931 downward, as a result of being struck by the ejected atom. The lattice wave superposition 'rebound', associated with the oscillation of interstitials at Sites 1084, 949 and 804, moved the center of lattice excitation to the close-packed line above Site 922. The leftward movement of atom 1220 in this line, as indicated in Fig. 4b at t = 7.4, is the beginning of this new excitation center. Between t = 7.4 and t = 60.8, the excitation energy in this line was dissipated along other close-packed lines which intersect it and a [111]-split interstitial was formed above and behind the site (922) from which the original atom was ejected. This complex displacement process typifies displacement events associated with difficult directions.

This example illustrates that the basic aspects of single Frenkel pair production in a metal actually begin to fall within the realm of collision cascade theory when the difficult directions for displacement are considered. The only simple displacement events appear to be those associated with replacement chains along the low index directions. Here 'simple' means the existence of no more than one vacancy and one interstitial at any time during the displacement process.

The concept of a displacement energy threshold becomes complex whenever several intermediate, transient Frenkel pairs are produced before a single stable Frenkel pair finally evolves.

Fig. 4. A complex Frenkel pair production process in fcc iron. This process involves production of transient multiple defects as an intermediate stage. As such, it is basically an interference effect, atomic collision cascade process determined by the conjunction of several close-packed line excitations.

Acknowledgments

    The authors are indebted to John McCarter for making the first
hard direction runs and to Thomas A. Wells for improving DYNAM's
continuation run sequence efficiency. DYNAM was written by R. H.
Evans, Jr. as part of a Ph.D. Thesis. All of the work was supported
by the U.S. Atomic Energy Commission under Contract AT-(40-1)-3912.
The work was importantly motivated by a request from T. C. Reuther,
Jr., of the USAEC.

References

[1] P. Jung and W. Schilling, Phys. Rev. B $\underline{5}$, 2046 (1972).
[2] P. Jung, R. L. Chaplin, H. J. Fenzl, K. Reichelt and
     P. Wombachen, Phys. Rev. B $\underline{8}$, 553 (1973).
[3] A. Bourret, Phys. Stat. Sol. (a) $\underline{4}$, 813 (1971).
[4] GRAPE Program, Arlene Larsen.
[5] J. B. Gibson, A. N. Goland, M. Milgram and G. H. Vineyard,
     Phys. Rev. $\underline{120}$, 1229 (1960).
[6] C. Erginsoy, G. H. Vineyard and A. Englert, Phys. Rev. $\underline{133}$,
     A595 (1964).
[7] R. A. Johnson, Phys. Rev. $\underline{134}$, A1329 (1964).
[8] R. von Jan and A. Seeger, Phys. Stat. Sol. $\underline{3}$, 465 (1963).
[9] D. G. Doran, Hanford Engineering Development Laboratory,
     private communication, 1972.

ENERGY DISSIPATION BY RANDOM COLLISIONS IN
COMPOUND TARGET MATERIALS

N. ANDERSEN and P. SIGMUND
Physical Laboratory II, H. C. Ørsted Institute
DK-2100 Copenhagen, Denmark

ABSTRACT

The paper gives a brief account of our calculations on the sharing of energy between the constituents of a polyatomic, in particular a binary medium
bombarded with heavy energetic ions. With a view
at radiation damage and sputtering in composite
targets we calculate displacement efficiencies as
a function of concentration, and particle fluxes
as a function of energy and concentration for a few
binary compounds. Pronounced deviations from stoichiometric behavior are found when the partial stopping
powers of the constituents differ from each other
because of different atomic number and mass.

We have investigated random atomic collision cascades in a
polyatomic, homogeneous, infinite medium by means of the integral
equations of transport theory [1]. A full report of this work is
being published [2].

We consider the sharing of kinetic energy [3] among the constituents of the medium [4-6], but neglect spatial variations [7].
Quantities investigated are the recoil density [8] $F_{ij}(E,E_o)$, defined as the average number per energy interval (E_o, dE_o) of recoiling
j-atoms in a cascade initiated by an i-atom of energy E, and the
slowing-down density [1,9] $G_{ij}(E,E_o)$ defined as the average number
per energy interval (E_o, dE_o) of moving j-atoms under steady-state
conditions during bombardment of ψ i-atoms per unit time of energy
E.

The slowing-down density obeys the set of integral equations

$$\sum_k \alpha_k \int d\sigma_{ik} \{ G_{ij} - G'_{ij} - G''_{kj} \} = \frac{\psi}{Nv_o} \delta_{ij} \delta(E-E_o) \quad , \quad (1)$$

where the α_k are the atomic concentrations ($\Sigma\alpha_k = 1$), $d\sigma_{ik}$ the differential cross section for scattering into the energy interval
(E', dE') and the recoil-energy interval (E'', dE''), $G'_{ij} = G_{ij}(E', E_0)$,
$G''_{kj} = G_{kj}(E'', E_0)$, N the atomic density, and v_0 the velocity of a
j-atom of energy E_0.
The recoil density follows from G_{ij} by the relation

$$F_{ij}(E, E_0) = \alpha_j \sum_{\ell} \int dE_1 \frac{Nv_1}{\psi} G_{i\ell}(E, E_1) \frac{d\sigma_{\ell j}(E_1, E_0)}{dE_0} \qquad (2)$$

The solutions quoted below have been found by neglecting electronic
stopping and inserting a power cross section [10]

$$d\sigma_{ij} = C_{ij} E^{-m_i} T^{-1-m_i} dT \qquad (3)$$

with constants from refs. 7 and 10 for $m_i = 0.333$ (Thomas – Fermi
interaction) and from refs. 9 and 11 for $m_i = 0.055$ (Born – Mayer
interaction). For E_0 in the eV-region, the former choice is assumed
to apply for light atoms up to atomic number ~ 10, and the latter
choice is used for heavier atoms [13].

Asymptotic solutions of eq. (1) for $E \gg E_0$ have been determined
by an extension of Robinson's [12] method involving Laplace transformation, and the deviations from asymptotic behavior have been studied
along the lines of ref. 13. We also included qualitatively the effect of electronic stopping along the scheme of ref. 3 and found
that in the limit $E \gg E_0$ only the absolute rather than the relative
recoil and slowing-down densities are affected.

The binding of atoms to their lattice sites was neglected in
the numerical evaluation, and the limitations to this approximation
have been investigated. Widely different binding energies do affect
the relative (and absolute) densities near threshold.

In the limit $E \gg E_0$, the nature of the bombarding ion turns
out to be immaterial as far as the relative densities are concerned.
Hence, the index "i" in eqs. (1) and (2) can be dropped. For a
binary target, we find the ratio of particle fluxes

$$\frac{v_0 G_1(E, E_0)}{v_0 G_2(E, E_0)} \sim \frac{\alpha_1}{\alpha_2} \cdot \frac{S_{21}(E_0)}{S_{12}(E_0)} \qquad (4)$$

where $S_{ik} = \int T \, d\sigma_{ik}$, and recoil densities

$$\frac{F_1(E, E_0)}{F_2(E, E_0)} \sim \frac{\alpha_1}{\alpha_2} \cdot \frac{S_{21}(E_0)}{S_{12}(E_0)} \cdot \frac{\alpha_1 S_{11}(E_0) + \alpha_2 S_{12}(E_0)}{\alpha_2 S_{21}(E_0) + \alpha_2 S_{22}(E_0)} ; \qquad (5)$$

For strictly stoichiometric behavior, we would just have α_1/α_2 on the right-hand sides of both eq. (4) and (5). Results for ternary targets are given in ref. 2.

It is seen that the deviation from stoichiometry is independent of composition in case of the slowing-down density, and dependent on composition in case of the recoil density. The magnitude of the deviations depends primarily on the mass ratio M_2/M_1.

It also follows that the deviation from stoichiometry may depend on energy E_0 in case of the slowing-down density (for $m_1 \neq m_2$), while all energy dependence drops out in eq. (5), irrespective of the choice of m_1 and m_2.

We have evaluated G_i and F_i for a few binary compounds with widely different masses. Figs. 1-5 show results for tungsten oxides. The choice of the scattering parameters m_i is specified in the figures.

Figs. 1 and 3 show the absolute quantities G_i and F_i as a function of spectral energy E_0. The absolute uncertainty especially in fig. 1 is considerable (at least a factor of 2) because of the uncertainty of the C_{ij} in eq. (3). Note that, according to fig. 1, it is the lighter constituent (0) that dominates at high E_0, and the heavy one (W) at low E_0. This is caused by the cross-over in stopping powers that can be deduced from the results of ref. 10.

Figs. 2, 4, and 5 show suitable relative quantities as a function of concentration. These three graphs visualize in a very direct way the deviation from strictly stoichiometric behavior.

The variation of G_i/G (fig. 2) near $\alpha_i = 1$ is much weaker than the stoichiometric variation. This effect can be understood qualitatively : G_i is determined both by the number of atoms set in motion and the time for slowing-down. An impurity of very different mass causes a decrease in the former quantity, and an increase in the latter.

The displacement efficiency K_i (fig. 4) has been defined by

$$N_i = \int_{E_{d,i}}^{\infty} F_i(E,E_0)\ dE_0 = \frac{E}{E_{d,i}} \cdot \alpha_i \cdot K_i \quad , \tag{6}$$

where N_i is the average number of displaced i-atoms and $E_{d,i}$ is the displacement threshold energy for atoms of type i (replacements and electronic energy loss have been ignored). The relatively low displacement efficiency of dilute impurities is due to the relatively inefficient energy transfer in collisions with host atoms and the small probability for impurity-impurity collisions.

Comparison of figs. 2 and 5 indicates that deviations from stoichiometric behavior tend to be more pronounced in the recoil density than in the slowing-down density.

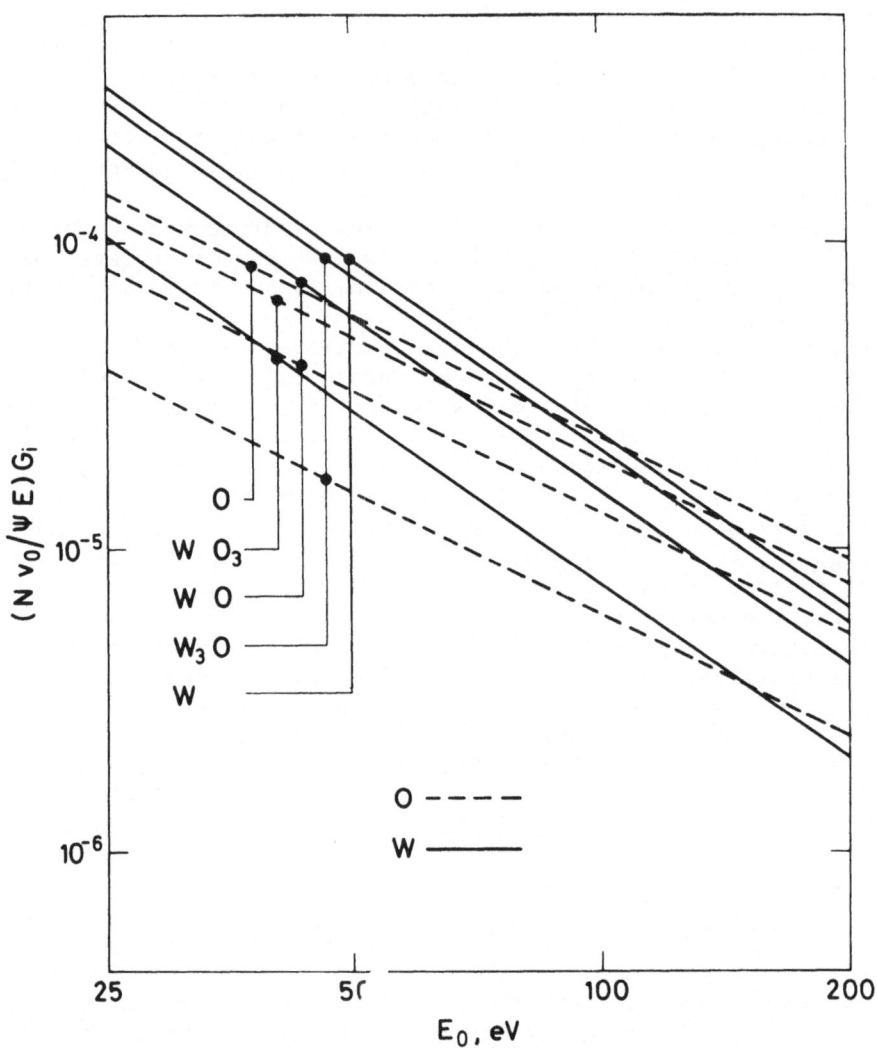

Fig. 1. Slowing-down densities of each of the two constituents of a tungsten-oxygen compound as a function of spectral energy E_0. Parameters m_O and m_W refer to the scattering law, eq. (3). Here $m_O = 0.333$ and $m_W = 0.055$. Note the different energy dependences of the spectra.

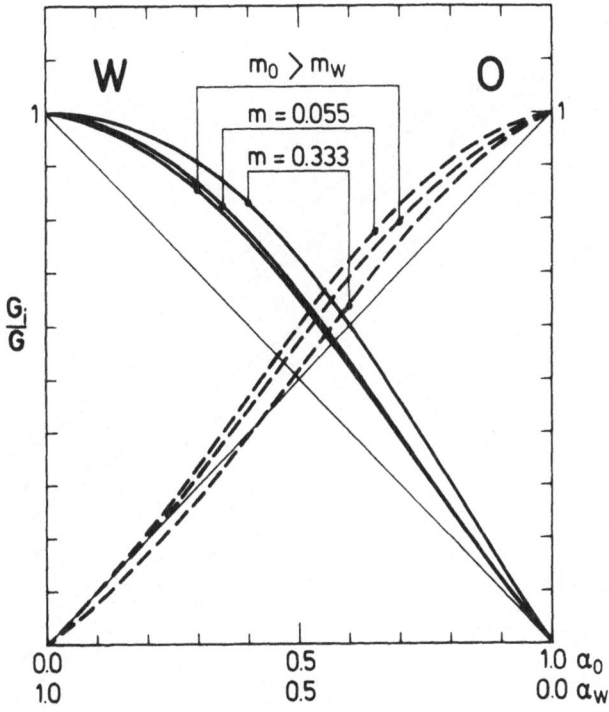

Fig. 2. Slowing-down densities of each of the two constituents, normalized to the values of the respective pure media, G, as a function
of concentration. The ratios G_i/G do not depend on spectral energy
E_o. In addition to the combination of scattering parameters m_i used
in fig. 1, two other combinations have been included for illustration.
The thin full-drawn lines refer to stoichiometric variation.

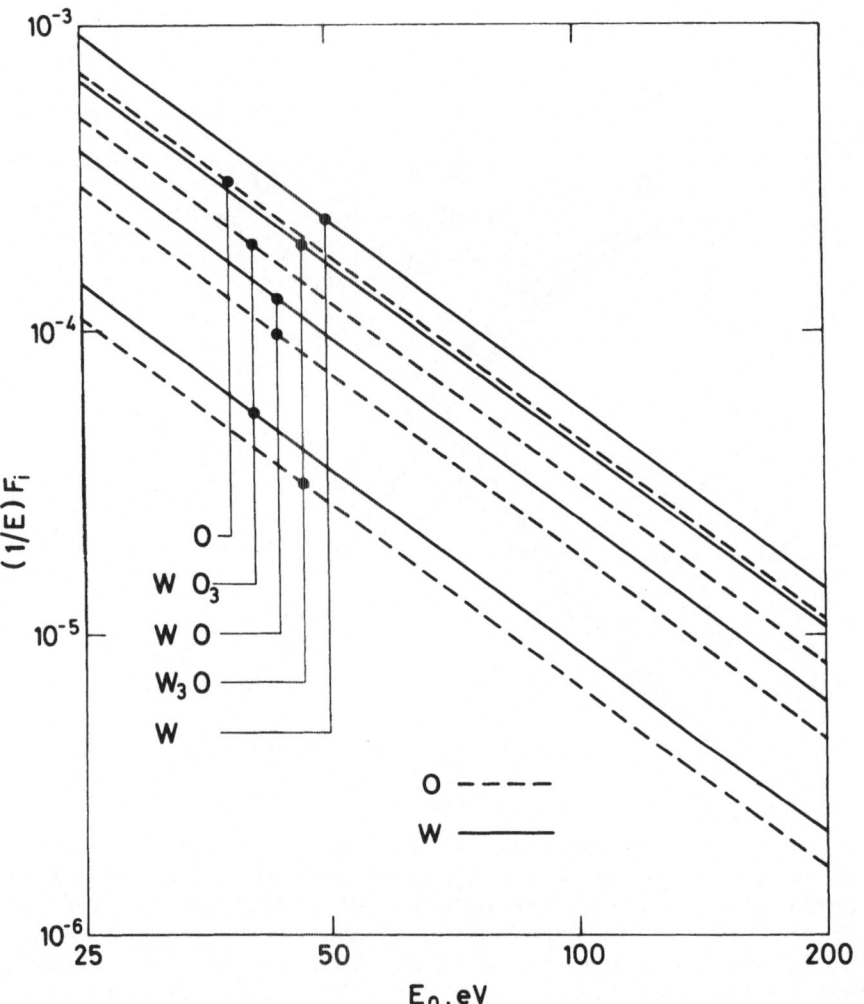

Fig. 3. Same as fig. 1 for the recoil densities. Note that all energy dependence goes as E_o^{-2}.

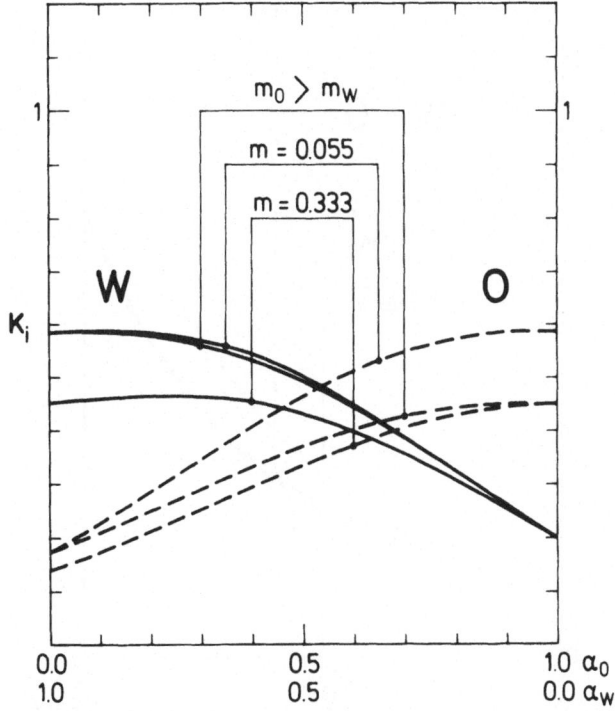

Fig. 4. Displacement efficiencies K_O and K_W of the two constituents [eq. (6)], as a function of concentration. Scattering parameters m_i as in fig. 2. Stoichiometric behavior would correspond to straight <u>horizontal</u> lines.

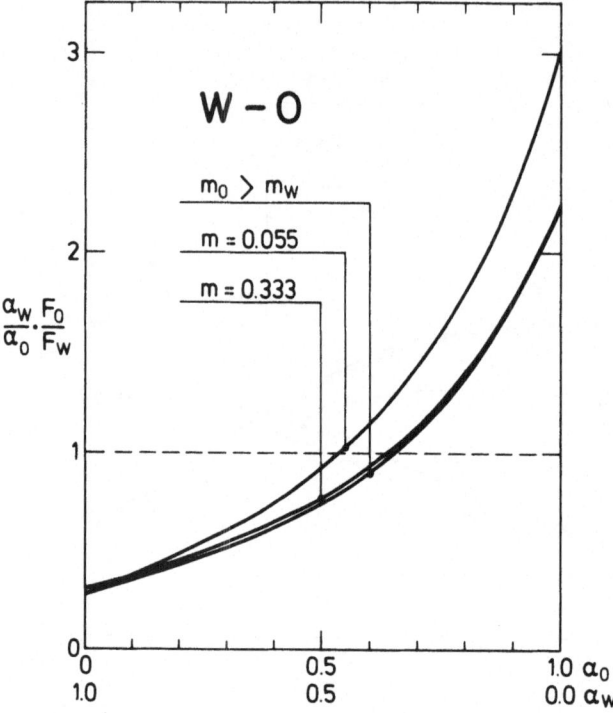

Fig. 5.   Ratio of the recoil densities, normalized so that stoichiometric behavior would correspond to the dashed horizontal line.
Scattering parameters m_i as in fig. 2.

References

[1] P. Sigmund, Rev. Roum. Phys. 17, 823, 969, 1079 (1972).
[2] N. Andersen, P. Sigmund, Mat Fys. Medd. Dan. Vid. Selsk. 39, No. 3 (1974).
[3] J. Lindhard, V. Nielsen, M. Scharff, P. V. Thomsen, Mat. Fys. Medd. Dan. Vid. Selsk. 33, no. 10 (1963).
[4] E. M. Baroody, Phys. Rev. 112, 1571 (1958).
[5] R. M. Felder, J. Phys. Chem. Sol. 28, 1383 (1967).
[6] M. D. Kostin, J. Appl. Phys. 37, 3801 (1966).
[7] K. B. Winterbon, P. Sigmund, J. B. Sanders, Mat. Fys. Medd. Dan Vid. Selsk. 37, no. 14 (1970).
[8] P. Sigmund, Appl. Phys. Lett. 14, 114 (1969).
[9] P. Sigmund, Phys. Rev. 184, 383 (1969).
[10] J. Lindhard, V. Nielsen, M. Scharff, Mat. Fys. Medd. Dan. Vid. Selsk. 36, no. 10 (1968).
[11] H. H. Andersen, P. Sigmund, Nucl. Inst. Meth. 38, 238 (1965).
[12] M. T. Robinson, Phil. Mag. 12, 741 (1965).
[13] P. Sigmund, Rad. Eff. 1, 15 (1969).

ON THE APPLICATION OF BOLTZMANN TRANSPORT EQUATIONS TO ION BOMBARDMENT OF SOLIDS

J. B. SANDERS

*FOM-Instituut voor Atoom- en Molecuulfysica
Kruislaan 407, Amsterdam, The Netherlands*

Introduction

The behavior of energetic particles in solid target materials has been the subject of many papers in the last 5 to 10 years [1,2,3]. However, most of these papers consider the spatial distribution of these particles when they have reached equilibrium with their surroundings. It is, however, of interest also to consider the state of the projectiles before they have reached this equilibrium, in order to obtain information about the speed with which they lose their kinetic energy and the depth where they lose it, which will be the subject of this paper. The starting point of the present paper is the derivation by Mazur and Sanders [4] of the range-equation for energetic projectiles in amorphous target material. The most essential points of this derivation will be briefly recapitulated here.

A probability function $P(\vec{v}_0|\vec{v},\vec{r},t)$ is introduced, which gives the probability density that a projectile, with velocity $\vec{v}_0$ at the origin will be at point $\vec{r}$ with velocity $\vec{v}$ an interval of time t later. (Of course $\int P(\vec{v}_0|\vec{v},\vec{r},t)\,d\vec{v}d\vec{r} = 1$.)

This definition implies that

$$P(\vec{v}_0|\vec{v},\vec{r},o) = \delta(\vec{v}-\vec{v}_0)\delta(\vec{r}). \tag{1}$$

For this function P we can now derive an equation. The essential hypothesis on which this derivation is based is that the process defined by the transition probability P is a Markoff-process, so that the Smoluchowski-equation is satisfied. We can write down this equation in two equivalent forms:

$$P(\vec{v}_0|\vec{v},\vec{r},t+\tau) = \int P(\vec{v}_0|\vec{v}',\Delta\vec{r},\tau)$$

$$\cdot P(\vec{v}'|\vec{v},\vec{r}-\Delta\vec{r},t)\,d\vec{v}'d(\Delta\vec{r}) \tag{2a}$$

$$P(\vec{v}_o|\vec{v},\vec{r},t+\tau) = \int_P P(\vec{v}_o|\vec{v}',\vec{r}-\Delta\vec{r},t)$$

$$\cdot P(\vec{v}'|\vec{v},\Delta\vec{r},\tau)d\vec{v}'d(\Delta\vec{r}) \ . \qquad (2b)$$

The form (2a) leads to the well-known equation for the range of amorphous material. Here we shall consider the form (2b). After some manipulations, analogous to those in ref. [4], this equation can be reduced to the form:

$$\frac{\partial P(\vec{v}_o|\vec{v},\vec{r},t)}{\partial t} + \vec{v} \cdot \frac{\partial}{\partial \vec{r}} P(\vec{v}_o|\vec{v},\vec{r},t)$$

$$= \int \{\omega(\vec{v}'|\vec{v})-A_2\delta(\vec{v}'-\vec{v})\} \ P(\vec{v}_o|\vec{v}',\vec{r},t) \ d\vec{v}' \qquad (3)$$

Here

$$\omega(\vec{v}'|\vec{v}) = \lim_{\tau\to o} \frac{\omega(\vec{v}'|\vec{v},\tau)}{\tau} \ , \qquad (4)$$

where $\omega(\vec{v}'|\vec{v},\tau)$ is the transition probability in velocity space in time τ and

$$A_2 = \int \omega(\vec{v}'|\vec{v})d\vec{v}' \ .$$

As in ref. [4], the derivation is based on a model of binary collisions only. We shall justify this model for the case of solid target material which of course has a very high number density. Eq.(3) can be reduced to a linear Boltzmann equation by specifying the transition probability ω. To this end we put

$$\omega(\vec{v}'|\vec{v}) = gd\sigma(\vec{v}'|\vec{v}) \ f^{(o)}(\vec{v}_1')d\vec{v}_1' \qquad (5)$$

In this formula $f^{(o)}$ is the Maxwell-Boltzmann velocity distribution of the target particles and $\vec{v}_1'$ is the velocity of a target particle such that in collision with a projectile, moving with velocity $\vec{v}'$, the latter will end up with velocity $\vec{v}$, $d\sigma$ is the differential cross section for this collision and g is the magnitude of the relative velocity of the two participants in the collision. If we substitute this we obtain the following equation:

$$\frac{\partial P(\vec{v}_o | \vec{v}, \vec{r}, t)}{\partial t} + \vec{v} \cdot \frac{\partial}{\partial \vec{r}} P(\vec{v}_o | \vec{v}, \vec{r}, t)$$

$$= \int g d\sigma d\epsilon d\vec{v}_1 \{f^{(o)}(\vec{v}_1') P(\vec{v}_o | \vec{v}', \vec{r}, t)$$

$$- f^{(o)}(\vec{v}_1) \ P(\vec{v}_o | \vec{v}, \vec{r}, t)\} \tag{6}$$

which can be recognized as a linear Boltzmann equation.

Justification of the Binary Collision Model

We have stated that the target material is supposed to be amorphous. This means that the target particles are distributed randomly in space, which is expressed mathematically by the Poisson distribution. Explicitly, the probability of finding k particles in a volume of magnitude V is given by

$$\frac{(NV)^k}{k!} e^{-NV},$$

where N is the average number of particles per unit volume. At this point we will make a short digression. Take a straight line with Poisson-distributed points on it, and consider the empty interval between the points. The length of the i'th interval we denote by X_i. Then the probability

$$P(x \le S_n = \sum_{i=1}^{n} X_i \le x+dx) = \alpha \frac{(\alpha x)^{n-1}}{(n-1)!} e^{-\alpha x} dx , \tag{7}$$

with α the average number of points per unit length. This is the same as the probability of n points in the interval between x and x+dx with the last one in dx. For the case n=2 the above probability becomes $\alpha^2 x \, e^{-\alpha x} dx$. This we multiply with the probability that the interval $x-\Delta x$ be empty, where Δx is situated at the beginning of the total interval x. This gives us $\alpha^2 x e^{-\alpha x} dx \cdot e^{-\alpha(x-\Delta x)}$. Taking the limit $\Delta x \to 0$ and normalizing, we obtain the expression

$$4\alpha^2 \, x e^{-2\alpha x} \tag{8}$$

as the probability that the nearest neighbor of a point on a line will be at a distance between x and x+dx. The argument for the three-dimensional case is the same and we take instead of x the volume $(2/3)\pi \ell^3$, by which we mean the forward directed spherical half space with the projectile in the center. The probability that the target particle nearest to the projectile will be a a distance between ℓ and $\ell+d\ell$ is then

$$16/3 \ \pi^2 N^2 \ell^5 e^{-(4/3)\pi N\ell^3} \ d\ell \ . \tag{9}$$

In an analogous way we calculate the probability that there will be one other particle somewhere in the above mentioned half-sphere. For this we find the expression

$$2^7/3^4 \ \pi^4 N^4 \ell^9 \ \cdot \ e^{-(4/3)\pi N\ell^3} \ d\ell \ . \tag{10}$$

In order to get some insight in the physical meaning of these results we shall consider a numerical example. We take the case of carbon projectiles, penetrating into a silicon target and we describe the interaction with Lindhard's well-known form of the differential cross section for a power potential [5]. For the power we shall take 2. This cross section is then given by $d\sigma = 2\pi bdb =$ 0.1515 keV $\text{Å}^2 \cdot E^{-1/2}T^{-3/2}dT$, b being the impact parameter and T the energy transferred by the projectile. A very simple calculation produces the relation $T = \gamma E\{(\pi b^2/2C)\gamma^{1/2} E + 1\}^{-2}$, where C = 0.1515 keV Å^2 and $\gamma = 4 \ m \ m_1/(m+m_1)^2$, m = mass of projectile, m_1 = mass of target.

In Table 1 we give a few numerical results. The number density of Si,N = 0.04997 Å^{-3}. If we now identify b with the radius of the forward half-sphere we find the probability that the projectile will undergo a collision with impact parameter $\leq$ b. We can also find the probability that, given the fact that a target particle is at distance between ℓ and $\ell+d\ell$ from the projectile, another one is still closer to it. These are given by (9) and (10) respectively. For the case of C on Si, their ratio is 0.0073 ℓ^4. We conclude from this that the probability for effective 3-body interactions is very small indeed for kinetic energies down to 0.5 keV at least.

Table 1

E = 50 keV			E = 95 keV		
b(Å)	T(eV)	T/γE	b	T	T/γE
1	0.185	$4.4\text{-}10^{-6}$	1	12.7	0.030
0.1	1270	0.03	0.1	382	0.910
0.05	8771	0.209	0.05	410	0.977
0.01	38246	0.911	0.01	419	0.998

Treatment of the Eigenvalue Problem of the Modified Boltzmann
Collision Operator

We return to eq. (6) in which we shall make a slight modification. This modification consists of the fact that in our binary
collision model the projectile interacts with the target particle
that is closest by, in contrast with the usual gas theoretical
treatments of the Boltzmann-equation, where the interaction takes
place with the particle which has the smallest impact parameter.
The modified Boltzmann-equation becomes

$$
\frac{\partial P(\vec{v}_0|\vec{v},\vec{r},t)}{\partial t} + \vec{v} \cdot \frac{\partial P}{\partial \vec{r}}
$$

$$
= \int_0^\infty f(\ell)d\ell \int_0^\ell g d\sigma \{ f^{(o)}(\vec{v}_1')P(\vec{v}_0|\vec{v}',\vec{r},t)
$$

$$
- f^{(o)}(\vec{v}) \; P(\vec{v}_0|\vec{v},\vec{r},t) \} \tag{6a}
$$

The meaning of this generalization is obvious. A target particle
can be struck with any impact parameter between 0 and ℓ, and this
contribution to the function P is multiplied with the probability
that the nearest target particle is at distance ℓ. Finally we
integrate over all $\ell \cdot f(\ell) = 16/3 \; \pi^2 N^2 \ell^5 e^{-(4/3)\pi N \ell^3}$.

Let us take as interaction potential a so-called Maxwell
potential, i.e., $V(r) \sim r^{-4}$. We can simplify eq. (6a) straightaway
by putting $P(v_0 v, \vec{r}, t) = f^{(o)}(\vec{v})\Psi(\vec{v}_0|\vec{v},\vec{r},t)$. This results in the
following equation for Ψ

$$
\frac{\partial \Psi(\vec{v}_0|\vec{v},\vec{r},t)}{\partial t} + v \frac{\partial \Psi}{\partial \vec{r}} = \int_0^\infty f(\ell)d\ell \int_0^\ell g b db d\varepsilon \; f^{(o)}
$$

$$
\cdot (\vec{v}_1)[\Psi(\vec{v}_0|\vec{v}',\vec{r},t) - \Psi(\vec{v}_0|\vec{v},\vec{r},t)] \tag{11}
$$

Use has been made of the fact that $f^{(o)}(\vec{v}')f^{(o)}(\vec{v}_1') = f^{(o)}(\vec{v})f^{(o)}(\vec{v}_1)$,
due to conservation of kinetic energy in elastic collisions.

For the Maxwell interaction the eigenvalues and eigenfunctions
of the Boltzmann collision operator are well-know [6] and we shall
use this technique. The operator is defined by the equation

$$
L_M \phi_{k,\ell} = \int g b db d\varepsilon \; f^{(o)}(\vec{v}_1)[\phi_{k,\ell}(\vec{v}') - \phi_{k,\ell}(\vec{v})] = \lambda_{k,\ell}\phi_{k,\ell}
$$

$$
\tag{12}
$$

with $f^{(o)}(\vec{v}_1) = N(\frac{\beta}{\pi})^{3/2} e^{-\beta v_1^2}$, $\beta = \frac{m}{2kT}$. The eigenvalues of L_M have the following form.

$$\lambda_{k,\ell} = -2\pi N \int_0^{\pi/2} \{1-(1-4\mu\mu_1 \cos^2\theta)^{\ell/2+k}$$

$$\cdot P\left(\frac{1-2\mu_1 \cos^2\theta}{\sqrt{1-4\mu\mu_1\cos^2\theta}}\right)\} B(\theta)d\theta \tag{13}$$

where $B(\theta)d\theta = gd\sigma$. ($B(\theta)$ does not depend on g in the Maxwell case.) $\theta = \frac{\pi}{2} - \chi/2$, χ deviation in the center-of-mass system and $\mu_{(1)} = \frac{m_{(1)}}{m+m_1}$ and P the Legendre polynomial. We have used the notation of ref. [6]. The eigenfunctions are

$$\phi_{k,\ell}(\vec{v}) = S_{\ell+1/2}^{(k)}(\beta v^2) v^\ell P_\ell(\cos\theta) \tag{14}$$

$S_{\ell+1/2}^{(k)}$ is the Sonine-polynomial [6] and θ the angle between $\vec{v}_o$ and $\vec{v}'$, so that $\vec{v}_o$ determines the coordinate system. It is worth pointing out that all eigenvalues are negative, except λ_{oo} which is 0. The corresponding eigenfunction $\phi_{oo}=1$. It is in fact the only collisional invariant of this problem, the particle number which is 1. In accordance with the results of the previous section we modify the operator L_M as in eq. (6a). This does not influence the eigenfunctions, but is causes a small change in the eigenvalues. The expression for the eigenvalues becomes in this case

$$\lambda'_{k,\ell} = -2\pi N \int_0^\infty f(\ell)d\ell \int_0^{\chi(\ell)} \{1 - (1 - \gamma\sin^2\chi/2)^{\ell/2+k}$$

$$\cdot P\left(\frac{1-2\mu_1\sin^2\chi/2}{\sqrt{1 - \gamma\sin^2\chi/2}}\right)\} B(\chi)d\chi , \tag{15}$$

where $\chi(\ell)$ is the deflection angle in the c of m system, belonging to impact parameter ℓ. In the case of C on Si the difference turns out to be negligible, and henceforth we shall omit the first integration in the expression for $\lambda_{k,\ell}$.

However, we can drastically simplify the whole calculation by neglecting the thermal motion of the target particles, which in any
case is very much slower than the motion of the projectiles which
have kinetic energies of the order of some keV. In that case
eq. (6a) becomes

$$\frac{\partial P}{\partial t} + \vec{v} \cdot \frac{\partial P}{\partial \vec{r}} = N \int_0^\infty f(\ell) d\ell$$

$$\int_0^\ell v'bdbd\varepsilon [P(\vec{v}_0|\vec{v}',\vec{r},t) - P(\vec{v}_0|\vec{v},\vec{r},t)] \quad (16)$$

v' is now the relative velocity between projectile and target particle, with which it collides. The eigenvalue problem has then a
form, corresponding to (12). The effect of this neglect is that we
are left only with the eigenfunctions $\phi_{k,\ell}(\vec{v})$ with the index $k=0$.
The Sonine polynomial $S_{\ell+1/2}^{(0)} = 1$, therefore

$$\phi_\ell(\vec{v}) = \phi_{0,\ell}(\vec{v}) = v^\ell P_\ell(\cos \theta). \quad (17)$$

That the $\phi_\ell(\vec{v})$ are orthogonal is obvious from the factor $P_\ell(\cos \theta)$.
The eigenvalues are not changed and therefore $\lambda_\ell = \lambda_{0,\ell}$ from (13).
The eigenvalues can now be calculated with the results of ref. [5]
and the result is

$$\lambda_\ell = -2\pi NC\left(\frac{m}{2}\right)^{-1/2} \int_0^1 \{1-(1-\gamma t^2)^{\ell/2}$$

$$P_\ell\left(\frac{1-\mu_1 t^2}{\sqrt{1-\gamma t^2}}\right)\} \frac{dt}{t^{3/2}} \quad (18)$$

The eigenvalues have the dimension $(\text{time})^{-1}$, so their inverses can
be interpreted as relaxation times.

Solution of the Boltzmann Equation in Moment Form

We wish to apply the results of the preceding section to the
solution of eq. (16) (Eq. (6) or (6a) can be solved in almost pre-
cisely the same way.) The obvious procedure is an expansion in
normalized eigenfunctions

$$P(\vec{v}_0|\vec{v},\vec{r},t) = e^{-v^2} \sum_\ell c_\ell(\vec{r},\vec{v}_0,t)\phi_\ell(v). \quad (19)$$

The normalization of the eigenfunctions can be found when we realize that the solutions for the "non thermal" model of the target are special cases of the Sonine polynomials. We normalize them therefore as follows:

$$\langle \phi_\ell | \phi_\ell \rangle = \frac{4\pi}{2\ell+1} \int_0^\infty e^{-v^2} v^{2\ell+2} dv = \frac{2\pi}{2\ell+1} \int_0^\infty e^{-v^2} (v^2)^{\ell+1/2} d(v^2)$$

$$= \Gamma(\ell+3/2) \cdot \frac{2\pi}{2\ell+1} \cdot$$

Hence the normalized eigenfunctions become:

$$\phi_\ell(\vec{v}) = \left(\frac{2\ell+1}{2\pi\Gamma(\ell+3/2)}\right)^{1/2} v^\ell P_\ell(\cos\theta) \tag{20}$$

When we substitute the expansion (19) into (16), we get

$$\frac{\partial}{\partial t} \sum_\ell c_\ell(\vec{r},\vec{v}_o,t) \phi_\ell(\vec{v}) e^{-v^2} + \frac{\vec{v}}{v} \cdot \frac{\partial}{\partial \vec{r}} \sum_\ell c_\ell(\vec{r},\vec{v}_o,t) v \, \phi_\ell(\vec{v}) e^{-v^2}$$

$$= \sum_\ell \lambda_\ell c_\ell(\vec{r},\vec{v}_o,t) \, \phi_\ell(\vec{v}) e^{-v^2} \tag{21}$$

where we have used eq. (12) with k=o. As in previous treatments we expand the c_ℓ into Legendre Polynomials

$$c_\ell(\vec{r},\vec{v}_o,t) = \sum_{\bar\lambda} (2\bar\lambda+1) c_\ell^{\bar\lambda}(r,v_o,t) P_{\bar\lambda}(\cos\eta), \tag{22}$$

where η is the angle between $\vec{r}$ and $\vec{v}_o$. We multiply (21) with $P_\lambda(\cos\eta)\, \phi_\ell(\vec{v})$ and integrate over η, ϕ and the entire velocity space. This produces the equation

$$4\pi \frac{\partial c_\ell^\lambda(r,v_o,t)}{\partial t} + \langle \lambda,\ell \left| \frac{\vec{v}}{v} \frac{\partial}{\partial \vec{r}} \right| \lambda+1, \ell+1 \rangle \langle \phi_\ell |v| \phi_{\ell+1} \rangle c_{\ell+1}^{\lambda+1}$$

$$+ \langle \lambda,\ell \left| \frac{\vec{v}}{v} \frac{\partial}{\partial \vec{r}} \right| \lambda+1, \ell-1 \rangle \langle \phi_\ell |v| \phi_{\ell-1} \rangle c_{\ell-1}^{\lambda+1}$$

$$+ \langle \lambda,\ell \left| \frac{\vec{v}}{v} \frac{\partial}{\partial \vec{r}} \right| \lambda-1, \ell+1 \rangle \langle \phi_\ell |v| \phi_{\ell+1} \rangle c_{\ell+1}^{\lambda-1}$$

$$+ \langle \lambda,\ell \left| \frac{v}{v} \frac{\partial}{\partial \vec{r}} \right| \lambda-1, \ell-1 \rangle \langle \phi_\ell |v| \phi_{\ell-1} \rangle c_{\ell-1}^{\lambda-1}$$

$$= 4\pi\lambda_\ell c_\ell^\lambda(r,v_o,t) \tag{23}$$

The matrix elements, occurring in this equation can be calculated, where we use the fact that the operator $\frac{\vec{v}}{v}\frac{\partial}{\partial\vec{r}}$ is the scalar product of two tensor operators of order 1. The elements can be factorized in two integrals, one over the angular coordinates of $\vec{r}$-space and one over those of $\vec{v}$-space. This last one results in $4\pi \left(\begin{smallmatrix} \ell & 1 & \ell\pm1 \\ 0 & 0 & 0 \end{smallmatrix}\right)^2$. This is a Wigner coefficient, the values of which have been tabulated [7]. The other integral can be performed with methods, given in ref. [7] and (23) becomes

$$
\frac{\partial c_\ell(r,v_o,t)}{\partial t} + \left(\begin{smallmatrix} \ell & 1 & \ell+1 \\ 0 & 0 & 0 \end{smallmatrix}\right)^2 \{ (\lambda+1) \frac{\partial c_{\ell+1}^{\lambda+1}}{\partial r} + (\lambda+1)(\lambda+2) \frac{c_{\ell+1}^{\lambda+1}}{r}
$$

$$
+ \lambda \frac{\partial c_{\ell+1}^{\lambda-1}}{\partial r} - \lambda(\lambda-1) \frac{c_{\ell+1}^{\lambda-1}}{r} \} \, <\phi_\ell |v| \phi_{\ell+1}>
$$

$$
+ \left(\begin{smallmatrix} \ell & 1 & \ell-1 \\ 0 & 0 & 0 \end{smallmatrix}\right)^2 \{ (\lambda+1) \frac{\partial c_{\ell-1}^{\lambda+1}}{\partial r} + (\lambda+1)(\lambda+2) \frac{c_{\ell-1}^{\lambda+1}}{r}
$$

$$
+ \lambda \frac{\partial c_{\ell-1}^{\lambda-1}}{\partial r} - \lambda(\lambda-1) \frac{c_{\ell-1}^{\lambda-1}}{r} \} \, <\phi_\ell |v| \phi_{\ell-1}>
$$

$$
= (2\lambda+1) \, \lambda_\ell c_\ell^\lambda(r,v_o,t) \tag{24}
$$

A trivial calculation gives for the matrix elements

$$
<\phi_\ell |v| \phi_{\ell+1}> = \frac{2\ell+3}{4\pi} \sqrt{\frac{2\ell+1}{2}} \tag{25a}
$$

and

$$
<\phi_\ell |v| \phi_{\ell-1}> = \frac{2\ell+1}{4\pi} \sqrt{\frac{2\ell-1}{2}} \tag{25b}
$$

The next step, as in previous treatments is the introduction of spatial moments

$$
{}^n c_\ell^\lambda (v_o, t) = 4\pi \int_o^\infty c_\ell^\lambda(r,v_o,t) \, r^{n+2} \, dr, \tag{26}
$$

for which we obtain the recurrence relations:

$$\frac{d^n c_\ell^\lambda (v_o, t)}{dt} + \begin{pmatrix} \ell & 1 & \ell+1 \\ 0 & 0 & 0 \end{pmatrix}^2 \{(\lambda+1)(\lambda-n)^{n-1} c_{\ell+1}^{\lambda+1}$$

$$- \lambda(\lambda+n+1)^{n-1} c_{\ell+1}^{\lambda-1}\} <\phi_\ell |v| \phi_{\ell+1}>$$

$$+ \begin{pmatrix} \ell & 1 & \ell-1 \\ 0 & 0 & 0 \end{pmatrix}^2 \{(\lambda+1)(\lambda-n)^{n-1} c_{\ell-1}^{\lambda+1}$$

$$- \lambda(\lambda+n+1)^{n-1} c_{\ell-1}^{\lambda-1}\} <\phi_\ell |v| \phi_{\ell-1}>$$

$$= (2\lambda+1) \lambda_\ell {}^n c_\ell^\lambda (v_o, t) . \tag{27}$$

For the solution of this system of differential equations we need the initial condition (1) and the expansion (19). We have

$$P(\vec{v}_o | \vec{v}, \vec{r}, o) = \delta(\vec{v}_o - \vec{v}) \, \delta(\vec{r}) = \sum_\ell c_\ell(\vec{r}, \vec{v}_o, 0) \, \phi_\ell(\vec{v}) \, e^{-v^2}$$

$$\tag{28}$$

The orthonormality of the ϕ_ℓ gives us

$$\int P \phi_\ell \, d\vec{v} = \phi_\ell(\vec{v}_o) \, \delta(\vec{r}) = c_\ell(\vec{r}, \vec{v}_o, 0) . \tag{29}$$

With the help of the orthogonality of the $P_\lambda(\cos \eta)$ we find

$${}^o c_\ell^o (v_o, 0) = 4\pi \int_0^\infty c_\ell^o (r, v_o, 0) \, r^2 dr = \int c_\ell(\vec{r}, \vec{v}_o, 0) \, d\vec{r}$$

$$= \int \phi_\ell(\vec{v}_o) \, \delta(\vec{r}) \, d\vec{r} = \phi_\ell(\vec{v}_o) = \phi_\ell(v_o) \tag{30}$$

(no angular dependence). A similar argument shows that all higher momenta are 0 at t=o. This enables us to obtain the formal solution to the system (27). We begin with the case $\lambda=n=0$. In that case all coupling between different ℓ disappears, and we are left with the simple equations

$$\frac{d^o d_\ell^o}{dt} (v_o, t) = \lambda_\ell {}^o c_\ell^o (v_o, t), \tag{31}$$

which have the solution

$$^{o}c^{o}_{\ell}(\vec{v}_{o},t) = \phi_{\ell}(\vec{v}_{o})e^{\lambda_{\ell}t}, \tag{32}$$

and in particular

$$^{o}c^{o}_{o}(\vec{v}_{o},t) = 1 \tag{33}$$

Next we take the case $\lambda=n=1$. Then (27) takes the form:

$$\frac{d^{1}c^{1}_{\ell}(\vec{v}_{o},t)}{dt} - 3 \begin{pmatrix} \ell & 1 & \ell+1 \\ 0 & 0 & 0 \end{pmatrix}^{2} {}^{o}c^{o}_{\ell+1} <\phi_{\ell}|v|\phi_{\ell+1}>$$

$$- 3 \begin{pmatrix} \ell & 1 & \ell-1 \\ 0 & 0 & 0 \end{pmatrix}^{2} {}^{o}c^{o}_{\ell-1} <\phi_{\ell}|v|\phi_{\ell-1}>$$

$$= 3\lambda_{\ell} {}^{1}c^{1}_{\ell}(\vec{v}_{o},t) . \tag{34}$$

Taking into account (32) we see that these ordinary linear differential equations whose solutions are

$$^{1}c^{1}_{\ell}(\vec{v}_{o},t) = e^{3\lambda_{\ell}t} \int_{o}^{t} \{\alpha^{+}\phi_{\ell+1}(\vec{v}_{o}) e^{(\lambda_{\ell+1}-3\lambda_{1})t'}$$

$$+ \alpha^{-}\phi_{\ell-1}(\vec{v}_{o}) e^{(\lambda_{\ell-1}-3\lambda_{\ell})t'}\}dt' \tag{35}$$

α^{+} and α^{-} are the coefficients of resp. $^{o}c^{o}_{\ell+1}$ and $^{o}c^{o}_{\ell-1}$ in (34). From this example the method of calculation of the higher moments is obvious.

Discussion of the Results

The geometrical interpretation is as could be expected analogous to the one in the earlier treatment.

Consider the integral $\int P(\vec{v}_{o}|\vec{v},\vec{r},t)r^{n+2}P_{\lambda}(\cos \eta)d\vec{r}$. This can be interpreted as the probability that projectiles, whose average value of $r^{n}P_{\lambda}(\cos \eta)$ at time t is given by this integral, will then have a velocity $\vec{v}$. From (19), (22) and (26) we see that

$$\int P(\vec{v}_{o}|\vec{v},\vec{r},t) r^{n+2}P_{\lambda}(\cos \eta)d\vec{r} = \sum_{\ell} {}^{n}c^{\lambda}_{\ell}(\vec{v}_{o},t) \phi_{\ell}(\vec{v}) \tag{36}$$

and we can calculate all these moments (for $n+\lambda$ even and $n \geq \lambda$) for every $\vec{v}$ and t. In view of the fact that $|\lambda_\ell|$ increases with ℓ, the series (36) is quickly convergent and only very few terms are sufficient. Of special interest is the case $\ell=0$. We can see from (35) that, because $\lambda=0$ and $\lambda_\ell \neq 0$ for $\ell \neq 0$ that only for $\ell=0$,

$$\lim_{t \to \infty} {}^1c_o^1(v_o,t) \neq 0, \quad {}^1c_o^1(v_o,t) = \int P(\vec{v}_o|\vec{v},\vec{r},t)rP_1(\cos \eta)\phi_o(\vec{v})d\vec{r}d\vec{v}.$$

Its interpretation is the average value of $r \cos \eta$ (penetration depth) of a projectile at time t, <u>irrespective of its velocity</u>. Hence $\lim_{t \to \infty} {}^1c_o^1(v_o,t)$ is the average penetration depth when it has come to rest, as calculated in [2], and obviously $\neq 0$. Let us calculate ${}^1c_o^1(v_o,t)$. From (34) and (25a) we have

$$\frac{d {}^1c_o^1(v_o,t)}{dt} = 3\begin{pmatrix} 0 & 1 & 1 \\ 0 & 0 & 0 \end{pmatrix}^2 {}^o c_1^o <\phi_o|v|\phi_1>$$

$$= 3\begin{pmatrix} 0 & 1 & 1 \\ 0 & 0 & 0 \end{pmatrix}^2 \phi_1(v_o) \, e^{\lambda_1 t} \frac{3}{4\pi\sqrt{2}} \tag{37}$$

and from (20) we get

$$^1c_o^1(v_o,t) = \frac{9}{4\pi} \sqrt{\frac{3}{4\pi\Gamma(5/2)}} \begin{pmatrix} 0 & 1 & 1 \\ 0 & 0 & 0 \end{pmatrix}^2 v_o \frac{e^{\lambda_1 t}-1}{\lambda_1}. \tag{38}$$

Now we take the limit $t \to \infty$ and use the result (18) for the eigenvalue λ_1. We obtain then.

$$\lim_{t \to \infty} {}^1c_o^1(v_o,t) = -\frac{9}{4\pi} \begin{pmatrix} 0 & 1 & 1 \\ 0 & 0 & 0 \end{pmatrix}^2 \sqrt{\frac{3}{4\pi\Gamma(5/2)}} \frac{v_o}{\lambda_1}$$

$$= \frac{9}{4\pi} \begin{pmatrix} 0 & 1 & 1 \\ 0 & 0 & 0 \end{pmatrix}^2 \sqrt{\frac{3}{4\pi\Gamma(5/2)}}$$

$$\cdot \frac{3}{2} \frac{v_o}{2\pi NC(\frac{m}{2})^{-1/2} \frac{m_1}{m+m_1}} \tag{39}$$

When we compare this with the results of the average penetration depth in ref.[2] for the case s=4, we see that we have the correct dependence of the initial energy, namely $E^{1/2}$. Like in [2], it is inversely proportional to the product NC, only the numerical factor is different. This discrepancy can be removed by carrying out a renormalization of the eigenfunctions ϕ_o and ϕ_1, which is permissible, because we have introduced an element of arbitrariness with the weight function e^{-v^2}. For ϕ_o the renormalization is simple.

We know that always the normalization $\int P(\vec{v}_o|\vec{v},\vec{r},t)d\vec{v}d\vec{r} = 1$ is valid. Substitute (19) in this and we find

$$\int e^{-v^2} \sum_\ell c_\ell(\vec{r},\vec{v}_o,t)\; \phi_\ell(\vec{v})d\vec{r}d\vec{v} = \sum_\ell \int c_\ell d\vec{r} \cdot \int e^{-v^2}\phi_\ell(\vec{v})d\vec{v}$$

$$= \int c_o(\vec{r},\vec{v}_o,t)\; d\vec{r} \cdot \int e^{-v^2}\phi_o(\vec{v})d\vec{v} = 1 \tag{40}$$

Both factors in this product must be 1 and because ϕ_o is independent of $\vec{v}$ we have $A \cdot \int e^{-v^2} v^2 dv \cdot 4\pi = 1$. This gives for the normalization constant $A = \pi^{-3/2}$, instead of $\pi^{-3/4}$, which would follow from (20). In order to find the new normalization factor for the function $\phi_1(\vec{v})$ we write down the result for the average penetration depth $^1c_o^1(v_o)$ from ref. [2] for the case s=4 (Maxwell interaction). There we find

$$^1c_o^1 = \lambda E_o^{1/2} = \frac{3}{2}\frac{m}{m+m_1}\gamma^{-3/4}\frac{E_o^{1/2}}{NC} \tag{41}$$

where E is the initial kinetic energy $E_o = 1/2\; mv_o^2$. We now change the normalization factor of $\phi_1(\vec{v})$ in such a way that the results (39) and (41) become identical. Via the coupling between different ℓ in eq. (27), this renormalization influences also the higher $\phi_1(v)$. Explicitly the renormalized eigenfunction becomes

$$\phi_1(\vec{v}) = Bv\cos\theta = \frac{2\pi}{\sqrt{3}}\gamma^{1/8}\; v\cos\theta. \tag{42}$$

To conclude we shall give a few numerical results for the case of silicon, bombarded by carbon ions. The eigenvalue λ_1, according to (18) has the value $-8.5502 \cdot 10^{12}$ sec^{-1}. Hence we have $^oc_1^o(v_o,t) = \phi_1(v_o)e^{-8.5502 \cdot 10^{12}t}$. We give an interpretation of this result. The renormalized eigenfunction $\phi_1(v) = Bv\cos\theta$. Using the orthogonality properties of the different functions occurring in this treatment we find:

$$\int P(\vec{v}_o|\vec{v},\vec{r},t)\phi_1(\vec{v})d\vec{v}d\vec{r} = Bv\cos\theta = {}^oc_1^o(v_o,t)\; <\phi_1|\phi_1>$$

$$= {}^oc_1^o(v_o,t)\frac{B^2\pi^{3/2}}{3}. \tag{43}$$

Therefore the average value of $v \cos \theta$, or the component of $\vec{v}$ in the direction of $\vec{v}_0$ at time t is given by $\frac{1}{3} B \pi^{3/2} \, {}^o c_1^o(v_o,t) = \frac{B}{3} \pi^{3/2} \phi_1(v_o) e^{-8.5502 \cdot 10^{12} t}$. The number λ_1^{-1} gives the corresponding relaxation time for the case of C on Si and has the value $2\pi\gamma^{1/8}/\sqrt{3} = 3.549$. We give also the average penetration depth as a function of time from (38) and (41). The result is

$$
{}^1 c_o^1(v_o,t) = -1.503 \, v_o \, \frac{e^{-8.5502 \cdot 10^{12} t} - 1}{8.5502 \cdot 10^{12}} . \tag{44}
$$

Finally we consider the moment ${}^1 c_1^1(v_o,t)$. For this moment we have the interpretation, of course also following from orthogonality, that it is the value of the moment $r \cos \eta \, v \cos \theta$ at time t. Its numerical value for the case of C on Si, calculated from (34) turns out to be

$$
{}^1 c_1^1(v_o,t) = -e^{-25.6506 \cdot 10^{12} t} \cdot \frac{0.565 \, v_o^2 (e^{17.3127 \cdot 10^{12} t} - 1)}{17.3127 \cdot 10^{12}}
$$
$$\tag{45}$$

In this particular case the second term disappears, which can be seen if we take the l=0 term of eq. (21) separately on both sides and then make a scalar multiplication with $\phi_1(\vec{v})$. We have used the relation $\dfrac{\partial c_o(\vec{r},\vec{v}_o,t)}{\partial t} + \vec{v} \dfrac{\partial}{\partial \vec{r}} c_o(\vec{r},\vec{v}_o,t) = 0$ which is an expression of Liouville's theorem [8].

Acknowledgments

        This work has been made possible by the financial support of the "Stichting F.O.M." (Foundation for Fundamental Research on Matter).
        The author thanks Drs. A. Tip and H. E. Roosendaal for helpful discussions.

References

[1] J. Lindhard, M. Scharff, H. Schiøtt, Mat. Fys. Med. Dan. Vid. Selsk. 33, 14 (1963).
[2] J. B. Sanders, Can. J. Phys. 46, 455 (1968).
[3] D. K. Brice, Rad. Effects 11, 227 (1971).
[4] P. Mazur and J. B. Sanders, Physica 44, 444 (1969).
[5] J. Lindhard, V. Nielsen and M. Scharff, Mat. Fys. Med. Dan. Vid. Selsk. 36, 10 (1968).

[6] R. Jancel and Th. Kahan, Electrodynamique des Plasmas, Dunod, Paris (1963).

[7] A. R. Edmonds, "Angular Momentum in Quantum Mechanics," Princeton University Press, 1957.

[8] R. Tolman, "The Principles of Statistical Mechanics," Oxford University Press.

INDICATION FOR AN IONIZATION DAMAGE PROCESS IN LIGHT ION IRRADIATION DAMAGE IN SILICON

H. J. PABST and D. W. PALMER
School of Mathematical and Physical Sciences
University of Sussex
Falmer, Brighton BN1 9QH, Sussex, England

ABSTRACT

Rutherford backscattering channeling experiments analysed in the framework of single scattering theory, taking into account the effects of flux peaking, lattice location of defects (through dechanneling cross section), and number of scattering centres per defect, indicate defect introduction rates for 300 keV H^+ and 275 keV He^+ in the ratio 1 to 2 which is similar to the electronic stopping power ratio but differs considerably from Kinchin and Pease predictions (ratio 1 to 20). A multiple ionization damage process is proposed.

Introduction

The study of irradation damage in silicon has in the past been centred upon electrical [1] and electron spin resonance [2] measurements after electron and γ-irradiation on the one hand and neutron and heavy ion irradiation damage as monitored by the above techniques [3] and several others including analysis with channeled ions [4]. This means a considerable amount of data is available on point defects in low concentrations and also on the behaviour of highly disordered silicon.

The present work [5] was aimed at defect location in n- and p-type silicon below and at room temperature for as low a defect concentration as was feasible with the channeling technique. In particular the different behaviour of the intrinsic silicon interstitial in n- and p-type material at low temperature [6] was hoped to be elucidated. Because of the discovery of the flux peaking effect for channeled ions [7] a novel method of analysis had to be developed [5,8,9] to obtain approximate defect concentrations. As channeling methods yield only scattering centre concentrations, appropriate defect models have to be taken into account to obtain defect concentrations [5,8,9]. These can then be compared with known models of defect introduction efficiencies, for instance, as in the present case, with the Kinchin and Pease model [10] for the elastic displacements, and a close pair model [11] for spontaneous recombination. The discrepancies found suggest an additional displacement mechanism [5] which will be described in some detail below.

Experimental set-up

Experiments were conducted in the single alignment (SA) mode (275 keV He$^+$; damage in <110>, <111>, and misaligned; analysis in <110> and <111>) and double alignment (DA) mode (300 keV H$^+$, <110> - <1$\bar{1}$0> - 90° geometry, with simultaneous recording of SA spectra) in a manner described previously [5,8,12]. The SA experiments were conducted on 1.2 Ω cm p-Si and 2.7 Ω n-Si of (111) orientation with an ion flux of 1.0 x 10^{13}/cm^2 sec, and the DA experiments employed 0.3 Ω cm p-Si and 0.003 Ω cm n-Si of (100) orientation with an ion flux of 2.1 x 10^{13}/cm^2 sec. In all cases the ion beam cross section was 1.0 mm^2. The SA He$^+$ experiments were performed with a goniometer cryostat similar to the one described in [8], the main difference being cooling by means of a copper braid attached to a liquid nitrogen container instead of by flexible stainless steel tubing carrying liquid nitrogen. The sample temperature was measured by means of a copper-constantan thermocouple attached to the copper sample holder 2 mm from the sample. The single alignment (surface barrier) detectors (3.0 x 10^{-3} sr) recorded ions scattered by 150° with an energy resolution of 25 keV and 13 keV for He$^+$ and H$^+$, respectively. The scattering spectra were recorded using conventional preamplifiers, amplifiers and multichannel pulse-height analysers.

Analysis of backscatter data

It has been shown elsewhere [13,5,9] that for an even distribution of damage both the minimum yield and the depth dependence of the backscatter yield contain information on defect related scattering centres, and this case is particularly simple to analyse in the framework of single scattering theory [13,5,9]. Uniform distribution of defects with depth can be assumed to hold true for energetic light ion damage when only small depths compared with the ion range are considered, i.e. when the energy loss of the ions is small compared with the ions' energy (compare formulas in reference 10). As the effects of plural and multiple scattering may become important at large penetration depths and also scattering cross sections and displacements cross sections do increase with diminished ion energy, it was found convenient to analyse the experimental data in terms of an extrapolated minimum yield χ_o and dechanneling parameter $1/\Lambda_o$ at the sample surface, which were found from least square fits (LSF) to the portions of the SA channeling spectra corresponding to the depth intervals 660 Å to 3800 Å (linear LSF) and 400 Å to 2100 Å (logarithmic LSF) for H^+ and He^+ analysis, respectively, using stopping power data kindly supplied by F. Eisen [14] and avoiding depth intervals masked by surface peaks from surface contamination (0 and C).

The dechanneling parameter $1/\Lambda$ (d), where d is the irradiation dose, is defined by

$$1 - \chi_o \,(x, \, d) = (1 - \chi_o \,(0, \, d)) \exp \,(-x/\Lambda \,(d)) \qquad (1)$$

where x is the penetration depth. According to single scattering theory, for a uniform defect concentration C, the normalised extra yield due to defects is given [5,8,9] by

$$\Delta Y_{SA}(x) = CF + CFN \,\sigma_D x \qquad (2)$$

where F is the average channeled ion flux at the scattering centre locations, N is the number of silicon atoms per unit volume, and σ_D is the scattering centre position dependent dechanneling cross section [5,9]. Equation (2) is valid when direct scattering yield (i.e. large angle collisions) and dechanneling (i.e. small angle collisions) are due to the same scattering centres in single collisions: the first term arises from direct scattering by the defects in the depth interval considered, the second term from ions dechanneled up to depth x.

At the sample surface only the first term of equation (2) contributes, i.e.

$$\Delta Y_{SA}(0) = \Delta\chi_o = CF \, , \qquad (3)$$

whereas the second term is linked to the dechanneling parameter changes due to defects [5,9] through equation (1) and

$$\Delta 1/\Lambda_o = CFN \,\sigma_D \, . \qquad (4)$$

Hence from (3) and (4)

$$\sigma_D = \Delta 1/\Lambda_o / (N\Delta\chi_o) \, . \qquad (5)$$

A relation similar to (5) has previously been derived [15].

Obviously the quantities employed in equations (3), (4) and (5) are not amenable to direct experimental measurements as they constitute extrapolations of bulk properties to the ion energy at the sample surface through equation (1). The scattering centre location dependence of the dechanneling cross section is due to ions with considerable initial transverse energy requiring collisions with scattering angles less than the critical channeling angle to acquire sufficient total transverse energy to be considered as dechanneled [5,9]. As the channeled ion flux is also a function of the maximum potential transverse energy the ions can achieve (through the theory of accessible area in the channel [7]), experimental dechanneling cross sections can be employed to estimate the ion flux at the defect location with the help of analytical models [7] or computer simulation calculations [7,5,9]. Through equation (3) estimates of the scattering centre concentration can then be made, which in conjunction with appropriate defect models consistent with the experimental dechanneling cross section then yields defect concentrations.

Oscillations in the ion flux [7,5,9] and nuclear encounter probability [16] are thought to be smeared out by the energy resolution of the detecting system and the LSF procedure.

The effective dechanneling cross sections for plural and multiple scattering by defects can be estimated to be smaller than the single scattering dechanneling cross section for the depths and defect concentrations analysed in this work [5].

Experimental results

The general features of yield changes due to 300 keV H^+ irradiation in a <110> direction at 80K and 300K have been presented elsewhere [8] for the depth interval 4000-8000 Å perpendicular to the sample surface. These features are seen also in the extrapolation to the surface described above, and hence the corresponding graph is omitted here. The relevant parameters obtained are incorporated in Table 1.

The yield changes for 275 keV He^+ irradiation in a <110> direction (simultaneous damage and analysis) in n- and p-type silicon at room temperature and low temperatures are presented in Fig. 1. As can be seen the extra yield introduction rates are similar, particularly at low doses. The differing slopes are partly due to the varying values for the random fraction of ions contributing to the displacements. A similar graph for damage and analysis in the <111> direction has been omitted for shortness. The relevant parameters are again incorporated in Table 1.

Results as monitored by <110> analysis for damage in a misaligned direction 5.4° from <111>, are given in Fig. 2 for p- and n-type silicon at room temperature and low temperature. The <111> analysis generally shows slightly higher yield changes. The higher yield for <110> analysis for 275 keV He^+ damage reported in [12] was due to the assumption in the spectra normalization of defect scattering contribution from lattice distortions around the perpendicular circular damage cylinder of near-<111>-damage being contained in

Table 1

Defect introduction: results

Ion	Si Type	EXPERIMENTAL				ANALYSIS				INTERPRETATION
		Temp. K	$\chi_0(0) = f_r$	F	$\Delta\chi_0(d)$ μC %	$\dfrac{\Delta C_{KP}\cdot f_r}{\text{μC %}}$	F_{Si}	m	η	N = n-Si, I = intrinsic
H+	P	300	0.042	0.05	0.0030	0.00071	0.5	4	42	V
	N	300	0.048	0.08	0.0103	0.00081	N 0.22 / I 0.5	N 12-16 / I 8	N 60-45 / I 40	V and LD from I-A
	P	80	0.030	0.05	0.0025	0.00051	0.5	4	49	V
	N	80	0.031	0.8	0.0016	0.00053	0.023	2.5-3.2	65-51	I and V
He+	P	300	0.045	1.0	0.017	0.0153	0.5	1	2.2	I type A
	N	300	0.056	1.0	0.025	0.0190	N 0.22 / I 0.5	1 / 1	N 6.1 / I 2.7	I type A, intrinsic
	P	113	0.027	1.0	0.016	0.0092	0.5	1	3.6	I type A
	N	155	0.039	1.0	0.025	0.0133	N 0.1 / I 0.5	N 2.5 3.2 / I 1	N 7.4 5.2 / I 3.7	I and V (n-Si) or / I type A, intrinsic
	P	300	1.3	0.05	0.068	0.44	0.5	4	1.6	V type A
	N	300	1.3	0.10	0.098	0.44	N 0.22 / I 0.5	4	N 2.5 / I 4.4	V type A / V type A, intrinsic
	P	109	1.3	0.05	0.32	0.44	0.5	4	7.4	V type A
	N	155	1.3	0.05	0.20	0.44	N 0.1 / I 0.5	4	N 23 / I 4.5	V type A, intrinsic

V Vacancy
I Interstitial
A Agglomeration
LD Lattice Distortion

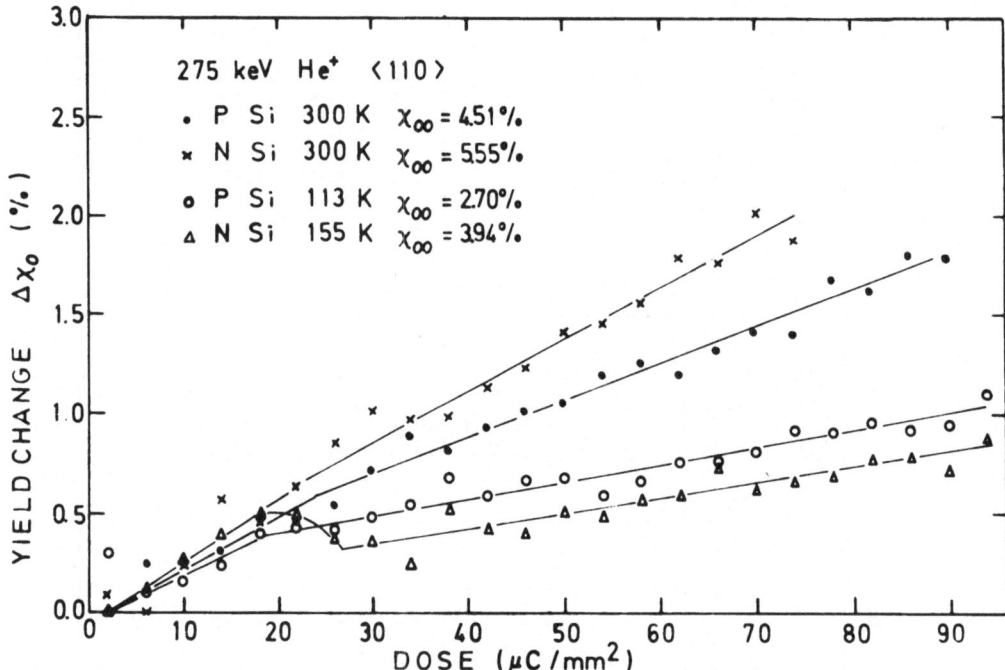

Fig. 1. Yield changes as a function of dose for <110> channeling irradiation with 275 keV He+ on p- and n-type silicon at and below room temperature.

the oblique elliptic analysis cylinder of <110>. This has been dis-proved by analysis of misaligned damage near <110> [5].

The particular misaligned direction near <111> employed in the present work results in a yield in damaging direction a factor of 1.3 larger than the yield from an amorphised sample, i.e. the mis-aligned defect introduction rates are expected to be approximately 30% larger than in a truly random direction.

It can be seen in Fig. 2 that for the misaligned damaging the effect of temperature is considerably more marked than for channeling damage (Fig. 1): the initial (i.e. low dose) scattering centre introduction rates are increased by lowering the temperature and that they are also considerably larger than for damage in channeling direction. Hence definite temperature and alignment effects can be observed for 275 keV He+ damage in silicon.

As may be seen from Table 1 the yield changes due to irradiation with 300 keV H+ are generally smaller in magnitude to those for the 275 keV He+ irradiations, but only by factors between 2 and 7. This is surprising in view of the fact that the Kinchin and Pease model [10] predicts reductions by a factor of 20. An attempt to explain this discrepancy between elastic collision theory and experiment is given below.

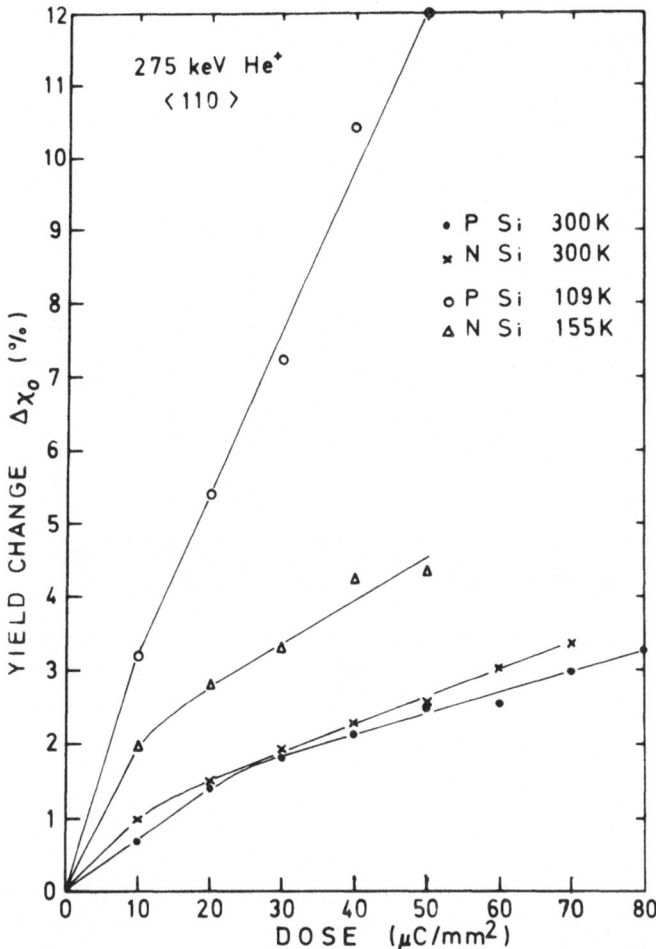

Fig. 2. Yield change as a function of dose for misaligned ("random") irradiation with 275 keV He$^+$ on p- and n-type silicon at and below room temperature as analysed in the <110> channeling direction.

Since the measured scattering yield depends on the concentration of defects, on the number of scattering centres per defect, and on the ion flux at the defect location, a more detailed analysis of the data is necessary. For this purpose experimental values for de-channeling cross sections were obtained according to equation (5). It can easily be deduced that the experimental dechanneling cross section will correspond in good approximation to the scattering centres subject to the highest flux.

An example of a graph according to equation (5), i.e. dechannel-ing parameter at the sample surface versus the extrapolated yield is given in Fig. 3 for damage in channeling directions. The example

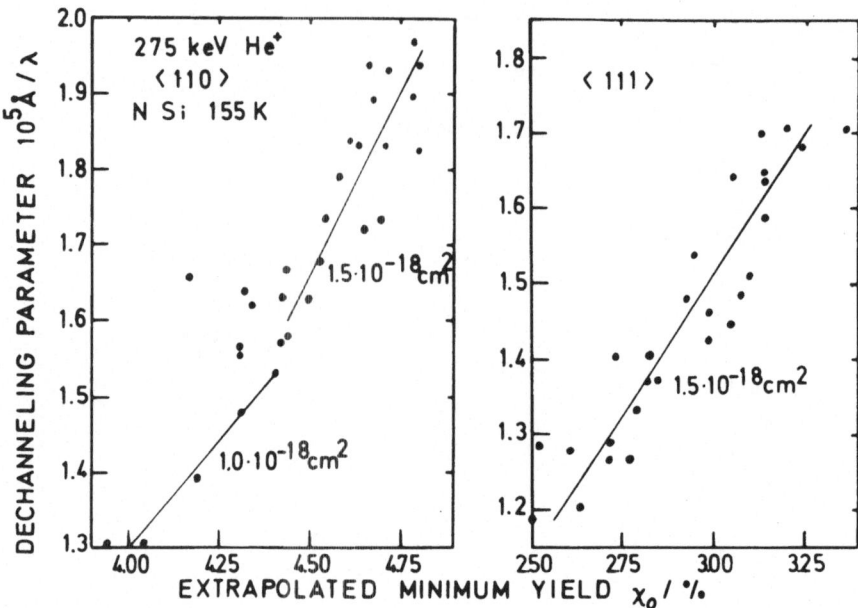

Fig. 3. Experimental dechanneling cross sections according to equation 5 for channeling irradiation (275 keV He[+] on n-Si at 155K) in <110> and <111> directions.

given is n-type Si at 155°K, which according to Fig. 1 exhibits a different behaviour for initial and later stages of the radiation damage in <110>. This is also seen in Fig. 3, where the initial damage shows a fairly smooth behaviour yielding an experimental dechanneling cross section of $(1.0 \pm 0.25) \times 10^{-18}$ cm^2 corresponding to 1.4 times the theoretical value at the channel centre of 0.71×10^{-18} cm^2, whereas the more advanced stages of damage, after the "annealing," show strong deviations from a simple relation, yielding on average an experimental dechanneling cross section of $(1.5 \pm 0.4) \times 10^{-18}$ cm^2 corresponding to 2.1 times 0.71×10^{-18} cm^2. The strong scatter of the experimental points, which is also seen in the case of <111> damage (Fig. 3) is larger than the experimental accuracy and is thought to be partly due to constant rearrangement of scattering centre location, hence also in dechanneling efficiency. In other words the scatter is thought to be partly due to the same effect which causes the large anneal step at 20-30 μCb (Fig. 1) where according to the interpretation of changes in the dechanneling cross section outlined in conjunction with equation (5) scattering centres near channel centre transform to and are superceded by scattering centres near atomic rows, in regions of small ion flux.

Some examples of similar plots for misaligned damage are presented in Fig. 4 for the case of <110> analysis. Slopes corresponding to 1.0 and 2.0 times the theoretical dechanneling cross section
for the channel centre are also included. p-type silicon at room
temperature exhibits both for <110> and <111> changes in the dechanneling cross section at low dose for the present experiments.
Generally the initial damage shows a dechanneling cross section
near twice the theoretical value for channel centre, i.e. scattering
centres near the atomic rows.

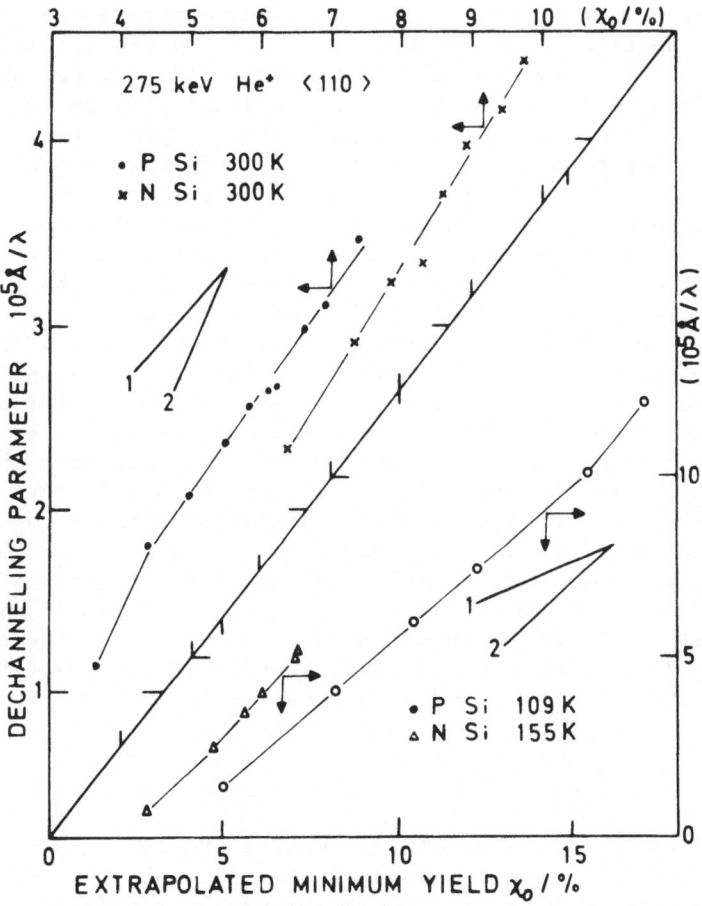

Fig. 4. Experimental dechanneling cross sections according to
equation 5 for misaligned irradiation (as Fig. 2) in <110> channeling
analysis.

The experimental dechanneling cross sections determined now in the same way for the 300 keV H$^+$ <110> channeling damage experiments described elsewhere [5,8] are consistent with the conclusions on defect locations as obtained from the double alignment analysis, i.e. dechanneling cross sections about twice the theoretical value for the channel centre in the three cases p-type Si at 80K and 300K and n-Si at 300K, and a dechanneling cross section of 1.2 times the value for the channel centre for 80K in n-Si damage. This reinforces the findings in [8] and supports the usefulness of experimental dechanneling cross section determination.

From the experimental dechanneling cross sections the ion flux at the scattering centre location can now be estimated to within a factor of two [5,9]. Because the channeled ion flux may vary by a factor of 100 over the channel area [7,5,8,9] this type of analysis is thought to be advantageous for obtaining good estimates of scattering centre concentrations in damaged crystals.

The experimental results (initial damage rates) are gathered in Table 1, section "Experiment": the first column specifies the damaging and analysing ion, the second the silicon type, and the third the irradiation and analysis temperature. Since in the following it will be assumed that the damage rate is proportional to the fraction of ions capable of small impact parameter encounters the fourth column of Table 1 shows the extrapolated minimum yield of the undamaged crystals χ_o (0) which is taken in the following to be the fraction f_r of ions on random paths. The ion flux at the scattering centre location as estimated from the experimental dechanneling cross section according to [5,9] is listed for each experiment in column 5. Finally the sixth column presents the changes in scattering yield per unit dose. Similar results for <111> analysis [5] are omitted here for shortness as they convey little additional information: with few exceptions the same figures are obtained. As can be seen from column 5 (ion flux F at scattering centre location) channeling damage with 300 keV H$^+$ and 275 keV He$^+$ result in different defects, an effect which may be caused by the different elastic collision damage efficiency. In addition misaligned damage for 275 keV He$^+$ produces scattering centre configurations (F $\leqslant$ 0.1, near atomic rows) different from those of channeling damage (F $\approx$ 1.0, well removed from atomic rows) which may be interpreted as an effect of the damage rate. And as could be seen in Figs. 3 and 4 scattering centre location rearrangements may occur at higher doses.

Discussion

According to the experimental results presented above the defect introduction rates are a function of the silicon type, the ion species, the fraction of ions on random paths, the irradiation temperature, and on the dose rate [4,5].

The Kinchin and Pease model [10] is thought to be a good approximation for considering defects introduced by means of elastic collisions and collision cascades between fast ions and target atoms. Disregarding any kind of annealing the concentration of Frenkel pairs according to this model (here for energetic light ions, i.e. coulomb recoil and hard sphere cascade) is given by

$$C_{KP} = \phi \; \frac{\pi}{2} \left(\frac{Z_1 Z_2 e^2}{4\pi\varepsilon_o} \right)^2 \frac{M_1}{M_2 E_1 E_d} \; \log \frac{\Lambda E_1}{E_d} \tag{6}$$

where the index 1 refers to the fast particle, 2 to the target material, Z_i e is the nuclear charge, M_i the nuclear mass, E_1 is the particle energy, E_d is the minimum energy for elastic atomic displacement, Λ the elastic energy transfer coefficient and ϕ is the number of energetic ions per unit area. Hence for non-channeled incidence and E_d = 20 eV [17] in the present case the calculated defect introduction rates are 0.017%/μCb and 0.34%/μCb for 300 keV H^+ and 275 keV He^+, respectively. The individual calculated defect introduction rates according to this model are contained in column 7 of Table 1.

Due to the nature of the defects, spontaneous recombination and thermal (and radiation) annealing are likely to take place. For electron damage in n-Si a close pair model has been introduced [11] which accounts for the reduction of surviving point defects in n-Si when lowering the temperature, as observed here for 300 keV H^+ channeling damage. The two charge state model [11] gives the fraction f^+ of close pairs in the more positive charge state as

$$f^+ = 1/1(1 + g \; \exp \; ((E_F - E_M)/kT) \; , \tag{7}$$

where g is the degeneracy, E_F the Fermy energy and E_M the energy of the electronic state of the close pair. If ΔE^+ and ΔE are the differences in energy barrier between recombination and dissociation jumps for the close pair in the more positive and the more negative charge state, respectively, the fraction of close pairs dissociating can be calculated to be [11]

$$F_{Si} = f^+/(1 + \gamma \cdot \exp \; (\Delta E^+/kT))$$
$$+ \; (1 - f^+)/(1 + \gamma \; \exp \; (\Delta E/kT)) \tag{8}$$

where γ is the ratio of the number of recombination paths to the number of dissociation paths. The electronic level of the close pair is assumed to be near the Fermi level of the n-type material

[11] (here $E_F(n) - E_M = 0.02$ eV). With $g = 1$ and $\gamma = 1$, $\Delta E^+ = 0.0$ eV and $\Delta E = 0.06$ eV (this choice is not very critical when considering relative relations, only affecting a scaling factor of the order of 1) the values F_{Si} in the eighth column of Table 1 were obtained, predicting no difference in defect introduction rates for p-Si in the temperature interval covered and some reduction for n-Si, depending on temperature.

The defect introduction rates can now be assumed to be governed by an equation of the form

$$\frac{dc_d}{d\phi} = \eta \cdot \frac{dc_{kp}}{d\phi} \cdot f_r \cdot F_{Si} \cdot (1 - c_d) - K \cdot f(c_d)/\frac{d\phi}{dt} \qquad (9)$$

where $\eta \cdot dc_{kp}/d\phi$ is the displacement rate for unit flux, f_r is the fraction of ions contributing to displacements, F_{Si} is the spontaneous recombination survival fraction, K is the form $K = k_o$ exp $(-E/kT)$ (K_o frequency factor, E migration energy of the defect), $f(c_d)$ is the function determining the thermal annealing law (i.e. proportional to c, or c^2 etc., depending on type of process) and $d\phi/dt$ is the dose rate.

Disregarding the thermal annealing term $Kf(c_d)$, the efficiency factor η in the above equation may be estimated using appropriate numbers in the scattering centres per defect. This has been realised in column 9 of Table 1. The principle applied was to assign 4 scattering centres near atomic rows to vacancies (Jahn-Teller distortions around vacancies [2]) and 2 to 3 scattering centres well removed from atomic rows to interstitials (split and bond centred, respectively). In addition it was assumed that silicon interstitials are mobile under irradiation conditions at all temperatures in p-type silicon and in n-Si above 140K [2,5,8], diffusing to the sample surface in the case of p-Si and building agglomerates in the case of n-Si. For the 275 keV He$^+$ irradiation results the possibility of the material turning intrinsic under irradiation, which would result, concerning recombination, in behaviour similar to p-Si has also been included. Column 11 of Table 1 gives defect interpretations.

The individual values of the displacement efficiency factor η calculated as outlined are listed in column 10 and are seen to be very similar for the same ion species as has to be expected but differing by a factor of about 10 when comparing H$^+$ and He$^+$. The consistency of preanneal defect introduction rates may be improved by taking into account thermal and radiation annealing [5].

With average values on η (300 keV H^+) = 45 and η (275 keV He^+) = 4.6 (the latter including <111> data [5]) the preanneal displace-

ment efficiency $\eta \dfrac{dc_{kp}}{d\phi}$ can now be calculated to be $6.0 \cdot 10^5$/(cm

and ion on random path) for 300 keV H^+ and $1.2 \cdot 10^6$/(cm and ion on random path) for 275 keV He^+. In other words 275 keV He^+ ions are only twice as efficient as 300 keV H^+ ions in producing defects in silicon. This ratio is close to the electronic stopping power ratio of the ions concerned, as opposed to the ratio 20 expected from energy into nuclear processes [10]. A simple M-shell ioniza-tion effect has to be ruled out to be responsible for the defect introduction rates because of the experimentally secured dependence on the fraction of ions on random paths (column 6 of Table 1 shows ratios of damage scattering yields of 4 to 20 between misaligned and aligned damage for He^+ irradiation and after taking f_r and the ion flux at the scattering centre location into account, the values of η (column 9) show no systematic difference). Following up a suggestion by Kistemaker [18] the ionization cross section of the L-shell was taken into consideration. The theory of Garcia [19] (binary encounter impulse approximation) yields L-shell ionization cross sections of $2 \cdot 10^{-17}$ cm^2 and 4.10^{-17} cm^2 for 300 keV H^+ and 275 keV He^+, respectively, which are sufficiently large to account for the observed defect introduction rates (according to the model below) and are also sufficiently small to account for reductions in the defect introduction rates when the ion beams are incident in channeling directions.

The model proposed [5] is a multiple ionization damage (MID) process caused by the coulomb repulsion of the high ionization of one atom after an Auger process and the ionization of nearest neigh-bours through M-shell vacancies produced along the ion path. It is thought of importance that M-shell vacancies must cause weakening of the bonding of the atoms.

The L-shell excitation of a light atom like silicon results in nearly 100% Auger recombinations [20], producing doubly or trebly charged Si-atoms with as many broken bonds (M-shell). This in it-self may not suffice to displace an atom: a net force on the atom is obtained when considering the high ionization of M-shells along the ion path of (in the present case) one or more electron-hole pairs per Angstrom (electronic energy loss 8 eV/Å and 22 eV/Å for 300 keV H^+ and 275 keV He^+, respectively, random [14]) hence it can safely be assumed that at least one nearest neighbour of the highly charged Si-atom introduced above is almost simultaneously ionized, Fig. 5. Neglecting polarization and dielectric effects a coulomb repulsion of about 5 eV per charge pair involved can be calculated. For three charge pairs (one M vacancy from direct ionization and two from Auger process paired with one M-shell vacancy on an adjacent

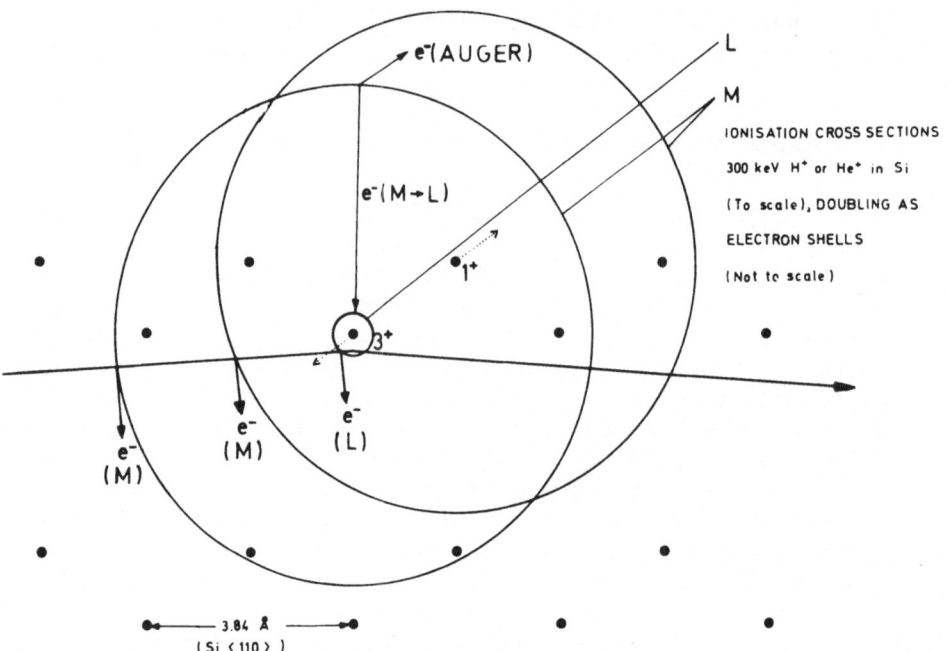

Fig. 5. Schematic diagram of multiple ionization damage process;
time sequence is: inner and outer shell ionization (+ bond breaking), Auger recombination of inner shell vacancy (+ bond breaking),
and coulomb repulsion resulting in atomic displacement.

atom) the Coulomb repulsion will amount to about 15 eV which compares favourably with an elastic displacement energy of 20 eV [17]
and is thought to be strong enough to cause the displacement of the
atom(s), since in one case only one bond remains to be broken, into
a quasi-stable close-pair configuration. Since the times for Auger
recombination and thermal vibration (10^{-13} secs) are short compared
to electron-hole pair recombination times (10^{-6} secs [21]) and formation energies of vacancies and interstitials in silicon have been
estimated to be in the order of 5 eV [22] the MID process here proposed is thought to be operational as a modified Varley mechanism
[24]. Of course it cannot be ruled out that the concurrent subthreshold elastic energy transfer is a prerequisit for damage creation.
        Because of the requirement of simultaneous L and M-shell ioni-
zation on two nearest neighbours (respectively), this model (MID)
does not allow for x-ray production of defects. On the other hand
the MID process may contribute to heavy ion damage in silicon (hence
also in Si collision cascades), due to the large cross sections for

L-shell excitation by heavy ions [20]. As the MID mechanism produces point defects (generally highly mobile under irradiation conditions in silicon) which may be trapped at the defect cascades produced by elastic collision displacements, observed damage distributions at room temperature [23] will generally agree with predictions of theories on energy into nuclear processes.

Concerning damage by electrons a Varley mechanism based on inner-shell ionization has been proposed by Zaikovskaya et al [25] to explain subthreshold irradiation effects in Si epitaxial films with 2 keV to 150 keV electrons, the maximum defect introduction rate having been found for electron energies of 6 keV, about three times the K shell ionization energy.

Regarding other materials, Table 2 gives the maxima in the L and K shell ionization cross section [19] with the corresponding energies for Si, Ge, C, N and O the latter because they are used in organic materials employed for track formation the former because of their approximation value for (compound) semiconductors. The ionization cross sections only halve for changes in energy by a factor of 3, hence permanent damage may be centred around lower energies as those of Table 2 because of the concurrent necessity for sufficient high probability of outer shell ionization for the suggested MID processes to be operational.

Table 2

Light ion inner shell ionization
Maximum cross section σ_{Im}, and corresponding energy E_m.
Theory of Garcia [19].

ELEMENT		Si	Ge	Ge	C	N	O
EL. SHELL		L	L	M	K	K	K
H^+	$\dfrac{\sigma_{Im}}{cm^2}$	$2 \cdot 10^{-17}$	$5 \cdot 10^{-19}$	$2 \cdot 10^{-17}$	$2 \cdot 10^{-18}$	$1 \cdot 10^{-18}$	$6 \cdot 10^{-19}$
	$\dfrac{E_m}{MeV}$	0.3	2.6	0.3	0.6	0.8	1.1
He^+	$\dfrac{\sigma_{Im}}{cm^2}$	$8 \cdot 10^{-17}$	$2 \cdot 10^{-18}$	$8 \cdot 10^{-17}$	$9 \cdot 10^{-18}$	$5 \cdot 10^{-18}$	$2.5 \cdot 10^{-18}$
	$\dfrac{E_m}{MeV}$	1.2	10.4	1.2	2.4	3.2	4.4

Acknowledgments

The authors wish to express their appreciation to J. W. Corbett for pointing out the electron irradiation data, to the Mullard Research Laboratory for providing some of the silicon samples and to Mr. E. G. Turpin for operation of the 300 kV accelerator. This work was supported in part by the Ministry of Defense (Procurement Executive).

References

[1] Recent review: H. J. Stein in Radiation Effects in Semiconductors, Ed. J. W. Corbett, G. D. Watkins, Gordon and Breach, 1971; p. 125.

[2] Recent review: G. D. Watkins in Radiation Damage and Defects in Semiconductors, Conference Series No. 16, The Institute of Physics, 1972; p. 228.

[3] Recent review: F. L. Vook in Radiation Damage and Defects in Semiconductors; Conference Series No. 16, The Institute of Physics, 1972; p. 60.

[4] Recent review: W. L. Brown in Radiation Damage and Defects in Semiconductors; Conference Series No. 16, The Institute of Physics, 1972; p. 416.

[5] H. J. Pabst, Thesis, 1973, University of Sussex.

[6] G. D. Watkins in Radiation Damage in Semiconductors, Ed. P. Baruch, Paris, Dunod, 1964, p. 97.

[7] D. van Vliet, Rad. Eff. $\underline{10}$, 137 (1971) and J. U. Andersen, O. Andreasen, J. A. Davies, E. Uggerhøj, Rad. Eff. $\underline{7}$, 25 (1971).

[8] H. J. Pabst, D. W. Palmer in Radiation Damage and Defects in Semiconductors, Conference Series No. 16, The Institute of Physics, 1972, p. 438.

[9] H. J. Pabst, this conference.

[10] G. H. Kinchin, R. S. Pease, Rep. Progr. Phys. $\underline{18}$, 1 (1955).

[11] B. L. Gregory, C. E. Barnes in Radiation Effects in Semiconductors, Ed. F. L. Vook, Plenum Press, 1968, p. 124.

[12] P. Baruch, F. Abel, C. Cohen, M. Bruneaux, D. W. Palmer, H. J. Pabst in Radiation Effects in Semiconductors, Ed. J. W. Corbett, G. D. Watkins; Gordon and Breach, 1971, p. 147.

[13] K. Morita, N. Itoh, J. of the Physics Soc. of Japan $\underline{30}$, 1430 (1971).

[14] F. H. Eisen, private communication.

[15] S. Miyagawa, K. Morita, N. Matsunami, K. Tachibana, N. Itoh, Rad. Eff. $\underline{13}$, 271 (1972).

[16] J. H. Barrett, Phys. Rev. B$\underline{3}$, 1527 (1971).

[17] S. B. Fisher, P. C. Banbury in Atomic Collision Phenomena in Solids, Ed. D. W. Palmer, M. W. Thompson, P. D. Townsend; North-Holland 1970; p. 232.

[18] J. Kistemaker, Comment after [17], p. 244.

[19] J. D. Garcia, Phys. Rev. A1, 280 (1970) and Phys. Rev. A3, 955 (1971).

[20] F. W. Saris, Thesis, Leiden 1971.

[21] Nakashime, Y. Innshi, Radiation Effects in Semiconductors, Ed., F. L. Vook, Plenum Press, 1968, p. 124.

[22] A. Seeger, K. P. Chik, Phys. Stat. Sol. 29, 455 (1968).

[23] F. H. Eisen, B. Welch, J. E. Westmoreland, J. W. Mayer, Collision Phenomena in Solids, Eds., D. W. Palmer, M. W. Thompson, P. D. Townsend, North-Holland 1970, p. 111.

[24] H. J. O. Varley, J. Phys. Chem. Solids 23, 985 (1962).

[25] M. A. Zaikovskaya, A. E. Kiv, O. R. Niyazova, Phys. Stat. Sol. (a)3, 99-104 (1970) and Phys. Stat. Sol. (a)8, K133-135 (1971).

RECOIL IMPLANTATION OF ^{18}O FROM SiO$_2$ BY HEAVY PROJECTILES

R. A. MOLINE, G. W. REUTLINGER and J. C. NORTH

Bell Laboratories

Murray Hill, New Jersey 07974

ABSTRACT

The recoil implanation yield of oxygen atoms recoiled from thin, ^{18}O enriched SiO$_2$ layers into silicon substrates has been studied using the ^{18}O$(p,\alpha)^{15}$N nuclear reaction. For 24 keV Kr projectiles the ^{18}O yield peaked at a thickness ($\sim$150 Å) which approached the expected range of the Kr in SiO$_2$, and $\sim$2.5 oxygen atoms recoiled into the Si for each projectile. The cross section for recoiling into the Si was $\sim$5 x 10^{-17} cm^2/oxygen atom in the near-linear region stated above. For a fixed SiO$_2$ thickness, the yield increased slightly with decreasing projectile energy until the projectile range was no longer greater than the oxide thickness.

The amount of ^{18}O which was backsputtered from the SiO$_2$ layers during the bombardment by Kr was determined. For a fixed Kr energy and a thin SiO$_2$ layer, this amount increased linearly with increasing SiO$_2$ thickness and was nearly 10 times the recoil implantation yield.

A simple cascade model does not describe the Kr data, but the primary knock-on yield from a 1/r^2 potential, which is a reasonable approximation to the Thomas-Fermi potential for impact parameters of interest here, predicts the observed yield to better than a factor of two. Making a first order correction for recoils caused by the primary knock-ons, the calculations give excellent agreement to the yield vs. oxide thickness data.

When using photoresist as a masking material for heavy projectiles, maximum energy carbon recoils have a mean range much larger than the projectile (a factor of $\sim$1.8 for As). Thus, for

159

high dose implantations, where a significant number
of carbon atoms are recoiled, the desired photoresist thickness is best dictated by the range and
range straggle of the recoils, not the projectiles.

Introduction

The quantity of a thin surface layer which is swept along with
an ion beam into the sample has been the topic of several papers
[1-5]. In analogy to the removal of surface layers using "backsputtering," the effect measured in this paper results from the
"forward sputtering" of atoms from the film. Forward sputtering
is concerned with all recoils which have sufficient energy to exit
a free standing foil in the direction of the incident beam. Recoil
implanation includes only those recoils which have sufficient energy
to leave a thin layer supported by a substrate and penetrate into
the substrate a sufficient distance such that when the thin surface
layer is removed the recoil atoms remain in the substrate, as shown
schematically in Fig. 1.

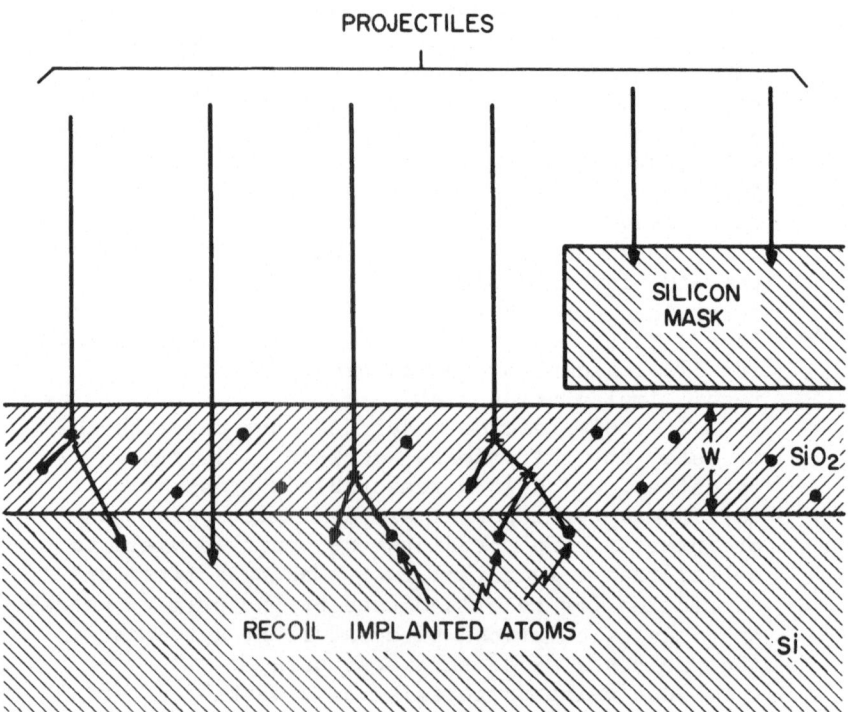

Fig. 1.   Schematic illustration of experimental technique.

    Considerable experimental data is available for backsputtering
yields, but very little is available for transmission sputtering or
recoil implantation yields. The recoil implantation yield is interesting not only as a tool for investigating the nature of recoil
cross sections, but also as it relates to applications of ion implantation. Photoresist is commonly used as a mask to block ion
penetration from specific device regions. If the carbon recoils
from the photoresist enter the silicon in any significant numbers,
precipitation might occur. In other situations, where the ions
penetrate a surface layer to enter the sample, some fraction of the
surface layer which the ion beam penetrates will be swept into the
substrate. A quantitative knowledge of this effect is important.
    Nelson [2] discussed the possibility of using recoil implanta-
tion as an implantation technique, where a film of the desired material is put on the sample to be doped, and then the structure is
bombarded with a beam of an ion species which lends itself to large
beam currents. The analysis treated in Ref. 2 used a cascade model
which yielded a nonintegrable (e.g., infinite) number or recoils.
    This paper describes a method for obtaining experimental results
for recoil implantation. The technique is described, typical results
are given, and the results are compared to numbers obtained from
integrating the $1/r^2$ potential for binary, elastic collisions.

Experimental technique

The goal of this work was to measure the number of atoms which recoil
in a forward direction out of a thin surface layer target, and are
implanted into the underlying substrate. Since the energy profile
of the recoils is peaked near zero energy, it is imperative that
the target/substrate system lend itself to quantitative measurements
of the target atoms in the substrate and complete separation of the
target layer from the substrate. Also, since for thin targets ($\sim$10 Å)
the number of recoil atoms is small, the experimental method used to
measure the recoil implantation concentration must be selectively
sensitive to the recoil atoms.
    High resistivity (100 Ω cm), 2 cm by 0.5 cm, n-type silicon
substrates were used. To remove any organic films and particulate
matter, the samples were rinsed in an ultrasonically agitated chloroform bath, dried and immediately mounted on the anode fixture. To
ensure that a minimum native ^{16}O oxide layer was incorporated into
the SiO$_2$ layer, the samples were held in dilute HF until the surface
was hydrophobic, rinsed (in running D.I. water), dried, and immediately placed in the anodization cell. A portion of each sample
was anodized by holding it in the electrolyte to a depth of $\sim$1/2".
A 5% solution of KCl in ^{18}O enriched water in which 42 atom % of the
oxygen was ^{18}O served as the electrolyte. The anodization was
carried out at room temperature with 40 volts applied to the cell
through a one megohm resistor. The voltage drop between the silicon
sample and a platinum cathode was monitored and the process stopped

at a predetermined voltage which was related to the desired thickness of SiO_2.

After anodization, the samples were rinsed, dried and baked in dry N_2 for 10 minutes at 630°C. This was done to dry the anodic oxide and thus eliminate the possibility of losing ^{18}O as H_2O from the layer during the experiment.

One half of the anodized region of each sample was then uniformly bombarded with 1×10^{15} Ne/cm^2, 1×10^{15} Ar/cm^2 or 2×10^{14} Kr/cm^2. The $^{18}O(p,\alpha)^{15}N$ nuclear reaction was used to measure the number of ^{18}O per unit area in both the bombarded and nonbombarded regions of the anodic oxide. The difference in the two results gave a measure of the amount of backsputtering which occurred during the implantation. The measurement technique and apparatus used has been discussed elsewhere [6].

After the initial ^{18}O analysis, the samples were cleaned in chloroform, rinsed, dried and then baked at 630°C for 10 minutes in dry N_2. This was done to remove the gross radiation damage produced by the bombardment before the oxide strip. The oxide was stripped by immersing the sample in a 1:8 solution of 48% HF in H_2O for 30 seconds. The samples were rinsed, dried, and immediately mounted in the target chamber of the Van de Graaff accelerator for analysis of the ^{18}O recoil implanted into the silicon substrate. Measurement made in the unbombarded areas served as control measurements to ensure that all of the $Si^{18}O_2$ was removed and to correct for any signal due to naturally occurring ^{18}O in any existing oxide on the silicon wafer.

Results

Figure 2 displays the α-particle yield from the unbombarded section of each oxide layer for a constant proton dose as a function of the oxide thickness as measured ellipsometrically. Each data point was corrected for the energy spectrum analyzer dead time, a non-negligible correction only for the thicker samples. As can be seen, the results give a near-linear relationship of 525 counts/Å of enriched SiO_2. The statistical uncertainty in the α-particle yield is very small. The data deviates from a linear relationship for the thinnest oxides (t $\lesssim$ 40 Å). The maximum deviation is 8 Å. This is consistent with previous results [7] obtained by the same techniques using oxides having the naturally occurring isotopic abundance of ^{18}O (0.2%). For the analysis below, the oxide thicknesses were determined from the (p,α) results.

The α-particle yield was converted to the number of ^{18}O atoms/cm^2 by comparison with the α-particle yield from a silicon sample implanted with 1.0×10^{15} ^{18}O/cm^2 at 3 keV. From the slope of the data in Fig. 2, it was found that the oxygen in the anodic oxide layer is in fact enriched to 36% ^{18}O. The source of the slight decrease from the 42% ^{18}O enriched water used as the electrolyte is unknown.

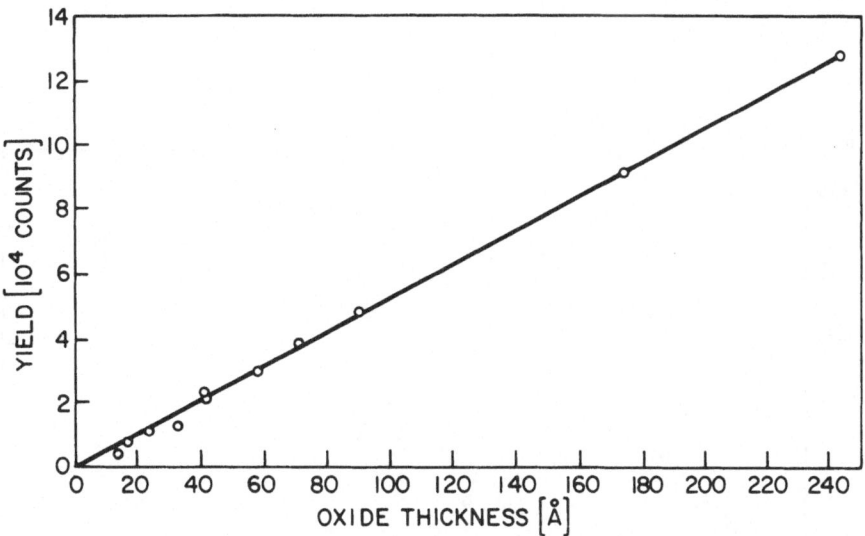

Fig. 2. Alpha-particle yield for constant proton dose from unim-planted oxides vs. oxide thickness as measured by ellipsometry.

We can conclude from Fig. 2 that the ^{18}O is uniformly distrib-uted through the oxide film. Therefore, the α-particle yield before and after bombardment can be used to account quantitatively for the backsputtering loss of SiO$_2$ during the bombardment. Because some of the SiO$_2$ layer is removed during the heavy ion bombardment, the thickness values used subsequently (Figs. 3 and 4) are the average thicknesses of the bombarded and unbombarded portions of each sample.

The number of ^{18}O atoms recoiled into the silicon is propor-tional to both the volume concentration of ^{18}O atoms in the target (SiO$_2$) and the dose of the heavy ion bombardment. The α-particle yield obtained from the silicon after the oxide was stripped was thus divided by the α-particle yield obtained from the oxide, per Å of oxide, and the ion dose (ions/Å^2). The resulting number gives a reduced yield in units of Å^3 of oxide per incident primary projec-tile. If one Si atom was recoiled from the SiO$_2$ into the silicon substrate for each two oxygen atoms recoiled, the thickness (Å) of SiO$_2$ recoiled into the silicon would be given by the product of this reduced yield times the ion dose (ions/Å^2).

Boron is a source of interference in these measurements due to the ^{11}B(p,α)^{8}Be* nuclear reaction. The energy spectrum of the α-particles from this reaction is much broader than (but overlaps) the spectrum from the ^{18}O reaction. Thus, the α-particle contribu-tion from the boron reaction to the α-particle energy region of the ^{18}O reaction can be subtracted as discussed elsewhere [6] This cor-rection typically corresponded to a reduced yield of <0.5 Å^3/ion.

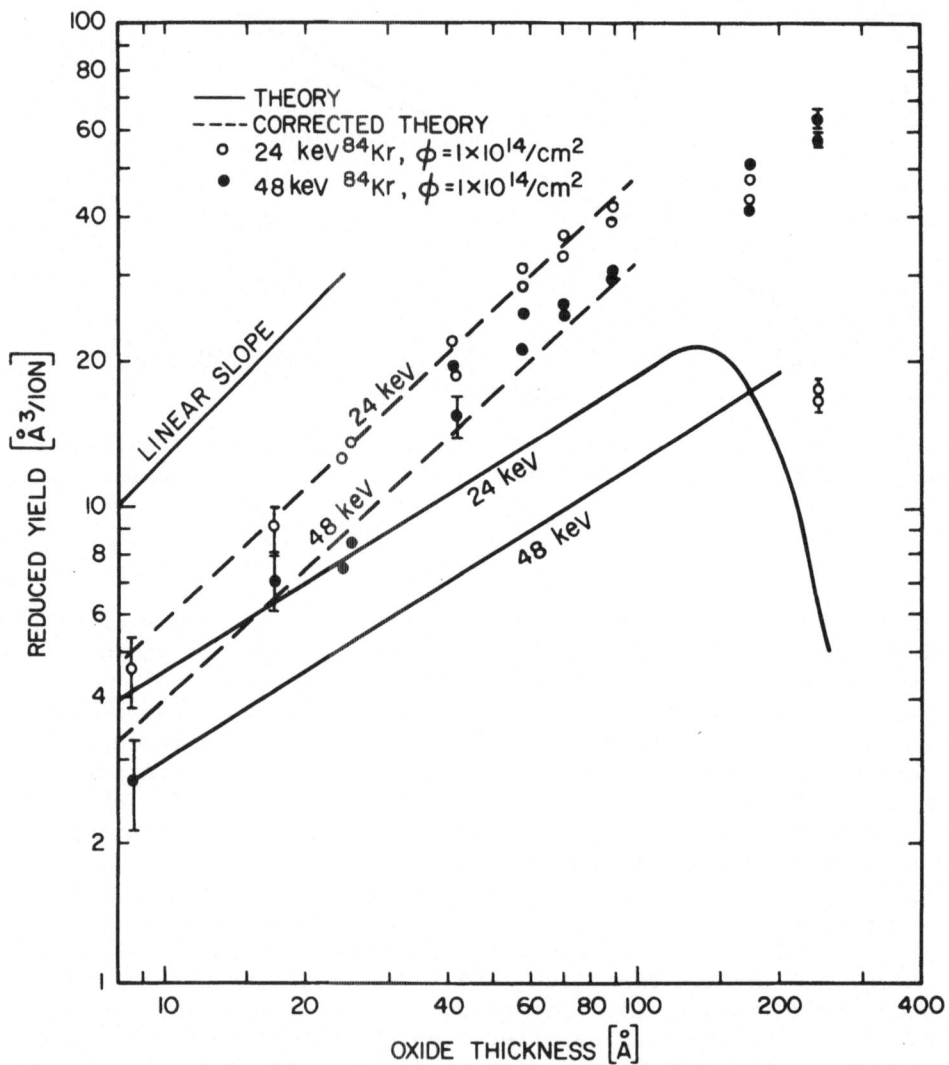

Fig. 3. Reduced yield vs. oxide thickness, theoretical curves and
experimental data, for 24 and 48 keV ^{84}Kr projectiles. Representative error bars are shown for several data points.

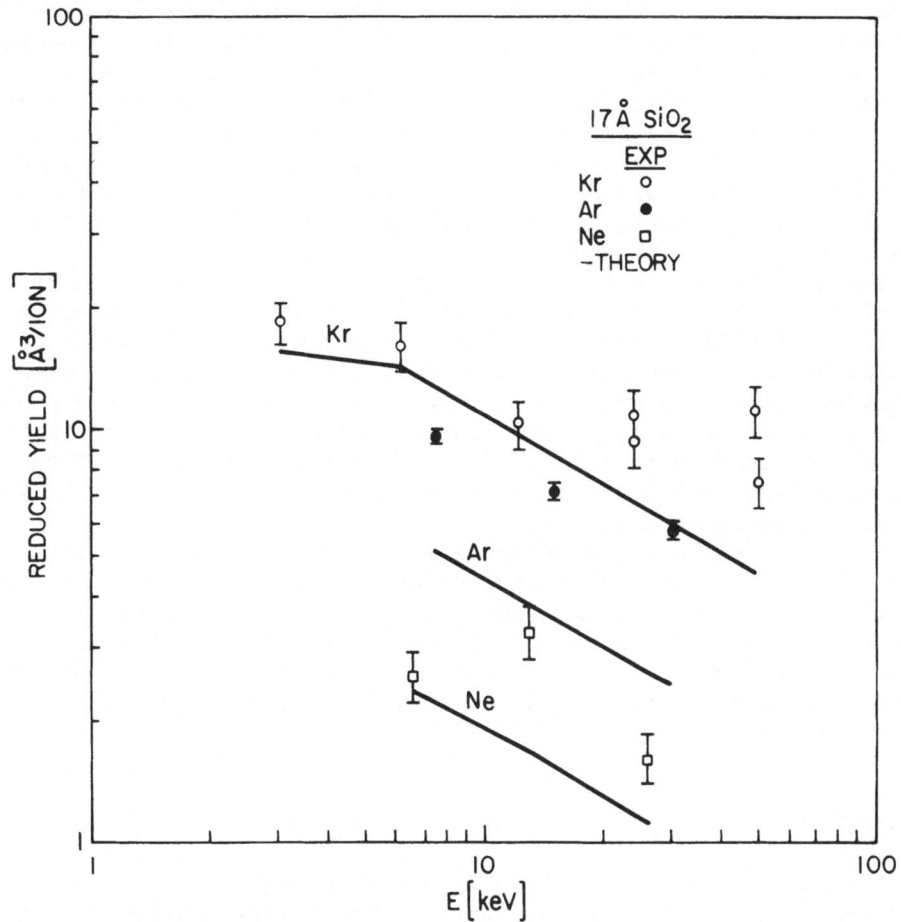

Fig. 4. Reduced yield vs. projectile energy for a 17 Å oxide layer. Theoretical curves and experimental data for ^{84}Kr, ^{40}Ar and ^{20}Ne projectiles are included.

The reduced yield in the sample region which was anodized but not bombarded was also measured as a control after the SiO₂ was stripped. This number represented a contamination from some unknown source and was subtracted from the reduced yield measured in the bombarded regions of the sample after the SiO₂ was stripped. This correction typically amounted to ∿1/40 Å of the enriched SiO₂ or a reduced yield of ∿1.4 Å³/ion.

The data points in Fig. 3 display the reduced yield as a function of oxide thickness for two different Kr energies. The solid and dashed curves are the results of calculations discussed below. Note that the yield varies nearly linearly with the oxide thickness over a one order of magnitude range. The expected projected range of 24 keV Kr in SiO_2 is 166 Å. This compares to the observed maximum in the recoil implantation yield for 24 keV Kr at an oxide thickness of ∿150 Å. Under these conditions ∿2.5 oxygen atoms are recoiled into the Si for each projectile. The yield is also seen to increase with decreasing Kr energy except for thick SiO_2 layers where the Kr range is less than the oxide thickness.

The data points in Fig. 4 display the reduced yield as a function of projectile energy for Kr, Ar and Ne projectiles impinging on a 17 Å thick SiO_2 target. The solid curves are the results of calculations discussed below. Note that the yield increases with increasing projectile mass. By a linear interpolation of the projected range of Kr quoted above, the projected range of 3 keV Kr in SiO_2 is estimated to be ∿21 Å. This is near the thickness of the oxide used here, and hence it is expected that the yield for Kr would drop off at energies near 3 keV.

Figure 5 displays the decrease in the reduced yield of the target as a result of backsputtering by the Kr bombardment, before the SiO_2 was etched off. Although there is no statistically significant difference between 24 and 48 keV Kr, the amount lost is linearly dependent on the oxide thickness up to a thickness of ∿90 Å. For thin targets, approximately 10 times more oxygen is backsputtered off than is recoiled into the substrate.

Calculations

For conditions where only a minute fraction of the forward sputtered atoms have ranges as large as the target thickness, the transmission sputtering yield has been calculated by Sigmund [1]. For the case of a high energy projectile, where the projectile emerges from the target with most of its original energy, both the calculations by Sigmund [1] and by Nelson [2] predict a transmitted sputtering yield which is independent of target thickness.

The ratio of the projected range of a maximum energy recoil divided by the projected range of the projectile tends to increase as the projectile mass (M_1) increases for a fixed recoil mass (M_2). For the case of an ^{84}Kr projectile and an ^{18}O recoil, with the target being SiO_2, the recoil from a 40 keV projectile can penetrate to a mean depth 2 times [8] that of the projectile itself. Thus, for $M_1 \gg M_2$, the condition that the target be "thin" for the projectiles and "thick" for the recoils cannot be satisfied.

The primary knock-on yield from a $1/r^2$ and a $1/r^3$ potential has been calculated in an attempt to find an analytic representation for the results measured here. Consider the recoil atom displayed in Fig. 6. If the range of the recoil atom projected in the direction of the incident projectile (for normal incidence) is greater than,

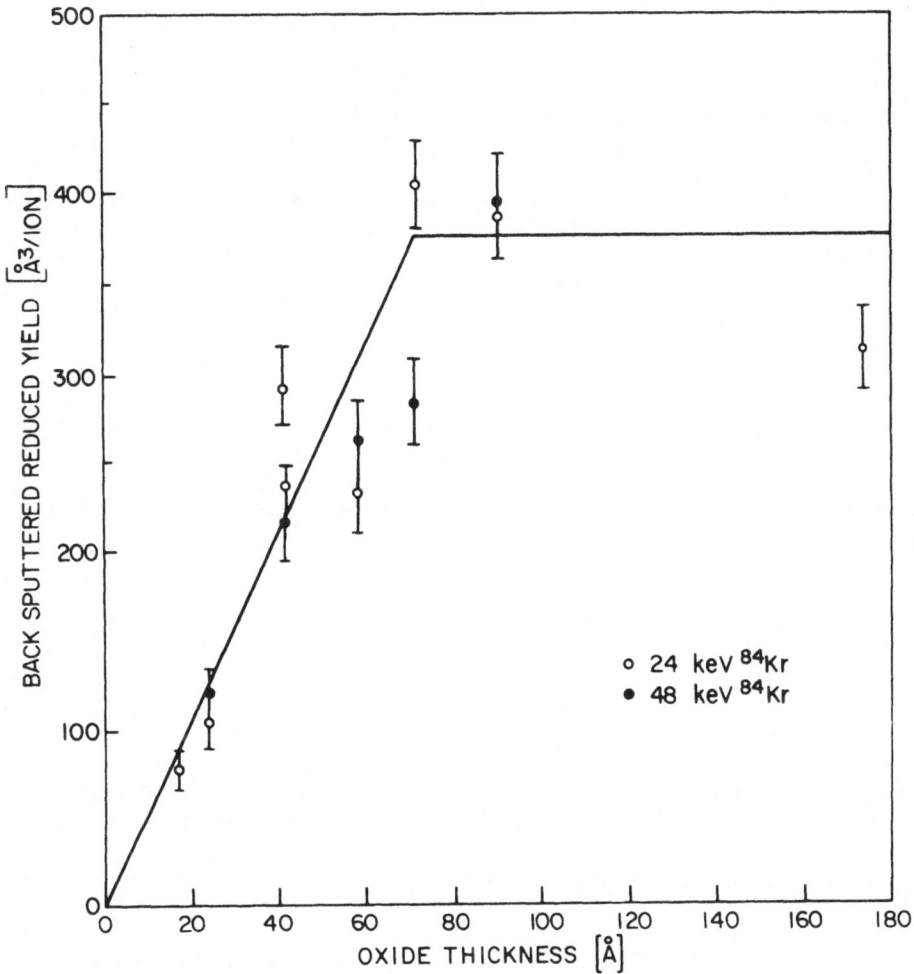

Fig. 5. Backsputtered reduced yield vs. oxide thickness for 24 and 48 keV ^{84}Kr projectiles.

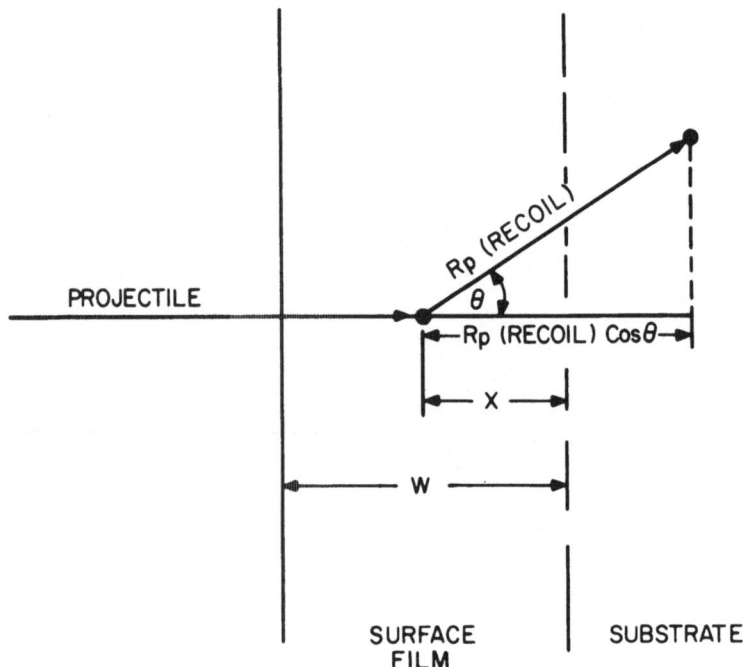

Fig. 6.　Schematic drawing of recoil atom defining analytical variables.

or equal to, x, the distance to the target/substrate interface, the recoil will escape from the target.　For a collision where the energy transfer to the recoil is T, the above condition requires that

$$R_p(T) \cos \theta \geqslant x , \tag{1}$$

where R_p is the projected range of the recoil along its initial direction.　The total number of recoils from the depth x (per projectile) is given by

$$\frac{dn}{dx} = N \int_{T_1}^{T_m} d\sigma \ (T) , \tag{2}$$

where N is the number of target atoms/cm^3, dσ is the differential cross section for a collision with an energy transfer T, T_m is the maximum energy transfer possible, and T_1 is the minimum energy transfer for which Eq. 1 is satisfied.

The lower limit of the integration above can be rewritten by substituting the expression [1] $\cos \theta = (T/T_m)^{1/2}$ in Eq. 1. The total yield per projectile is then obtained by integrating Eq. 2 over all values of x which allow the recoils to reach the substrate.

For an elastic $1/r^2$ potential, the differential cross section in Eq. 2 is given by [9] (for $T < T_m$)

$$d\sigma(T) = \frac{0.33 \ \pi a^2 T_m^{1/2} \ dT}{2 \ \varepsilon_1 T^{3/2}} \ , \tag{3}$$

where $a = a_o * 0.8853 \ (Z_1^{2/3} + Z_2^{2/3})^{-1/2}$, and ε_1 is the projectile energy in dimensionless units [9]. Using the $1/r^2$ potential, Eq. 2 yields

$$\frac{dn}{dx} = \frac{0.33 \ \pi a^2 N}{\varepsilon_1} \left[\left(\frac{T_m}{T_1} \right)^{1/2} - 1 \right] . \tag{4}$$

If the stopping powers of the target are known, Eq. 4 can be integrated using the functional relationship between ε_1 and the energy lost by the projectile as it slows down in the target. Here, the case has been treated where the stopping power for the projectile (S_1) and recoil (S_2) are constant. This is a good approximation for moderately low values of ε_1 (c.f. Ref. 9). For this simplified case

$$E_1 = E_o - (W - x)S_1 \ , \tag{5}$$

where E_o is the energy of the projectile at $x = W$, the front surface of the target. Thus

$$n = 0.33 \ \pi a^2 N \left(\frac{E}{\varepsilon} \right) \int_o^W \frac{u}{E_1} \ dx \ , \tag{6}$$

where

$$u = \left\{ \left[\frac{\gamma E_1}{S_2 x} \right]^{1/3} - 1 \right\} \quad \text{for } (\gamma E_1) > S_2 x,$$

$$u = 0 \qquad \qquad \text{otherwise,}$$

$\gamma \equiv T_m/E_1$, and E/ε is a dimensionless constant [9].

For a thin sample, where the energy loss of a projectile is much less than its original energy, E_1 can be taken as constant (E_o) and Eq. 6 yields

$$n_t = 0.33 \ \pi a^2 N \ \frac{E}{\varepsilon} \ \frac{W}{E_o} \left[\frac{3}{2} \left(\frac{\gamma E_o}{S_2 W} \right)^{1/3} - 1 \right] \underset{\alpha}{\sim} \left(\frac{W}{E_o} \right)^{2/3} . \tag{7}$$

Equation 6 was numerically integrated over x for a distribution of stopping powers, S_1. This was done to simulate the effect of the range straggle of the projectiles. Solid lines in Figs. 3 and 4 show the result of these calculations, with no adjustable parameters, for the primary recoil yield.

The agreement for Kr is excellent at small thickness but the theory predicts a $W^{2/3}$ slope to the yield while a near linear slope is observed. The dashed curves in Fig. 3 were obtained by assuming the primary yield calculated above causes secondaries with a multiplication factor given by $1 + (W/W_o)^{2/3}$, where $W_o = 50$ Å for the fit shown. This is clearly a gross oversimplification of the development of a cascade, but for small multiplication factors a $W^{2/3}$ dependence might be expected from the primary recoil calculations, (c.f. Eq. 7), and the recoil flux approaches the projectile flux at thicknesses not far different from 50 Å, as can be seen from the data in Fig. 3.

For values of ε_1^2 (T/T_m) below 4×10^{-4}, a $1/r^3$ potential gives a better fit to the Thomas-Fermi cross section than the $1/r^2$ potential [10]. At a value of 4×10^{-6} the difference in the two cross sections approaches a factor of 2. For $\varepsilon > 0.1$ the $1/r^3$ potential is a poor approximation to the Thomas-Fermi potential for large values of T/T_m. This corresponds to 20 keV Kr on ^{18}O. Assuming that recoil implantation is insensitive to recoils with a range less than 1 Å, the only calculations presented here which might be affected by using the $1/r^3$ potential are those for the lowest energy Kr projectiles studied.

Equation 2 was solved for a $1/r^3$ potential [10] in a fashion identical to the $1/r^2$ potential with the exception that S_1 was taken to be proportional to $E^{1/3}$ [10]. For Kr energies equal to 3 or 6 keV the difference between the yield for the $1/r^2$ and $1/r^3$ potential was less than 10%. Thus, the assumption that the $1/r^2$ potential is a valid form to use for the impact parameters of interest here seems justified.

Discussion

The dashed curve in Fig. 3 fits the data far better than could be wished. The treatment neglects the contribution from a developed cascade, where a significant number of low energy recoils "boil" over the potential between the target and substrate. The thin target agreement between the data and theory for other elements and energies is not as spectacular, as shown in Fig. 4, but the deviations are less than a factor of 2.

It is difficult to compare the above calculations with those obtained by Sigmund [1] or Nelson [2] since their calculations are based on the assumption that $M_1 < M_2$, which is opposite to the assumption made here. One notable difference with the calculation given here is that it predicts a profile in the substrate much less sharply peaked at the interface. If the stopping power of the substrate equals that of the target, Eq. 6 can be solved for the number of recoils reaching a depth D or more in the substrate by integrating from D to D + W instead of from 0 to W. This was done for 50 keV Kr projectiles passing through a 10 Å SiO$_2$ layer. This result, along with the concentration profile obtained by differentiation of the integral curve, is shown in Fig. 7. The range straggle of the recoils has been neglected. It is notable that 1% of the projectiles penetrate to a depth beyond ∿2 times the Kr range. The profile, for small D, diverges as $D^{-1/3}$. Thus the integral (the total number of recoil implanted oxygen atoms) does not diverge as $D \to 0$, in contrast to the calculation by Nelson [2] for the case of forward sputtering.

Because some recoils penetrate deeper than the projectiles when $M_1 > M_2$, it is important when using light atomic weight materials for masking heavy projectiles to choose the mask thickness such that the recoils are stopped. This thickness can be a factor of two greater than that needed for the projectile itself. For example, when heavy projectiles are implanted through an oxide layer, many oxygen atoms will recoil into the silicon. The significance of this fact will depend on the particular device and thermal treatments involved. Under certain conditions oxygen does alter the electrical properties of silicon. [11]

An additional point to be noted is that for a thin, free standing target, the forward flux of recoils (forward sputtering) must be greater than the backward flux of recoils (backsputtering). This is due to the fact that all primary recoil atoms recoil forward. Also, it was found in this experiment that the backsputtering yield was ∿10 times greater than the recoil implantation yield. Therefore, the forward sputtering yield must be more than 10 times greater than the recoil implantation yield.

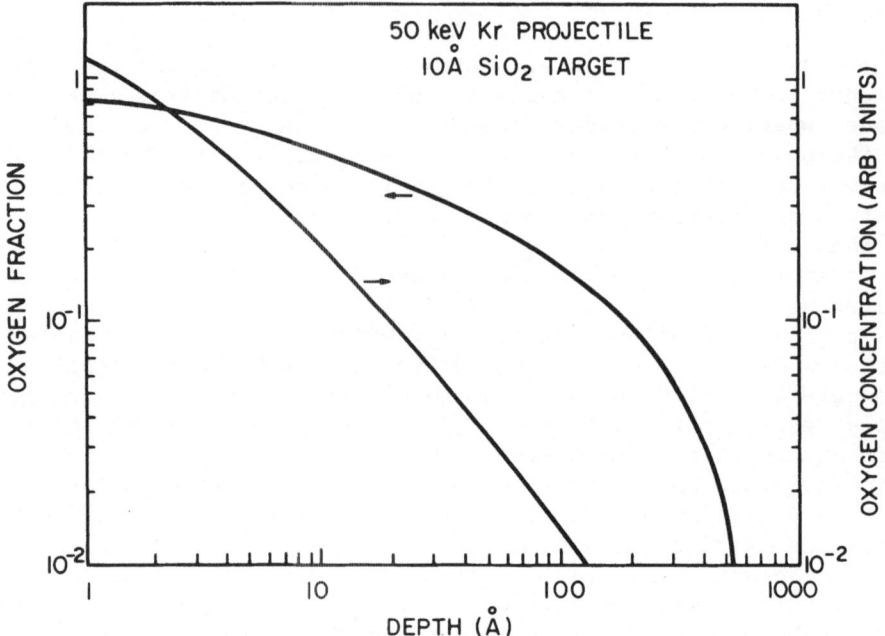

Fig. 7. Calculated fraction of the recoil implanted oxygen atoms
which are deeper than the indicated depth and the oxygen concentration vs. depth obtained for 50 keV ^{84}Kr projectiles into a 10 Å SiO$_2$
target.

Conclusions

It has been found that calculations based on the $1/r^2$ interatomic potential agree well with the observed recoil implantation
yield of Kr projectiles on an ^{18}O target. For thin oxide layers the
yield is nearly linear with thickness and from the results shown in
Fig. 3 the cross section is $\sim$5 x 10^{-17} cm^2/oxygen atom for 24 keV Kr.

The backsputter yield is approximately 10 times the recoil implication yield.

Acknowledgments

Thanks are due R. H. Kaiser for assistance with the ellipsometer
measurements, D. E. Post and R. W. Treible for the bombardments,
W. M. Augustyniak for assistance with the Van de Graaf accelerator,
and P. Sigmund for valuable discussions.

References

[1] Peter Sigmund, Physical Review 184, 383-416 (1969).

[2] R. S. Nelson, Radiation Effects 2, 47-50 (1969).

[3] L. E. Collins, J. G. Perkins and P. T. Stroud, Thin Solid Films 4, 41-45 (1969).

[4] L. E. Collins, P. A. O'Connel, J. G. Perkins, F. R. Pontet and P. T. Stroud, Nuc. Inst. Methods 92, 455-459 (1971).

[5] J. G. Perkins and P. T. Stroud, Nuc. Inst. Methods 102, 109-115 (1972).

[6] E. C. Lightowlers, J. C. North, A. S. Jordan, L. Derick and J. L. Merz, J. Appl. Phys. 44, 4758-4768 (1973).

[7] J. C. North and E. C. Lightowlers, Unpublished.

[8] William S. Johnson and James F. Gibbons, "Projected Range Statistics in Semiconductors," Stanford Univ. Bookstore (1970). The published projected range was adjusted for the densty of thermal SiO₂.

[9] J. Lindhard, M. Scharff and H. E. Schiøtt, Kgl. Danske Videnskab, Mat.-Fys. Medd. 33, No. 14 (1963).

[10] P. Sigmund, Rev. Roum, Phys. Tome 17, pp. 823-870 (1972).

[11] W. R. Runyan, "Silicon Semiconductor Technology," Texas Instruments Electronics Series, McGraw-Hill Book Co., New York, pp. 256-259 (1965).

EFFECTS OF LATTICE DEFECTS ON DECHANNELING AND ON CHANNELED-PARTICLE DISTRIBUTION

NORIAKI MATSUNAMI and NORIAKI ITOH
Department of Nuclear Engineering
Nagoya University, Nagoya, Japan

ABSTRACT

The diffusion equation was applied to obtain the dechanneling caused by lattice defects in Si. It is shown that the predominant effect of defects on dechanneling for large surface concentration or large depth is the modification of the particle distribution in transverse energy and enhancement of dechanneling by lattice vibrations and electronic collisions. It is pointed out that this effect modifies considerably the depth profile of lattice defects extracted from backscattering experiments.

Introduction

Dechanneling has been used to investigate the defect concentration in ion-implanted layers in semiconductors [1-7]. In these investigations, the theoretical relations between the defect concentration and dechanneling have been used. The previous treatments used to derive the relation between the defects and dechanneling include only single scattering and multiple scattering between the channeled particles and defects, and do not take it into consideration that the defects modify the particle distribution in transverse energy and enhance the dechanneling caused by lattice vibrations and the electronic collisions. Very recently the latter effect, which may be called the interference effect between dechanneling by defects and that by electrons and thermal vibrations, has been studied by employing the diffusion equation and was shown to play an important role in dechanneling by defects [8].

In the present paper, the modification of the particle distribution in channels by defects was calculated for Si. Also the dechanneling caused by defects was obtained and compared with the results obtained with the method of single scattering and multiple scattering.

Method of Calculation

The method of calculation was described in detail elsewhere [8]. The distribution function $g(E_\perp, z)$ in transverse energy $E_\perp$ at a depth z was obtatained by solving the diffusion equation:

$$\frac{\partial g(E_\perp, z)}{\partial z} = \frac{\partial}{\partial E_\perp} \left[A(E_\perp) D(E_\perp) \cdot \frac{\partial}{\partial E_\perp} \frac{g(E_\perp, z)}{A(E_\perp)} \right] , \qquad (1)$$

where $A(E_\perp)$ is the area of the region where the transverse energy is larger than the transverse potential, and $D(E_\perp)$ is the total diffusion coefficient which is described later. In the present calculation the transverse potential was approximated by one string potential. The initial distribution $g(E_\perp, 0)$ was calculated, assuming that the particles are incident uniformly on the crystal surface.

The defects were simulated by atoms displaced from the string nearly to the center of the channel. The component of the diffusion coefficient $D_d(E_\perp)$ originating from scattering by the displaced atoms was derived as the second moment of the change in the transverse energy. The diffusion coefficient $D_p(E_\perp)$ for scattering by electrons and lattice vibrations has been calculated by Lindhard [9]. The total diffusion coefficient $D(E_\perp)$ is given by

$$D(E_\perp) = D_p(E_\perp) + c\, D_d(E_\perp) , \qquad (2)$$

where c is the fractional concentration of the displaced atoms. The equation (1) was solved numerically [10] and the random fraction $\chi(z)$ was calculated by the following integral:

$$\chi(z) = \int_{E_\perp^C}^{\infty} g(E_\perp, z) \, dE_\perp , \qquad (3)$$

where $E_\perp^C$ is the critical transverse energy, which is equal to the particle energy E times the square of the critical angle.

Results of Calculation and Discussion

The calculation was made for 1 MeV He$^+$ incident along a <110> string of Si, in which the displaced atoms are distributed uniformly in depth. Figure 1 shows the components of the diffusion coefficient $D(E_\perp)$, caused by electrons, lattice vibrations and displaced atoms for c = 0.3%. At lower transverse energies, the defect contribution as well as the electronic contribution are prominent.

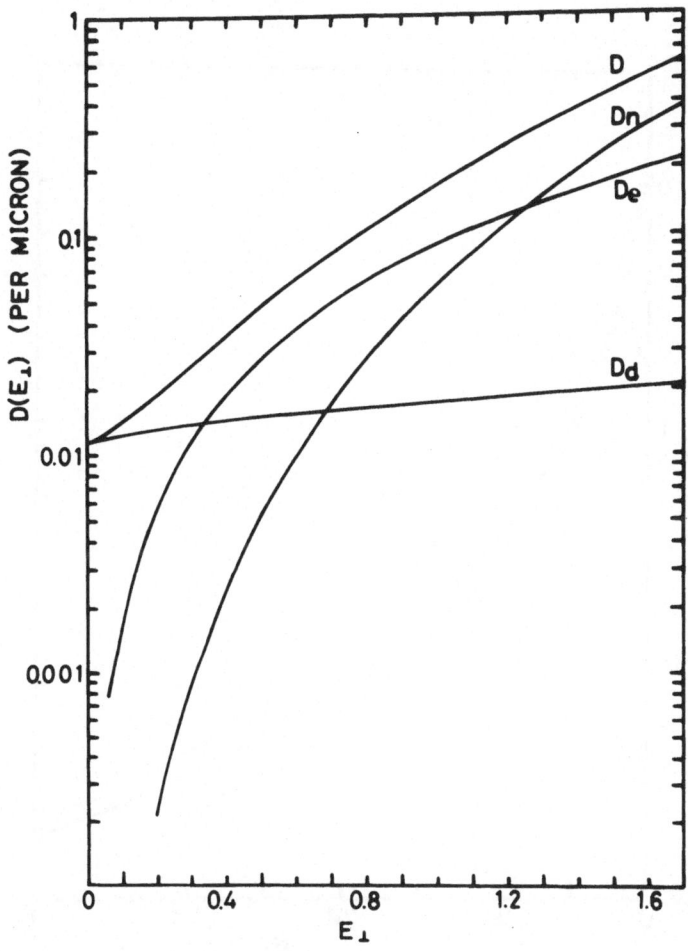

Fig. 1. Components of the diffusion coefficient resulting from scattering by thermal vibrations (D_n), electronic collisions (D_e) and scattering by displaced atoms with c = 0.3% (D_d), for 1 MeV He$^+$ incident along a <110> channel of Si.

Figure 2 shows the initial distribution used in the present calculation and the distribution at $z = 1$ μ in channels without defects and with defects (c = 0.3%). The surface concentration of defects in this depth is 1.5×10^{16} cm^{-2}, and from the figure, it is seen that the depression of the distribution function by an order of magnitude takes place around $E_{\perp} = 0$. This effect would explain why the experimental heights [11] of the flux peak in several ion-implanted crystals are smaller than theoretically predicted [12].

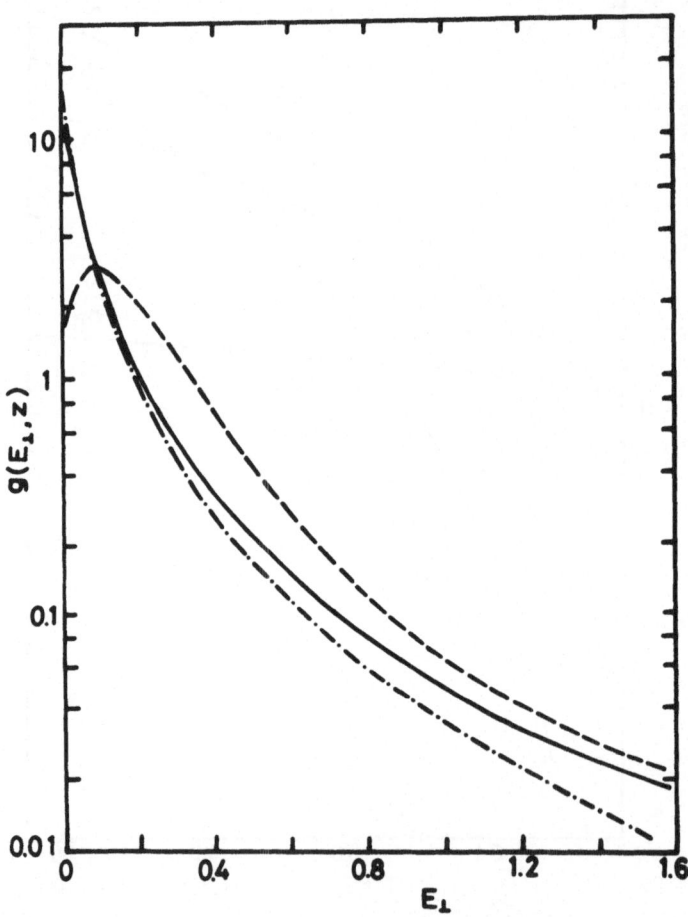

Fig. 2. Distribution in transverse energy of particles at $z = 1$ μm in <110> channels in Si without defects (solid line) and with defects of c = 0.3% (broken line). The initial distribution is shown by the dot-dash line.

Figure 3 shows the difference $\Delta\chi$ between the random fraction for crystals with defects and that for crystals without defects obtained by the diffusion equation (this method is referred to as the diffusion model) and also by single scattering and multiple scattering models, as a function of surface concentration cNz, where N is the number of host atoms per cm^3. The result for the diffusion model shown in Fig. 3 was obtained with c = 0.3%, but the relation between $\Delta\chi$ and cNz shows only a small dependence on c,

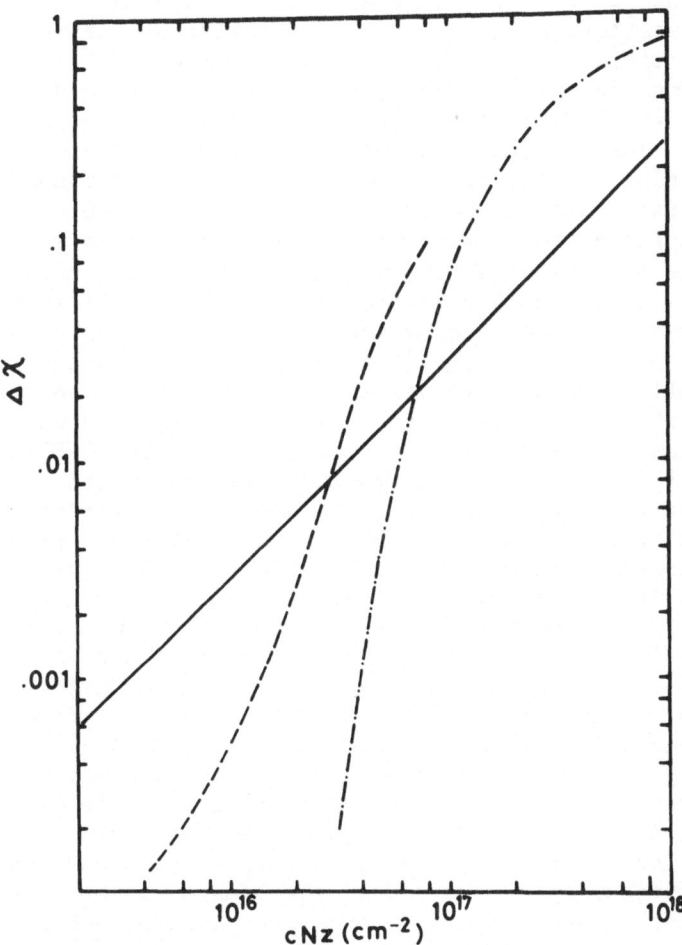

Fig. 3. Defect-induced increase in random fraction, calculated by the single scattering model (solid line), by the multiple scattering model (dot-dash line) and by the diffusion model (broken line). The calculation was made for 1 MeV He$^+$ incident along a <110> string of Si with displaced atoms of c = 0.3%.

as obtained for KCl [13]. The differences $\Delta\chi$ obtained by the diffusion model are smaller than those obtained by the single scattering model at low surface concentration of defects but the former increases much more rapidly than the latter as the surface concentration increases. This indicates that the modification of the particle distribution by defects results in the enhancement of the dechanneling caused by lattice vibrations electronic collisions at high surface concentration or at large depth. The reason that $\Delta\chi$ obtained by the diffusion equation is smaller than that obtained by the single scattering model at low surface concentration would be partly because the diffusion equation underestimates the violent collisions and partly because the single scattering model oversimplifies such effects as the flux distribution and the correlation between the scattering angle and the direction of the motion of channeled particles. Since the underestimation of the violent collisions is considered to be more important, one can assume that the single scattering model is a good approximation at low surface concentrations. The results shown in Fig. 3 also indicate that multiple scattering process with defects, suggested by Lindhard [9], is not important until $\Delta\chi$ becomes nearly equal to one.

The results of the diffusion model could be used to extract the depth profile of defects in implanted layers. It is clear that the total surface defect concentrations derived by the diffusion model is smaller than that derived by the single scattering by a factor of 4. The depth profiles [4-7] extracted by the single scattering model generally show a small value near the surface and increases suddenly as cNz exceeds 10^{17} cm^{-3}. These profiles do not agree with the theoretical results [14,15] or with the result of EPR and optical experiments [16,17]. According to the diffusion model $\Delta\chi$ exceeds the value for the single scattering model at a certain depth. Therefore, if the diffusion model is used to extract the depth profile, the sudden increase in the apparent defect concentration extracted from the single scattering model may disappear. The $\Delta\chi$ vs cNz curve obtained by the present study was used to derive the depth profile, but the profile depends very critically on the shape of the $\Delta\chi$ vs cNz curve and on the magnitude of $\Delta\chi$, and that some modification of the $\Delta\chi$ vs cNz curve was necessary to obtain the depth profile in accordance with theoretical predictions.

In applying the diffusion model to obtain the depth profile $\Delta\chi$ reaches values as high as 0.9 and, comparing with the experimental results [5], it is necessary to assume that the channeled fraction increases after the particles pass through the damaged layer. Since the flux distribution at the end of the damaged layer is highly distorted, the rechanneling effect as discussed by Morita and Itoh [18] would take place.

In conclusion, the contribution of displaced atoms on dechanneling is predominantly due to the modification of the particle distribution, followed by the enhancement of dechanneling by lattice

vibration and electronic scattering. It is shown that the dechanneling by defects becomes larger with increasing penetration depth and
it is pointed out that this effect should be taken into consideration
when extracting the depth profiles from backscattering experiments.

References

[1] J. A. Davies, J. Denhartog, L. Eriksson, and J. W. Mayer,
    Canadian J. Phys. 45, 4053 (1967).
[2] J. W. Mayer, L. Eriksson, S. T. Picraux and J. A. Davies,
    Canadian J. Phys. 46, 663 (1968).
[3] E. Bøgh, Canadian J. Phys. 46, 663 (1968).
[4] L. C. Feldman and J. W. Rodgers, J. Appl. Phys. 41, 3776 (1970).
[5] F. H. Eisen, B. Welch, J. E. Westmoreland and J. W. Mayer,
    Atomic Collision Phenomena in Solids (North-Holland, Amsterdam,
    1970) p. 111.
[6] J. M. Ziegler, J. Appl. Phys. 43, 2973 (1972).
[7] K. Schmid, Radiation Effects 17, 201 (1973).
[8] N. Matsunami and N. Itoh, Phys. Letters 45, 435 (1973).
[9] J. Lindhard, Mat. Fys. Medd. Dan. Vid. Selsk. 34, No. 14 (1965).
[10] E. Bonderup, H. Esbensen, J. U. Andersen and H. E. Schiøtt,
    Radiation Effects 12, 261 (1972).
[11] J. U. Andersen, O. Andreasen, J. A. Davies and E. Uggerhøj,
    Radiation Effects 7, 25 (1971).
[12] D. Van Vliet, Radiation Effects 10, 137 (1971).
[13] N. Matsunami, to be published.
[14] D. K. Brice, Appl. Phys. Letters 16, 103 (1970).
[15] T. Tsurushima and H. Tanoue, J. Phys. Soc. Japan 31, 1695 (1971).
[16] H. J. Stein, F. L. Vook and J. A. Borders, Appl. Phys. Letters
    16, 106 (1971).
[17] K. L. Brower, F. L. Vook and J. A. Borders, Appl. Phys. Letters
    16, 108 (1970).
[18] K. Morita and N. Itoh, J. Phys. Soc. Japan 30, 1430 (1971).

SECTION III
COMPUTER SIMULATION

COMPUTER SIMULATION OF ATOMIC COLLISIONS IN SOLIDS

D. P. JACKSON
Chalk River Nuclear Laboratories
Atomic Energy of Canada, Limited
Chalk River, Ontario, Canada

ABSTRACT

A discussion of the application of computer
simulation to atomic collision phenomena in solids
is presented, organized under three main headings:
(a) technique, (b) the applications of simulation,
and (c) the advantages and limitations of simulation. The aim is not to be comprehensive but
rather the emphasis is on the nature and uses of
the simulation process with examples from the
field of particle-crystal interactions.

Introduction

When high speed digital computers became available, computer
simulation became possible; this roughly coincided with the beginning of a period of burgeoning interest in atomic collisions in
solids. So that although the simulation technique is relatively
new, it has had a significant influence on our theories of radiation damage and directional effects in crystals. Because of the
intimate connection between computer simulation and the experimental and theoretical development of these fields for the past dozen
years, an historical review would mean reviewing the evolution of
ideas in these areas. This is not feasible in the space available
here; rather, we will try to convey the essential concepts of simulation to the reader. A description of the simulation technique
will be followed by a discussion of its applications with examples
relevant to radiation damage and directional effects in single
crystals. In the latter the emphasis will be on results which
illustrate the particular contribution of simulation to the theoretical study of these processes. Next we will summarize the
inherent advantages and limitations of simulation in order to

dispel certain misconceptions surrounding it. In conclusion, we will indicate briefly its future role in view of current trends in the atomic collisions field.

The Computer Simulation Technique

   It was in the early 1960's that Vineyard and his collaborators [1] and Robinson and Oen [2] pioneered the two basic types of simulation programs. All subsequent work in the simulation of atomic collisions in solids has been based on one or the other of these fundamental prototypes.
   Vineyard's group [1,3,4] introduced the many-body integration (MBI) approach in what was the original application of computer simulation to radiation damage. They wrote the now classic simulation codes called "GRAPE" which are still used. GRAPE considers a finite crystallite of unit cells (f.c.c. [1] or b.c.c. [3,4]) containing in total a few hundred to a thousand atoms. Each atom interacts with each of the other atoms in this crystallite through a two-body central interatomic potential. The atoms comprising the boundaries of the crystallite are subjected to special forces applied to hold the assembly together. Collision sequences inducing large perturbations at the crystallite boundaries are therefore to be avoided. Assuming classical mechanics, one can now write down the Newtonian equations of motion for each atom. An event is initiated by starting one atom in motion with a specified velocity vector, simulating a primary knock-on (PKA) by a bombarding electron, neutron or ion. All other atoms are assumed initially at rest. The equations of motion of all the crystallite atoms are integrated numerically until an equilibrium position is reached, using damping at the boundaries if necessary, or until the collision process of interest has reached one of the crystallite surfaces. In this manner GRAPE was used to study both collision sequences of various types and the production of defects such as interstitials and vacancies.
   The GRAPE model involves the simultaneous integration of several thousand non-linear and non-separable differential equations. This is accomplished in GRAPE by a central difference scheme based on time stepping. At each time step in the numerical integration several thousand force components must be calculated as a function of interatomic distances. An energy balance check is used to ensure the numerical stability of the integration, the time step being adjusted accordingly. All this results in lengthy computer runs-- Gibson et al [1] quote a figure of three and a half hours for a typical run on the now obsolete IBM 704 computer. Modern computer speeds would reduce this by perhaps a factor of ten. The other fundamental problem with GRAPE is the size of the crystallite; all the ramifications of events with PKA energy greater than a few hundred eV cannot be reasonably contained within the boundaries.

Beeler [5] estimates that at least 1500 differential equations are
required per keV of PKA energy, whereas radiation damage in the
reactor field involves PKA's in the range 1-1000 keV. This situation can be improved somewhat employing "leap frog" methods where
the crystallite is moved along with the zone of interaction by
geometric continuation [3]. Erginsoy et al [4] were able to study
channeling with GRAPE in this way. GRAPE has had a very great
influence on our concepts of radiation damage and many later investigations have followed its lead, often using the original GRAPE
computer programs generously supplied by the Vineyard group to
other workers.

The binary-collision approximation (BCA) is an alternative to
MBI which circumvents many of the latter's limitations. In this
approximation the overall interaction is represented by an ordered
sequence of binary (or two-body) collisions. An early contribution
of Yoshida [6] used the BCA to study damage in an amorphous solid;
Oen et al [7] also treated the same problem. However, the work of
Robinson and Oen [2] is the key to subsequent simulations of directional effects. They recognized the phenomena of channeling as the
explanation for the anomalously large penetrations of ions into
crystals which had been observed experimentally at about the same
time.

It is convenient to discuss this model in terms of simulating
an ion penetrating into a single crystal lattice. We assume a
two-body central interatomic potential acting between the ion and a
lattice atom. But instead of constructing equations of motion we
trace the path of the ion through the lattice, treating each encounter with a lattice atom as an isolated binary collision. It
is not necessary to store a large representation of a lattice in
the computer; the symmetry properties of the lattice in conjunction
with the spatial coordinates of the ion allow one to determine the
next collision with relative ease. The scattering of the ion from
a lattice atom is calculated in the free-space centre-of-mass formulation for two-body collisions. Techniques are available for
doing this quickly on the computer which we cannot discuss here
[8,9,10]. In between collisions the ion travels on a straight
path undisturbed by the lattice, termed "free-flight". The ion is
allowed to continue in this way through the lattice until its
energy falls below a pre-set value. The trajectory of the ion can
thus be traced rapidly by the BCA without the spatial limitations
characteristic of MBI.

Gibson et al [1] were interested in studying events with
GRAPE. Robinson and Oen [2], on the other hand, simulated the fate
of a beam of ions impinging on a crystal. In this connection a
useful term to describe the initial geometric parameters of the
ion before its interactions with the crystal is system impact
parameter. This, in general, is a set of quantities giving the
initial spatial relationship of the ion to the crystal, analogous
to the function of the impact parameter in a binary collision. A

beam of ions is a directed pencil of ions with an angular width
given by a collimation angle. There may also be a mosaic spread
in the crystal surface. And very important is the distribution of
ion entry points at the crystal surface.

To simulate a beam, many ions with initial conditions sampled
from the overall distribution of possible system impact parameters
are simulated individually in their interaction with crystal. And
then a distribution of outcomes is constructed by adding together
the outcomes (e.g., depth to which the ion penetrated) of these
independent trials. (Notice here the concept of independent trials
which pervades computer simulation in many forms.) Clearly the
sample must be uniform so that the outcome distribution is not
distorted and it must be representative so that no possibility in
the spectrum of outcomes is neglected. Furthermore, the number of
ions involved must be sufficient to achieve an outcome distribution
with meaningful statistical confidence. An unfortunate characteristic of our sampling procedure, which is usually called Monte
Carlo, is its very slow convergence ($\sqrt{1/N}$) as a function of the
number of trajectories (N). Therefore, even with a BCA model relatively lengthy computer runs may be required to simulate experimental situations.

We will now discuss in turn a few of the more significant
topics concerned in the computer simulation of physical effects
in our area of interest.

The Interatomic Potential

There is space here for only a few comments about interatomic
potentials, a subject with an extensive literature (see, for example, the recent monograph of Torrens [11] and for current applications, Ref. 12.) An interatomic potential is an approximation
which hopefully simulates the behavior of interacting particles
for a particular range of parameters. Therefore, simulations,
since they include interatomic potentials, give results directly
related to the potential used. In short, simulations are model-
dependent with respect to the interatomic potential. Quantitative
comparison of computer simulations with experiment become very
difficult for this reason (and, of course, other difficulties are
involved in other model approximations). It is sometimes more
fruitful to search for an empirical potential giving good agreement
and then use it to correlate experimental data or to extrapolate
the experiment results to other but similar parameter ranges.

For relatively high interaction energies the potentials
commonly used consist of a Coulomb repulsive core surrounded
by screening electrons, the latter modeled by a screening function.
The screened Coulomb, Bohr, Born–Mayer, Thomas–Fermi (and the
Molière) potentials belong to this category. In high-energy, low-
impact parameter collisions the core repulsion dominates and the
details of screening are relatively unimportant. However, for

glancing angle collisions screening dominates. In a solid the
screening function is expected to be modified somewhat by the
presence of surrounding atoms and the free space collision concept
of the BCA will not strictly apply. In practical BCA models, the
"potential" is considered in parts. A relatively short range repulsive potential is used for the elastic collisions and the inelastic or electronic part of the collision is modeled independently.
This division is somewhat more difficult in MBI models.

At relatively low energies ($\lesssim$ a few hundred eV), where many-
body effects are known to be significant, the situation is much
more complex because calculations are very much more model-dependent.
Especially difficult is the radiation damage problem of starting
PKA's at relatively high energies eventually producing final defect
configurations which are very sensitive to the potential. Various
potential matching methods have been attempted with mixed results
[12]. Finally, we should mention that the MBI method uses the
summation of pairwise potentials (SPP) approximation. It is sometimes assumed that the summation of two-body pairwise free-space
potentials gives a somehow more accurate representation of the
potential experienced by a particle in a solid. There is no guar-
antee of this. In fact, it is not plausible that the summation
process per se can account for the complicated changes in electronic
screening that occur in solids. A potential to be used in SPP
calculations must be specially derived for this purpose, based on
experimental properties of the solid.

Comparison of the MBI and BCA Models

It should be obvious at this point that the validity of the
BCA is closely related to the interatomic potential. At first
sight it would appear easy to use both methods for the same physical situation and then to infer the accuracy of the BCA by comparing the results. However, we are confronted with the question:
what is the equivalent BCA potential for a given MBI potential?
No convincing and unambiguous recipe is available for making this
correspondence. Certainly one can show that an MBI calculation is
behaving essentially like a binary collision process but there is
a nebulous area in a strict comparison. Erginsoy et al [4] observed that damage cascades in α-iron with PKA of greater than a
few hundred eV proceeded approximately as a sequence of two-body
collisions. Beeler [13] obtained a similar result. Gay and
Harrison [14] essentially agreed with this for damage events in a
copper lattice. They concluded that the BCA will slightly under-
predict the true energy transfers at all times, the largest errors
being for large impact parameters. But further efforts of Harrison
and his co-workers [15,16] pointed out the impossibility of making
meaningful comparisons because the assigning of a range of interaction for the BCA potential corresponding to the equivalent MBI,

was uncertain. Eltekov et al [17,18] emphasized that there are in fact any number of MBI models possible depending on how many lattice atoms are included in the integration. They found by comparison between different MBI models that particular trajectories were very sensitive to this number. This makes the problem even more intractable.

Morgan and van Vliet [19] gave the following justification for the BCA in connection with 5–500 keV protons in copper. Consider the displacement δ of a lattice atom in a collision:

$$\delta \sim 2a(m_1/m_2) \; \theta \; . \tag{1}$$

θ is the scattering, a is the screening radius and m_1 and m_2 are the proton and copper masses respectively. Suppose we change the impact parameter by small amount Δb, then the corresponding change in the scattering angle may be approximated:

$$\Delta\theta \sim (\Delta\theta/\Delta b)\cdot\delta \sim (\theta/a)\cdot 2a \; (m_1/m_2) \; \theta = 2(m_1/m_2) \; \theta^2 \tag{2}$$

Since $(m_1/m_2) = 1/63.5$ we obtain $\Delta\theta<<\theta$ concluding that small changes in impact parameter due to the influence of surrounding atoms will not have a very large effect on the main collision. This argument is intuitively attractive for high-energy light-ion propagation; however, the troublesome thought remains that phenomena such as channeling involve the cumulative effect of many gentle (very small θ) collisions.

A danger in BCA models concerns the treatment of (nearly) simultaneous collisions if the projectile interacts with two or more lattice atoms in highly symmetric positions, e.g., in a pair or ring. This, in effect, is a many-body situation. BCA models step the projectile forward after each collision and thus may miss the atom symmetric with the atom just collided with. Torrens and Robinson [20] solve this problem by testing for simultaneity and if it is present by vector-adding the results of the two-body collisions with all the symmetric atoms to arrive at a total scattering angle. Neglect of simultaneity can obviously produce significant errors in BCA simulations of directional phenomena.

To conclude, there is apparently no simple and satisfactory criterion for the validity of BCA. One uses it because the MBI is not feasible for many practical problems; as we shall see below it has achieved some success in simulating experimental results.

Thermal Vibrations

The thermal displacements of solid atoms represent slight departures of the lattice from geometrical perfection which are significant in axial channeling and other directional phenomena. The

Debye-Waller theory [21] is the basis for modeling thermal effects; it gives the mean-square displacement of an atom in the ith direction at temperature T:

$$<\rho_i^{\,2}> = \frac{3\hbar^2}{Mk\theta} [\frac{1}{\alpha^2} \int_o^\alpha \frac{tdt}{e^t-1} + \frac{1}{4}] \tag{3}$$

where θ is the Debye temperature, h, k and M have their usual meanings, and $\alpha = \theta/T$. The probability of finding the displacement to be ρ in the ith direction is given by:

$$P(\rho) = 2\pi<\rho_i>^{-1/2} \exp[-\rho^2 4<\rho_i^{\,2}>] . \tag{4}$$

Displacements are usually sampled from the distribution (4) using the technique used by Robinson and Oen [2] to approximate a Gaussian Distribution:

$$P(\rho) = \{(6<\rho_i^{\,2}>)^{1/2} - |\rho|\}/<\rho_i^{\,2}> . \tag{5}$$

An advantage in program speed is obtained and occasional very large displacements are avoided in this formulation. It is also feasible to assume an harmonic oscillator with the appropriate amplitude and sample from a sine or cosine distribution. The calculated displacements are then applied in the BCA model to modify the position of the lattice atom involved in the collision. In the MBI model the assumption is usually made that the lattice atoms are "discovered" in their displaced condition by the penetrating ion.

The standard method just discussed does not include the possibilities of directional anisotropy or of correlated displacements. The former could be included in a simulation by sampling from a more elaborate distribution of ρ_i's. Correlated thermal lattice motions were discussed by Nelson et al [22] who derived limiting expressions for the normalized correlation between the displacement of two lattice atoms in the same direction. Using their model, based on the Debye phonon distribution, the full expression is:

$$<r_1 r_n>/<r_1^{\,2}> = \frac{2\alpha \int_o^\alpha \frac{\sin(kt/\alpha)dt}{k(e^t-1)} + \frac{a^2}{k^2}(1-\cos k)}{2 \int_o^\alpha \frac{tdt}{e^t-1} + \frac{\alpha^2}{2}} . \tag{6}$$

In this expression α has the same meaning as in (3) and $k = n\nu\pi$; n is the number of length units the atom "n" is displaced from atom 1 and ν is a parameter characteristic of the lattice.

This function indicates a surprising degree of correlation between lattice atom motions. Ryabov [23-25] has included thermal correlations in his computer simulations of blocking, although he does not compare "uncorrelated" with "correlated" simulations. So it is not clear from this work exactly how correlations affect the situation. Agranovich and Kirsanov [26] have also used correlations to extend the original considerations of Nelson et al [22], done for isolated strings, to lattices showing the modification in focusson propagation due to the lattice. It would be interesting to have more accurate experimental data on correlations to verify that the effect is as large as predicted by the Debye theory.

At this point, if space permitted, we could also go into a discussion of the electronic stopping effects incorporated in computer models. In general, a trend is seen, as in the case of thermal vibrations, to more and more elaborate treatments of inelastic effects in simulations. In particular, attempts are made to relate impact-parameter-dependent stopping formulas to distance-integrated stopping powers (for example, see References 15 and 17-18). However, we cannot discuss the details here and the reader is referred to other contributions in this volume.

The Uses of Computer Simulation

To review the applications of computer simulation under such headings as "radiation damage," "channeling," etc., would be futile since a reasonable discussion of the contents of each contribution cannot be given here. Furthermore, a balanced assessment of the role of simulation in the development of any particular topic would require an explanation of the full experimental and theoretical picture of that phenomenon. Our purpose here is to convey the essential features of computer simulation and hence, a functional approach will be adapted. Somewhat artificial categories are used in this classification--obviously substantial overlap is present-- but they serve to illustrate our points.

The Recognition of New Effects

We have seen how the simulation process can be regarded in a general way, as a method of solving a large system of non-linear differential equations for a specified set of initial conditions (system impact parameters). However, all the possible consequences of these equations are not known a priori. Experimental clues indicating the effect to be explained are the usual starting points for non-simulation theories whereas simulation offers the possibility of "discovering" in one's results a phenomenon not hitherto suspected. This, as we have seen, was how channeling was identified by Robinson and Oen [2]. They were also the first to recognize the type of channeling known as proper or hyperchanneling. Gibson et al [1] observed a variety of phenomena in the propagation of radiation damage, notably various replacement and focussing sequences. They also saw a new type of defect configuration--the split interstitial. The anisotropy of displacement thresholds was observed to be a significant factor in the evaluation of radiation damage in crystals. The total impact of these papers [1,2] was in the full recognition of the significance of the lattice order.

The flux peaking effect is an example of the feasibility demonstration function of computer simulation. An effect is seen experimentally and it is hypothesized to be due to some type of interaction. Moreover, this may not be easily demonstrated to be feasible by elementary means. Alexander et al [27] were able to show flux peaking in their computer simulation which went a long way to establishing the now accepted mechanism, and opening the topic of spatial variations in ion flux as an application of simulation [28]. The application of classical simulation methods to describe the high energy interactions of electrons with crystals resulted in the identification of weavon motion by Nip and Kelly [29] and rosette motion by Bell et al [30]. These examples illustrate the role of computer simulation, in the words of Beeler [5], as a "synthetic imagination."

The Resolution of Complexity

There are some situations by nature so complex that they defy elementary description. A good example can be seen in the work by Beeler [5] on a BCA model of radiation damage cascades, started at the same time Robinson and Oen [2] were developing it for directional effects. The whole radiation damage problem is tackled in these simulations. First, it is necessary to simulate the deposition of damage in the solid by the bombarding particles. This gives a group of points from which cascades can be initiated. The cascades are then followed in all their ramifications; overlap of neighboring cascades must be taken into account in the production of this primary damage picture. From this point, one attempts to calculate the final damaged configuration using a variety of defect interaction processes. This is an enormous task impossible without the use of computer simulation. A good example of the increase in complexity resulting from relatively small changes in the physical situation studied can be seen in the work of Chadderton, Torrens, and Morgan [31-34]. They used essentially the same approach as Gibson et al [1] to simulate damage events in alkali halides. The introduction of two atomic species with different potentials made for a great increase in complexity in the results as compared with those of Gibson et al [1]. In many single crystal problems, there seems to be this irreducible complexity which requires the use of simulation techniques.

The Testing of Analytic Theory

The approximations included in Lindhard's [35] theory of channeling have been widely tested by computer simulation. Morgan and
van Vliet [19,36] have done a group of simulations concerned with
the validity of the string and plane approximations used in the
Lindhard theory. They concluded that these were generally correct
and suggested certain modifications to the Lindhard equations for
the critical angle and minimum yield. Barrett [37] has performed
an extensive series of comparisons of these quantities with corresponding experimental data for a large number of ion-solid combinations. He was also able to modify the basic Lindhard formulas to
give good agreement with experiment. A function called "the nuclear encounter probability" was introduced which proved to be a
convenient way of expressing the total effect of elastic collisions
as a function of depth. These are a few illustrations of the widespread use of computer simulations as a test of the approximations
used in non-simulation theories; at this point in time any simplifying assumption used by an analytic theory is almost invariably
tested by simulation if this is possible.

Finally we should add that computer simulations perform the
usual function of theoretical models in that they provide a basis
for extrapolating known experimental data to unknown situations,
are used to predict the likely experimental results of parameter
changes, and to correlate a variety of experimental data.

Remarks on the Simulation Process

Harsh criticism from its oponents and exaggerated claims from
its proponents are frequent in discussions of computer simulation.
It is therefore worthwhile to reflect on the nature of simulation
and hopefully clear up some misunderstandings about it.

It should be clear from the foregoing sections that the computer
is merely a calculating machine employed to solve a well-defined
mathematical problem specified by the user. No mystical element is
added by the computer. The adage of the computer age "Garbage in,
Garbage out!" applies to simulations (as it does to all theories).
Results are direct consequences of model assumptions. Every result
from a simulation should be understandable in an "inductive" way
as Robinson and Oen [2] pointed out. By this they meant that if
something surprising or interesting occurred it should have a sensible
physical explanation. Too often authors use phrases such as "the
computer said," "ask the computer," etc., causing confusion in the
reader about these very points. Similarly we have followed the
common practice in this review writing "ion" and "atom" referring
to mathematical objects intended only to simulate the properties of
real ions and atoms. It is clumsy to write "computer ion" or some
such phrase every time to underline this distinction. Nevertheless
the lack of such terminology leads easily to a state of mind which

blurs the difference between simulation and reality. Another problem arises from the description "Monte Carlo" applied to computer simulation. It seems to imply a negligent gambling game with unpredictable results. This, of course, is a misconception. Sampling from distributions of impact parameters or thermal vibrations is a well-defined process for modeling stochastic variations present in nature, with the proviso that a sufficient, uniform and representative sample is taken.

There is no fundamental conflict between computer simulation and nonsimulation or traditional theoretical methods. Both assume similar physical pictures of interatomic potentials, electronic stopping, etc. And as we have seen above simulations incorporate many theoretical descriptions. The traditional approach is basically to isolate what is significant in a problem and discard what is not. This requires prejudgment of all the various interaction possibilities--such an assessment can be very difficult for complex many-body interactions in solids. We have noted that computer simulation is very valuable for making these judgments and indeed for recognizing previously unsuspected phenomena. Certainly it is feasible to do a great deal of simplification in cases where symmetry can be exploited. For example, Lindhard [35] was able to treat the channeling problem with great success by modeling lattice order in terms of strings and planes. Similarly interactions in amorphous solids are amenable to analysis due to a complete lack of symmetry. Computer simulation has found wide application in testing the validity of the approximations of nonsimulation methods. Conversely when the main features of a physical effect are clarified by simulation it is often possible to epitomize these features in some convenient nonsimulation theory. But for some phenomena it is necessary to consider the details of the individual collisions--details which do not survive the averaging approximations needed for convenient analytic treatments. Simulation can achieve this level of detail. However, it may not be feasible to obtain any further simplifications. Therefore, simulation and nonsimulation methods complement and supplement each other.

Finally, some cautionary comments about simulation computer programs: the construction of a simulation code with a reasonable degree of "reality" is no mean undertaking. The level of detail to ensure correct handling of many standard and special interactions is very high indeed. To completely debug these complex codes is very difficult. Eliminating obvious flaws which give dramatically wrong results is relatively straightforward. Subtle errors, perhaps occurring only occasionally, may elude the programmer for some time or conceivably are never detected. Many tests for consistency and accuracy are and should be applied to simulations; however, the achievement of complete accuracy is far from trivial. Simulation is an expensive and time-consuming business compared to the inspired use of traditional approximation methods. One should always be sure that computer simulation has some useful purpose and that a simpler model cannot be applied.

Conclusion

The power of computer simulation lies in its generality. As Barrett [37] points out, "As the complexity of a set of calculations increases, the difficulties of analytical methods tend to rise more sharply than those of numerical methods." This statement summarizes the future role of computer simulation. For this reason it will become increasingly important as progress in the atomic collisions field goes forward. More and more attention is being paid to the details of various physical effects; as the level of detail increases so will the application of computer simulation.

References

[1] J. B. Gibson, A. N. Goland, M. Milgram and G. H. Vineyard, Phys. Rev. 120, 1229 (1960).

[2] M. T. Robinson and O. S. Oen., Phys. Rev. 132, 2385 (1963).

[3] C. Erginsoy, G. H. Vineyard and A. Englert, Phys. Rev. 133, A595 (1964).

[4] C. Erginsoy, G. H. Vineyard and A. Shimizu, Phys. Rev. 139, A118 (1965).

[5] J. R. Beeler, Jr. in Radiation Damage in Reactor Materials, Vol. II, IAEA, Vienna, p. 3 (1969).

[6] M. Yoshida, J. Phys. Soc. Japan 16, 44 (1961).

[7] O. S. Oen, D. K. Holmes and M. T. Robinson, J. Appl. Phys. 34, 302 (1963).

[8] C. Lehmann and G. Leibfried, Z. Physik. 172, 465 (1963).

[9] F. J. Smith, Physica 30, 497 (1964).

[10] M. T. Robinson, "Tables of Classical Scattering Integrals," ORNL-4556 (1970).

[11] I. M. Torrens, "Interatomic Potentials," Academic Press, New York (1972).

[12] "Interatomic Potentials and Simulation of Lattice Defects," Eds. P. C. Gehlen, J. R. Beeler, Jr. and R. I. Jaffee, Plenum Press, New York (1972).

[13] J. R. Beeler, Jr., Phys. Rev. 150, 470 (1966).

[14] W. L. Gay and D. E. Harrison, Jr., Phys. Rev. 135, A1780 (1964).

[15] D. E. Harrison, Jr., R. W. Leeds and W. L. Gay, J. Appl. Phys. 36, 3154 (1965).

[16] D. E. Harrison, Jr. and D. S. Greiling, J. Appl. Phys. 38, 3200 (1967).

[17] V. A. Eltekov, D. S. Karpuzov, Yu. V. Martynenko and V. E. Yurasova, "Proceedings of the III International Conference on Atomic Collisions in Solids" (Brighton), p. 657 (1969).

[18] V. A. Eltekov, D. S. Karpuzov, Yu. V. Martynenko, E. A. Rubakha, V. A. Simonov and V. E. Yurasova, "Proceedings of the IV International Conference on Atomic Collisions in Solids" (Gausdal), p. 113 (1971).

[19] D. V. Morgan and D. van Vliet, Can. J. Phys. 46, 503 (1968).

[20] I. M. Torrens and M. T. Robinson in Ref. 12, p. 423 (1972).

[21] See for example: A. A. Maradudin, E. W. Montroll and G. M. Weiss,"Solid State Physics," Suppl. 3, Academic Press, New York (1963).

[22] R. S. Nelson, M. W. Thompson and H. Montgomery, Phil. Mag. 1, 1385 (1962).

[23] V. A. Ryabov, Phys. Stat. Sol. 38, 63 (1970).

[24] V. A. Ryabov, Phys. Stat. Sol. 49, 469 (1972).

[25] V. A. Ryabov, Sov. Phys. Sol. State 12, 2216 (1971).

[26] V. M. Agranovich and V. V. Kirsanov, Sov. Phys. Sol. State 11, 540 (1969).

[27] R. B. Alexander, G. Dearnaley, D. V. Morgan, and J. M. Poate, Phys. Letters 32A, 365 (1970).

[28] D. van Vliet, Rad. Effects 10, 137 (1971).

[29] H. C. H. Nip and J. C. Kelly, "Proceedings of the III International Conference on Atomic Collisions in Solids" (Brighton), p. 50 (1969).

[30] F. Bell, H. J. Kreiner and R. Sizmann, Phys. Letters 38A, 373 (1972).

[31] L. T. Chadderton and I. M. Torrens, Nature 208, 880 (1965).

[32] L. T. Chadderton, D. V. Morgan and I. M. Torrens, Phys. Letters 20, 329 (1966).

[33] I. M. Torrens, L. T. Chadderton and D. V. Morgan, J. Appl. Phys. 37, 2395 (1966).

[34] I. M. Torrens and L. T. Chadderton, Phys. Rev. 159, 671 (1967).

[35] J. Lindhard, Kgl. Danske Vid. Selskab, Mat.-Fys. Medd 34, No. 14 (1965).

[36] D. V. Morgan and D. van Vliet, Rad. Effects 8, 51 (1971).

[37] J. H. Barrett, Phys. Rev. B 3, 1527 (1971).

THE EFFECT OF STRAGGLING IN ELECTRONIC STOPPING ON RANGE DISTRIBUTIONS

K. B. WINTERBON

Theoretical Physics Branch, Atomic Energy of Canada Ltd.
Chalk River, Ontario, Canada

ABSTRACT

Previous calculations of range and damage distributions assumed electronic stopping to be a continuous
frictional term, independent of the path of the particle.
In this paper the electronic stopping is allowed to depend
on the distance of closest approach in collisions with
target atoms, following approximately Firsov's results.
Moments calculated by the two models are compared.
Mean range and straggling do not differ appreciably.

COMPUTER SIMULATION OF THE MULTIPLE SCATTERING
OF HIGH ENERGY HEAVY IONS IN THIN FILMS

K. GÜTTNER and G. MÜNZENBERG
II. Physikalisches Institut
6300 Giessen, Germany

ABSTRACT

A Monte-Carlo method is applied to describe
the multiple scattering of fast heavy particles
(MeV-region) by amorphous metal films. Different
potentials are fitted to a screened Coulomb potential supplying rather simple connections between
impact parameter and deflection angle for the
single interactions. The energy loss of the charged
particles is taken into account. The calculations
yield the angular and energy distribution of the
scattered particles together with the percentage
distribution of the complete particle histories.
The results are compared to experimental multiple
scattering distributions of fission fragments in
thin films. Furthermore differences and agreement
concerning other theoretical results are discussed.
As an example for the further application of the
computational program we calculated the scattering
of high-energy ions (4 MeV/nucleon) by an edge-
shaped slit.

Introduction

The multiple scattering of high-energy heavy ions by thin foils
has been carefully studied the last few years. Most of these studies
have been done with ions in the keV-region. Because of the development of heavy ion accelerators the higher energies become more and
more interesting, but very few observations exist about the reflection of these ions by the walls and slits of an apparatus. Some
experiments were performed concerning the backscattering of fission
fragments from solid materials [1,2] and Monte-Carlo calculations are

in satisfactory agreement with the experimental results [3]. But only one experiment exists, to the authors knowledge, in which the multiple scattering of fission fragments was measured, having uniform mass and energy [4]. The maximum value of the Born parameter $\alpha = Z_1 Z_2 / 137 \beta$ involved in these experiments was 1 or 2 orders of magnitude larger than previous experimental values. The measured distributions were compared with those predicted by the theories of Molière [5] and Nigam, Sundaresan and Wu [6], using two different models to account for the additional screening of the ions. In the former case the theoretical distributions were calculated with an effective charge of the fission fragments, in the latter the potential was fitted by a suitable screening parameter. With only the theory of Molière it was possible to obtain reasonable and consistent agreement with experiment. Additional comparison with the theory developed later by Meyer [7] has not been made because only the distribution of the projected angle has been measured in the experiments.

In the following results of Monte-Carlo calculations are discussed, which we utilized to study the scattering of fission fragments by thin foils. Different potentials are used to describe the single interactions and the energy loss of the ions in the foils is taken into account. The results are compared with the preceeding experiments and theory in order to check the fundamental range of these simulations for the description of more complicated scattering processes by walls and slits.

Calculation Procedures

The fundamental method which has been explained in more detail elsewhere [3] consists of following single particle histories in metal films of known thickness. It is supposed that the atoms in the film are in no regular order, an assumption satisfactorily fulfilled by polycrystalline foils. Each ion traverses several layers without any deflection but with continuous energy losses by excitation and ionization of the atomic electrons. After each of these paths the particle suffers a deflection interacting with one atom at that moment. The length of the paths are selected by random numbers as a function of the total cross section. The scattering angles during the single interactions are selected by two different methods depending on the particular potential and fitting method.

For the selection of the scattering angles a differential cross section is needed. This can be found by integration of the well known classical relation between the angle χ in the centre-of-mass system and the impact parameter ρ

$$\chi = \pi - 2 \int \frac{(\rho/r^2)\, dr}{\sqrt{1 - (\rho/r)^2 - 2\, V(r)/m_o v^2}} \quad . \tag{1}$$

Our calculations are based on interaction potentials of the form

$$V(r) = \frac{Z_1 \cdot Z_2}{r} \cdot f(r/a) \quad . \tag{2}$$

In one part the exponential function from Bohr [8] is used as the screening function

$$f(r/a) = \exp(-r/a_B) \quad , \tag{3}$$

with $\quad a_B = a_o / \sqrt{z_1^{2/3} + z_2^{2/3}}$, where $a_o = 0,529.10^{-8}$ cm.

For comparison the Thomas-Fermi screening function is taken in the approximation of Molière [5]

$$f(r/a) = \sum_{i=1}^{3} a_i \cdot \exp\left(-\frac{b_i \cdot r}{a_{TF}}\right) \quad , \tag{4}$$

with $\quad a_{TF} = 0,885 \cdot a_o / \sqrt{z_1^{2/3} + z_2^{2/3}}$.

A screened Coulomb potential of the form

$$\bar{V}(r) = E_c \left(\frac{a_c}{r} - 1\right) \tag{5}$$

is fitted to the interaction potential (2), where the constants E_c and a_c are dependent of the particular fit.

In our former calculations [3] we used only the Bohr potential. For simplification the two potentials (2) and (5) were equalized in value and inclination at a distance b, where

$$b = \frac{Z_1 \cdot Z_2}{\frac{1}{2} m_o v^2} \quad , \tag{6}$$

m_o = reduced mass of the interacting particles and v = their relative velocity.

This approximation leads to closed expressions, a finite total cross section, and an energy dependent selection of the scattering angles by a rather simple formula. The possibility also exists to reduce the total cross section by introducing a suitable lower limit for the scattering angle, for which we chose 0.1° in the laboratory system. This older approximation will be denoted by (YBN) in the following.

In the new calculations the potentials in question were adjusted
for that value of the impact parameter, for which the scattering
angle must be computed by integration of (1). Considering many
values the connection between impact parameter and scattering angle
is given by energy dependent tables. From these tables the relation
between random numbers and the scattering angle is calculated piece
by piece and the result is again given by energy dependent tables.
Intermediate values must be interpolated. The total cross sections
are established by a lower limit of the scattering angle. For this
angle θ_g we chose $0.1°$ and $0.3°$ in the laboratory system. The distributions for the two different potentials will be denoted by (TF)
in the case of the Thomas-Fermi potential and by (YB) for the Bohr
potential.

Energy Losses

Two different energy losses must be considered. The overwhelming
part, due to electron excitations, is calculated using the expression

$$\Delta E_e = \left| dE/dx \right|_e \cdot s \ , \tag{7}$$

where s is the distance travelled by the particle. The differential
energy loss $\left| dE/dx \right|_e$ is taken by interpolation from the tables of
Northcliffe and Schilling [9].
The energy losses due to nuclear collisions which are only a
small amount of the total losses can be calculated following classical rules for elastic collisions

$$\Delta E_c = C \cdot E \cdot \sin^2(\chi/2),$$

where C is a constant.

Results and Discussion

For comparison with known experiments the distributions were
calculated for the following particles: $Z_1 = 53$, $M_1 = 135$, $E_O =$
81 MeV. The gold films ($Z_2 = 79$, $M_2 = 197$) had thicknesses of 0.315
mg/cm^2 and 0.630 mg/cm^2, which mean 1630 Å and 3260 Å, if the bulk
density is assumed.
The experimental results and the scattering distributions calculated in the different approximations are shown in Figures 1 and
2. The values of the lower limit of the scattering angle θ_g were
$0.1°$ and $0.3°$ in the case of 0.63 mg/cm^2, whereas for the thickness
of 0.315 mg/cm^2 only $0.1°$ was used because of the low number of
scattering events.

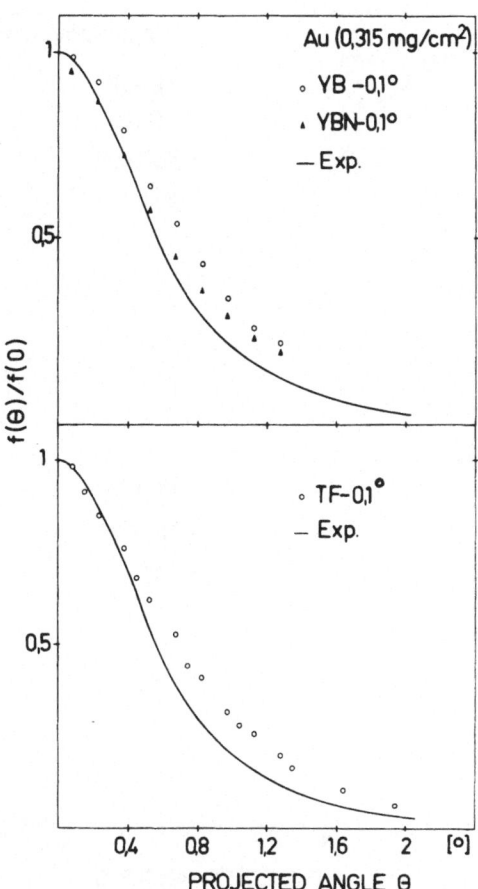

Fig. 1. The experimental multiple scattering distributions and the calculated distributions as a function of the projected scattering angle; $\theta_g = 0.1°$, film thickness $t = 0.315$ mg/cm^2.

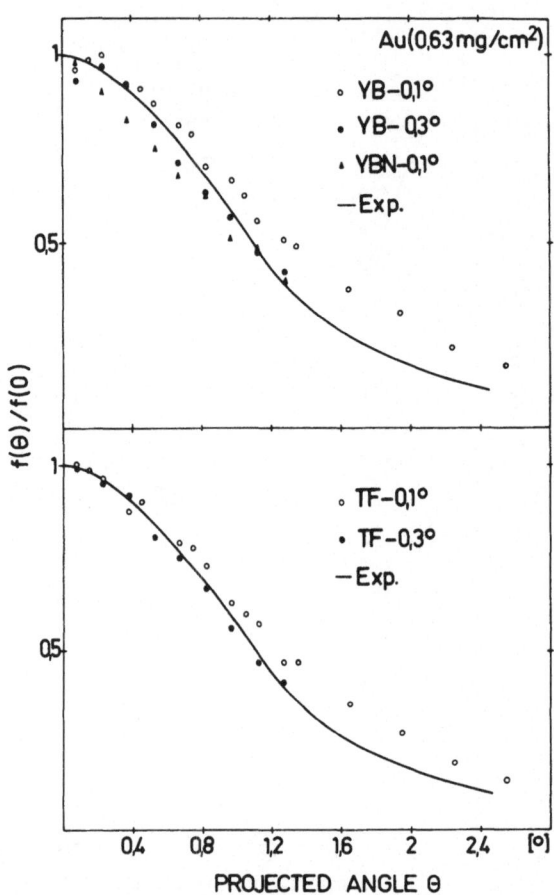

Fig. 2. The experimental multiple scattering distributions and the calculated distributions as a function of the projected scattering angle; $\theta_g = 0.1°$ and $0.3°$, film thickness t = 0.63 mg/cm^2.

The differences between the two potentials are not very pronounced, if the same potential fit is applied. This is not astonishing, because the potentials only differ slightly over the range of impact parameters considered. For $\theta_g = 0.1°$ the TF-distribution is somewhat smaller, but both distributions exceed the experimental width for the two thicknesses. Better agreement is found between calculation and experiment for $\theta_g = 0.3°$ and the thickness 0.63 mg/cm^2. With the older approximation YBN contrary results are found for both foils. The corresponding distributions for the number of scattering events are shown in Figures 3 and 4.

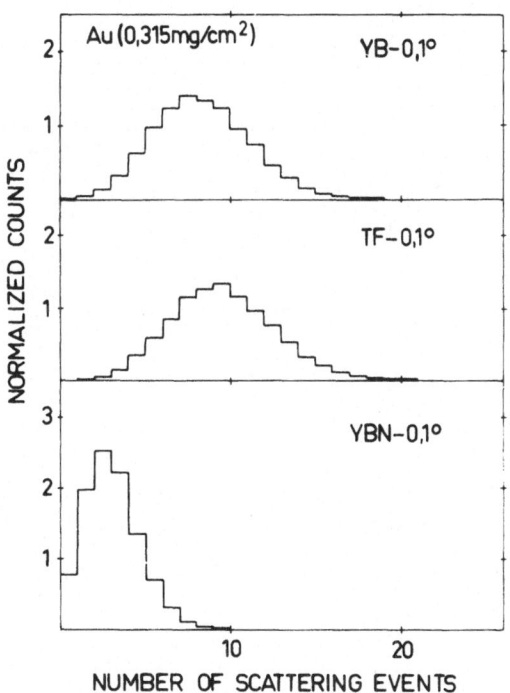

Fig. 3. The distributions of the collision numbers for the film thickness t = 0.315 mg/cm^2.

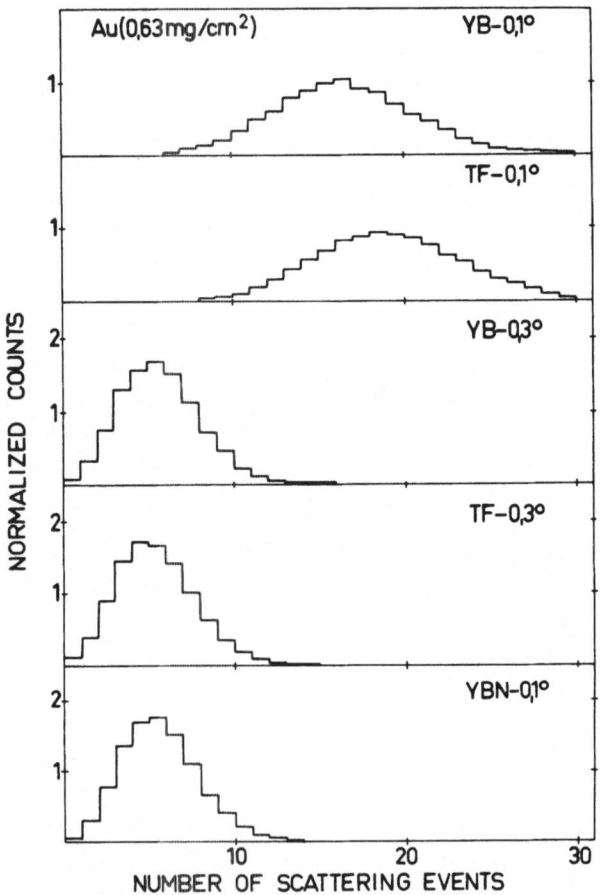

Fig. 4. The distributions of the collision numbers for the film
thickness t = 0.63 mg/cm².

These distributions are determined by the upper limits of the impact parameter given by Table I.

Table 1

Maximum impact parameter (10^{-9} cm) depending on the lower limit of the scattering angle θ_g (E = 81 MeV)

θ_g (°)	TF	YB	YBN
0.1	1.712	1.602	0.918
0.3	0.886	0.914	

For θ_g = 0.1° the average scattering numbers of the TF-simulation are somewhat higher than for YB, whereas for θ_g = 0.3° just the opposite result is found, but less distinct. The YBN-distribution is comparable to that for TF-0.3°.

There are two opposing effects which influence the width of a scattering distribution. As the screening increases the scattering angle in a single scattering event becomes generally larger tending to broaden the distribution. At the same time the average number of collisions, depending on the maximum impact parameter, decreases, tending to make the distribution narrower. If the number of collisions is reasonable large the latter effect dominates. This can be clearly seen if one examines the distributions for a particular potential and the two different limits of the scattering angles. Decreasing collision number reduces the width of the distribution. The more complicated interaction of the two effects causes the differences between the two potentials and approximations, especially for θ_g = 0.3°, where the average number of scattering events is rather small.

Besides the projected distributions our scattering distributions can be compared with the calculations of Meyer [7]. This has been done in Figures 5 and 6.

The differences between the two potentials and the same fit are less distinct than in the case of the projected angles. The deviations between the theoretical distributions are just opposite to the result of the comparison between experiment and our calculations. In the latter case the best agreement is found for θ_g = 0.1°, whereas large deviation appear for θ_g = 0.3° and the approximation YBN-0.1°. It is the same complicated reciprocal action of the collision number and the values of the scattering angles, which causes these differences, depending on the rather arbitrary value of θ_g, which is not yet quite understood.

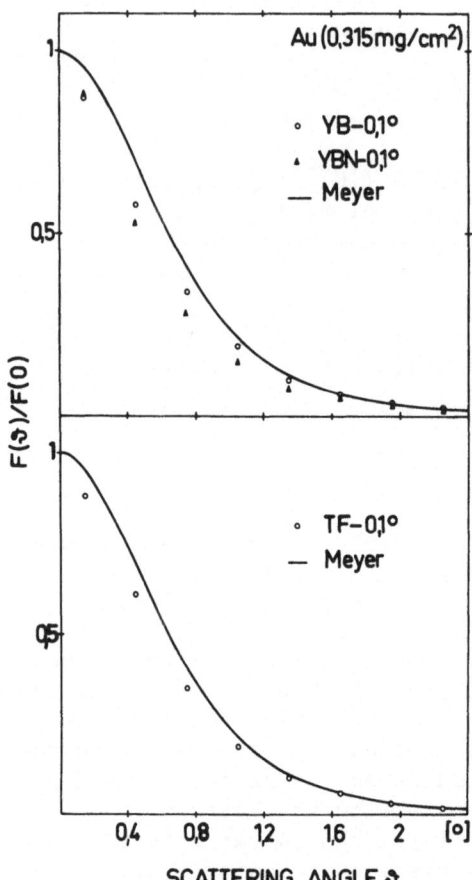

Fig. 5. The distributions calculated according to the theory of Meyer and to the Monte-Carlo simulations for t = 0.315 mg/cm^2.

Despite the agreement for θ_g = 0.1° one must conclude from the results concerning the projected distributions that the widths of Meyer's theoretical distributions are too large.

Slit Scattering

An example for a possible application is given in Figure 7, where a beam of high-energy heavy particles strikes the edge of a wedge-shaped slit.
    The following values are chosen:

Beam: Z_1 = 54, M_1 = 136, E_o = 567 MeV (Xe).

Slit: Z_2 = 79, M_2 = 197 (Au); α = 80°.

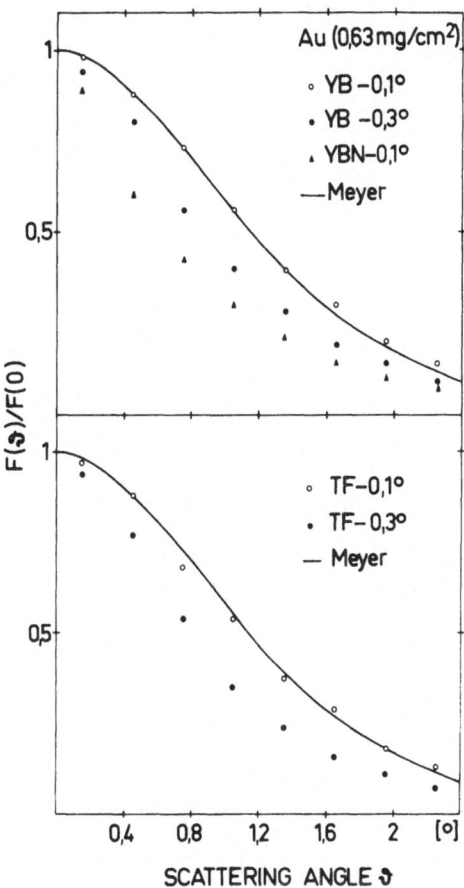

Fig. 6. The distributions calculated according to the theory of Meyer and to the Monte-Carlo calculations for t = 0.63 mg/cm².

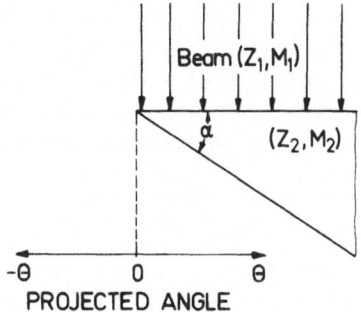

Fig. 7. Example for a beam striking the edge of a wedge-shaped slit.

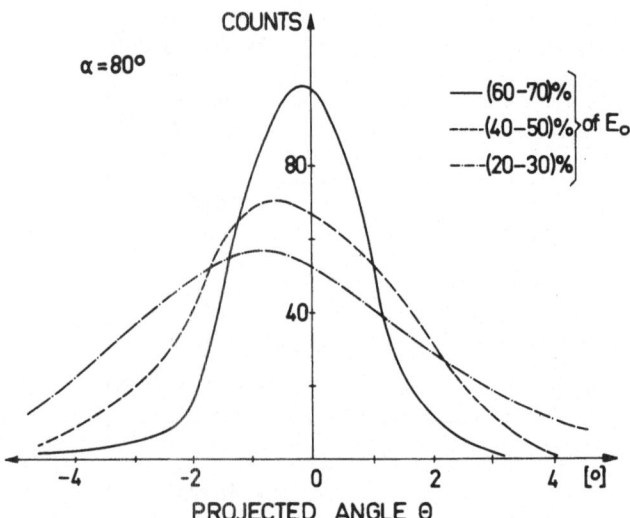

Fig. 8. Projected distributions of particles scattered by a slit for different energy intervals.

The projected distributions of the scattered particles in the plane of incidence is shown in Figure 8 for several energy intervals. With decreasing energy the width of the distribution increases becoming more and more asymmetric in relation to the edge of the slit.

Conclusion

The present work shows that these computer simulations are appropriate for the description of multiple scattering processes. Uncertainties are given by the rather arbitrary choice of the lower limit of the scattering angle, which may influence the distributions considerably.

Acknowledgments

We would like to express our gratitude to Prof. H. Ewald for encouraging to do those calculations and to Mr. H. Schmidt for valuable discussions.

References

[1] D. Engelkemeir, Phys. Rev. 146, 304 (1966).
[2] J. Albrecht and H. Ewald, Z. Naturf. 26a, 1296 (1971).
[3] K. Güttner, Z. Naturf. 26a, 1290 (1971).
     K. Güttner, H. Ewald and H. Schmidt, Radiation Effects 13, 111
     (1972).
[4] D. Kerr, G. Siegert, K. Kürzinger, E. Konecny and H. Ewald,
     Z. Naturf. 22a, 1799 (1967).
[5] G. Molière, Z. Naturf. 2a, 133 (1947).
                          3a, 78 (1948).
[6] B. P. Nigam, M. K. Sundaresam and Ta-You Wy, Phys. Rev. 115,
     491 (1959).
[7] L. Meyer, Phys. Stat. Sol. (b) 44, 253 (1971).
[8] N. Bohr, Mat. Fys. Medd. Dan. Vid. Selsk. 18, 8 (1948).
[9] L. C. Northcliffe and R. F. Schilling, Nucl. Data A7, 233 (1970).

NUMERICAL SIMULATION OF RANGE AND BACKSCATTERING FOR keV PROTONS INCIDENT ON RANDOM TARGETS[*]

J. E. ROBINSON and S. AGAMY
Department of Engineering Physics
and
Institute for Materials Research
McMaster University, Hamilton, Ontario Canada

ABSTRACT

Using a Monte-Carlo technique, projected range distributions and backscattering coefficients have been calculated for keV protons normally incident on heavy targets. For an incident reduced energy range of $1 < \varepsilon_o < 20$, both the projected range distributions and backscattering coefficients have been found to be in good agreement with a third order Edgeworth range approximation. Backscattered energy and angular distributions have also been calculated and are compared to available theoretical and experimental data.

Introduction

Recent interest has been shown in the interaction of light keV ions with matter in the intermediate energy range where both nuclear and electronic stopping are important [1,2,3]. This energy range is also of particular importance for predicting surface radiation damage and sputtering in fusion research [4]. Weissman and Sigmund [5], and Bøttiger and Winterbon [6] have recently calculated the projected range distributions and reflection coefficients of light keV ions incident on solids by extending Schiøtt's integral equation method [7]. Ideally, these predictions should be checked experimentally. However, an accurate measurement of the projected range distribution is difficult due to diffusion of the implanted ions in the solid. Also, a direct measurement of the scattering is difficult because of (a) the scattered particles charge dependence (H^+, H^0, H^-)

on surface contaminants, and (b) the lack of a suitable light isotope
(of H, D, T, He) for utilization of a radioactive tracer technique
[8,9]. In this study, projected range distributions and reflection
coefficients are calculated by following the histories of 1.5×10^2
to 2×10^3 particles using a high speed computer. A comparison is
then made to the third order Edgeworth approximation.

Computation Procedure

To calculate the projected range and scattering of a particular ion incident on a solid, we make use of three random numbers at
each collision (N) to calculate the distance between collisions
$(X_{N+1} - X_N)$ and two scattering angle $(\theta_{N+1}, \phi_{N+1})$. The use of a
random number to select a collision distance, rather than directly
using the mean free path concept, more accurately models the backscattering of ions near the surface. The nuclear collision cross
section used in the calculation is based on the Thomas-Fermi potential approximation [10,11]. Continuous electron slowing down is
also assumed [12], i.e.

$$\frac{dE}{dx} = - kE^{1/2} \tag{1}$$

where E is the particle energy and x is a distance along the particle
path.

The technique for calculating the distance between collisions
can be summarized as follows. If we assume that the total distance
at which an ion is stopped is x_{max}, which corresponds to ion energy
E_{min} = 25 ev, the probability that at least one nuclear scattering
event occurs is p_n and it is given by

$$p_n = 1 - \exp \left\{ - \int_0^{x_{max}} N' \, \sigma(E) \, dx' \right\} , \tag{2}$$

where N' is the atomic density of the target material, and $\sigma(E)$ is
the scattering cross section as a function of energy (and consequently a function of distance). This probability p_n varies from
0 to 1. If we choose a random number q at a particular x, and if
$q \geqslant p_n$, no nuclear scattering will occur. If $q < p_n$, we will have
at least one nuclear scattering event before the particle is stopped

by electronic energy loss. Hence, the probability that we have a nuclear scattering at distance x ($< x_{max}$) is given by,

$$q = 1 - \exp\left\{-\int_0^x N'\sigma(E) \, dx'\right\} \quad . \tag{3}$$

Substituting equation (2) into equation (3), we have the form

$$q = 1 - (1 - p_n) \exp\left\{\int_x^{x_{max}} N'\sigma(E) \, dx'\right\} \quad . \tag{4}$$

Changing variable in equation (4) using the relation of equation (1), we have

$$\int_{E(x)}^{E(x_{max})} -\frac{N'\sigma(E') \, dE'}{K(E')^{1/2}} = \ln\left[\frac{1-q}{1-p_n}\right] \equiv F(E) \tag{5}$$

or

$$q = 1 - (1 - p_n) \exp\{F(E)\}$$

$$= 1 - \exp\{-F(E_o)\} \exp\{F(E)\} \quad ,$$

and

$$F(E) - F(E_o) = \ln[1 - q] \quad . \tag{6}$$

E_o is the initial energy of the particle and E is the energy at which the scattering occurs at a distance x.

Now we list the steps to be followed for determining the nuclear scattering distance between collisions:

1. tabulate and store F(E) for energies ranging from 25 ev to the maximum energy of an incident particle,

2. determine $F(E_o)$ from step 1 and p_n from equation (6) with F(E) = 0,

3. choose a random number q and check $q < p_n$,

4. from equation (6), determine $F(E_1)$,

5. use equation (1) with the table of F(E) to determine $(X_1 - X_o)$,

6. repeat steps 2 through 5 at each successive collision.

To calculate the scattering angle, θ, in the center of mass system, we choose a second random number, $o < q' < 1$, and evaluate the integral for the differential cross section, i.e.

$$q' = \frac{T_{min}\int^{T} d\sigma}{\int_{T_{min}}^{T_{max}} d\sigma} \tag{7}$$

where $\sin^2\theta/2 = T/T_{max}$, and T is the energy transferred. The center of mass angle can then be converted to a lab angle ψ, knowing θ and the masses of the incident and target particles, M_1 and M_2

The azimuthal scattering angle, ϕ, is found by choosing a third random number, $0 < q'' < 1$, and evaluating $\phi = 2\pi q''$.

For a given initial energy, E_o, we determine the first nuclear collision location and evaluate the energy and direction after the collision. In the same way we proceed to the second and third collision, ... etc. At each point we calculate the projected range. Also, at each location we check whether the projected range is negative or positive. In the former case, the particle is backscattered. If this occurs, we determine the scattered energy and the angle of the outgoing particle.

Discussion and Summary

Projected range distributions of H^+ normally incident on Al and Nb, for reduced energies of 1,2,5 and 10 keV, are shown in Fig. 1. For the case of aluminum, 300 particles were followed for 50-100 collisions to a minimum energy of 350 eV. With niobium as the target material, 162-186 particles were followed for 50-200 collisions to a minimum energy of 500 eV. As can be seen in Fig. 1, the computer simulation is in good agreement with the third order Edgeworth expansion [5]. However at small values of projected range, the accuracy of the simulation is limited by a small number of events recorded. Since the reflection of particles is sensitive to this

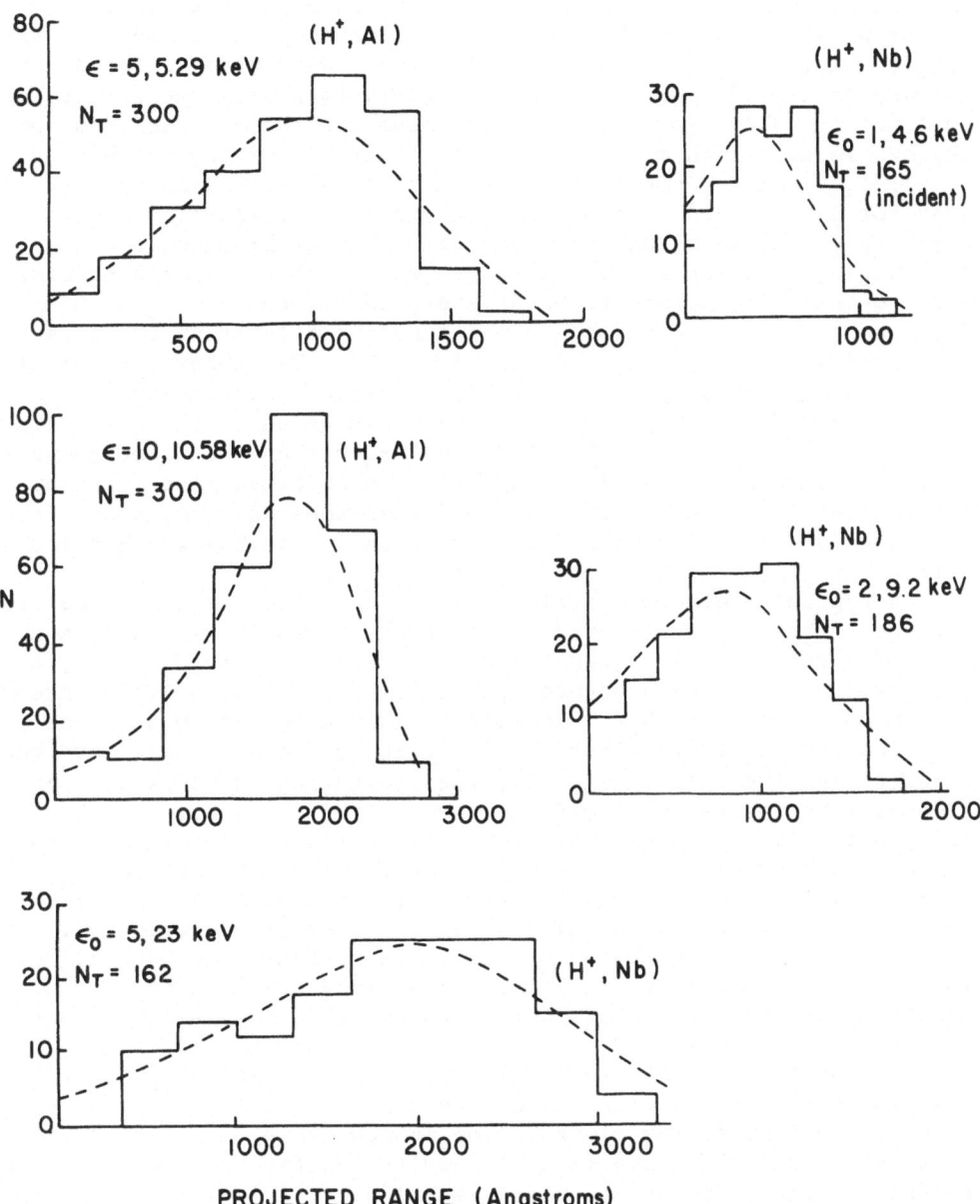

Fig. 1. Projected range for protons incident on niobium and aluminum.

part of the projected range distribution, the third order expansion can be more accurately checked by comparing Monte-Carlo reflection coefficients with those found from Edgeworth expansions. To increase the computing speed for the reflection calculation, particle histories were terminated when a particle's depth penetration into the target was too large to permit reflection. With this modification, as many as 2×10^3 particles were followed to obtain the reflection coefficients shown in Fig. 2. Also shown are the coefficients for the single collision model [2] and the third order Edgeworth expansion [5]. Good agreement with the third order expansion can be seen for $1 < \varepsilon_o < 10$. At higher energies, the reflection coefficient is larger than the Edgeworth value but still much less than the single collision model. If a transition to the single collision model occurs, it appears to be at energies higher than $\varepsilon_o = 20$. Also shown for completeness are the results of Ishitani et al using 10^3 particles and a variety of targets [13]. These earlier computer results are in good agreement with the present calculations.

It should be noted that the calculated value of the reflection coefficient at $\varepsilon_o = 1$, $0.18 \pm .02$, is larger than the value computed from the Edgeworth expansion. A similar result was also found by Ishitani et al for $\varepsilon_o \simeq 1.5$. These differences are consistent with the "surface correction" not included in the Edgeworth calculation.

A typical scattered particle energy spectrum and the normalized average energy of a scattered particle are shown in Fig. 3(a) and Fig. 3(b) respectively. For the cases studied, where $M_1 \ll M_2$, the average energy of the scattered particles appears to be independent of the mass ratio (M_2/M_1). However, a marked variation in slope with energy can be seen. At small ε_o, the average energy curve has a large negative slope. That is, the scattering cross section increases with decreasing incident energy with the result that both the penetration depth and energy losses decrease. For $\varepsilon_o > 5$, the normalized average energy appears to approach a constant value: $\bar{E}_s/E_o \simeq 0.2$. This is in qualitative agreement with the single collision theory which has a value $\bar{E}_s/E_o \simeq 0.1$ at high energy [2]. The lower average value of the single collision theory can be attributed to the peak in the single collision scattered spectrum at $E_s = 0$. A peaking at zero scattered energy was not observed in any of our computer simulations.

The scattered particle angular distribution is shown in Fig. 4(a) for the superposition of the two cases, $\varepsilon_o = 1,2$. As can be seen in the figure, the distribution is approximately a cosine. This is in agreement with experimental results observed at lower incident energies using radioactive tracer techniques [9]. As shown in Fig. 4(b), a shift to larger scattering angles is expected from the single collision theory at higher incident energies. However, for all cases evaluated with the computer simulation, a shift from the cosine distribution was not observed. This observation could be due to the large statistical error associated with the small number of scattered particles at higher incident energies.

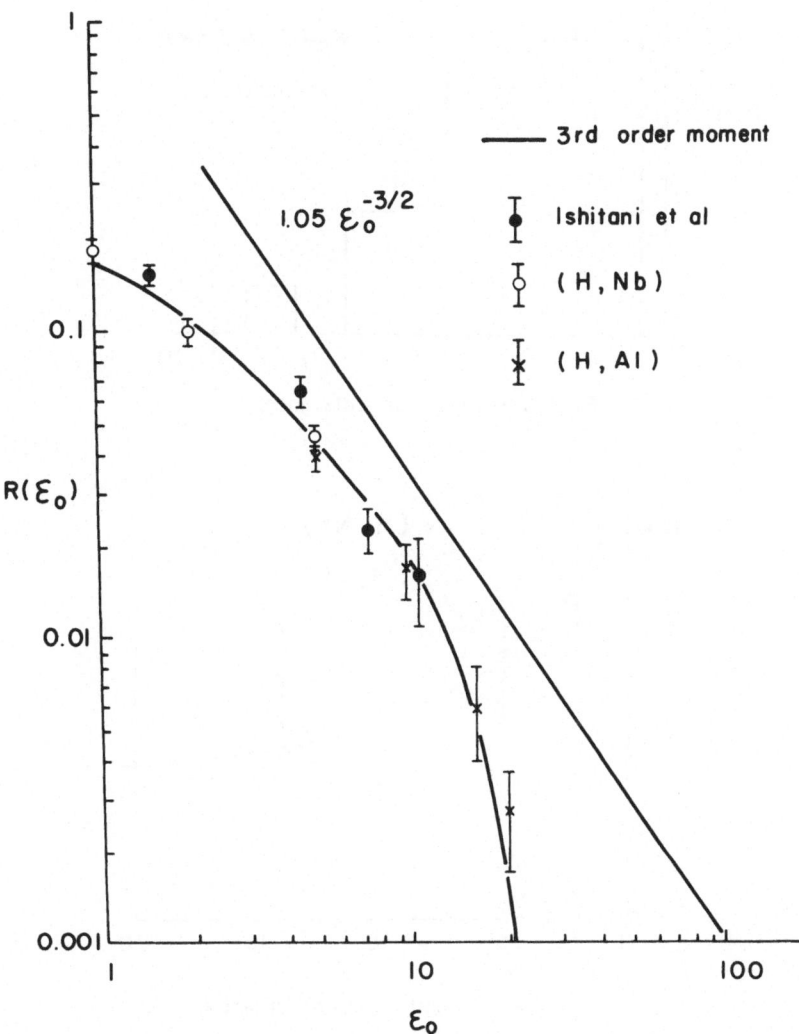

Fig. 2. Reflection coefficient as a function of reduced incident energy.

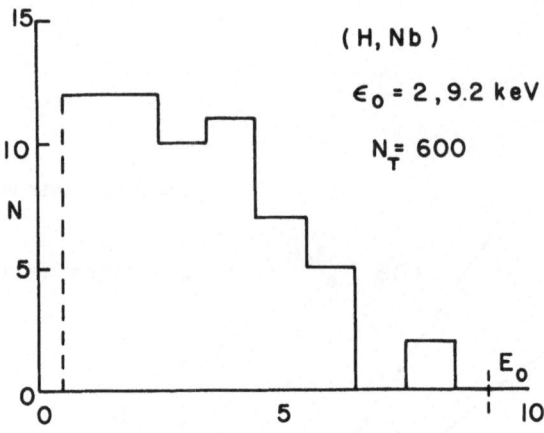

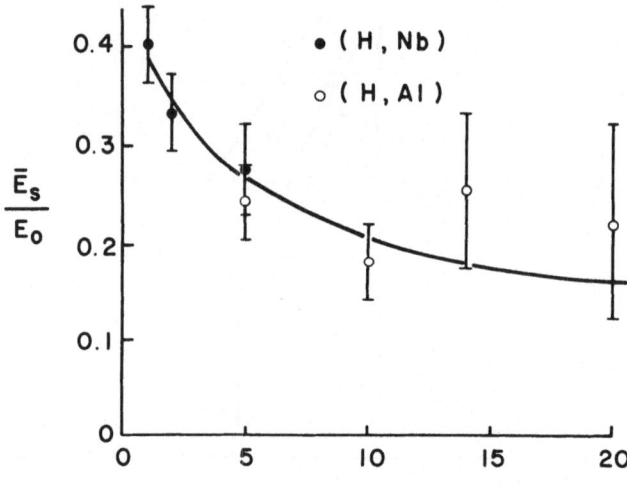

Fig. 3.   (a) Scattered particles energy distribution for protons
normally incident on niobium, E_0 = 9.2 keV.   (b) Normalized scattered particle average energy as a function of incident energy
(solid line is the best fit).

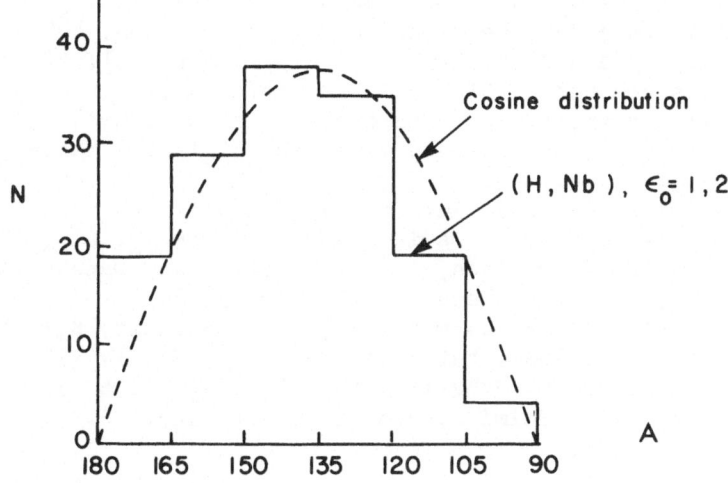

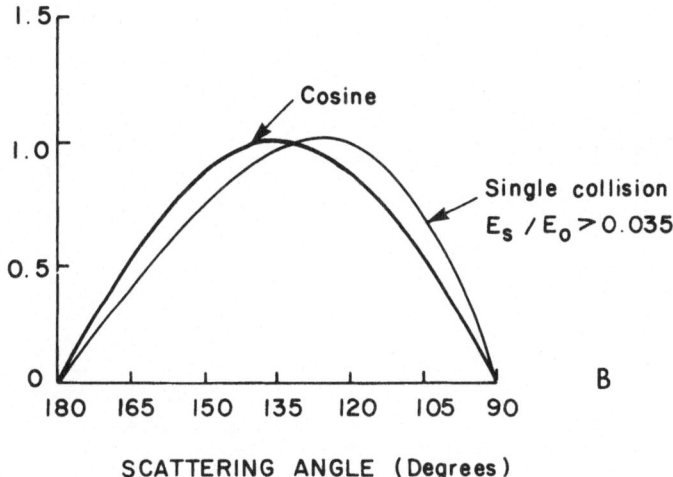

Fig. 4. (a) Scattered particle angular distribution (particles per angular interval). (b) Comparison of the cosine and the single collision angular distributions.

References

   *Research supported by the National Research Council of Canada.
[1] E. S. Mashkova and V. A. Molchanov, Rad. Effects 16, 143 (1972).
    (Review Article)
[2] G. M. McCracken and N. J. Freeman, J. Physics B, 2, 661 (1969).
[3] R. Behrisch and R. Weissman, Phys. Letters 30A, 506 (1969).
[4] M. Kaminsky and S. K. Das, Appl. Phys. Lett. 21, 443 (1972).
[5] R. Weissmann and P. Sigmund, Rad. Effects (in press).
[6] J. Bøttiger and K. B. Winterbon (to be published).
[7] H. E. Schiøtt, Kgl. Danske Videnskab. Selskab, Mat.-Fys. Medd.
    35, No. 9 (1966).
[8] J. A. Phillips, Phys. Rev. 97, 404 (1955).
[9] J. Bøttiger and J. A. Davies, Rad. Effects 11, 61 (1971).
[10] J. Lindhard, V. Nielsen, and M. Scharff, Kg. Danske Videnskab.
     Selskab., Mat.-Fys. Medd. 36, No. 10 (1968).
[11] K. B. Winterbon, P. Sigmund, and J. B. Sanders, Kgl. Danske
     Videnskab. Selskab., Mat.-Fys. Medd. 37, No. 14 (1970).
[12] J. Lindhard and M. Scharff, Phys. Rev. 124, 128 (1961).
[13] T. Ishitani, R. Shimizu, and K. Murata, Japan J. Appl. Phys.
     11, 125 (1972).

COMPUTER STUDIES OF REPLACEMENT SEQUENCES IN SOLIDS ASSOCIATED WITH ATOMIC DISPLACEMENT CASCADES*

D. K. HOLMES and M. T. ROBINSON
Solid State Division
Oak Ridge National Laboratory
Oak Ridge, Tennessee 37830, U.S.A.

ABSTRACT

Focussed sequences of replacement collisions have been studied, especially in association with radiation damage cascades. Radiation
damage events are simulated using a computer
program based on the binary collision approximation. The results are averaged over a complete
set of initial directions of the primary knock-
on atom for each primary energy to obtain quantities of physical interest for which the effects
of replacement sequences may be made explicit.
The calculations are compared with recent experimental results.

Introduction

The concept of a focussed sequence of collisions along a close-
packed row of lattice atoms was introduced by Silsbee [1] in 1957.
Since then the theory has been discussed by many investigators [2,
3,4,5] but until very recently there has been no firm establishment
of this physical phenomenon on an experimental basis (at least, not
in metals). The purpose of the present work is to extend the theoretical study of the effects of such focussing sequences (especially
those which involve mass transport along the line, i.e. replacement
sequences) in association with radiation damage cascades, and to
attempt to compare the results with recent experiments.

The calculations reported here are based on a computer simulation of radiation damage events. This computer program (called
Marlowe) has been discussed in greater detail elsewhere [6,7]. The
most important features of the program for the present discussion

are: (i) the full symmetry of the crystalline lattice is used but
(ii) only binary collisions between moving atoms and stationary
lattice atoms are allowed. The first point is clearly necessary
if replacement sequences are to be studied; on the other hand the
restriction to binary collisions makes the approximation poorer for
such sequences of collisions. It is in the low energy range in
which focussed replacement sequences are important; these are energies of motion of the order of 10-50 eV. In this energy region the
lattice atoms have a strong interaction over distances of the order
of the equilibrium lattice distance, so that a moving atom will, in
general, experience significant forces from several stationary
atoms at once. Nevertheless, it is felt that with careful treatment the binary collision approximation can give valuable insight
into the role of focussing events. The advantage of the binary
collision approximation is that the full detail of relatively large
displacement cascades in the lattice can be treated. The new feature of the Marlowe computer program which has been incorporated to
assist in the present study is a statistical analysis of replacement sequences accompanying each radiation damage cascade.

A brief outline of the use of the Marlowe program for these
calculations follows. For a given substance the lattice geometry
is specified and the interatomic potential is chosen. Generally
the latter is the Moliére approximation to the Thomas-Fermi potential with an appropriately chosen screening radius. For each binary collision an inelastic energy loss is also permitted (based
on the Firsov model [8]). One atom, the primary, is started with
a given kinetic energy from a lattice site and the full development
of the subsequent collision cascade is followed until all energies
of motion have fallen below a specified value. This process is repeated for a number of cascades with the starting direction of the
primary allowed to vary randomly over the full unit sphere. For
the present calculations the number of such initial directions for
each primary energy was usually taken to be one thousand. After
the set of cascades has been calculated the program presents as
output a statistical summary of the important results. For the
present the most significant quantities are: (i) the total number
of vacancies and interstitials produced, along with the distribution of the separations of vacancy-interstitial pairs, (ii) the
total number of replacement events (both proper and improper, see
later discussion), and (iii) the distribution of lengths of replacement sequences along the principal lattice directions.

Some results of these calculations will now be discussed along
with possible implications for experimental observations.

The Total Length of Replacement Sequences as a Function of Primary
Energy

In the Marlowe program a replacement event is said to occur
for a collision in which the energy of the moving atom falls below

a certain value while the energy transferred to the stationary atom
is above a specified value. These two limiting energies are not
necessarily the same, but have been taken to be the same in the present calculations. The value of that energy, called the displacement energy, E_d, has mostly been taken as 5 eV for the calculations
reported here. If the momentum is sufficiently low and is fairly
well directed along a close-packed lattice row, a sequence of replacement collisions will result which the computer program follows
in detail. The sequence terminates when the energies of both of
the colliding atoms falls below E_d after the collision. At this
point in the lattice the final atom in the sequence becomes an interstitial. The replacement sequences discussed in the present
work are generated under very favorable circumstances. Both fcc
(copper) and bcc (iron) lattices have been used, with potentials
leading to good focussing along <110> and <111> rows, respectively.
Qualitatively, the results reported here apply equally well for both
substances. The parameters chosen for the calculations are such
that the interaction with lattice rows neighboring the given replacement sequence is neglected. The energy loss to neighboring
rows has been separately calculated, and is about one-tenth of an
electron volt per collision in the replacement chain. Further, the
lattice is taken as perfectly static for these calculations (except
for special purposes to be discussed later, for which very large
thermal vibrations were introduced.) Again the effect of thermal
vibration has been calculated separately; it is estimated that as
much as 1.5 eV may be lost per collision in the chain due to zero-
point vibrations alone.

As the cascade of displaced atoms develops from a given primary event a number of replacement sequences may result, moving in
various lattice directions, mostly away from the region containing
the vacancies produced. This is illustrated schematically in Fig.
1 which shows a two-dimensional array of atoms containing the displacement cascade and some resulting replacement sequences. Such
replacement sequences carry an atomic mass down the close-packed
lattice line, and when the sequence terminates an interstitial is
formed at the end of the chain. Thus, these events carry interstitials far from the vacant lattice sites. The replacement sequences
produced will vary in length depending on the energy and direction
of motion of the first atom in the chain. One of the quantities
which can be calculated is, then, the total length (in total number
of replacement events) of such replacement sequences associated with
a given cascade. When the result is averaged over a statistical
distribution of primary directions, the value calculated is characteristic of the primary energy alone (for the chosen crystal). The
total length itself is not, of course, physically observable in a
monatomic lattice, but the discussion of this quantity brings out
some of the important aspects of replacement sequences associated
with cascades, and, further, the quantity itself is of importance
in consideration of order-disorder alloys. For the present

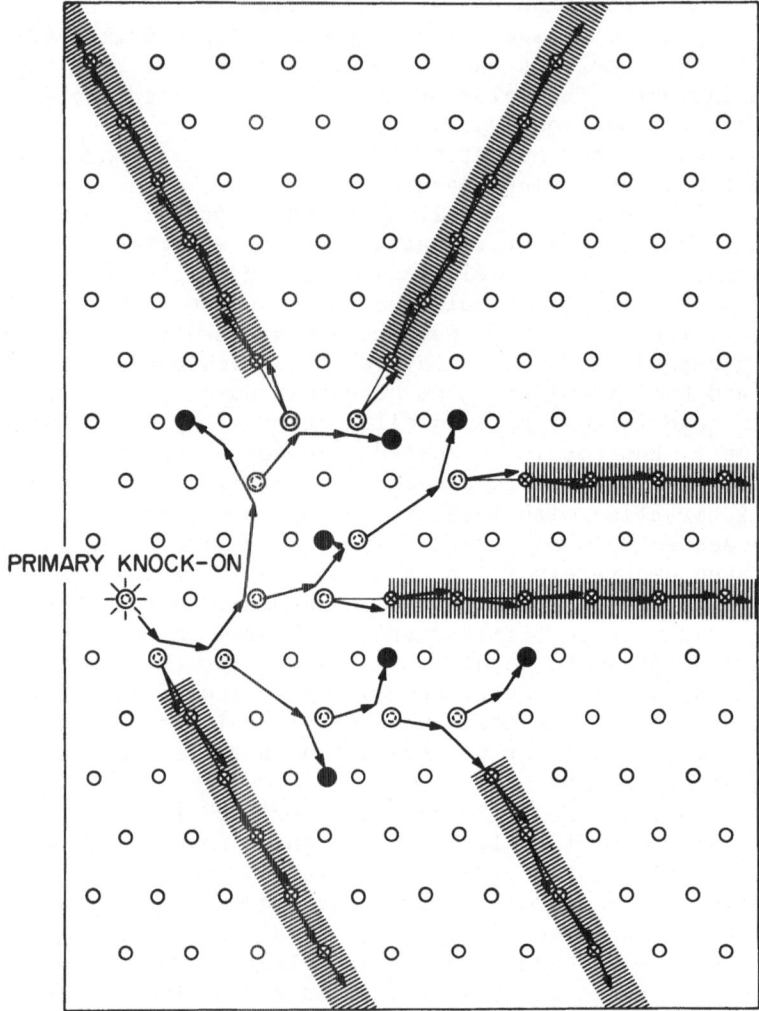

Fig. 1. Schematic representation of a radiation damage cascade
including vacancies ⊘, interstitials ●, and replacement sequences
(shaded lines of atoms). At the end of each replacement sequence
(not shown) is an interstitial atom.

discussion the "total length" will be taken to mean the total length
of all sequences which involve more than two atoms. This is an
attempt to concentrate on those sequences which are definitely focussed along a lattice row.

The dependence of the total length of replacement sequences on the primary energy is presented in Fig. 2. The example chosen is copper (fcc) with a screening radius [6] of 0.0738 Å for the Moliére potential. The length reported in this case is mostly (98.6%) due to <110> type sequences. From the results plotted it may be seen that there are two regions in energy on each side of a distinct minimum at about 50 eV. A closer examination of the details of individual cascades shows that the high and low energy regions do, indeed, involve different sets of replacement sequences. In the regions below 50 eV the principal events will be called here, for convenience, "primary replacement sequences." These sequences result directly from the action of the primary knock-on atom on a close-packed row which contains the original position of the primary. Such an event is schematically presented in Fig. 1 which shows one replacement sequence directly stemming from the primary. In the energy range above 50 eV the replacement sequences arise from collisions between moving cascade atoms and stationary lattice atoms, as represented in Fig. 1 by all of the sequences shown except the "primary replacement sequence." These secondary replacement sequences will be called here "cascade replacement sequences."

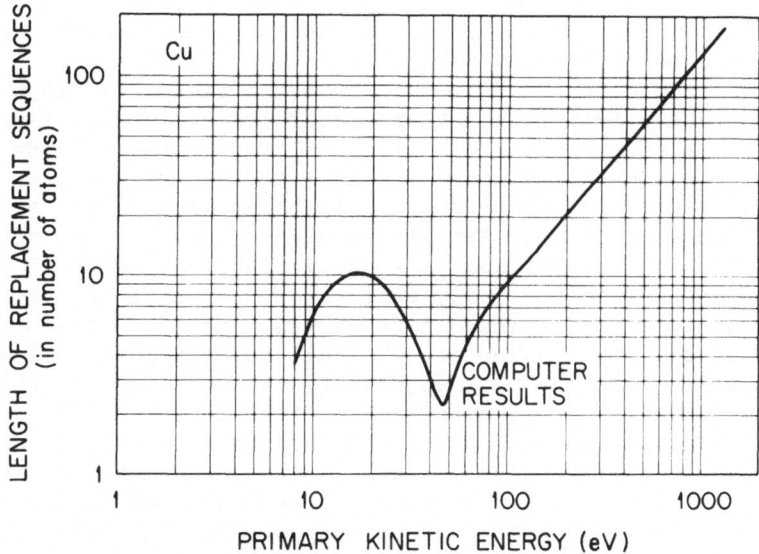

Fig. 2. The total length of replacement sequences associated with cascades resulting from primaries of various energies. The total length is measured in total number of atoms replaced and is given for all sequences greater than two atoms in length. The results are for copper (fcc).

The processes occurring in the energy range below 50 eV are just those which have been extensively discussed in the literature in connection with focussing events [9]. The primary event tends to produce a single replacement sequence and not much of an associated cascade. The most favorable energy region for long sequences is in the general region of the focussing energy which is about 13 eV for the present choice of potential for copper (for the screening radius chosen for iron, 0.0781 Å, for the calculations reported later, it is about 17 eV). In the upper part of the low energy range (20-50 eV) the sequences are initially defocussed and thus lose a large portion of their energy in the early collisions with a consequent reduction in the total length. In the lower part (below 15 eV) the initial energy is so low that the primary replacement sequence is cut off by the energy criterion before it achieves much length.

In the energy region above 50 eV it may be seen that the total length tends to rise linearly with the initial energy of the primary. This is simply a reflection of the fact that the higher energies produce larger displacement cascades with more resulting cascade replacement sequences. A simple analytical theory for this region may be constructed by assuming that the energy distribution of collisions in the cascade is essentially that for the hard sphere model ($\sim \frac{2E_o}{E^2}$) and that a cascade replacement sequence has a certain (low) probability of being produced whenever a struck lattice atom is given an energy below, say, 50 eV. This theory shows the expected linear behavior, and may be fitted to the computer results by the choice of the probability for production of a cascade replacement sequence and of its total length.

The minimum near 50 eV results from the facts that in this energy region the primary replacement sequences are defocussing and short, while the cascades are small resulting in few cascade replacement sequences. The minimum calculated by the computer program is, in fact, much lower than would be predicted by the simple analytical model fitted to the linear high energy portion of the curve of Fig. 2. This is due to a physical effect which is not taken into account in the analytical theory. When a moving cascade atom approaches a close-packed lattice row, the very close-packed positions of the atoms makes it difficult for the moving atom to produce a cascade replacement sequence. At the higher energies, (say 100 eV or above) this merely means that the probability for sequence production is low. However, when the energy gets down near the focussing energy range, the loss of energy, which is a consequence of the interaction of the moving atom with the atoms of the row, makes it impossible to produce long sequences. This is illustrated in Fig. 3 which shows in the upper part the easy production of a long cascade sequence by a moving 50 eV atom when one atom is missing from the row. On the other hand for a perfect row, as shown in the lower part of the figure, the same moving atom loses too much energy to be able to initiate a long sequence.

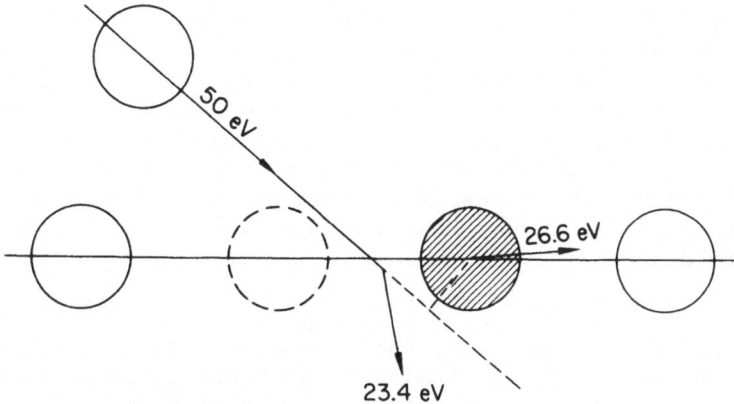

(a) COLLISION **WITHOUT** PREVIOUS ATOM
IN FOCUSSING CHAIN

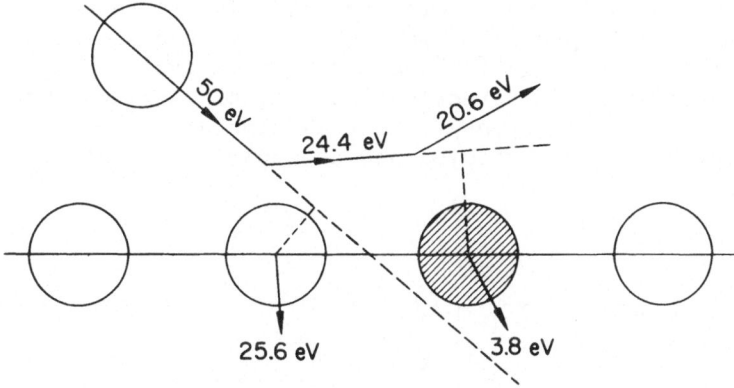

(b) COLLISION **WITH** PREVIOUS ATOM
IN FOCUSSING CHAIN

Fig. 3. The collision of a moving cascade atom with a close-packed
line of lattice atoms. (a) The chain is incomplete and a long replacement sequence results. (b) The same conditions but with a complete chain and no focussed replacement sequence results.

The Total Damage as a Function of the Primary Energy

     Focussed replacement sequences also have an effect on the total
damage remaining as a result of a radiation damage cascade. The damage remaining, for purposes of discussing this effect, will be taken
to be the number of vacancy-interstitial pairs. In order to make
the results physically more realistic it is important to allow for
the collapse of close pairs which will not be counted in the total
damage remaining (see reference 7 for a discussion of this point).

For this discussion we will concentrate on the Fe (bcc) lattice and
allow the collapse of all vacancy-interstitial pairs within a radius,
r_v, which is taken, for this case to be three lattice distances (cubic cell edges). The resulting damage as a function of primary energy is shown in Fig. 4 as the curve labeled "total". It may be
seen that the results again show different behaviors on either side
of a minimum, this time located at about 25 eV. (The different positions of the minima in energy·in Figs. 2, 4, and 5 is not completely understood.) The higher energy range shows the expected behavior
in that the damage rises linearly with primary energy. If the relationship is taken to be of the form $\frac{E_o}{2E_d}$, at very high energies,
then $E_d \cong 33$ eV. On the other hand the damage does not fall smoothly to zero at low energies. Below 25 eV there is a distinct maximum
in the damage remaining. From the previous discussion it may be suspected that this effect is due to primary replacement sequences.
This is confirmed by the results shown in the second curve of Fig.
4. In order to generate a damage curve without the effects of long
replacement sequences, very large thermal vibrations were introduced
into the Marlowe program (see reference 7 for a discussion of thermal vibrations as used in the computer simulation). The parameters
were chosen so that the effective RMS thermal amplitude was 0.3
lattice distances. As a result of this high amplitude the lattice
rows were effectively destroyed and no replacement sequences as long

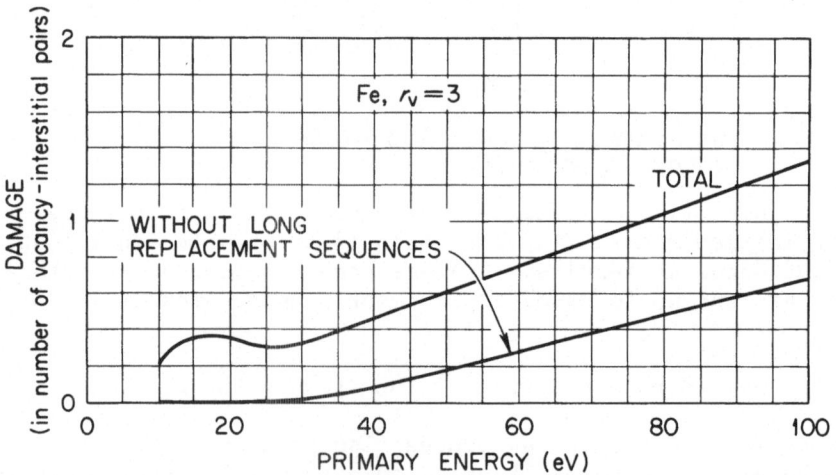

Fig. 4. The total damage remaining as a function of primary energy.
The particular case chosen is iron (bcc) with a pair collapse radius, r_v, of three lattice distances (cubic cell edges). The damage
is measured in terms of vacancy-interstitial pairs per cascade.
The upper curve is for the perfect lattice, while the lower curve is
for the disordered lattice (large amplitude thermal vibrations).

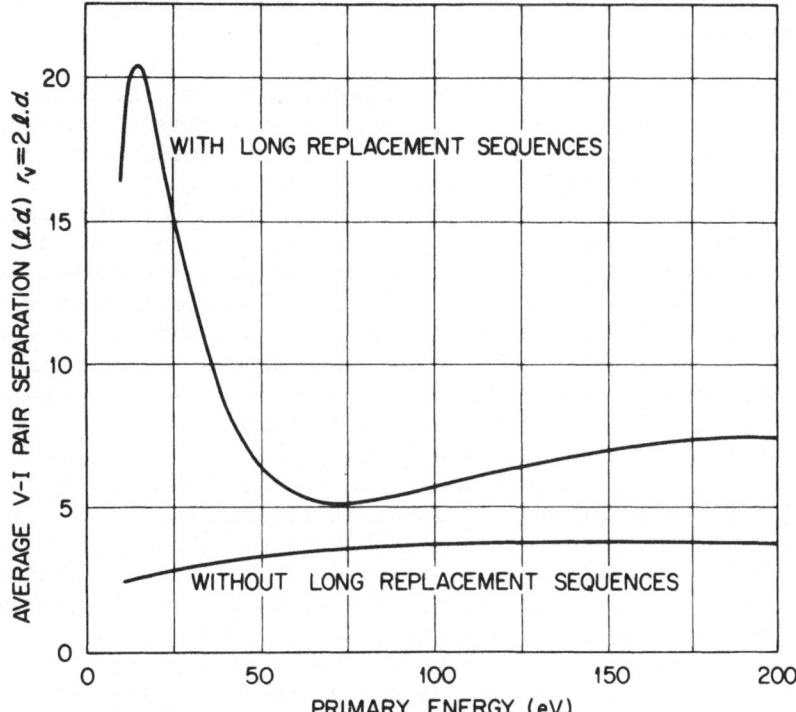

Fig. 5. The average vacancy-interstitial pair separation as a function of primary energy. The particular case chosen is iron (bcc)
with a pair collapse radius, r_V, of two lattice distances (cubic
cell edges). The separation is measured in lattice distances. The
upper curve is for the perfect lattice, while the lower curve is
for the lattice disordered by large amplitude thermal vibrations.

as three lattice distances were possible. The resulting curve in
Fig. 4 shows that the damage below 25 eV is almost completely eliminated. This indicates that the onset of residual damage in the
low primary energy range can be due to long (primary) replacement
sequences which carry the interstitial far away from the vacancy at
the lattice site of the primary atom. The difference between the
two curves in the energy range above 25 eV indicates that the cascade replacement sequences are effective in adding to the stable
damage produced by separating pairs to distances greater than the
collapse radius. (It is to be noted that the remaining damage in
either case in the low energy range is well below one pair per primary atom.)

The Average Separation of Vacancy-Interstitial Pairs as a Function
of Primary Energy

As mentioned above, the Marlowe program gives a statistical
analysis of the separations of the vacancy-interstitial pairs resulting from a given primary atom. This is of special interest because of recent experimental results obtained by Becker et al. [10]
As in the case of the total damage, pairs below a certain separation
radius, r_v, were allowed to collapse and thus were not counted in
the final results. Fig. 5 shows the average pair separation as a
function of primary energy. For this case the calculations were
made using the Fe (bcc) lattice and the collapse radius was taken
as two lattice distances. The upper curve of Fig. 5 shows the average pair separation when long replacement sequences are allowed.
As before, it may be seen that the results are different in the two
energy ranges above and below a minimum at about 75 eV. Again it
is reasonable to suppose that the pronounced maximum in the low energy range is due to primary replacement sequences which carry the
interstitial far from the vacancy. This is confirmed by the lower
curve in Fig. 5 for which large amplitude (0.2 lattice distances)
thermal vibrations were introduced so that replacement sequences
larger than three lattice distances did not occur. As may be seen,
destroying the primary replacement sequences greatly reduces the
average pair separation in the low energy range. Again there is
also a reduction in the average separation in the high energy range
indicating that the cascade replacement sequences contribute some
well separated pairs to the total damage remaining.

The results shown in the upper curve of Fig. 5 agree well with
the conclusions of Becker et al. [10] from their experimental work
(using electron irradiation of copper crystals). (As previously
mentioned the differences between the results for copper (fcc) and
iron (bcc) are believed to be negligible, at least qualitatively,
for the potentials used here.) These are (i) the average pair separation increases greatly as the maximum primary energy is reduced
below 30 eV. (ii) The average pair separation rises very slowly
(or not at all) with rising primary energy in the range above 75
eV. (iii) The large separations observed in the vicinity of 20 eV
are due to the production of long focussed replacement sequences.

Replacement Sequences in Order-Disorder Alloys

In the ordered state of an order-disorder alloy a focussed replacement sequence can produce disorder by being directed along a
lattice line which is itself an ordered arrangement of more than
one atomic species. Such replacement disorder can add to the observed damage produced by irradiation. Again such events are of
special interest because of recent experimental work on Ni_3Mn by
Becker et al. [11] and Blewitt et al. [12] The damage events were

simulated in Ni_3Mn using the Marlowe program. The damage calculated (after allowing pair collapse within two lattice distances) falls
into three categories: (i) the surviving vacancy-interstitial
pairs, (ii) "improper" replacements, i.e. replacements of a nickel
atom by a manganese atom or vice versa, which occurred during the
development of the cascade, and include single and double replacement events as well as longer replacement sequences, and (iii) the
"improper" replacements which result from the pair collapse process
when the "wrong" type of interstitial atom falls into a vacant lattice site. For a simplified approach, remaining pairs and "improper"
replacements were taken to give equal contributions to the observable damage (e.g. electrical resistivity) (see the discussion on
this point by Becker et al. [11]). Also the weighting given to each
atomic type for purposes of starting them as primary atoms was taken
to be just their atomic percentages in the lattice. This does not
quite correctly reflect the experimental situation, especially in
the case of thermal neutron capture damage (Blewitt et al. [12]).
With these assumptions the Marlowe program can be used to calculate
the ratio of the damage expected in ordered Ni_3Mn to the damage expected in pure nickel. Under the conditions of the calculation (no
thermal vibrations, no interaction of replacement sequences with
neighboring rows, and interatomic potentials and lattice spacing
the same as for copper [fcc]) large effects of long replacement sequences may be expected.

In order to compare the calculational results with the experimental data two different primary energies were used. Becker et
al. [11] irradiated with 1.0 to 3.0 MeV electrons; to simulate
events in this range a primary energy of 50 eV was chosen. Blewitt
et al. [12] used thermal neutron irradiation with a reported average
recoil energy of 400 eV; for comparison with this experiment, then,
a 400 eV primary was chosen. As may be seen from Fig. 2 these two
primary energies correspond to quite different total lengths of replacement sequences. A 50 eV primary is near the minimum and gives
a total length of about two atoms, while a 400 eV primary is well
up in the "cascade replacement sequence" region and gives a total
length of about 50 atoms.

The resulting damage ratios, however, do not differ so greatly.
For a 50 eV primary the damage ratio (ratio of damage in Ni_3Mn to
that in Ni) is about 2.6. This value agrees well with the results
reported by Becker et al. [11] For a 400 eV primary the ratio is
about 4.7. This result does not agree well with that reported by
Blewitt et al. [12] who give a damage ratio of 60. Consider an
arbitrary increase in the total length of all cascade replacement
sequences of length two atoms or more for the 400 eV cascade as an
attempt to obtain a damage ratio of 60. The total length required
would be of the order of 1000 atoms instead of the 50 atom total
length calculated. (If the resistivity contribution of a vacancy-
interstitial pair is taken to be twice that of an improper replacement, as may be suggested by the discussion in reference 11, a total

length of the order of 2000 atoms would be required.) This is particularly striking when it is recalled that the 50 atom length was
calculated under very favorable conditions for long replacement sequences; namely no losses to neighboring rows or thermal vibrations
were taken into account.

References

*Research sponsored by the U. S. Atomic Energy Commission under
contract with Union Carbide Corporation.

[1] R. H. Silsbee, J. Appl. Phys. 28, 1246 (1957).

[2] G. Leibfried, J. Appl. Phys. 30, 1388 (1959).

[3] P. H. Dederichs and G. Leibfried, Zeit. f. Physik 170, 320
(1962).

[4] J. B. Sanders and J. M. Fluit, Physica 30, 129 (1964).

[5] C. Erginsoy, G. H. Vineyard, and A. Englert, Phys. Rev. 133,
No. 2A, A595 (1964).

[6] (a) I. M. Torrens and M. T. Robinson, "Radiation Induced Voids
in Metals," edited by J. W. Corbett and L. C. Ianniello (U.S.A.
E.C., CONF-710601, 1972) p. 739.
(b) idem, "Interatomic Potentials and Simulation of Lattice
Defects," edited by P. C. Gehlen, J. R. Beeler, Jr., and R. I.
Jaffee (Plenum Publ. Corp., New York, 1972) p. 423.

[7] M. T. Robinson and I. M. Torrens, Phys. Rev. B 9 (1974), in
press.

[8] O. B. Firsov, Zh. Eksperim. Teor. Fiz. 36, 1517 (1959) [English
translation: Soviet Phys.-JETP 36, 1076 (1959)].

[9] G. Leibfried, Bestrahlungseffeckte in Festkörpern, B. G.
Teubner, Stuttgart, 1965, Section 7.2.

[10] D. E. Becker, F. Dworschak, and H. Wollenberger, Rad. Eff. 17,
25 (1973).

[11] D. Becker, F. Dworschak, C. Lehmann, K. T. Rie, H. Schuster, H.
Wollenberger and J. Wurm, Phys. Stat. Sol. 30, 219 (1968).

[12] T. H. Blewitt, A. C. Klank, T. L. Scott, and S. T. Macrander,
private communication.

RADIATION DAMAGE IN TRANSITION METAL HEXAHALO COMPLEXES:
THE APPLICATION OF ATOMIC COLLISION DYNAMICS IN HOT ATOM CHEMISTRY

K. RÖSSLER

*Institut für Nuklearchemie der Kernforschungsanlage Jülich GmbH
D-5170 Jülich, FRG*

and

M. T. ROBINSON

Solid State Division, Oak Ridge National Laboratory
Oak Ridge, Tennessee 37830, U.S.A.*

ABSTRACT

In transition metal hexahalo complexes
$K_2[MeX_6]$, the chemical consequences of "hot" atom
reactions can be observed in relatively simple
systems. The primary recoils, generated by nuclear
reactions or implanted as radioactive ions, can be
studied by radiochemical methods. The halogen recoils may appear either as free halide $^*X^-$ or as
labeled complex anion $[MeX_5{}^*X]^{2-}$. The hot atom
chemistry of these systems can be accounted for
in largely solid state physical terms, using computer simulation techniques. In general, excellent
agreement between the experimental data and the
model calculations can be obtained using reasonable
model parameters. This agreement shows that the
product distribution in these systems is controlled
by atomic collision dynamics and simple reactions
of a well-defined number of simple defects.

Radiation damage studies in metals and alkali halides do not
generally include chemistry, since the chemical states of displaced
atoms in such substances do not change significantly. In contrast,
hot atom chemistry makes use of radiochemical analysis to determine
the fates of the primary recoils generated by nuclear reactions or
by implantation of radioactive ions (for a review cf. [1]). Un-
fortunately, most of the systems studied hitherto are too complex

for a simple physical treatment. Transition metal hexahalocom-
plexes $K_2[MeX_6]$ (Me = Re, Os, Ir, Pt), however, lend themselves to
relatively simple observation of the chemical consequences of recoil reactions. Hot halogen distribution studies in $K_2[ReX_6]$ exhibit only two products: free halide $^*X^-$ and labeled parent complex $[ReX_5{}^*X]^{2-}$ ("retention") [2,3]. Electronic defects do not
seem to be involved in these processes [2,4]. Thus, the discussion of mechanisms can be restricted to the reactions of atomic
point defects. Kinetic analysis of hot halogen distribution in
the $K_2[ReBr_6]$-$K_2[ReCl_6]$ mixed crystal system reveals three different mechanisms: primary retention (the recoil atom is bound in
its original complex ion), direct replacement, and recombination
after ligand exchange, which correspond to zero, first and second
order chemical kinetics, respectively [5,6]. The simple product
spectrum of only monosubstituted complexes indicates that no "hot"
or "molten" zone is formed, but only comparatively isolated defects. The radiation damage consists mainly of ligand vacancies
and interstitials, probably $[ReX_5\square]^-$ and X^-, respectively.

The hot atom chemistry of these systems can be accounted for
in largely solid state physical terms, using computer simulation
techniques [7]. Binary collision cascades have been calculated
with the machine program "Marlowe" [8] originally developed for
radiation damage studies in metals [9]. The atoms are treated as
mass points; chemical bonds and ionic interactions are neglected.
Collisions of the atoms are governed by repulsive interatomic potentials of the screened Coulomb type. A displacement threshold
of 5 eV is assumed. Recombination of the nascent point defects
takes place when the interstitial-vacancy separation is < 0.8 a_0,
where a_0 is the edge of the cubic unit cell. In general, excellent
agreement between experimental data and model calculations is obtained [7]. This work presents a more detailed comparison of computed and radiochemical results with the aim of revealing the connection between atomic collision dynamics and hot atom chemistry
in complex ionic crystals.

$K_2[ReCl_6]$ and $K_2[ReBr_6]$ are used as standard systems, both of
which crystallize in a f.c.c. structure of the $K_2[PtCl_6]$ type with
a_0 = 9.843 Å and 10.387 Å, respectively; for further details see
[7]. Energetic halogens with some 100 eV recoil energy are created
by the nuclear processes listed in Table 1. A detailed comparison
of the fate of hot halogens as obtained from experiments [2-6] and
computer simulation [7] shows that the observed retention in poly-
and monocrystalline samples of pure $K_2[ReCl_6]$ (84,90 %) and
$K_2[ReBr_6]$ (88,91 %) is almost the same as the computed value of
94% including direct replacements and the collapse of close Frenkel
pairs. A similar agreement is found for the primary retention and
for the mean vector range.

The 50% probability for the direct replacement of halogen ligands points to an interesting mechanistic aspect. The final
product-determining steps seem to proceed in the low energy parts

Table 1
Comparison of experimental and calculated data
for recoil halogen cascades in $K_2[ReX_6]$.

	$^{37}Cl(n,\gamma)^{38}Cl/K_2[ReCl_6]$		$^{81}Br(n,\gamma)^{82}Br/K_2[ReBr_6]$	
	E_r = 300 eV		E_r = 80–100 eV	
	exp. [3,5]	sim. [7]	exp. [3,5]	sim. [7]
retention, %	84,90	85–94	88,91	85–94
primary retention, %	5	4	14	12
displacement, %	56	50	61,66	50
ligand vacancies	≥ 2.4	20.6, 5.8	> 0.66	7.1, 2.2
mean vector range, Å	> 15	20	$\approx$ 10	10

of the cascade. In this context the old billiard ball model proposed by Libby [10] should be called to mind. This conceived the
hot reactions to proceed via hard sphere collisions leading to 50%
retention. (A more accurate calculation [11] gives a retention of
$10^2 \ln 2$ = 69.3% for a hard sphere model and lower values for more
realistic interactions.) This model can only work in systems in
which most of the atoms are of the same kind as the recoil. In
addition the recombination of close defects has to be taken into
account. Experimentally, the 50% replacement probability is reflected in the first order terms from mixed crystal-system kinetic
analysis [5,6]: 56% (Cl→Cl), 61% (Br→Cl), 66% (Br→Br). These values are somewhat greater than the calculated 50% since some rapid
combination processes must probably also be included in these yields.
     According to the calculations about 7 transient ligand vacan-
cies are formed per 100 eV recoil energy (21 for $K_2[ReCl_6]$ and 7
for $K_2[ReBr_6]$), about 2 of which survive the mechanical collapse of
close pairs within the 0.8 a_0 range. The values agree reasonably
with the experimental findings [5,6]. The number of stable vacancies in this calculation is similar to that derived from the classical 25 eV displacement threshold. Thus, the choice of the low 5
eV threshold followed by spontaneous recombination of the close
pairs is an alternative procedure which finally leads to the same
results as the traditional approach. However, a lot of information
on the processes involved in retention formation is obtained by this
procedure. A linear dependence of the number of vacancies on the
primary energy is shown in Fig. 1. However, this holds only for
energies up to $\sim$ 1 keV. For higher energies the cascade becomes so
dense that the single collision sequences overlap each other. An
increased annealing of defects must be taken into account and the
curves in Fig. 1 may finally reach a plateau value.

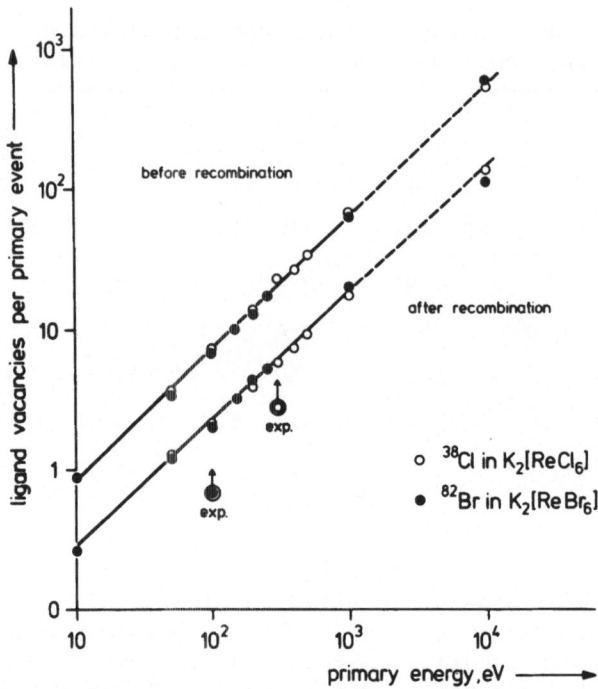

Fig. 1. Number of ligand vacancies created by energetic halogens in hexahalorhenates according to computer simulation.

Table 2
Radiohalogen retention after (n,γ) and
$(n,2n)$-reactions in hexahalometalates

System	Retention, %		
	X = F	X = Cl	X = Br
$K_3[RhX_6]$	–	$88.6 \pm .05$[b]	–
$K_2[ReX_6]$	81.1 ± 0.5[a]	84.5 ± 0.4[b]	87.8 ± 0.4[b]
$K_2[ReX_6]$[f]	–	89.7 ± 0.3[c]	91.0 ± 0.3[c]
$K_2[OsX_6]$	88.6 ± 0.4[a]	87.2 ± 0.5[c]	88.5 ± 1.0[e]
$K_2[IrX_6]$	81.3 ± 0.5[a]	88.7 ± 0.5[d]	83.0 ± 1.0[e]
$K_2[PtX_6]$	85.7 ± 1.5[a]	88.5 ± 0.5[c]	89.0 ± 1.0[e]

a) $^{19}F(n,2n)^{18}F$, [6]   b) (n,γ) and $(n,2n)$, [2,3]

c) (n,γ), this work   d) $^{37}Cl(n,\gamma)^{38}Cl$, [12]

e) $^{81}Br(n,\gamma)^{82}Br$, [13]   f) monocrystals

From Table 1 it is obvious that hot halogen reactions in hexahalorhenates can be satisfactorily interpreted by a collision dynamic treatment. However, it still remains questionable whether this
statement can be extended to other systems, e.g. the rest of the
hexahalo complexes. The small variation of the mass, when going to
other metal atoms (Os, Ir, Pt) certainly will not affect the results
of the calculation; furthermore, the similarity of the computed data
for $K_2[ReCl_6]$ and $K_2[ReBr_6]$ exhibits the fact that gross altera-
tions of the product spectrum can not be expected for a variation of
ligands from F to Br. But the influence of different structures,
bond energies and recoil energies for various nuclear processes can
not be ruled out so easily. Therefore, the recoil halogen product
distribution has been studied in several hexahalometalates with
classical chromatographic techniques, cf. [2,3]. The whole set is
compiled in Table 2 together with older data from our [2,3,6] and
other laboratories [12,13]. In all cases, even for $K_3[RhCl_6]$,
rather similar and high retentions ranging from 80 to 90 % have
been observed. A small increase is observed in going from fluoro-
to bromocomplexes; monocrystalline samples (grown from solution)
show an only slightly increased retention as compared to polycrystalline ones.

Thus, it may be concluded that neither the type of metal atom
or halogen ligand, the differences in chemical binding and bond energies, the different structures (hexagonal in the case of the hex-
afluorocomplexes and f.c.c. for the chloro- and bromocomplexes),
nor the different primary energies (ranging from some 10 to some
10^4 eV for the nuclear processes studied) influence the recoil retention greatly. Quite obviously, it is the general arrangement of
the atoms in the lattice, the stoichiometry, which governs the hot
halogen yield. A comparison of the volume fractions of the various
ions in the unit cell, according to Pauling's ionic radii [14], is
presented in Fig. 2. This shows that in all the complexes a large

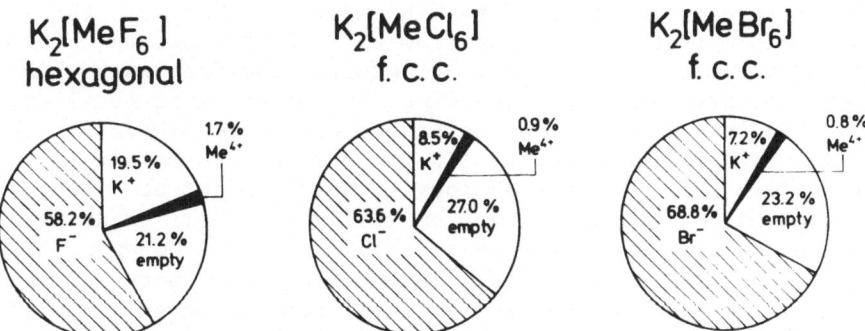

Fig. 2. Average volume fractions for ions in $K_2[MeX_6]$ (Me = Re,
Os, Ir, Pt), according to Pauling's ionic radii.

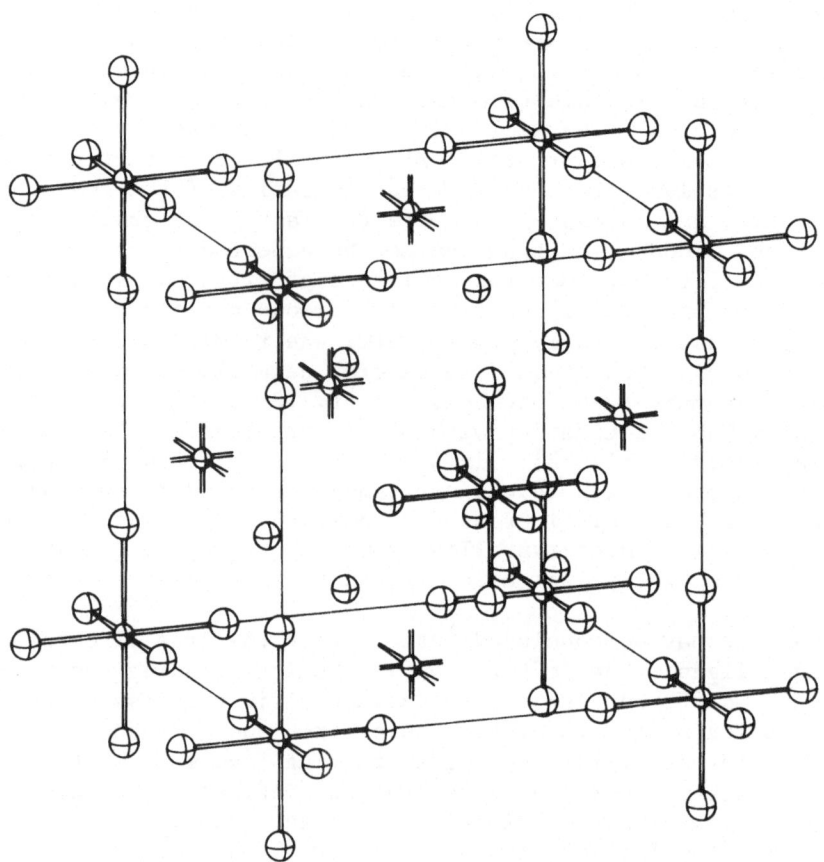

Fig. 3.   The unit cell of f.c.c. $K_2[MeX_6]$.

and rather similar volume of the lattice is occupied by the halogens. Only a slight increase is observed when going from fluoro-
(58%) to bromocomplexes (68%), which might be responsible for the
small increase in retention (cf. Table 2). Furthermore, the large
fraction of empty space ranging from 21 to 27 % shows that these
compounds possess a rather open structure. The frequency of halogen ions in these lattices resembles that of atoms in metal crystals. Also, the number of sites in the defect recombination volume
is rather similar for $K_2[MeX_6]$ (50 for 0.8 a_0) and metals (e.g., 48
in Pt [15]); cf. [7]. This analogy to metals might be a reason for
the rather high and equal retention yields and their quantitative
simulation by collision cascade calculations. The homogeneous distribution of defects, especially in the low energy parts of the
cascade, gives rise to the similar retentions observed for different primary energies.

Next to the calculation of hot atom distributions among different products, the spatial distribution of the interstitials

about the vacancies is interesting. A determination of the
interstitial-vacancy pair separations could give more information
on the distance at which rapid recombination occurs; an analysis
of the positions of defects which escape the collapse of close
pairs might help to elucidate the thermal annealing and solid state
exchange processes observed in these ionic compounds [2,3].

The lattice of the hexahalometalates does not allow a random
distribution of interstitials, but the primary recoil-vacancy separation should reflect the structure. The schematic representation
of the unit cell of f.c.c. $K_2[MeX_6]$ in Fig. 3 shows empty spaces
between the halogen ligands, which may serve as sites for interstitials. Empty regions within easy reach are found at distances 0.26
a_0, 0.56 a_0 and 0.75 a_0 from the ligand sites. Interstitials at
those positions as well as halogen ions on potassium sites (0.34 a_0,
less probably 0.61 and 0.66 a_0) are likely to recombine easily with
nearby vacancies. From the next farther open sites at 0.91 a_0 and
1.04 a_0 the vacancy cannot be reached so directly.

An extended calculation of 5000 cascades has been carried out
for each of the typical recoil processes (300 eV Cl in $K_2[ReCl_6]$ and
100 eV Br in $K_2[ReBr_6]$) to determine the distribution of the closest
primary recoil-vacancy distances with good statistics. The results
are shown in Fig. 4. The separation distribution can be resolved

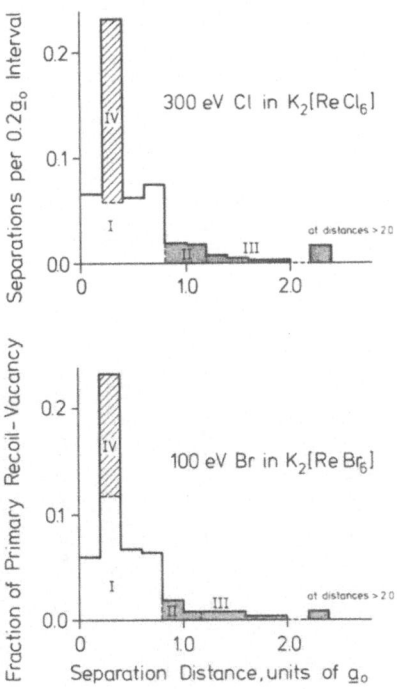

Fig. 4. Primary recoil-vacancy separations; average of 5000 events.

Table 3

The fate of recoil halogens in computer simulation
and experiments in monocrystalline $K_2[ReX_6]$

	position of halogen and possible reactions	distance to nearest vac., units of a_0	300 eV Cl in $K_2[ReCl_6]$		100 eV Br in $K_2[ReBr_6]$	
			sim.	exp.	sim.	exp.
	direct replacement	0.0	49%		53%	
I	interstitials, collapse of close pairs	≤ 0.8	27% (93%)	90%	31% (95%)	91%
	halogens on K-sites,					
IV	collapse of close pairs	0.34	17%		11%	
II	interstitials, thermal recombination	0.8–1.2	3.6%	2–3%	2.6%	3–4%
III	interstitials, annealable by bulk reactions	≥ 1.2	3.4%	3–4%	2.5%	3–4%
	trapped halogens, nonannealable	---	---	4.5%	---	4.0%

into four fractions by comparing the calculations with experiments. This is done in Table 3. Fraction I is attributed to the close interstitials at distances < 0.8 a_o. Fraction II is attributed to those interstitials which survive rapid recombination, but on thermal annealing recombine nevertheless. Fraction III are those interstitials at distance ≥ 1.2 a_o which anneal by bulk reactions such as solid state exchange of ligands, cf. [2,3]. Finally, fraction IV is assigned tentatively to halogen ions on potassium sites (0.34 a_o). The comparison with experimental data from annealing and solid state exchange studies in $K_2[ReX_6]$ monocrystals shows a reasonable agreement for the sum of fractions I, IV and direct replacement, as well as for the fractions II and III. In the experiments a fraction of nonannealable halogen is observed. This is due to irreversible trapping at surfaces, dislocations etc., and is not considered in the machine program. The calculated retention is slightly greater for the hexabromo- than for the hexachlorocomplex, again reflecting the greater volume fraction of bromine atoms in the lattice and, thus, the greater cross section. The difference in the K replacement by Cl and by Br can be explained by the difficult energy transfer between atoms of different mass. Although the assignment of fractions II and III has to be treated with great care, at least some stronger evidence for the postulated 0.8 a_o recombination radius can be obtained from this study.

The computer simulation has been applied so far to events whose primary energies far surpass the displacement threshold. At lower recoil energies, the influence of the displacement energy increases considerably and it is interesting whether the calculations will still fit the experimental data. Low energy recoils are encountered as a consequence of the isomeric transitions of metastable nuclides, which in general deliver sufficient energy for the breakage of bonds. However, the $^{80m}Br(IT)^{80g}Br$ process in $K_2[ReBr_5{}^{80m}Br]$ [16] and in $K_2[ReBr_5{}^{80m}Br]$ – $K_2[SnCl_6]$ mixed crystals [17] leads to 100% retention. Similarly, equal retentions have been found for the $^{79}Br(n,\gamma)^{80g}Br$ and the $^{81}Br(n,\gamma)^{82m}Br(IT)^{82g}Br$ processes in $K_2[ReBr_6]$ [2,3]. This may indicate either that the bonds are not broken at all or that immediate recombination takes place after transient bond rupture. The latter seems highly probable, since it reflects the restoring force on the interstitial ions, which already has been made responsible for the lack of ionizing radiation effects [2,4]. The computer simulation has been performed for 10 eV Br recoils in $K_2[ReBr_6]$. The results are shown in Fig. 5. No primary recoil vector ranges and no interstitial-vacancy separations greater than 0.8 a_o are found. Both distribution curves exhibit a similar dependence on the distance, indicating that the interstitials are all correlated with the primary vacancy and thus can undergo recombination. A problem is posed by the fraction of 49% ligand replacement calculated. Certainly, the assumption of lower primary energies than 10 eV would settle this problem. However, according to [17] the 10 eV represents the lower limit of

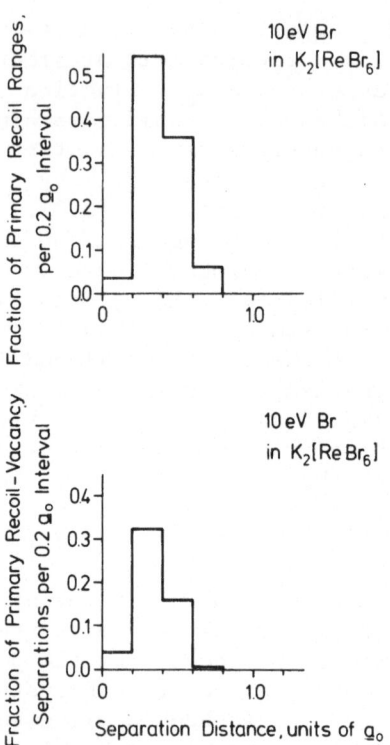

Fig. 5. Range distribution and separation to nearest vacancies for a 10 eV Br-recoil in $K_2[ReBr_6]$.

recoil energies generated by (IT)-processes with subsequent Auger effect. Thus, either the displacement threshold assumed is too low or the recoil atom replaces its neighbors in its own complex. This remains to be clarified by later calculations.

The calculations of low energy events reflect the general situation quite reasonably, but the simple nature of the program leaves at least some problems unsolved. In this context some obstacles should be mentioned which impede the extension of the rather successful calculations in the halogen recoil case to other systems. The extremely simple halogen recoil with its relatively mild disrupture of the molecular structure and its uncomplicated chemistry hides some of the difficulties which will be encountered when going to most of the classical hot atom systems. This can best be demonstrated in the case of the Re atom recoil in the hexahalorhenate complexes. A comparison of experimental and computed data is given in Table 4. It can be seen that after the $^{185}Re(n,\gamma)^{186}Re$ (E_R = 30 to 50 eV) nuclear reaction the Re-retentions increase from the hexachloro- to the hexaiodocomplex. The greater probability of

Table 4

30–50 eV Re recoils in $K_2[ReX_6]$ in experiments and computer simulation [7].

System	prim.retention %		retention %		mean vector ranges units,of a_0	
	exp.	sim.	exp.	sim.	exp.	sim.
$K_2[ReCl_6]$	$\sim$ 18 [18]	< 7	61±6[19-22]	$\sim$ 20	$\approx$ 0.5 [24]	1.0-1.3
$K_2[ReBr_6]$	$\sim$ 11 [18]	< 22	76±6[20,23]	$\sim$ 40	$\approx$ 0.5 [24]	0.8-1.1
$K_2[ReI_6]$	--	--	87±5 [20]	--	--	--

retention is certainly responsible for this increase, due to the better energy transfer from the metal ion to the ligands as their masses increase. The calculations show the right trend, but nevertheless do not agree with the absolute experimental values. Three effects are responsible for this dissimilarity: First, the greater disruption of the complexes when the central atom is removed; second, the complicated chemistry of the polyvalent Re, which even when on an interstitial site can reform its ligand shell by the attraction of ligands; third, and we think this to be a major point, the atoms in a chemical system do not occupy sites with equal energy states. Some atoms may be situated in energetically deep traps and exhibit greater displacement thresholds than others. This is the case for the complexed metal atoms such as Re. It is not only closely surrounded by its ligands and thus sterically hindered for motion more than other atoms in the lattice, but it is also connected to the ligands with six bonds of about 1.5 and 1.8 eV each in the case of hexabromo- and chlororhenate, respectively [3]. A Re recoil escaping from its ligand shell thus has to overcome about 9 to 11 eV in ligand binding energy. Thus, the retention and range calculated for the Re recoils correspond to those of particles more energetic than is really the case. Table 4 shows indeed that the recoil ranges calculated are greater than those obtained from mixed crystal experiments [24]. Future calculations must be modified to allow for these binding effects. Furthermore the rearrangement of ligands to interstitial Re atoms can be taken into account by a detailed calculation of Re-halogen interstitial separations and the recombination probability of the halogens. Thus, even the complicated central atom recoil may be treated satisfactorily in the future.

The calculations show clearly that hot atom reactions, at least in these simple systems, are governed entirely by collision steps and simple reactions of a well defined number of single defects. There is no need to invoke the concept of a "hot spot" ("thermal spike") surrounding the primary recoil in which chemical reactions take place. In fact, a simple analysis shows that the

primary recoil comes to rest in no more than about 10^{-13} sec after
its creation. Such times are far shorter than those for which it
is appropriate to speak of thermal motions, diffusive jumps, and
the like.

References

*Operated for the U. S. Atomic Energy Commission by the Union
Carbide Corporation.

[1] G. Stöcklin, Chemie heisser Atome, ("Chemische Taschenbucher,"
Bd. 6, Verlag Chemie, Weinheim/Bergstr., Germany, 1969);
"Chimie des atomes chauds" (Masson et Cie, Paris, 1972).

[2] R. Bell, K. Rössler, G. Stöcklin, and S. R. Upadhyay, Jülich
Report Jül-625-RC (1969).

[3] R. Bell, K. Rössler, G. Stöcklin, and S. R. Upadhyay, J. Inorg.
Nucl. Chem. 34, 461 (1972).

[4] R. Bell and G. Stöcklin, Radiochim. Acta 13, 57 (1970).

[5] K. Rössler, J. Otterbach, and G. Stöcklin, J. Phys. Chem. 76,
2499 (1972).

[6] J. Otterbach, Jülich Report Jül-832-RC (1972).

[7] M. T. Robinson, K. Rössler, and I. M. Torrens, J. Chem. Phys.
60, 680 (1974).

[8] I. M. Torrens and M. T. Robinson, "Radiation Induced Voids in
Metals," edited by J. W. Corbett and L. C. Ianiello (U.S.A.E.C.
publication CONF-710601, 1972), 739; "Interatomic Potentials
and Simulation of Lattice Defects," edited by P. C. Gehlen, J.
R. Beeler, Jr., and R. I. Jaffee (Plenum, New York, 1972), 423;
M. T. Robinson and I. M. Torrens, Phys. Rev. B 9 (1974), in
press.

[9] M. T. Robinson, "Radiation Induced Voids in Metals," edited by
J. W. Corbett and L. C. Ianiello (U.S.A.E.C. publication
CONF-710601, 1972), 397.

[10] W. F. Libby, J. Am. Chem. Soc. 69, 2523 (1947).

[11] P. H. Dederichs, C. Lehmann, and H. Wegener, Phys. Stat. Solidi
8, 213 (1965).

[12] W. J. van Ooij, Thesis, Technische Hogeschool Delft (1971).

[13] W. Herr, K. Heine, and G. B. Schmidt, Z. Naturforsch. 17a, 590
(1962).

[14] L. Pauling, "The Nature of the Chemical Bond" (Cornell Univ.
Press, Ithaca, N. Y., 1960), p. 514.

[15] H. J. Wollenberger, "Vacancies and Interstitials in Metals,"
edited by A. Seeger, D. Schumacher, W. Schilling, and J. Diehl
(North-Holland Publ. Co., Amsterdam, 1970), p. 215.

[16] G. B. Schmidt and W. Herr, "Chemical Effects of Nuclear Transformations," (IAEA, Vienna, 1961) Vol. 1, p. 525.

[17] H. Müller and D. Cramer, Radiochim. Acta, 14, 78 (1970).

[18] H. Müller, J. Inorg. Nucl. Chem. 27, 1745 (1965).

[19] W. Herr, Z. Elektrochem. 56, 911 (1952).

[20] G. K. Schweitzer and D. L. Wilhelm, J. Inorg. Nucl. Chem. 3, 1 (1956).
[21] D. J. Apers and A. G. Maddock, Trans. Faraday Soc. 56, 498 (1960).
[22] K. Rössler, to be published.
[23] H. Müller, Naturwissensch. 49, 182 (1962).
[24] H. Müller, "Chemical Effects of Nuclear Transformations," (IAEA, Vienna, 1965) Vol. 2, p. 359.

MONTE CARLO SIMULATION OF BACKSCATTERING PHENOMENA

D. K. HUTCHENCE and S. HONTZEAS
University of Saskatchewan
Division of Natural Sciences & Mathematics
Regina, Canada

ABSTRACT

A Monte Carlo simulation program (BACKS) has been written in such a way as to be efficient for the generation of backscattering data for particle energies of approximately one keV to two MeV. The program is several thousand times as efficient as conventional programs in generating backscattering data at higher energies due to an information return analysis routine involving particle generation over selected sub-areas of the crystal face.

The program takes into account energy losses and thermal motion. It allows for a nearly complete range of beam to crystal face angles. Most crystal forms can be used as well as crystals containing several different kinds of atoms.

The data output includes a depth of information distribution energy loss spectrum, a number of collisions, distribution, and a backscattering pattern.

Among the results obtained from the program is a quantitative prediction of a "classical" isotope effect. The results obtained also shed light on the effects of changing particle energy and crystal structure.

The program is seen as forming a useful bridge between theoretical approaches and experimental results since the effects of change in theoretical values can readily be used to generate backscattering information which can be compared with experimental results.

Description of the Program

A computer program has been written that is capable of producing backscattering results efficiently for a wide range of particle energies (approximately 1 keV to 2 MeV). The lower energy limit is due to only binary collisions being considered and the upper energy limit is determined by the non inclusion of relativistic effects. While the particle identity may be fairly freely chosen it is at present limited to having the mass of the particle less than the mass of target atom. Crystal type, composition, and temperature may be freely varied. Beam to crystal axis angles and crystal face slope are input determined. Sample size, the depth to which particles are followed, and the number of collisions through which a particle is followed are also input determined.

The program produces as output distributions of the number of collisions, angles of scattering events, and deepest lattice layer reached for the backscattered particles. An energy loss spectrum for the totality of backscattered particles is produced as well as a backscattering pattern. Because of limited availability of computer facilities samples sizes and hence output resolution have been limited. The resolution may, however, be readily improved. Statistics on crystal reflectivity, regions of backscattering information and program efficiency for the given run are also output.

The present program uses a Bohr potential (screened coulombic potential) as a particle-atom interaction potential. Evaluation of scattering is done by use of the method of matched truncated potentials developed by Lehmann and Robinson [1]. Collision energy losses are evaluated by standard classical methods. Electronic energy losses in the low and moderate energy regions are evaluated by a formula developed by Lindhard, Scharff and Schiott [2]. High energy electronic energy losses are evaluated using a Bethe-Block formula [3].

Temperature simulation for the crystal is done by randomly picking atomic displacements from a triangular distribution. The r.m.s. amplitude of the displacement is based on results based on and in agreement with x-ray work [4].

The feature which enables the program to function efficiently for high particle energies is the manner in which particles are generated. The particles are generated by randomly picking particle coordinates over the surface of the crystal within successive limited areas of generation. The initial area of generation is the area defined by the particle-atom separation at which ninety degree scattering of the particle would occur. The succeeding areas of generation are annulii of progressively greater radii. The particle coordinates are chosen so as to ensure a uniform areal density.

The logic of the above method is best understood by considering a beam of particles parallel to a crystal atomic string. Particles generated directly over the centers of the strings must necessarily

be backscattered. Particles generated in the centers of channels
will be channeled indefinitely so that as far as backscattering is
concerned there is no point in generating them. The program determines the areas in which it is profitable to generate particles
by starting with an area in which nearly all particles are backscattered and then proceeding to areas of progressively lower

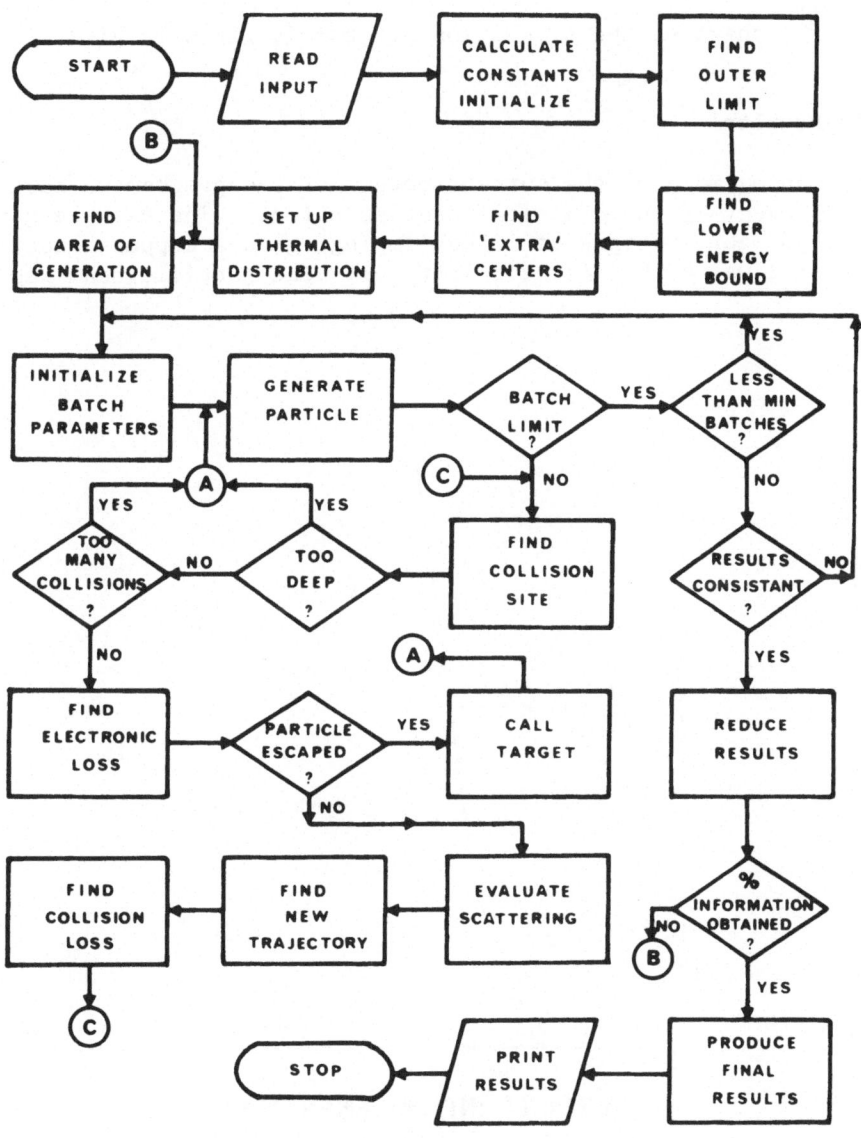

Fig. 1. Flowchart for BACKS

information return until the amount of information being returned
is negligibly small in comparison to that already obtained. Results from each area of generation are weighted according to the
information content of the area (information density expressed as
percentage particles returned times the area of the area of generation) and summed to produce the final results. In general the
area surveyed is only a small fraction of the total crystal surface. When the particle beam is not parallel to the crystal atomic
strings it is necessary to locate the additional atoms in the
strings that become visible to the beam and to generate particles
for them.

A flowchart of the simulation program is given in Figure 1.

Sample Results

As an example of the current output of the program we have
chosen a study of the classical isotope effect. The example given
is for hydrogen isotopes at 10 keV impinging on a copper monocrystal.
The crystal was at 0°K (static lattice). The particle beam was

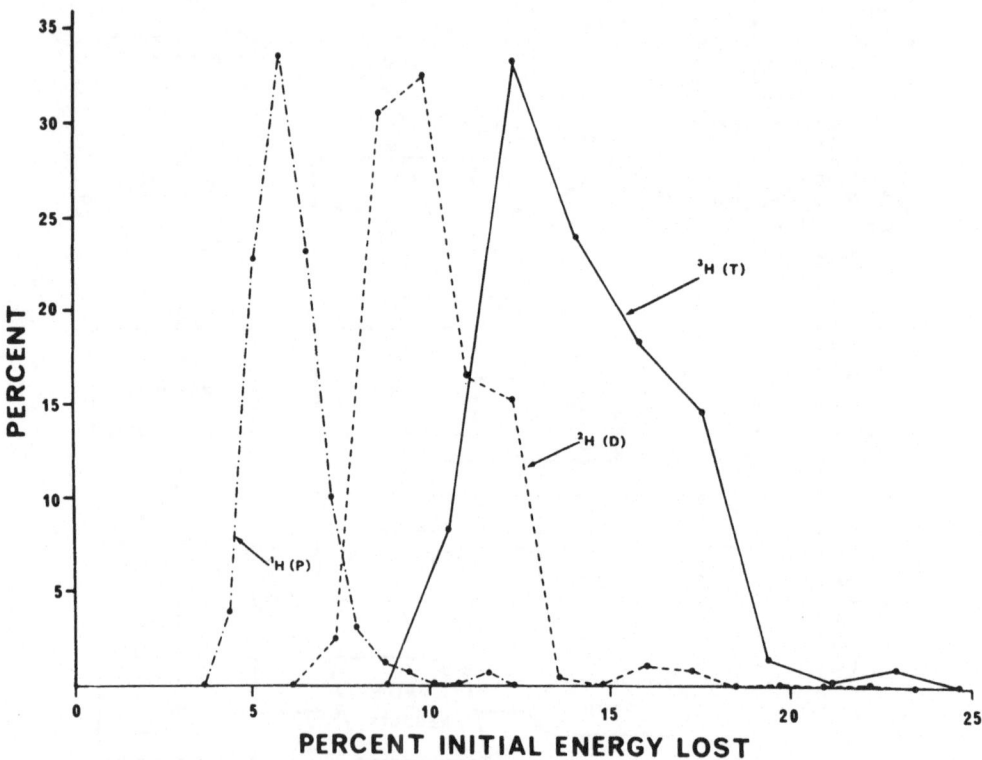

Fig. 2. Energy loss spectra for hydrogen isotopes in Cu (10 keV).

orthogonal to the (100) plane of the crystal. The same sequence
of random numbers was used for particle generation in all cases.
The number of backscattered particles was approximately 1000 particles in each case.

In the present model the energy losses and angle of scattering
for a given particle-atom separation are dependent upon the mass
of the particle. Since the isotopes of a given atomic number all
have different masses one may anticipate a difference in backscattering behaviour among the isotopes of a given species. As this
effect depends only on mass differences which are used in classical
computations we have called the effect the classical isotope effect.
It is not known at present what relationship this effect has with
the quantum mechanical isotope effect put forward by Chadderton [5]
and others [6].

Because of rather small sample sizes, the limited development
of the present program and the limited resolution of the outputs
it seems unwise to attempt any in depth analysis of the results.
In terms of establishing the existence of the effect all that is
required is that the results for each isotope be different. This
objective appears to have been obtained.

Figure 2 shows the energy loss spectra for the three hydrogen
isotopes. The spectra are all dominated by a single large peak.
The expected shift in principal peak position does occur. The
principal peaks are also quite different in width and shape. The
larger the isotope mass the greater the peak width. The H^1 peak
is quite sharp, whereas the H^2 and H^3 principal peaks have definite
shoulders on the high energy loss sides suggesting that one or
more additional peaks may be present.

Figure 3 shows the distributions of number of collisions
undergone by the backscattered particles for each of the isotopes.
There are clearly dissimilarities among the distributions. There
is a possibility of trends as indicated by the occurrences of three
and four collisions. The lighter the isotope the lower the prob-
abilitiy of having undergone three collisions and the greater the
probability of having undergone four collisions. The reason for
the oscillating appearance of the distributions is not known.
Statistical fluctuations are thought to be a likely contributing
factor.

Backscattering patterns are given for H^1 (Figure 4) and H^3
(Figure 5). The results are expressed as the ratio of particles
found in a given area to particles expected had the backscattering
distribution been uniform. The two backscattering patterns are
dissimilar in many details. The H^2 backscattering pattern (not
shown) was different from those of H^1 and H^3. It must be borne
in mind that certain features are predominantly due to the structure
of the crystal. Indeed a pattern of blocking and focusing is
apparent for all of the backscattering patterns. The blocking and
focusing occuring at low latitudes in the middle of the quadrant
is observed for all helium and hydrogen isotopes in copper.

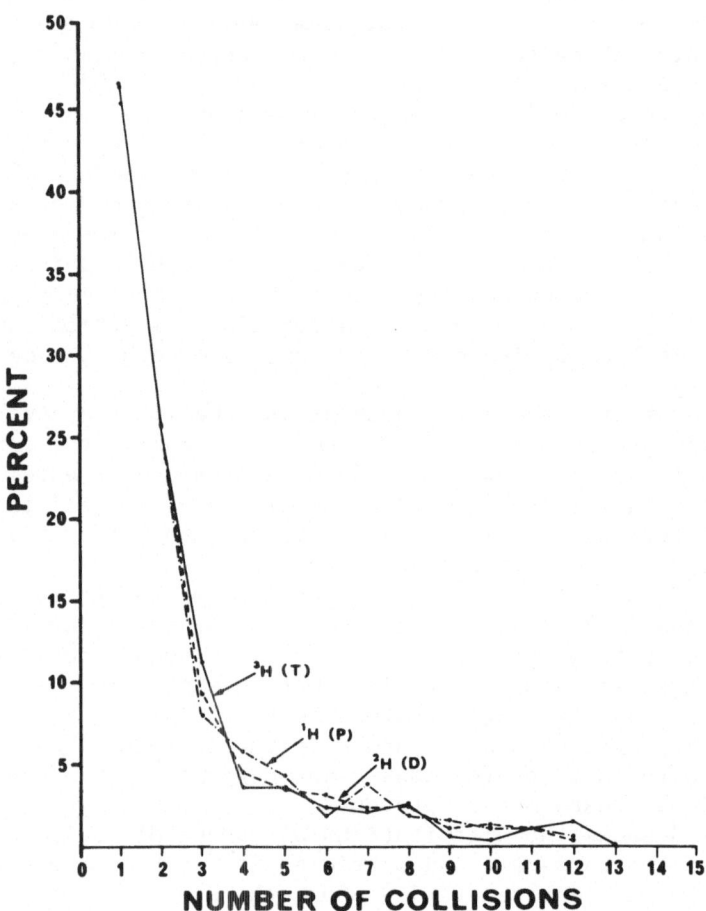

Fig. 3. Number of collisions distribution. Hydrogen isotopes in Cu (10 keV).

Use and Development of the Program

While it may ultimately be possible to produce backscattering results by use of a simulation program more quickly and more economically than by experimentation this appears to be a goal that is not likely to be reached soon. It is rather in conjunction with experimental results that the program is likely to be of most value. The accuracy of the program is dependent upon the expressions used for various model parameters such as the potential of interaction and electronic energy losses. By studying the changes produced in the program outputs by varying these parameters one may find the expressions yielding the best comparison with experimental results. Accurate expressions for such parameters are of

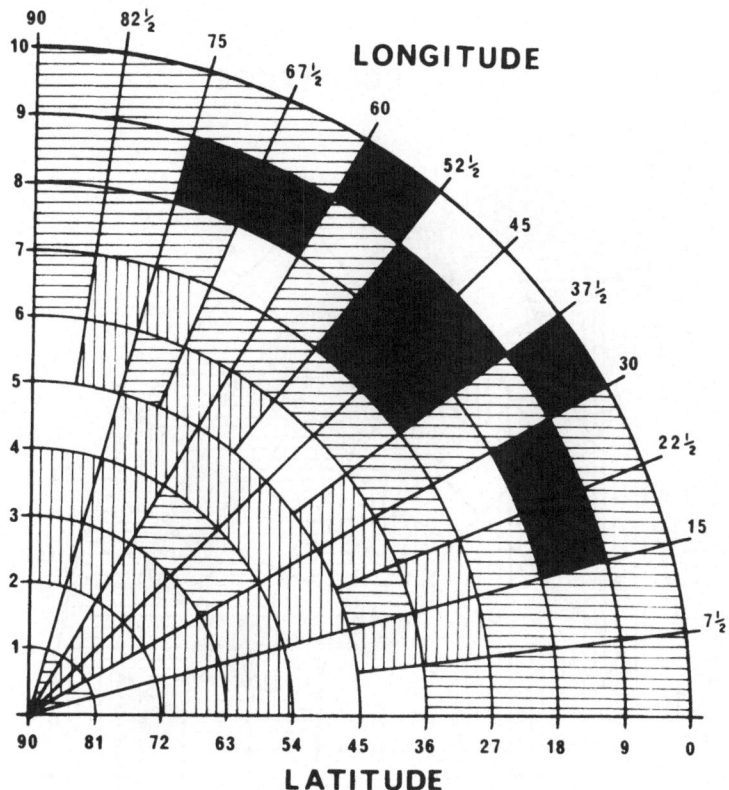

Fig. 4. Backscattering pattern of protons in Cu (10 keV).

widespread interest. The program is also capable of producing a
wide variety of descriptions of processes going on inside the
crystal and thereby aiding theoretical understanding of the backscattering phenomena.

Use of larger sample sizes and increased resolution of outputs
are goals to be sought shortly. Studies of model sensitivities
to model parameter changes should also be undertaken. Conceivably
some additional outputs aimed at further elucidation of the backscattering process may be added.

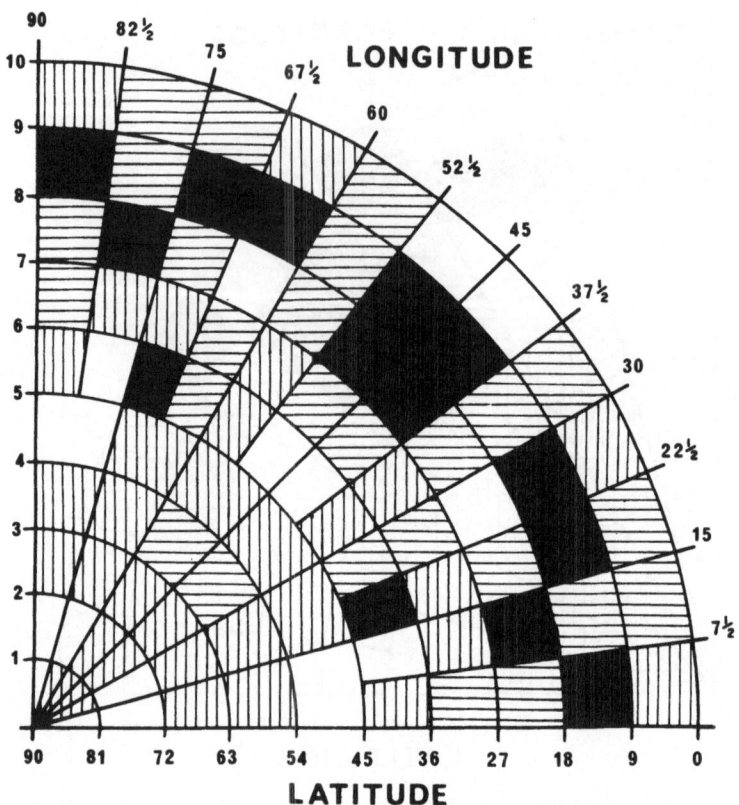

Fig. 5. Backscattering pattern of ^{3}H(T) in Cu (10 keV).

References

*This work is supported by the National Research Council of
Canada.
[1] C. Lehmann and M. T. Robinson, Phys. Rev. 134, A37-44 (1964).
[2] J. Lindhard, M. Scharff and M. E. Schiott, Mat. Fys. Medd. Dan.
    Vid. Selsk. 33 (14), 1-42 (1963).
[3] J. Lindhard, Mat. Fys. Medd. Dan. Vid. Selsk. 34 (14), 1-63
    (1963).
[4] R. W. James, "The Optical Principles of the Diffraction of
    X Rays," p. 236 ff, G. Bell and Sons Ltd., London (1967).
[5] L. T. Chadderton, Phys. Letters 23 (5), 303-4 (1966).
[6] G. Foti, F. Grasso et al., Phys. Letters A., 31 (4), 214-15
    (1970).

SECTION IV
SCREENING AND
CHARGE EXCHANGE

ION SCREENING IN SOLIDS[*][†]

WERNER BRANDT
Department of Physics
New York University
New York, New York 10003

Introduction

In 1950, Hall [1] reported the results of an investigation of charge states in proton beams emerging from metals with energies between 20 and 400 keV, as summarized in Fig. 1. If one defines the neutron beam fraction

$$\Phi_o = \frac{n_o}{n_+ + n_o} \tag{1}$$

in terms of the number of protons, n_+, and of hydrogen atoms, n_o, in the emerging beam, the ratio $R = (1 - \Phi_o)/\Phi_o$ plotted versus the projectile kinetic energy given in units of $E_0 = 25$ keV, or the projectile velocity, given in units of $v_0 = e^2/\hbar = 2.18 \times 10^8$ cm/sec (upper scale), rises rapidly. Hall summarizes: "For the four media studied, the beam composition is almost identical, being indistinguishable in the three lighter media and slightly more neutral in gold at ion speeds greater than two v_o. Existing theory offers no explanation for this lack of independence on medium."

In 1969, Groeneveld and Kaminsky [2] measured Φ_o in beams of deuterons emerging from foils of Al and Ni in the velocity range $1.4 < (v_1/v_o) < 2.2$. The exit surfaces were either clean or covered with various materials. The data shown in Fig. 2 exhibit no discernible differences. Last year, Betz [3] published a compilation of average projectile charges, $\bar{q}_S$, in beams of heavy ions emerging from solid foils and finds a universal trend, independent of the foil material, if the data are reduced in a suitable manner relative to the velocity, v_1, and the atomic number, Z_1, of the projectiles (Fig. 3). A similar plot obtains for the average charge number, $\bar{q}_G$, of heavy ions emerging from gas targets (Fig. 4) [3], but the v_1 and Z_1 dependences of $\bar{q}_S$ and $\bar{q}_G$ differ. Betz summarizes: "No quantitative theory is available for average equilibrium charge states which are produced by solid targets. For many light ions, $\bar{q}_S$ differs very little from $\bar{q}_G$, but for heavy ions, $\bar{q}_S$ exceeds $\bar{q}_G$ often by more than

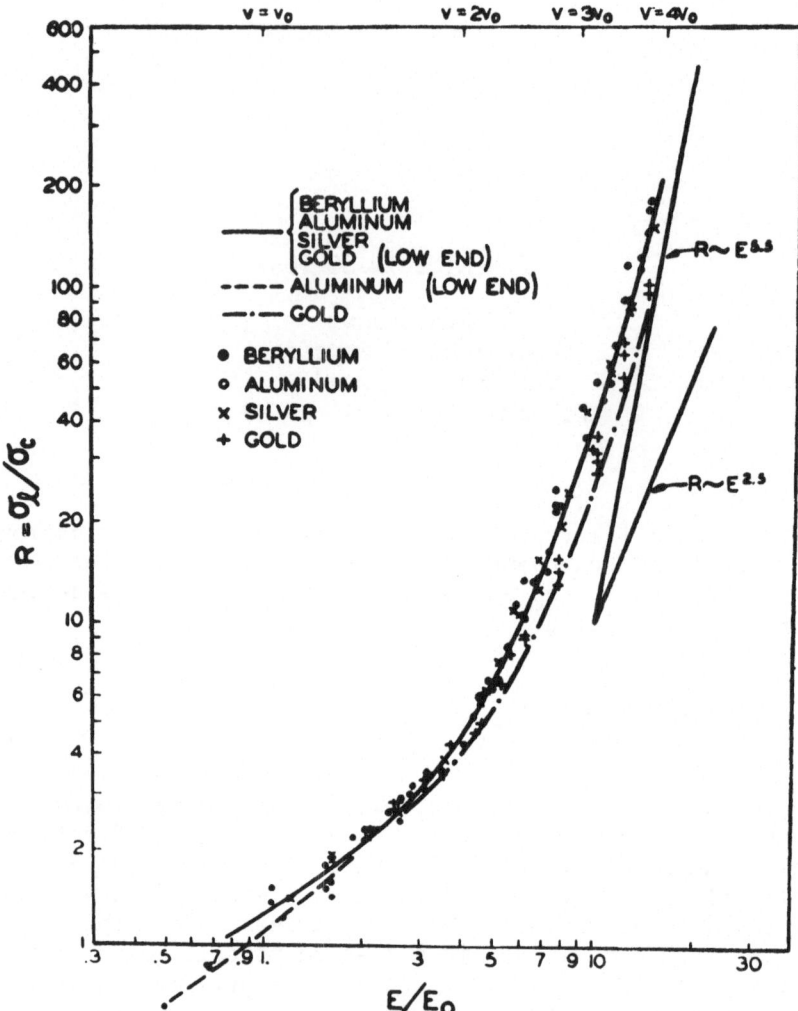

Fig. 1. Ratio $R = n_+/n_o$ in proton beams emerging from various
solids [1].

a factor two. Interestingly, the mechanism of that effect has not
been completely explained, and it is still being disputed whether
the large increase is produced <u>inside</u> or <u>outside</u> the solid."

    While charge states in emerging beams are well defined observ-
ables, no such conditions prevail inside solid targets. For example,
in first Born approximation the energy loss of a particle per unit
distance traversed, i.e., the stopping power $(-dE_1/dx)$, is given by
the product

$$- \frac{dE_1}{dx} = n \ q_1^2 \ S(v_1, Z_2), \qquad (2)$$

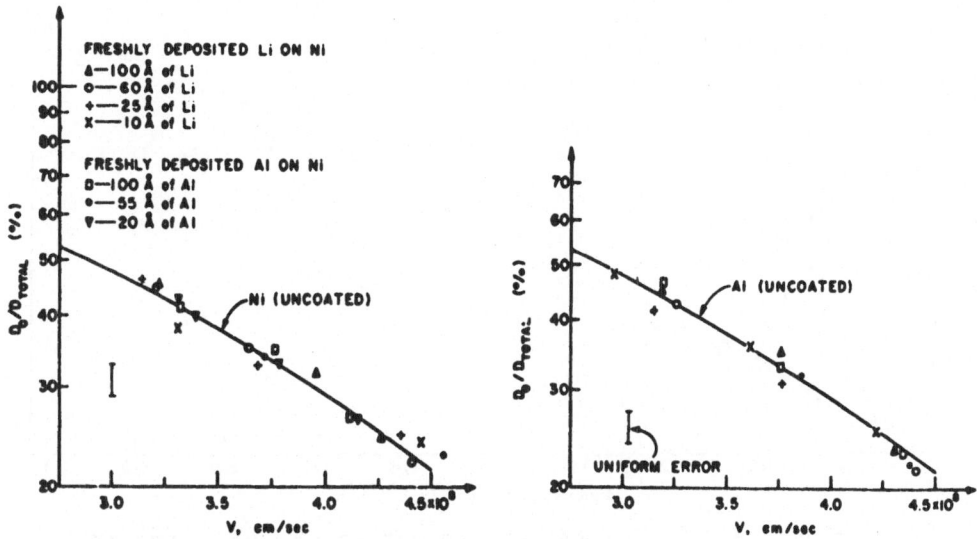

Fig. 2. Ratio R in deuteron beams emerging from Ni and Al foils with a variety of exit-surface conditions [2].

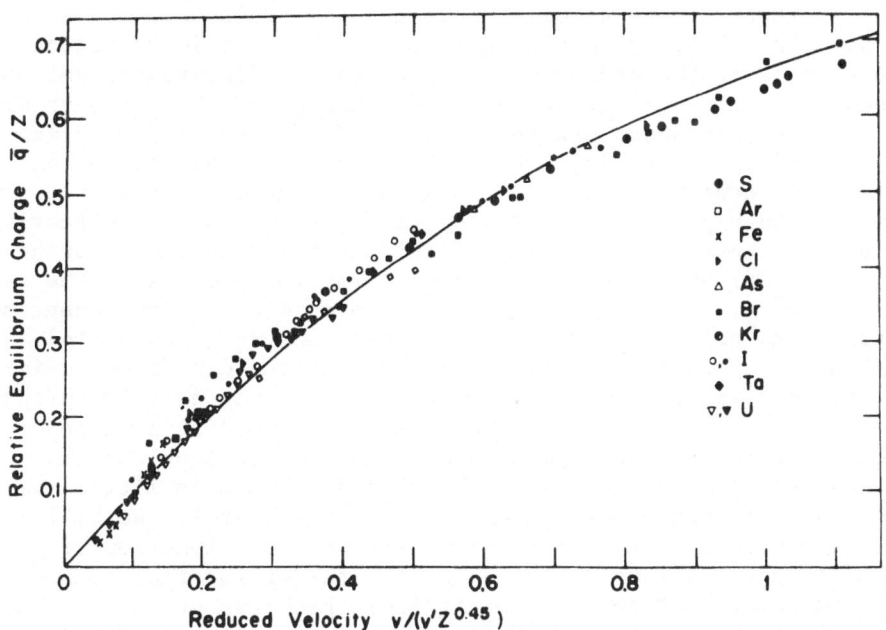

Fig. 3. Average projectile charge of heavy ions emerging from foils of carbon (full symbols) and other solids (open symbols) [3].

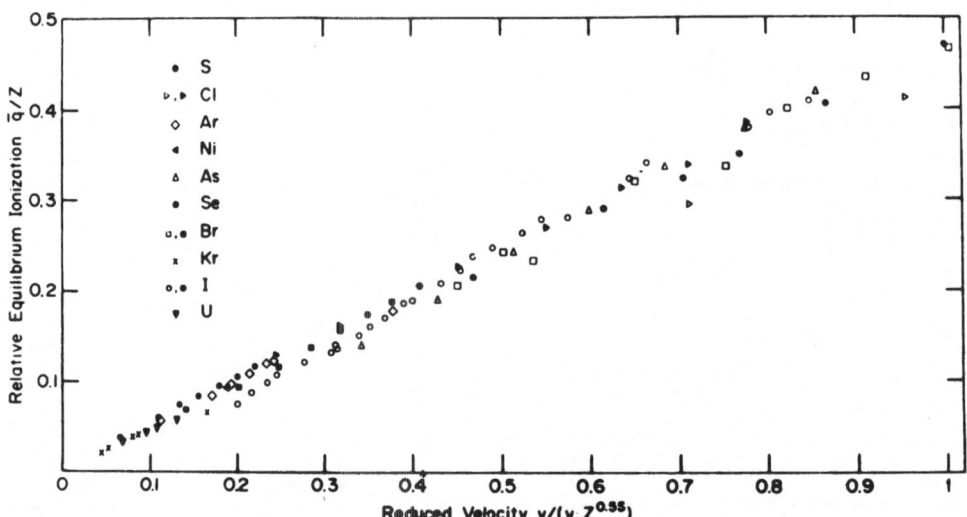

Fig. 4. As Fig. 3, but for gaseous targets of N_2, O_2, and air [3].

where q_1 is the charge number of the projectile, n the target atomic
density, and $S(v_1,Z_2)$ a function only of the projectile velocity v_1
and of the target atomic number Z_2. A residual density effect in
S becomes important at nearly relativistic projectile velocities.
When $v_1 \gg Z_1 v_0$, where $Z_1 v_0$ is the orbital velocity in the K shell
of the projectile with atomic number Z_1, the projectile moves as a
bare point particle, of charge number $q_1 = Z_1$. At lower velocities,
the electrons of the medium can respond to the disturbance set up
by the projectile and screen the moving charge, either through electron capture into bound states or by the formation of a polarization
cloud around the particle. This effectively reduces the Coulomb
interaction between the particle and the target electrons and,
hence, the energy loss. One can operationally define an effective
mean squared charge number q_1^2 by dividing the measured stopping
power $(-dE_1/dx)$ by the known stopping power for protons of the same
velocity, $S(v_1,Z_2)$. Remarkably, q_1^2 is practically independent of
the target material and depends only on v_1 and Z_1 (Fig. 5) [4].
      These and other observations imply that the dynamic behavior
of swift ions in dense media is insensitive to the detailed properties of the media. It is often assumed that the distribution of
charges measured in beams emerging from solids is equal to the effective charge distribution inside the solids. There is little evidence that this conjecture is correct or even useful. We will not
address here the question to what extent the wave function of a
projectile carries information about the target interior to an external detector. It touches on difficult problems of the quantum
theory of measurements. We shall take a pragmatic approach in describing the conditions encountered by a charged projectile in a

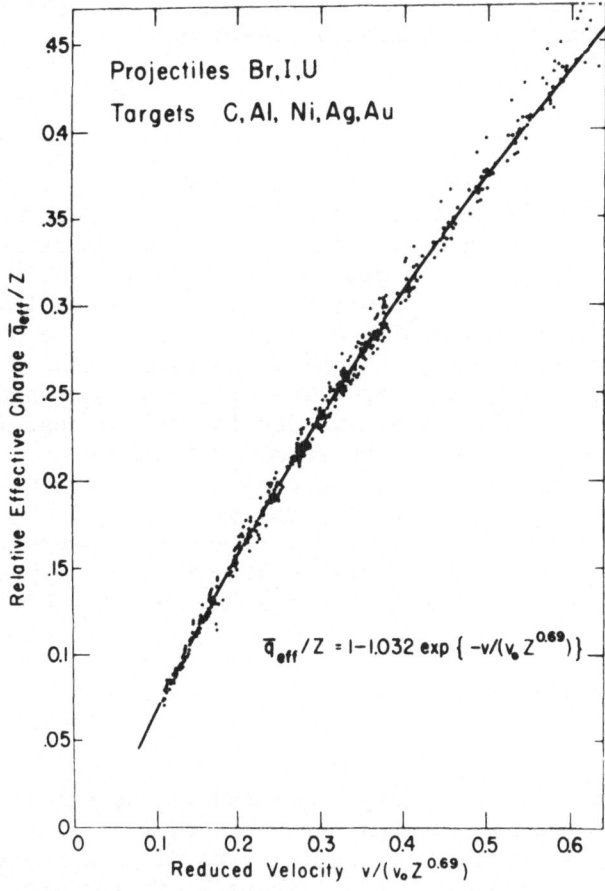

Fig. 5. Effective charge of Br, I, and U ions deduced from energy loss measurements in solids [3,4].

solid and summarize some of the detectable consequences of these conditions. The emphasis is on the nonscaling aspects of the phenomena in solids. By this we mean aspects that cannot be deduced adequately by merely scaling, according to density, data relevant to the same phenomena in gases. The most important solid-state effects in this context are the advent of plasma oscillations and the screening of the projectile charge, owing to the response of the valence electrons to the perturbation set up by the moving particle. The derivation of the response function of an electron gas at metallic densities is one of the most challenging many-body problems in solid state physics. Despite its inherent complexity, certain simple approximations can be employed to develop reliable guidelines for the important quantitative trends of the screening

of charged particles in solids. This may aid the planning of experiments and the interpretation of their results.

Static Screening

     To fix ideas, let us place a heavy point charge, of charge
number Z_1, at rest in a metal. The metal may be described by the
jellium model consisting of a uniform positive background, representing the ion cores of the metal atoms, and an electron gas of
density ρ_0. It is customary to characterize the density of the
electron gas by the radius of the Wigner-Seitz sphere occupied in
the mean by each electron of the gas, $r_s a_0$, in units of the Bohr
radius $a_0 = 1$ a.u. $= 0.53$ Å [5]. The values of r_s range between 2
and 6. The valence electrons have density values near 10^{-2} a.u. and
Fermi velocities v_F close to the atomic velocity v_0.
     The point charge attracts the conduction electrons. They are
free to respond by forming an electron cloud around the "impurity"
in which the electron density, $\rho(r)$, is enhanced over ρ_0 with increasing radial distances. Charge neutrality in the particle-target
system requires a density change $\Delta\rho(r) \equiv \rho(r) - \rho_0$ such that

$$\int \Delta\rho(r)d^3r = Z_1 \quad . \tag{3}$$

Consequently, at distances large compared to the extension of $\Delta\rho(r)$
the particle charge is screened out completely.
     In the present context it suffices to discuss the screening in
the linear-response approximation, where Z_1 is assumed to be sufficiently small that $\Delta\rho$ is proportional to Z_1. Figure 6 shows the
density enhancement around a proton in a metal like Au ($r_s = 3$), in
two different many-body approximations (B and H) and the linearized
Thomas-Fermi (TF) approximation [6]. From a finite value, at $r = 0$,
the enhanced density for the B and H approximations falls off nearly
exponentially and follows the TF approximation when $rk_F > 1$. At
large distances, the density change exhibits small oscillations of
amplitude $\propto \cos (2k_F r)/r^3$, the Friedel oscillations [7]. Considerations of nonlinear terms show that the random-phase approximation
seriously underestimates the density enhancement at small distances
[8]. As Z_1 increases, the response begins to change, bound states
on the impurity become possible [9], and an ion core develops. At
very high Z_1, the atomic shells overlap more and more and approach
the smooth density distribution of the TF atom [10], shown in Fig.
7, which is the exact distribution in the limit $Z_1 \rightarrow \infty$ [11], renormalized to finite Z_1.
     In the linearized TF approximation [12], the density increment
surrounding Z_1 becomes (dotted curve in Fig. 6)

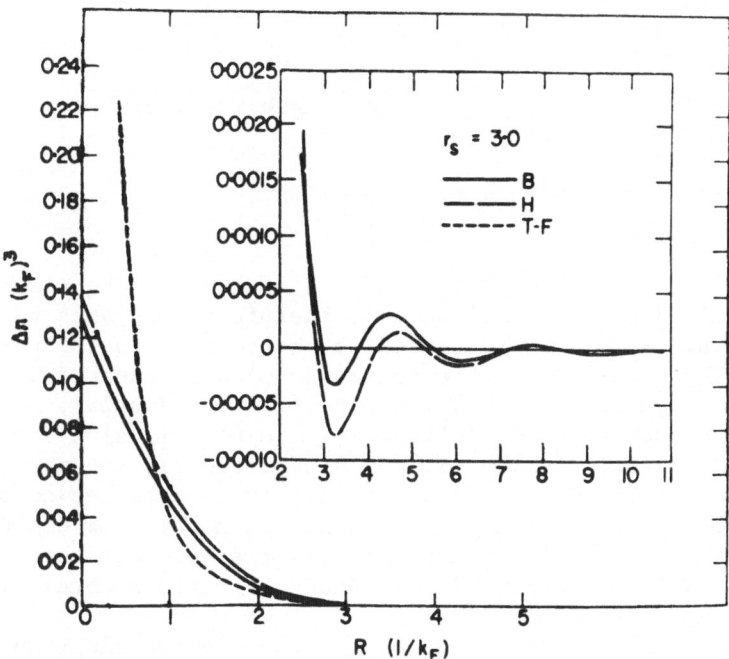

Fig. 6. Density enhancement at a unit positive point charge in an
electron gas of metallic density [6].

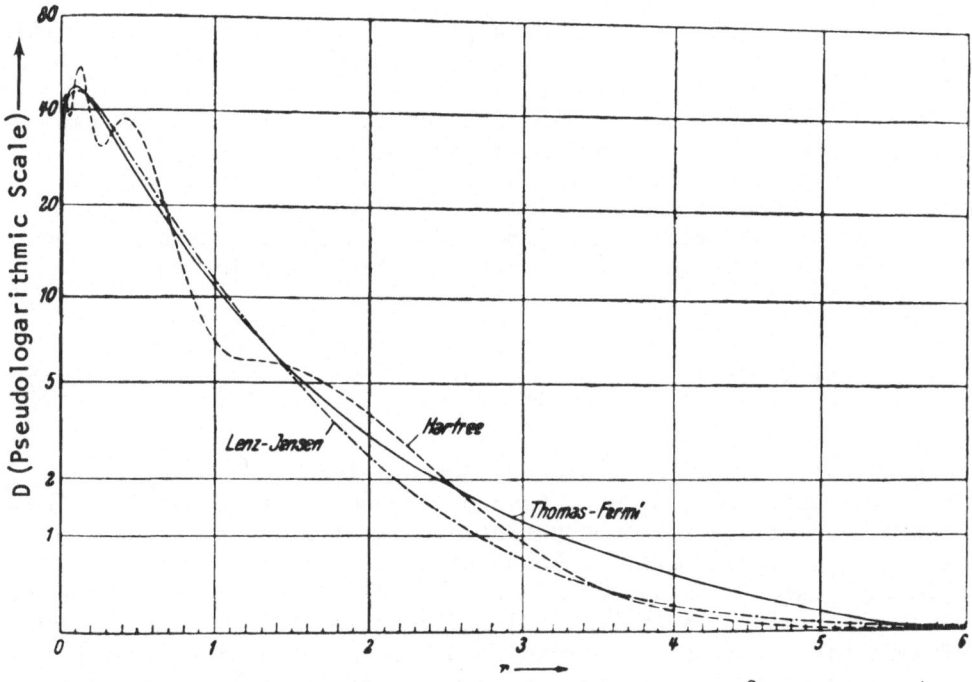

Fig. 7. Electron density distribution, $D = 4\pi\rho(r)r^2$, of the Rb^+
ion in various approximations; r is plotted in a.u. [10].

$$\Delta\rho_{stat} (r) = (Z_1/4\pi a_{TF}^2 r) \exp(-r/a_{TF}) \tag{4}$$

which obeys Eq. (3). The screening length is given by

$$a_{TF} = v_F/3^{1/2}\omega_p = r_s^{1/2} a_o/1.56 , \tag{5}$$

where $\omega_p = (4\pi\rho_o)^{1/2}$ a.u. is the plasma frequency. Equation (4) is a long-wavelength approximation and, therefore, exhibits no Friedel oscillations. This, however, need not concern us in the present context. The extension of the screening cloud, measured in terms of $r = a_{TF}$, depends only weakly on the electron density, viz., as $\rho_o^{-1/6}$. The mean value in metals, $a_{TF} \approx 0.6$ Å, is always smaller than the interatomic distances. As a consequence, the potential set up by an ion is essentially screened out over distances comparable to the interatomic spacing. By contrast in gases, the Coulomb interaction between the particle and the target atoms falls off as r^{-1} and, hence, is long range. Intermediate conditions obtain in semiconductors and insulators [13].

One may ask to what extent bound states of an electron can exist in a potential, $V(r)$, screened by the enhanced density of a form given by Eq. (4). This potential is

$$V(r) = (Z_1 e/r) \exp(-r/a_{TF}) . \tag{6}$$

The solution of the Schrödinger equation for an electron in this potential yields the binding energy as a function of a_{TF}. The result is shown in Fig. 8 [14]. The abscissa values relevant for metals lie between 1 and 1.6 times Z_1. In this calculation, no bound state exists when $a_{TF} < 0.84 a_o/Z_1$. While the number of bound states for an electron in the Coulomb field of a point charge ($a_{TF} \rightarrow \infty$) is infinite, only

$$n^* = 0.583 + 0.499 a_{TF} Z_1/a_o \tag{7}$$

bound states exist in the screened potential Eq. (6).

Alpha particles, $Z_1 = 2$, have one bound state in metals. They exist as He^+ with a screened potential, Eq. (6), proportional to $Z_{1eff} \approx (Z_1 - 0.3) = 1.7$. A bound state on protons, $Z_1 = 1$, is a borderline case in this approximation. Additional factors, such as the contribution of exchange [15] and the need for binding energies sufficiently large not to permit spontaneous ionization in an electron gas with Fermi velocities v_F [16], lead, on balance, to the conclusion that protons cannot bind an electron in a stable orbit

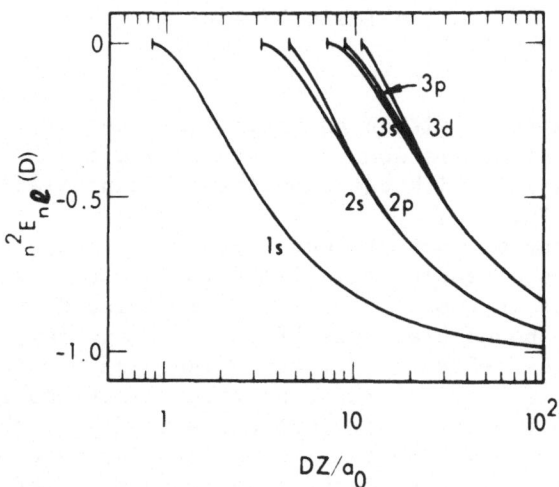

Fig. 8. Binding energy, in units of Ry = 13.6 eV, of electrons on a point charge Ze, as a function of the screening length D = a_{TF} [14].

in any metal. A detailed analysis for a real metal support this finding [17].

The most thoroughly investigated experimental evidence for the enhancement of the electron density at a positive point charge comes from studies of positron annihilation characteristics in metals. When positrons are injected into a metal, they thermalize in times short compared to their lifetime before annihilation with an electron into two gamma quanta. The annihilation rate γ (= inverse lifetime) is directly proportional to the density of electrons at the site of the positron at the time of annihilation. If the lifetime were determined by ρ_0, it would be given by $\gamma = (12/r_s^3)(nsec)^{-1}$ ("Sommerfeld approximation"). If it were determined by electron capture leading to positronium formation, the annihilation rate would be $\gamma = 2.0$ $(nsec)^{-1}$, independent of r_s. The observed annihilation rates, however, vary with r_s as predicted for the enhanced electron density at $r \to 0$,

$$\rho = \rho_o \, h \, (\rho_o, \, Z_1) \, , \tag{8}$$

where the density enhancement factor h can be approximated by [13]

$$h = 1 + \frac{r_s^3 + 10}{6} \tag{9}$$

for positrons. For nuclei this enhancement factor can be written as

$$h = 1 + Z_1 \frac{1.5r_s}{v_F/v_o} = 1 + 0.75 \ Z_1 r_s^{\ 2} \ . \tag{10}$$

Figure 9 compares the positron annihilation rate based on Eq. (9) with the many-body calculation by Bhattacharyya and Singwi [18]. The theory agrees well with the experimental annihilation rates of positrons in metals.

For atoms heavier than protons there will always be an ion core of bound electrons of radius r_c. The valence shell, where the effective atomic number is close to one, can be replaced by a screened distribution of unbound electrons of the form Eq. (4), where Z_1 is replaced by a number close to N_1, the valency of the ion. The Pauli exclusion principle prevents the conduction electrons from interacting strongly with the ion core. One can account for the interaction of the metal electrons with the ion by introducing a pseudopotential which simulates the effects of the condition of orthogonality between the conduction-electron and core-electron wave functions. The result is as follows. Outside the core, $r > r_c$, the ion sets up a Coulomb potential, $N_1 e/r_1$, of a point charge of charge number N_1 located at

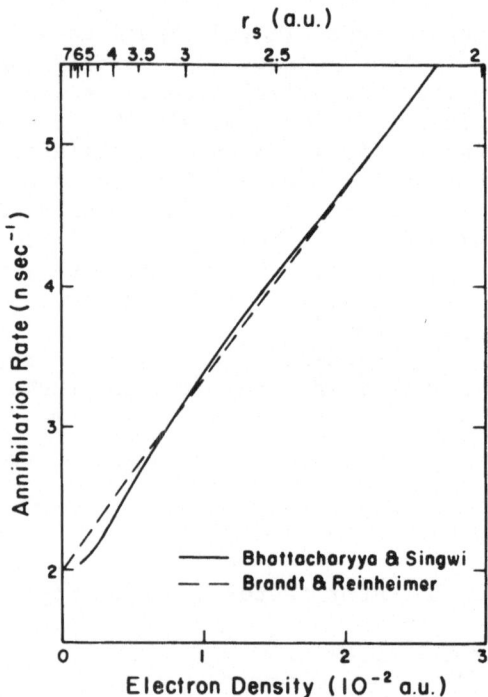

Fig. 9. Comparison of the many-body calculation of the positron annihilation rate [18] with the approximate form based on Eq. (9) [13].

the nucleus. It is screened by the conduction electrons to take the approximate form of Eq. (6) with Z_1 replaced by N_1. Inside the core, $r < r_c$, only a weak effective potential acts on the conduction electrons which is practically constant throughout the core volume $4\pi r_c^3/3$. For our purposes, it suffices to replace the pseudopotential by a constant model potential that can be adjusted through sample calculations or in some other suitable way. As a consequence, the density distribution surrounding a heavy ion at rest inside a metal always resembles closely that of a neutral atom immersed in the sea conduction electrons. At distances $r < r_c$, it is an atomic Hartree-Fock distribution, with only a small component from the valence gas. At distances $r > r_c$, the density distribution, Eq. (4), ranges over $\sim a_{TF} \sim 1$ a.u., much like that of an isolated neutral atom. The question of whether the outer electrons are actually bound or not is of importance only to the extent that the overlap between the wavefunctions of the valence electrons and the ion core can create conditions inside the ion which differ from those of an isolated atom. An example is the magnetic polarization near the nucleus of an ion imbedded in a magnetized metal.

Moving Particles

If we set the particle in motion, with velocity $v_1 \approx v_F$, the response of the electron gas begins to change, in accordance with the frequency and wave number dependence of the response function of the electron gas. The build up in time t of the charge cloud is approximately given by

$$\Delta\rho = \Delta\rho_{stat} \left[1 - \exp(-t/\tau) \right] , \qquad (11)$$

with a relaxation time of the electron gas, τ, of the order of ω_p^{-1}, where ω_p is the plasma frequency [5]. The characteristic time of of the perturbation is $t \approx a_{TF} v_1^{-1}$, so that $t/\tau \approx v_F/v_1$. Consequently by Eq. (11), when $v_1 < v_F$, $\Delta\rho \approx \Delta\rho_{stat}$, whereas for $v_1 > v_F$, $\Delta\rho \approx (v_F/v_1)\Delta\rho_{stat}$. Following Eq. (10), the enhancement factor becomes

$$h = 1 + Z_1 \frac{1.5r_s}{v_F/v_o} \text{ for } v_1 < v_F \qquad (12)$$

and

$$h = 1 + Z_1 \frac{1.5r_s}{v_1/v_o} \text{ for } v_1 > v_F . \qquad (13)$$

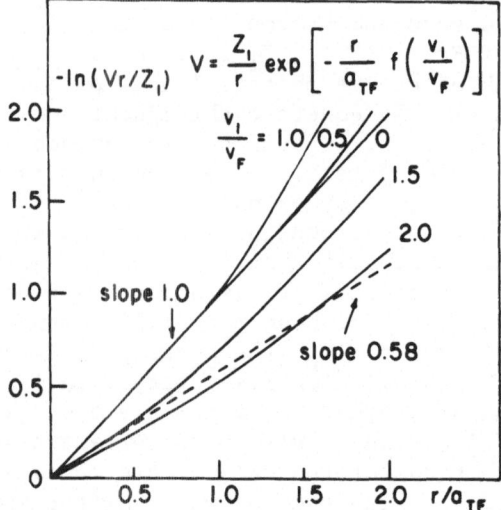

Fig. 10. Dependence of the screening function $f(v_1/v_F)$, defined through the potential given in the legend, on v_1 for r_s = 2 [20].

A wave mechanical discussion based on the Hulthén potential leads to similar results [19].

As the enhancement decreases with increasing particle velocity, the screening of the moving charge is reduced as well. Ebel [20] has calculated the leading spherical term to the screening using the frequency and wave number dependent dielectric function which describes the response of the electron gas to the longitudinal field of the charged particle. The result is sketched in Fig. 10. The function $f(v_1/v_F)$ in the potential cited in Fig. 10 is equal to 1 for $v_1 \ll v_F$, leading to Eq. (6). As $v_1 \to v_F$, the screening increases slightly at distances larger than a_{TF}. Only when v_1 exceeds v_F by a factor two does the screening diminish significantly, to $f \sim 0.6$, which is still not small enough to admit a stable H projectile at metallic densities. At even higher velocities collision broadening makes the bound electron state unstable. We conclude that protons cannot bind an electron inside a normal metal at any velocity.

Heavier projectiles, $Z_1 > 1$, carry the zero-velocity core as long as $v_1 < Z_1^{2/3}v_o$, where $Z_1^{2/3}v_o$ is the mean electron velocity in the core. Moreover, as long as $v_1 < v_o$, they are dressed like point particles with an electron cloud that screens dynamically the ion charge over a distance a few times a_{TF}. In the range $v_1 \approx Z_1^{2/3}v_o$, the dynamic screening subsides, and the projectile moves with bound electrons as an ion of charge $Z_{1 \text{ eff}}e \le Z_1e$. When $v_1 > Z_1v_o$, the ion-core electrons cannot follow adiabatically the interaction with the target electrons, and they strip off.

As a consequence, heavy projectiles moving with low velocities $v_1 \lesssim v_0$ in dense targets act, as regards their long-range interaction with the target electrons, effectively as neutral atoms. The stopping power is determined by momentum transfer in direct atom-atom collisions. The interaction potential is that of overlapping atomic clouds, and the question of charge states does not enter into the discussion. This is the basis for the low velocity stopping power theory for heavy projectiles in terms of the Thomas-Fermi statistical model of the atom developed by Firsov and by Lindhard and his collaborators [21].

At velocities $v_1 > v_F$ the screening length a_{TF} in Eqs. (4) and (6) increases from $\sim v_F/\omega_p$, Eq. (5), to $\sim v_1/\omega_p$, and the dynamic screening of the ion by target electrons subsides. The effective projectile charge rises monotonically toward $Z_1 e$ as v_1 approaches and becomes larger than $Z_1 v_0$. As a consequence, a broad stopping power maximum for heavy ions develops in a velocity range where $v_1^2/v_0^2 > K_B Z_2/4Ry \sim Z_2/4$ ($Ry = 13.6$ eV and $K_B =$ Bloch's constant) and, hence, with regard to the target, Bethe's stopping power theory applies. The maximum develops when the proton stopping power decline $\propto v_1^{-2} \log 2mv_1^2/K_B Z_2$ is balanced by an increase of $Z_{1\ eff}^2$ until it approaches Z_1^2. This is the physical basis for the observation that, when $(v_1/v_0)^2 \gtrsim Z_2/4$, the electronic stopping power for heavy ions can be correlated well through Eq. (2) via an effective charge $Z_{1\ eff} e = q_1 e$ which is a function predominantly only of v_1 and Z_1.

Proton Stopping Power Maximum

Figure 11 shows stopping power maxima for protons in various solid targets [22]. There does not seem to be a simple systematic trend with the atomic number, Z_2, of the target. The maxima sometimes have been attributed to the onset of electron capture and loss processes which do not depend in any clear way on Z_2. However, if protons cannot bind electrons in metals at any velocity, as argued in the previous sections, the maxima must be tied to the properties of the targets. Let

$$x \equiv \frac{v_1^2}{v_0^2 Z_2} \tag{14}$$

be the reduced proton energy parameter. We can write the stopping power in the reduced form

$$-\frac{dE_1}{dx} \frac{mv_0^2}{4\pi(Z_1 e^2)^2 n} \equiv s(x) = \frac{L(x)}{x} \ , \tag{15}$$

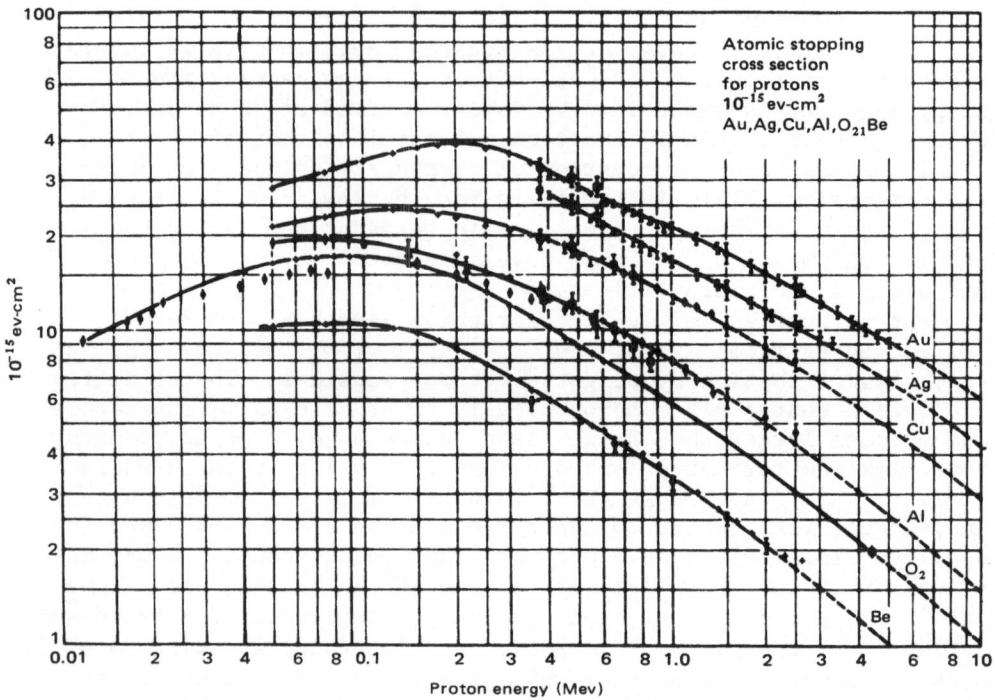

Fig. 11. Stopping power maxima of various targets for protons [22].

where the stopping number per target electron can be approximated by [23]

$$L(x) = \int_0^{\Omega = 2x} g_2(\Omega') \ln \frac{2x}{\Omega'} \, d\Omega' \tag{16}$$

in terms of the target oscillator strength distribution $g_2(\Omega)$ of re-reduced frequency $\Omega = \omega/2Z_2 Ry$ normalized such that $\int_0^\infty g_2(\Omega) d\Omega = 1$. For large x,

$$L(x) \to \int_0^\infty g_2(\Omega) \ln \frac{2x}{\Omega} \, d\Omega = \ln \frac{4Ry}{K_B} x \, , \tag{17}$$

where $K_B \sim$ Ry denotes the Bloch constant defined such that $I_2 = K_B Z_2$ is the mean excitation energy of stopping. A stopping power maximum implies the conditions

$$xs = L(x)$$

$$xs = x \, L'(x) \tag{18}$$

at some value $x = x_m$. Differentiation of Eq. (16) yields the function

$$\psi \equiv x \, L'(x) = Z_2^*(x)/Z_2 \int_0^{\Omega = 2x} g_2(\Omega') d\Omega' \, , \tag{19}$$

where $Z_2^*(x)/Z_2$ is the fraction of oscillator strength of the target electrons that contributes to the stopping of protons with velocity corresponding to x. By Eq. (18) the maximum, then, occurs when

$$x_m \, s(x_m) = \psi(x_m) \, . \tag{20}$$

Figure 12 tests Eq. (20) [24]. The curves represent various approximations for $g_2(\Omega)$. The dashed curve results from Eq. (19) for the Thomas-Fermi model of the atom [25,26]. The points are measured values of proton stopping power maxima in the reduced form of Eq. (15), $s(x_m) \equiv s_m$, multiplied by the x_m values, Eq. (14), corresponding to the proton velocities at which the maxima occur. The agreement between the curve and the experimental values $x_m s_m$ supports the assumptions built into Eq. (20). They are not tied to capture-and-loss processes, as in fact they cannot be if no bound states exist for electrons on protons moving in condensed targets. The proton stopping power maxima occur when, with decreasing proton velocity v_1, the stopping power rise $\propto v_1^{-2}$ in the Bethe theory is outweighed by the depletion of target oscillator strength available for stopping.

Effective Charge and Stopping Power Maxima for Heavy Ions

Atomic projectiles act as point particles when $v_1 > Z_1 v_o$. The treatment of stopping power maxima as presented in the previous section, therefore, applies to ions so light that $x_m > Z_1^2/Z_2$, where x_m is given by Eq. (18) and refers to the projectile velocity at which the proton stopping power has its maximum.

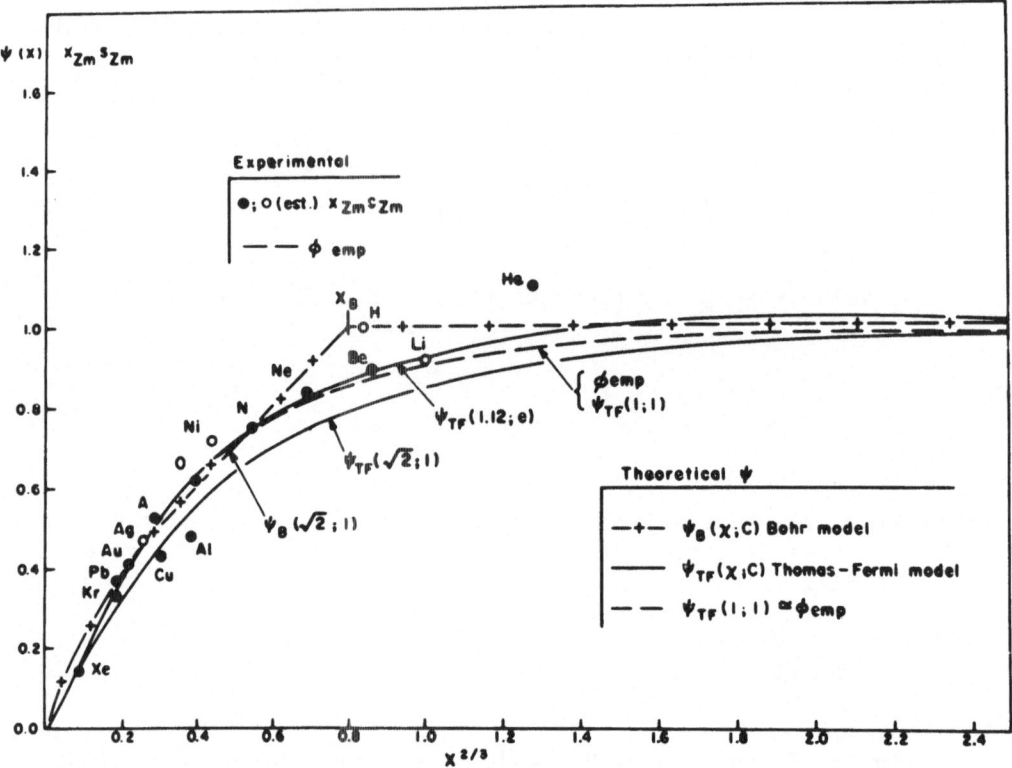

Fig. 12. Comparison of measured stopping power maxima with Eq. (19) [24].

Stopping power maxima for heavy particles occur at higher projectile velocities $\left[\sim (1 \text{ MeV/amu})^{1/2} \right]$ than the maxima for protons $\left[\sim (0.1 \text{ MeV/amu})^{1/2} \right]$. The depletion of target oscillator strength available for stopping makes itself felt when $x \leq x_m$ and should not be the primary cause for the maxima observed with heavy ions. When $x > x_m$, therefore, one can properly write

$$s(x) = (Z_{1 \text{ eff}}(x)/Z_1)^2 L(x)/x . \tag{21}$$

One can estimate the effective projectile charge number $Z_{1 \text{ eff}}(x)$ by invoking the Bohr criterion that all electrons bound to the moving ion in orbits with velocities less than v_1 are stripped off by the target. Let $\xi \equiv r/a$, with $a = 0.8853 a_0 Z_1^{-1/3}$, be the scaled distance from the nucleus of the TF model of the projectile, where the Fermi velocity in the shell between ξ and $\xi + d\xi$ is given by

$v_F(\xi) = \left[3\pi^2 \rho_{TF}(\xi) \right]^{1/3}$ a.u. The stripping criterion, then, considers all electrons as free except those of orbital velocity $bv_F(\xi) = v_1$ or larger, where b is a constant of the order of 1. Formally, this limit can be expressed in terms of the local resonance frequency, $\Omega(\xi)$, as it appears in Eq. (16), because it is related to the scaled local plasma frequency, $\Omega_o(\xi)$, as

$$\Omega(\xi) = \chi \Omega_o(\xi) = \chi \ z^{-1} \left[4\pi \rho_{TF}(\xi) \right]^{1/2} \text{ a.u.} \tag{22}$$

The values of the parameter χ range between 1 and $\sqrt{2}$ [23,31]. With

$$y \equiv v_1/z_1^{2/3} v_o , \tag{23}$$

we obtain

$$\frac{Z_{1\ eff}(y)}{Z_1} = \int_0^{\Omega_o = b^{-3/2}(4/3\pi)^{1/2}y^{3/2}} g_1(\Omega')d\Omega' . \tag{24}$$

In this cursory attempt, we assume that $g_1(\Omega)$ is independent of the state of ionization. The integral in Eq. (24) is then the same as the integral in Eq. (19). We have calculated it for the density distribution in the TF atom as a function of the upper integration limit $\Omega_o(\xi)$. Some values are tabulated in Table I. The universal curve in Fig. 13 shows Eq. (24) for the trial value $b = 2^{1/3} = 1.26$. For comparison, empirical fits to the data by Barkas et al., by Pierce and Blann, and by Brown and Moak, as reviewed in Ref. 3, are drawn as three different sets of points. The early treatments of this problem in the TF approximation by Brunings, Knipp, and Teller, in 1941, also were based on the Bohr stripping criterion but on other assumptions for the cutoff velocity. Equations (23) and (24) suggest the reasons why empirical analytical fits to Z_1 $_{eff}$ data have yielded nearly the same trends with projectile velocity and Z_1 if they are scaled approximately in terms of $v_1/Z_1^{2/3}v_o$. In fact, Eq. (24) gives good estimates of the magnitude and v_1 dependence of Z_1 $_{eff}(y)$. With Eq. (24), Eq. (21) accounts for the stopping power maxima for heavy ions: they are determined by the competition between the rise of $Z_1^2{}_{eff}(y)$, where $y \equiv (Z_2^{1/2}/Z_1^{2/3})x^{1/2}$, and the decline of $L(x)/x$ with increasing $x > x_m$. It is this circumstance which permits the factorization in Eq. (21) down to x values below the values at which the heavy-ion stopping power maxima are observed.

Table 1

Functions calculated for the Thomas-Fermi atom, as used in this and the previous section, and in Figs. 12 and 13. The last column lists the function which, with the third column, is $\psi(x) = Z_2^*(x)/Z_2$ in Eq. (19) and, with the fourth column, is $Z_{1\ eff}(y)/Z_1$ in Eq. (24).

ξ	$\Omega_o(\xi)$	x/χ	y/b	$\int_o^{\Omega_o(\xi)} g_{TF}(\Omega')d\Omega' = \frac{4\pi a^3}{Z} \int_{\xi(\Omega_o)}^{\infty} \rho_{TF}(\xi')\xi'^2 d\xi'$
		(1)	(2)	
25	0.0015	0.0008	0.018	0.012
15	0.0053	0.0026	0.040	0.034
10	0.0132	0.0066	0.074	0.070
8	0.0211	0.0106	0.102	0.101
6	0.0377	0.0189	0.150	0.155
5	0.0534	0.0267	0.189	0.197
4	0.0802	0.0401	0.248	0.256
3.4	0.106	0.0530	0.298	0.305
3.0	0.131	0.0655	0.343	0.344
2.6	0.165	0.0825	0.400	0.391
2.2	0.214	0.107	0.476	0.447
2.0	0.247	0.124	0.524	0.480
1.8	0.288	0.144	0.580	0.515
1.6	0.340	0.170	0.648	0.554
1.4	0.409	0.205	0.733	0.598
1.2	0.501	0.251	0.839	0.645
1.0	0.631	0.316	0.979	0.698
0.8	0.825	0.413	1.17	0.756
0.6	1.14	0.570	1.45	0.819
0.5	1.39	0.695	1.66	0.852
0.4	1.75	0.875	1.93	0.886
0.3	2.32	1.16	2.33	0.919
0.2	3.37	1.69	2.99	0.952
0.1	6.14	3.07	4.46	0.981

(1)   The dashed curve in Fig. 12 is drawn for $\chi = 1$.

(2)   The solid curve in Fig. 13 is drawn for $b = 2^{1/3} = 1.26$.

When the projectiles have atomic numbers $Z_1 \lesssim 10$, the Thomas-Fermi model does not apply and $Z_{1\ eff}(y)$ scales approximately as [27]

$$y = v_1/Z_1 v_o , \qquad (25)$$

where $Z_1 v_o$ is the mean electron orbital velocity in the projectile K-shell (Fig. 14). The reduced losses $s/Z_{1\ eff}^2$ in a given target

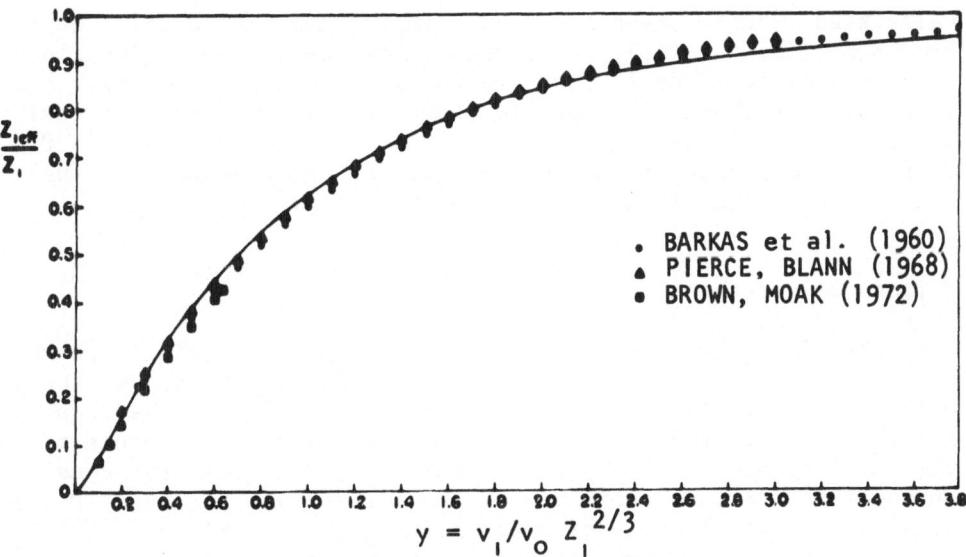

Fig. 13.   The effective charge of heavy ions calculated from Eq. (24)
(solid curve) as a function of velocity, compared with experimental
data [3,4].

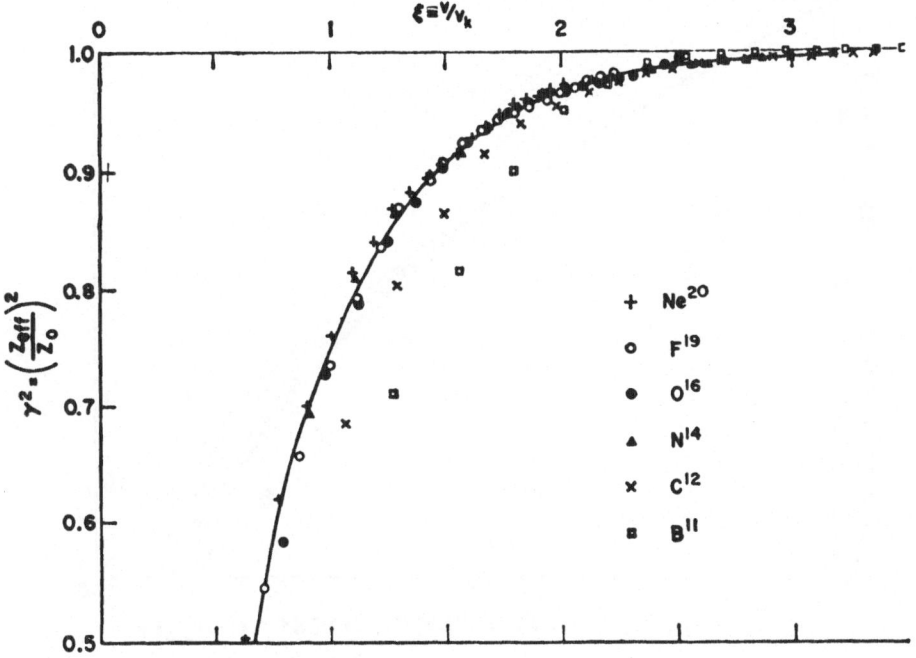

Fig. 14.   Effective charge of light ions as a function of velocity
[27].   The curve is empirical.

are nearly the same at the same projectile velocity, except at low velocities, where screening becomes important. As is to be expected from Eqs. (12) and (13), the screening of the moving projectile has terms proportional to Z_1 that are not contained in Eq. (21): heavy ions per unit charge attract electrons from the target more easily, at higher velocities, than light ions; therefore, as shown in Fig. 15 [27], the reduced ranges are longer for the heavier ions (Ne) than for the light ions (H) of equal velocity.

Effects of Screening

When Coulomb-excited nuclei are imbedded by recoil in ferromagnetic materials, precessions of the nuclear spin can be observed which correspond to magnetic fields at the nucleus in the megagauss range. These enormous local fields are a direct consequence of the velocity-dependent enhancement of the conduction electron density in the field of the recoil ion at the site of its nucleus. Figure 16 shows the total angle of precession $\varphi(v_1)$ of various nuclei Z_1 with initial velocity v_1 divided by the nuclear g-factor, for implantation in iron and gadolinium [19]. The curves are calculated with enhancement factors as given in Eqs. (12) and (13) and integrated over the entire recoil range. The results of this investigation clearly demonstrate the importance of the dynamic screening

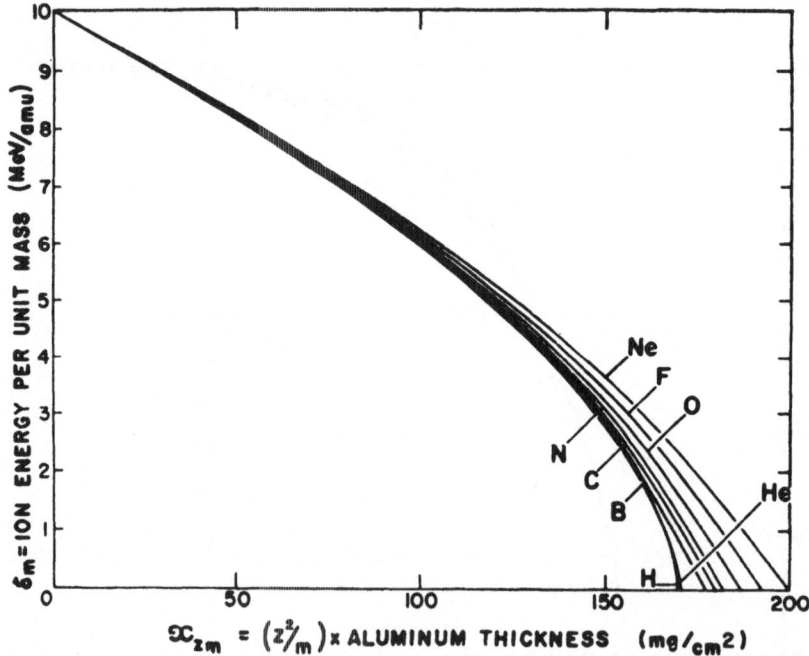

Fig. 15. Variations in ion energies per amu with universal absorber thicknesses of Al [27].

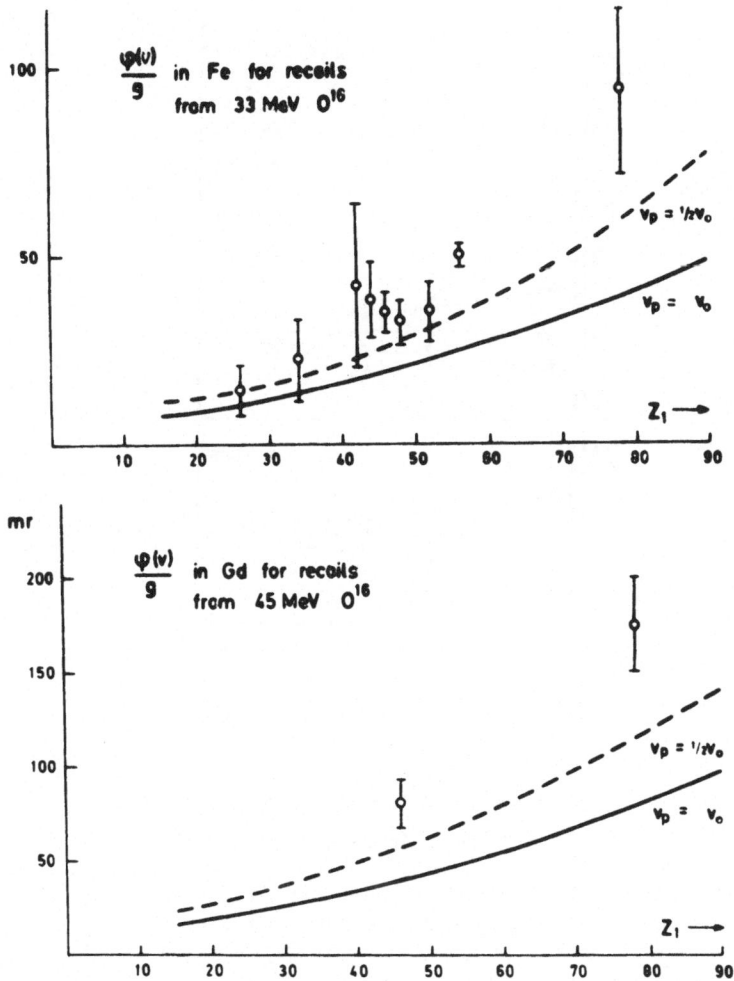

Fig. 16. Measured angles of precession of various nuclei of atomic number Z_1, compared with the theory based on the valence-electron density enhancement at the nuclei [19].

through the many-body response of the valence electron gas in solids to the charge of the moving ions.

Radiative electron capture of conduction electrons into K shells has recently been identified [28] and is discussed in this conference [29]. It can be treated as an inverse photoionization effect. In the attempt of the electron gas to screen out the disturbance set up by the moving projectiles, electrons are captured into bound states in inner atomic shells which carry holes created by Coulomb or Pauli excitation in collisions with target atoms.

As a corollary, the effective charge on moving ions for inner-shell target excitations depends on the response of the target

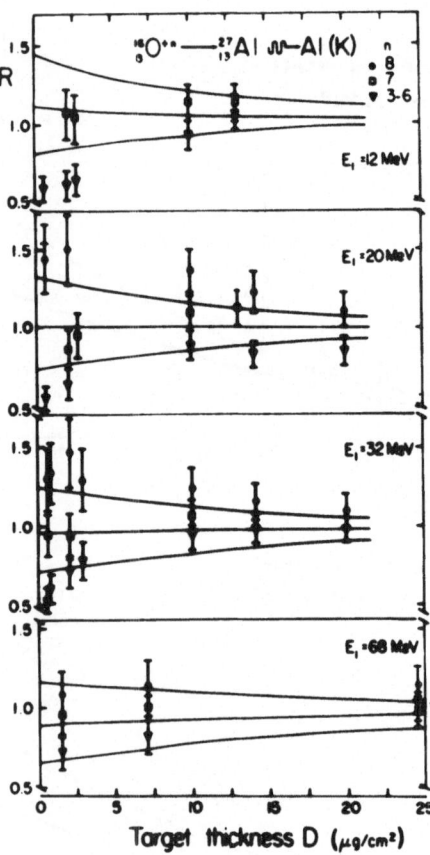

Fig. 17. Relative Al(K) x-ray yields for different incoming oxygen charge states as a function of Al target thickness [31].

electron gas. Figure 17 shows the Al(K) X-ray yields produced by O^{+n} ions in thin Al foils [30]. The curves are calculated under the assumption that the electron gas screens the field of the O^{+n} ions and responds at a rate that is equal to the damping rate of the plasma oscillations. The steady-state condition can comprise electrons which remain tightly bound throughout penetration, as demonstrated by the channeling stopping-power data for selected oxygen charge states reported by Datz et al. at this conference. The residual target polarization by stripped ions at high velocities makes itself felt, not through a screened and hence reduced interaction as at low velocities, but through $(Z_1 e)^3$ dependent increments, for example, of the stopping power, and K-shell ionization cross section of the target [31]. At lower velocities screening prevails but for effects where large-impact parameters are important, such as energy loss by target-electron excitation, it matters little for the mean target response whether the electrons that screen the moving ion

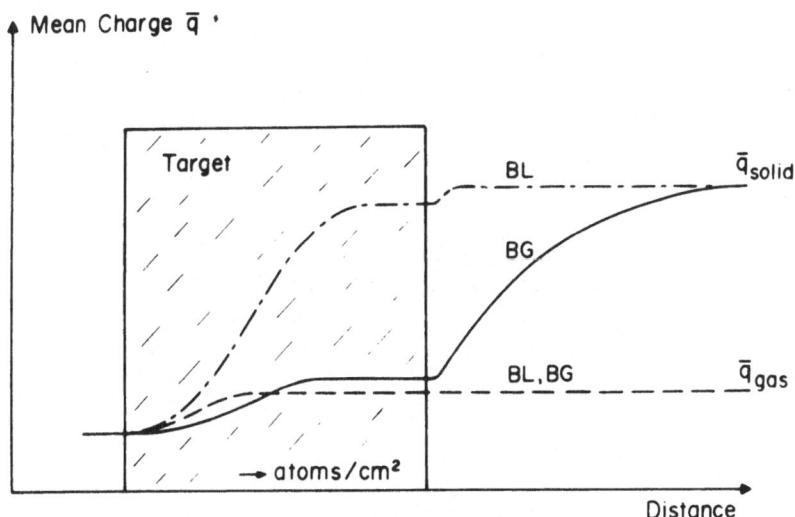

Fig. 18. The mean charge of ions moving in solid and gaseous targets (after Betz) [3,33].

are bound to the ion or follow it as a polarization cloud. One can discuss in these terms the differences between the descriptions by Bohr and Lindhard [32] (BL) and by Betz and Grodzins [33] (BG) of the mean charge of an ion in a solid (Fig. 18). In particular, inner-shell holes can be created in the moving ion by the onrushing electron gas which lifts the degeneracy in inner shells. In the statistical approximation this can be viewed as establishing a steady-state temperature distribution in the moving atomic cloud. Experimental evidence for inner-shell hole distributions comes from energy shifts of the characteristic x-rays owing to the change in the outer screening of deep atomic levels, and from changes in the fine structure of their spectra. On emergence of an ion from the surface and cessation of the electron impact excitation, the holes in the atomic shells are rapidly filled, predominantly through Auger cascades, leading to high ionization states of the projectile after leaving the foil. In gases, the moving ion core does not encounter this steady-state excitation by impact with a dense electron gas and, thus, the emergence from a gas target is not accompanied by additional Auger-induced ionizations.

Surface Effects

The emphasis shifts to the electronic selvage at the metal edge, if one returns to the question why the charge states in proton beams emerging from solids are to a large extent independent of the nature of the solid (cf. Figs. 1 and 2). For if no bound states

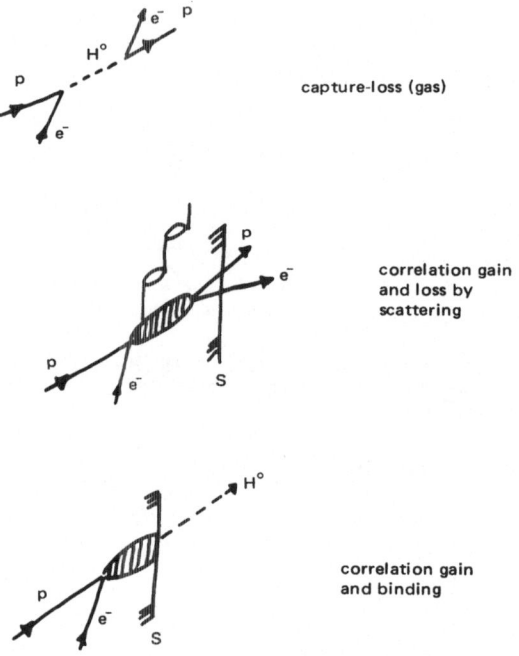

Fig. 19. Processes leading to H formation in proton beams emerging
from a surface S [34].

exist on protons at any velocity inside a metal and neutral fractions
are observed outside the metal, electron capture must occur at the
surface. One can discuss this phenomenon in the following terms
(Fig. 19) [34]. In a gas, a proton collides with a gas atom where
it can capture an electron. It moves as an H atom for a time long
compared to the orbital time. In a subsequent collision, it can
lose the electron and move again as a proton. In the dense electron gas of a solid, however, electrons cannot be bound to protons.
Instead electrons in scattering encounters with a moving proton can
gain correlation in speed and direction such that they follow the
proton. After a certain time, the electrons will have lost so much
spatial correlation by multiple scattering with the target electrons
that, when the surface arrives at the moving proton-electron system,
the proton will emerge with a coterie of electrons too far away to
be bound, but correlated in velocity. Indeed such velocity correlations have been observed (Fig. 20) [35]. The phenomenon is rem-
iniscent of the process which has been termed a "charge exchange to
a continuum state" [36]. However, if the surface arrives before
spatial correlation is lost, the proton can capture a correlated
electron when the density in the metal selvage has dropped so low
that screening is unimportant and bound states become stable. One
can describe these processes in terms of the statistical model of a

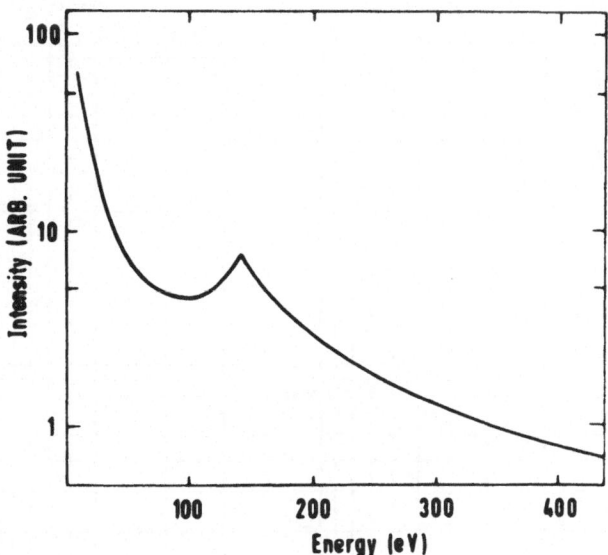

Fig. 20. Energy distribution of electrons emerging with a 275 keV proton beam from a carbon foil. The velocity at the peak coincides with the proton velocity [35].

solid, and derive the velocity dependent function Φ_0 [34]. It agrees in a comprehensive manner with the experimental data shown in Fig. 21, where Φ_0 values drop from about 1 to 10^{-4} over the range of $(v_1/v_0)^2$ values from 0.1 to 70 [37].

Summary

        The mean charge of an atomic projectile appears as a parameter which, in general, depends on the phenomena studied. The scattering of electrons on an ion moving slowly in condensed matter in effect cancels the long range Coulomb potential such that, over distances from the ion comparable to an interatomic spacing in the target, the charge is completely screened out. The projectile behaves like a neutral atom. Energy loss occurs by momentum transfer in collisions with interatomic penetration, where the scattering potential at inner-shell distances is insensitive to the electronic configuration in outer shells. At higher velocities the dynamic screening is reduced and the ion moves with a velocity-dependent coterie of bound electrons. The effective ionic charge rises with velocity. As one consequence, a broad stopping power maximum obtains in a velocity range where the Bethe theory of a monotonically declining target stopping power is valid. When the velocity exceeds all electron

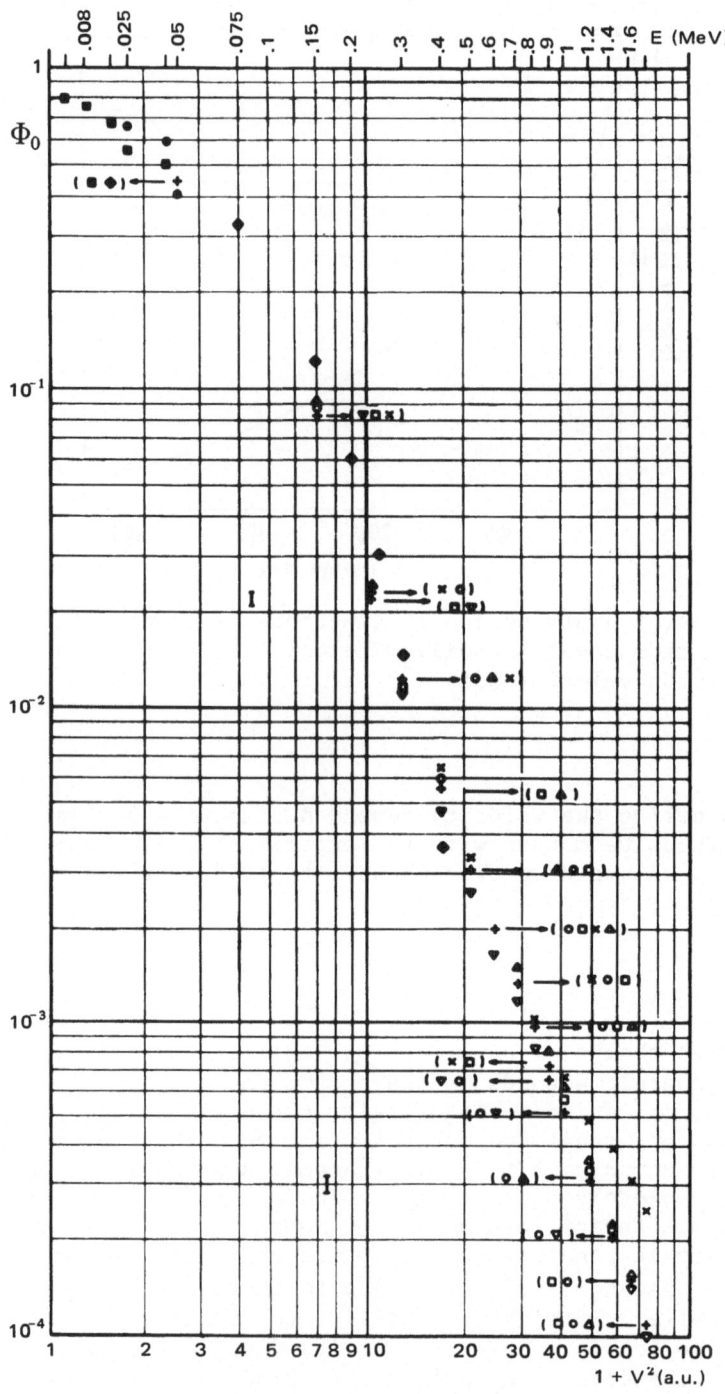

Fig. 21. Neutral charge fraction, $\Phi_0 = (1 + R)^{-1}$, in proton beams emerging from different solids with velocity v_1. Solid points are earlier data, open symbols are new data for C, Al, Ni, Au [37].

orbital velocities in the projectile, the ion moves as a stripped point charge.

If phenomena are observed which depend on the interior of the moving ion, screening can play different roles: it enhances the magnetic field at the nucleus in ferromagnetic targets, it fills inner-shell holes by radiative electron capture, and it affects the energy shifts, fine structure and fluorescence yields of collision-induced characteristic x-rays.

Charge states of ions after emergence from a solid are determined by the correlation gain of electrons inside the solid in scattering collisions with the moving ion, by the number of bound states available in the screened potential, by the inner shell excitation of the moving ion core, and by the effects of the electronic selvage at the target surface.

Acknowledgement

We are indebted to K. F. Stanton for a critical reading of the manuscript.

References

[*]Work supported by the United States Atomic Energy Commission.
[†]Invited paper, Third International Conference on Atomic Collisions in Solids, Gatlinburg, Tennessee, September 1973.

[1] T. Hall, Phys. Rev. 79, 504 (1950).

[2] K. O. Groeneveld and M. Kaminsky, Argonne National Laboratory Annual Report N. 7728, p. 177 (1969) and preprint.

[3] H. Betz, Revs. Mod. Phys. 44, 465 (1972).

[4] M. D. Brown and C. D. Moak, Phys. Rev. B 6, 90 (1972).

[5] The constant r_s can be calculated from the atomic weight of the target, $A(g)$, its specific gravity $d(g/cm^3)$, and the number of valence electrons per target atom, N_2, as $r_s = 1.389 \, (A/N_2 d)^{1/3}$ a.u. Correspondingly, $\rho_o = 8.92 \times 10^{-2} \, (N_2 d/A)$ a.u. $= 0.602 \times 10^{24} \, (N_2 d/A) cm^{-3}$. The electron gas has, in atomic units, the Fermi momentum $k_F = (3\pi^2 \rho_o)^{1/3} = (9\pi/4)^{1/3} r_s^{-1} = 1.917 \, r_s^{-1} \simeq 2 r_s^{-1}$; the Fermi velocity becomes $v_F = k_F = 2 r_s^{-1}$, the Fermi energy $E_F = v_F^2/2 = 2 r_s^{-2}$ and the plasma frequency
$$\omega_p = (4\pi \rho_o)^{1/2} = 3^{1/2} r_s^{-3/2}.$$

[6] J. S. Langer and S. H. Vosko, J. Phys. Chem. Solids, 12, 196 (1959).

[7] J. Friedel, Nuovo Cimento, Suppl. 2, 287 (1958).

[8] H. Payne, Phys. Rev. B 1, 3645 (1970).

[9] J. Friedel, Advanc. Phys. 3, 446 (1954).

[10] P. Gombás, Die statistische Theorie des Atoms und ihre Anwendungen (Springer-Verlag, Wien 1949).

[11] E. H. Lieb and B. Siman, Phys. Rev. Lett. 31, 681 (1973).

[12] N. F. Mott, Proc. Cambridge Phil. Soc. 32, 281 (1936).

[13] W. Brandt and J. Reinheimer, Can. J. Phys. 46, 607 (1968);
Phys. Rev. B 2, 3104 (1970); Phys. Lett. 35A, 109 (1971).

[14] F. J. Rogers, H. C. Graboske, Jr., and D. J. Harwood, Phys.
Rev. A 1, 1577 (1970); and references cited therein.

[15] I. Isenberg, Phys. Rev. 79, 736 (1950).

[16] J. Callaway, Phys. Rev. 116, 1140 (1959).

[17] J. Friedel, Phil. Mag. 43, 153 (1952).

[18] P. Bhattacharyya and K. S. Singwi, Phys. Rev. Lett. 29, 22
(1972).

[19] J. Lindhard and A. Winther, Nuclear Phys. A166, 413 (1971).

[20] M. Ebel (private communication). We are grateful to Professor
Ebel for the permission to show his results in Fig. 10 prior
to publication.

[21] O. B. Firsov, Zh. Eksp. Teor. Fiz [Sov. Phys. - JETP 36, 1076
(1959)]; J. Lindhard, M. Scharff and H. E. Schiøtt, Kgl. Danske
Videnskab. Selskab, Mat.-Fys. Medd. 33, No. 14 (1963).

[22] W. Whaling, "Handbuch der Physik," edited by S. Flügge (Springer-
Verlag, Berlin 1958) Vol. 34, p. 193 ff.

[23] W. Brandt, J. Lindhard and M. Scharff, "Proc. Conf. Electronic
and Atomic Collisions," Boulder, Colorado (1961).

[24] W. Brandt, Trans. New York Acad. Sci. 29, 210 (1966).

[25] W. Brandt and S. Lundqvist, Phys. Rev. A 139, A612 (1965).

[26] W. Brandt, L. Eder, and S. Lundqvist, J. Quant. Spectr. and
Radiative Transfer 7, 411 (1967).

[27] L. C. Northcliffe, Phys. Rev. 120, 1744 (1960).

[28] H. W. Schnopper, H. D. Betz, J. P. Delvaille, K. Kalata, A. R.
Sohval, K. W. Jones and H. E. Wegner, Phys. Rev. Lett. 29, 898
(1972).

[29] F. Bell and H.-D. Betz; H. W. Schnopper (these Proceedings).

[30] W. Brandt, R. Laubert, M. Mourino, and A. Schwarzschild, Phys.
Rev. Lett. 30, 358 (1973).

[31] J. Ashley, W. Brandt and R. H. Ritchie (these Proceedings) and
papers cited therein; G. Basbas, W. Brandt, and R. Laubert,
Phys. Lett. 34A, 277 (1971).

[32] N. Bohr and J. Lindhard, K. Dan. Vidensk. Selsk. Mat.-Fys.
Medd. 28, 7 (1954).

[33] H.-D. Betz and L. Brodzins, Phys. Rev. Lett. 25, 211 (1970).

[34] W. Brandt and R. Sizmann, Phys. Lett. 37A, 115 (1971) and to
be published.

[35] K. G. Harrison and M. W. Lucas, Phys. Lett. 33A, 142 (1970).

[36] J. Macek, Phys. Rev. A 1, 235 (1970).

[37] A. Chateau-Thierry and A. Gladieux (these Proceedings).

SURF-RIDING ELECTRON STATES: POLARIZATION CHARGE DENSITY EFFECTS ASSOCIATED WITH HEAVY ION MOTIONS IN SOLIDS*

V. N. NEELAVATHI[†] and R. H. RITCHIE

Health Physics Division, Oak Ridge National Laboratory
Oak Ridge, Tennessee 37830

Characterization of charged particle energy loss to electrons in condensed media in terms of polarization waves created by the particle is well known from the early work of Fermi [1], Bohr [2], and others. These polarization waves give rise to an oscillatory potential field extending behind and moving along with the particle [3]. We wish to point out that there may exist quasi-stationary states of electrons in the minima of such a field and that electrons may be captured into these surf-riding states.

A point charge Ze moving with a velocity $\vec{v}$ through a metal having a long-lived plasmon state with eigenenergy $\hbar\omega_p$ generates a wake of oscillating polarization charge density. It may be shown [3] that the scalar electric potential associated with such motion, referred to a frame moving with the ion, can be written as

$$\phi(\vec{r}) = Zek\left\{g(\vec{r}) + 2\,\sin(kz)K_o(k\rho)\exp\left[-\frac{\gamma kz}{2\omega_p}\right]\theta(z)\right\} \qquad (1)$$

where $k = \omega_p/v$, z and ρ are the cylindrical coordinates measured from the charge and where the z-axis is taken in the direction of $-\vec{v}$. $K_o(x)$ is the modified Bessel function of the second kind, $\theta(x)$ is the unit step function and

$$g(\vec{r}) = \int_o^\infty \left(\frac{x^2}{1+x^2}\right) J_o(k\rho x)\exp(-xk|z|)\,dx \qquad (2)$$

The quantity γ is the damping rate of the plasmon state in the long-wavelength limit. The first term on the right-hand side of Eq. (1) represents the potential of the incident particle screened by the polarization charge. For $r \ll v/\omega_p$, this term varies as r^{-1} and for $r \gg v/\omega_p$, it varies as r^{-3}. The second term represents the oscillatory contribution of the polarized waves. Since

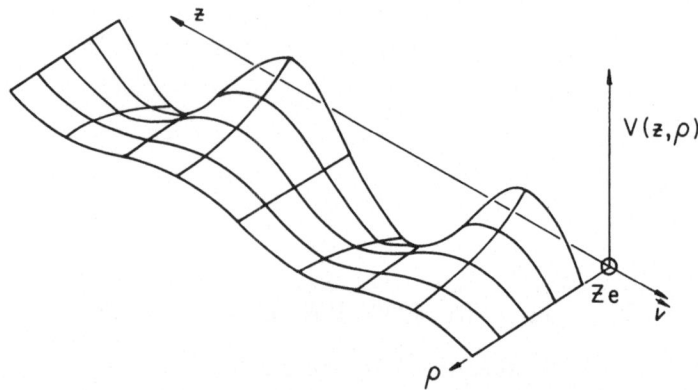

Fig. 1. A schematic representation of the oscillatory portion of
the potential energy of an electron as a function of position behind a swift charged particle moving in a medium which displays a
well-developed plasmon state. The coordinate z is measured from
the position of the particle and in a direction antiparallel with
its motion. The polar coordinate ρ is measured from the track of
the particle.

$\gamma/2\omega_p \sim .03$ for metals, it is clear that the polarization effect
may persist to many cycles of oscillation behind the ion. Figure 1
shows schematically the oscillatory portion of the potential energy
of an electron at rest in the ion frame, specified as a function
of position z and ρ relative to the ion.

Since this potential is fairly complex in form, we have made
some simplifying assumptions about it in order to facilitate the
study of the eigenstates of an electron moving in this potential.
We believe that our conclusions are qualitatively correct and will
not be changed substantially when more accurate calculations are
made.

We assume that the eigenfunction of a given trough does not
overlap appreciably with the neighboring ones. In other words, we
assume an atomic type of wave function rather than a Bloch type.
Also the variation of $g(\vec{r})$ over the extension of a given trough is
assumed negligible. The factor $\exp[-\gamma kz/2\omega_p]$ is treated in the
same way. The z-coordinates of the centers of the troughs are
taken to lie at kz_n , $(3 + 4n)\pi/2$ where $n = 0, 1, 2, \ldots$.

We have assumed a trial wave function for the ground state to be $U_o(\vec{r}) = A \exp[-\alpha\xi^2 - \beta\rho^2]$ where α and β are variational parameters and ξ is the z-coordinate measured from the center of a given trough. We have used standard Rayleigh-Ritz variational theory to estimate the binding energy E_o of an electron in a given trough behind an ion with arbitrary charge and speed. We find $E_o = -(k^2/8)(1/a + 2/b - 1/a^2)$ where $a = k^2/8\alpha$, $b = k^2/8\beta$ and a and b are solutions of the transcendental equations

$$W \, a^2 \exp(b - a) \, E_1(b) = 1 \tag{3}$$

$$Wb \exp(-a) \, E_1(b) = 2 \tag{4}$$

$W = 16 \, cZ/k$, and $c \equiv \exp[-\gamma k\xi/2\omega_p]$ and $E_1(x) = \int_x^\infty du \, e^{-u}/u$. Here energies are measured in units of $Ry = e^2/2a_o$ and lengths are measured in units of the first Bohr radius $a_o = \hbar^2/me^2$. When $W \gg 1$, we get from (2) and (3) $b \sim 2/W$, $a \sim (W \ln[W/2\Gamma])^{-1/2}$ and $\ln \Gamma \simeq 0.5772 \ldots$ is Euler's constant. Then it follows that

$$E_o \sim \frac{k^2}{8} \left[W \ln(\frac{W}{2\Gamma}) - W - \sqrt{W \ln \frac{W}{2\Gamma}} \right] \tag{5}$$

Figure 2 shows a plot of $k^{-2} E_o$ vs W obtained by solving for a and b numerically from (2) and (3).

Table I gives values of the binding energies of an electron in several of the potential minima behind the charged particle. Two different physical situations are considered: (1) a 120 MeV S^{+16} ion traversing an Al target, (2) a 40 MeV O^{+8} ion in the same target. Figure 3 gives the binding energies in the $n = 0$ troughs as a function of the incident energy of these ions.

As the ion emerges from the metal into vacuum, the polarization potential will disappear. We expect surf-riding electrons to appear as a group of electrons with velocity centered around $\vec{v}$. Under the assumption that $v \gg v_o$, v_o being a representative velocity in the polarization well, we find that the probability for observing the electron with momentum $\vec{q}$ is proportional to

$$\exp[-q^2 \sin^2 \Theta/2\alpha - (q \cos \Theta - p)^2/2\beta] \tag{6}$$

where Θ is the angle which $\vec{q}$ makes with $\vec{v}$, and $\vec{p} = m\vec{v}/\hbar$.

We have also made estimates of the photon distribution arising from radiative capture of electrons into the first trough in the

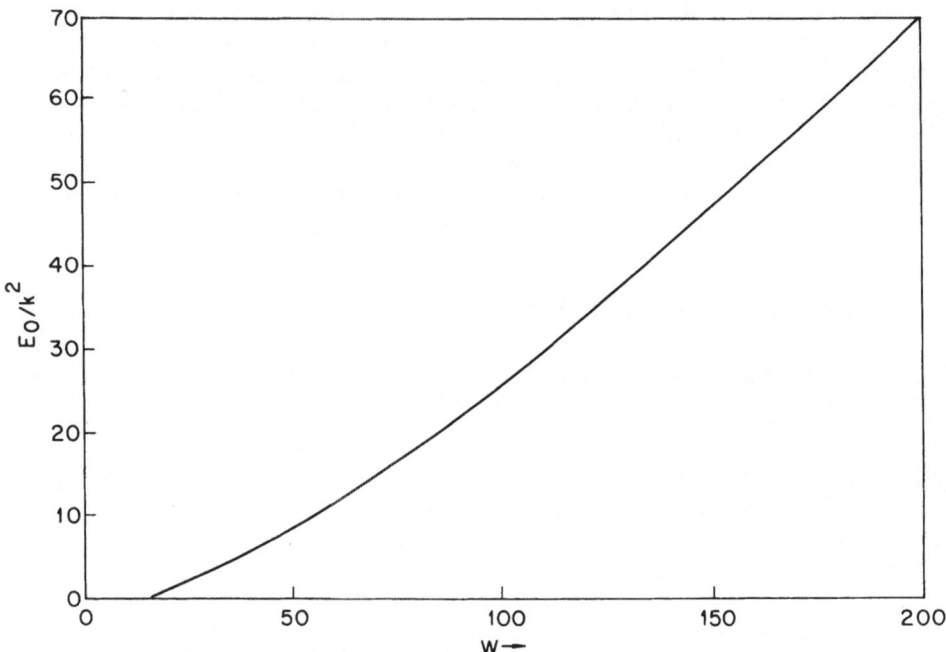

Fig. 2. Ground state binding energy E_0 of an electron in a surf-riding state plotted as a function of the energy variable W defined in the text. All energies are measured in units of Ry = $e^2/2a_0$ = 13.6 eV.

Table I

Values of $E_0^{(n)}$ for two different physical situations: (1) a 120 MeV S^{+16} ion traversing an aluminum target and (2) a 40 MeV O^{+8} ion in an aluminum target

$k_1 z_n$	$E_0^{(n)}$ (eV)	
	120 MeV S^{+16} ions in Al	40 MeV O^{+8} ions in Al
3/2	107 eV	52 eV
7/2	84	39
11/2	66	30
15/2	52	22
19/2	41	15.5
23/2	32	10.3
27/2	25	6.1

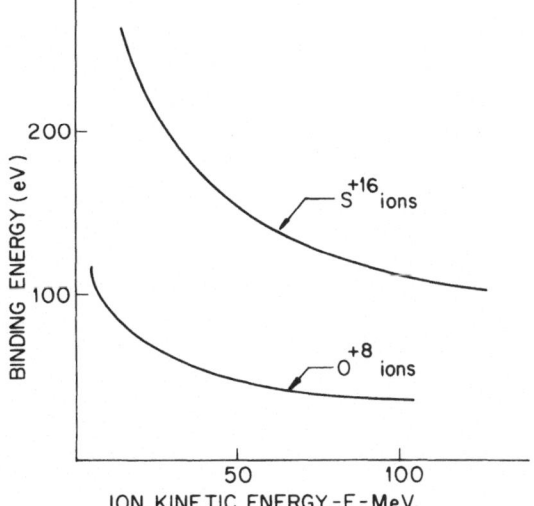

Fig. 3. Ground state binding energy (in eV) of an electron in the first (n = 0) trough of the polarization potential associated with the motion of S^{+16} and O^{+8} ions. This energy is plotted as a function of ion kinetic energy for both cases.

wake of the ion for the case of a 5 MeV S^{+16} ion in an Al target. The binding energy in this case is ∿379 eV. The photon distribution has a peak at ≈32 Ry and a width at half height of ≈67 eV. We note that these results are expected to be only qualitatively correct. The variational wave function is not expected to be highly accurate when used to study properties other than the ground state energy.

Helpful conversations with Werner Brandt, Bill Appleton, and Sheldon Datz are gratefully acknowledged.

References

*Research sponsored by the U. S. Atomic Energy Commission under contract with Union Carbide Corporation.
†Graduate student, University of Tennessee, Knoxville, TN 37916.
[1] E. Fermi, Phys. Rev. 56, 1242 (1939); Phys. Rev. 57, 485 (1940).
[2] A. Bohr, Kgl. Danski Vid Sels, Mat-Fys. Medd. 24, No. 19 (1948).
[3] J. Neufeld and R. H. Ritchie, Phys. Rev. 98, 1632 (1955).

ELECTRON SPIN POLARIZATION AT FERROMAGNETIC SINGLE CRYSTALLINE NICKEL SURFACES DETERMINED THROUGH ELECTRON CAPTURE BY SCATTERED DEUTERONS

CARL RAU
Sektion Physik der Universität München
Munich 40, FRG

and

RUDOLF SIZMANN*
Physics Division, Euratom
CCR-Ispra, Italy

ABSTRACT

A beam of 150 keV deuterons is scattered at grazing incidence on a nickel single crystal surface (hkl) placed in a magnetic field. Polarized electrons captured by the scattered deuterons can orient the deuteron nuclei via the hyperfine interaction. Their polarization is then a measure of the original electron polarization and is determined from the angular distribution of the alpha-particles emitted in a $T(d,n)^4He$ reaction.

The electron spin polarization is found to be parallel to the magnetizing field in the nickel at (100)[110], (110)[110], (111)[1$\bar{2}$1], and (111)[110], antiparallel in (120)[$\bar{2}$10]. The symbols in the brackets indicate the beam direction, the magnetization being in the plane and perpendicular to the beam direction. An electron spin polarization of (96.1±2.8)% is found at the (110) surface, other surfaces giving as yet lower values.

Introduction

Charged particles, passing through a solid can emerge as neutral atoms from the solid surface by capturing a target electron. At low particle velocities the order of a few times $v_o = e^2/h \approx 2.18 \cdot 10^8$ cm/s a high degree of neutralization can be

achieved, of order 25% [1]. In particular, if the target electrons
are polarized, they can transfer their polarization to the nuclei,
after being captured. Already in 1957 Zavoiskii proposed such a
possibility for the polarization of a proton beam [2]. The first
successful experiment was performed in 1969 by M. Kaminsky, who
used a well-channeled deuteron beam traversing a thin single
crystalline magnetized nickel foil [3]. In 1972 Feldman, Mingay
and Sellschop reported on another measurement of polarization of
deuterons channeled through a nickel foil, which confirmed qualitatively the large polarization observed by Kaminsky [4]. The
main difficulties in these experiments are the preparation of the
thin nickel foil and its quick deterioration through radiation
damage during the transmission of the swift particles. Recently,
however, it has been observed by Kaminsky [5] and independently
deduced by Rau and Sizmann [6] as a consequence of the physics of
the electron capture process that small angle scattering of e.g.
deuterons at single crystal surfaces of magnetized nickel targets
can yield polarized deuterium. In this manner intensive polarized
beams of protons, deuterons, ^{3}He can be produced without target
deterioration. Because of its simplicity the technique may become
of considerable interest to nuclear physicists.

     Here we report on the application of electron capture by
scattered deuterons for probing the spin polarization of electrons
at various (hkl) surfaces of a magnetized solid. This is of primary interest in solid state physics, particularly in the field of
electron band structure calculations. A beam of deuterons is
reflected at the surface of the target. The degree and direction
of the spin polarization of the captured target electrons is then
determined in the neutral beam fraction via the anisotropy of the
α-emission in a $T(d,n)^4$He reaction induced by the deuterium atoms
impinging on a tritium target.

Experimental

     The experimental arrangement is illustrated in Figure 1. A
heavy ion accelerator provides a beam of 150 keV deuterons with a
beam divergence of less than 0.025°. Differential pumping in the
beam line reduces the pressure of $\sim5\cdot10^{-6}$ Torr in the accelerator
tube to $\sim4\cdot10^{-9}$ Torr in the target chamber. The vacuum system
consists of two ion getter pumps and a titanium sublimation pump.
Initially, the gas in the target chamber is removed by zeolithe
absorption pumps.
     The deuteron beam is reflected with 0.3° grazing incidence
from a (hkl) surface of a nickel single crystal which is magnetized
to saturation parallel to the surface plane and perpendicular to
the beam direction. The magnetization in the 12x8x.5 mm nickel
sample is achieved by a permanent magnet, the strength of which
is characterized by 0.2T in the air gap. Pole pieces provide the
magnetic shunt between the nickel crystal and the magnet. Details
are shown in Figure 2.

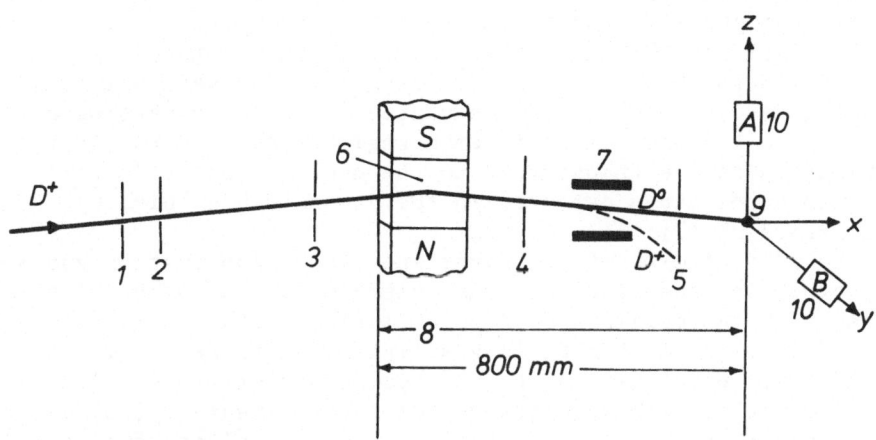

Fig. 1. Experimental arrangement. 1 - 5: collimating slits;
6: nickel target (12 x 8 x 1.5 mm), magnetized perpendicularly to
the beam direction; 7: electrostatic condenser; 8: weak magnetic
field (0.8 mT), parallel to the target magnetizing field; 9: T-Ti
target; 10: alpha solid state detectors A and B.

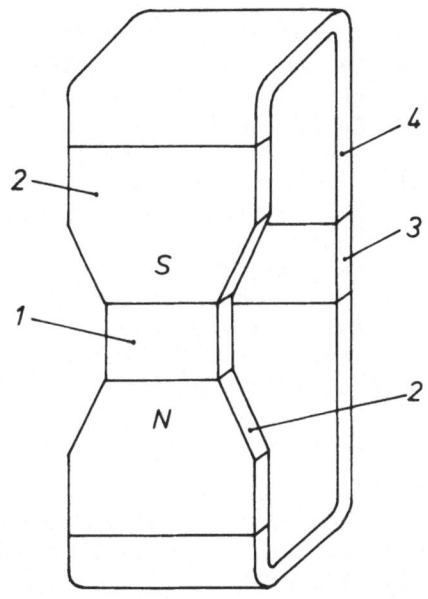

Fig. 2. Target holder. 1: nickel single crystal; 2: special iron
pole pieces used to magnetize the single crystal and to adjust the
required magnetic field gradient; 3: permanent magnet; 4: iron
connecting pieces for magnetic shunt.

After reflection at the nickel surface the beam passes from the strong magnetic field at the target surface into a weak external magnetic field of $\sim$0.8mT. Polarized electrons captured by the deuterons in the high magnetic field region represent selectively populated Paschen-Back hyperfine states in which by passing adiabatically to the weak field (Zeeman regime) the electron spin polarization can be transferred to the nuclei. For a deuterium atom with an electron captured in the ground state "high field" means H>11.7 mT [7].

The adiabaticity of the hyperfine transition must be guaranteed, otherwise the final nuclear polarization cannot be taken as a measure of the original spin polarization of the captured electron. This requires the field gradient $\Delta H/\Delta x$ to be smaller than $(\mu_B/h)(H^2/v_1)$ everywhere along the trajectory of the neutral deuterium moving with velocity v_1 [8]. Experimentally, by adjustment of the shape and the mounting of the pole pieces this condition on $\Delta H/\Delta x$ can be fulfilled.

After scattering at the crystal surface the beam passes through an electrical field to extract the residual charged deuterons. A system of slits provides a beam collimation such that only the specularly reflected particles can impinge upon a titanium-tritium target. A circular aperture covering the tritium target limits the half angle of acceptance to <0.025 degrees. The α-particles emitted in the $T(d,n)^4$He reaction induced in the target spot are counted by two solid state detectors A and B located at 90° to each other in a plane which is perpendicular to the beam direction; A is parallel to the direction of the magnetic field applied to the nickel sample, see Fig. 1. The half angles of acceptance of these detectors are 9 degrees.

The count ratio Z_A/Z_B of the two detectors measures the tensor polarization of the deuterons in the deuterium beam. To compensate for slight geometrical asymmetries the counters are calibrated by substituting for the nickel sample a (non-magnetized) polycrystalline copper sample of identical dimensions, assuming the analyzing power of the $T(d,n)^4$He reaction to be 1. Each measuring sequence consists of two runs with the copper target, four runs with the nickel sample and again two runs with the copper target. Z_A/Z_B is measured with such an accuracy that the electron polarization P_{el} can be calculated within ±0.015 from

$$P_{el} = (N\uparrow-N\downarrow)/(N\uparrow+N\downarrow) = 12(Z_A/Z_B-1)/(Z_A/Z_B+2) \qquad (1)$$

$N\uparrow$ is the number of electrons with spin parallel to the direction of the magnetic field applied to the nickel crystal. It follows from Eq. (1) that Z_A/Z_B=14/11=1.273 belongs to P_{el}=+100% and, therefore, is the largest count ratio measurable. On the other hand, Z_A/Z_B=10/13=0.769 corresponds to P_{el}= -100% and represents the smallest count ratio observable in these experiments. A point of considerable concern is the preparation of the nickel

single crystal surfaces. For the (110) nickel surface we have
found the following procedure satisfactory. The sample is prepared
by wire saw cuttting along (110) from a large lump of a nickel
single crystal. Then the surface is ground in successive steps
with diamond powder of 7, 3, and 1 μm grain size. Subsequently the
surface is etched electrolytically in a bath composed of nitric
acid, sulphuric acid, orthophosphoric acid and glacial acetic acid
at 75°C and a current density of 120 mA/cm^2 [9]. The final step
consists in thermal annealing for 5 hours at 1350° in hydrogen at
normal pressure followed by cooling at a rate of ∿100°/h. After
mounting in the goniometer the sample is again annealed at 200°C
in situ in the target chamber filled with 10^{-7} Torr H_2. Measurements of spin polarization are performed after evacuating the target chamber to $4 \cdot 10^{-9}$ Torr where, however, more than 90% of the
residual gas pressure is H_2. Under these conditions, the polarization measured remains constant over many days of operation.

Results

    Table I summarizes the experimental results obtained so far.
Two aspects of these data are of interest; the sign of the electron
polarization and its magnitude. It appears that the sign is a
characteristic of the (hkl) surface and perhaps of the beam direction <uvw> used in the scattering experiments. In a previous
investigation +32% electron spin polarization was deduced from the
measured Z_A/Z_B ratio. Confining the $T(d,n)^4$He reaction to specularly reflected deuterium atoms only brings the apparent electron
polarization up to ∿72%. Then, by eliminating those magnetic stray
field components which are perpendicular to the nickel surface and
at the same time taking care that the adiabaticity criterion is
fulfilled (cf. Section 2), the apparent electron polarization rises
to the reproducible value of 96% as given in Table I.
    Moreover, the magnitude of the measured electron polarization
can depend on the target preparation. The surface ought to be
atomically flat to avoid close particle-target atom collisions in
which unpolarized core electrons can be captured. The strong
reduction of polarization which results from such particle core
interactions is evident in the transmission experiments quoted in
the introduction: a high degree of polarization is only achieved
with a well-channeled deuteron beam, where the projectiles experience the largest impact parameters possible with the ion cores.
In the present investigation it was attempted to improve the
polarization in (120) [210] from the previously measured -9% in
the same manner as described above and proved to be successful for
(110) [110].
    It amounted, however, only to a change of the -9% to -16%
electron spin polarization. At this moment it is not yet clear
whether -16% is the extreme value obtainable for (120) [$\bar{2}$10] or a
more perfect surface would lead to a larger magnitude of polarization.

Table I

Electron spin polarization of nickel single crystal surfaces.

Reflection Plane	Beam Direction	Polarization Experimental[a] P_{el} (hkl) %
(110)	[110]	+96.1±2.8
(110)	[1$\bar{1}$2]	+23±2[b)c)]
(100)	[110]	+19±2[b)]
(111)	[1$\bar{2}$1]	+12±2[b)]
(111)	[110]	+10±2[b)]
(120)	[$\bar{2}$10]	-16±1[b)]

a) + sign: predominant electron spin polarization parallel (magnetic moment antiparallel) to the direction of the applied magnetizing field in the nickel sample.
b) These values are possibly lowest limits only and subject to experimental conditions. The error limit is taken from repeated experiments with different samples and is not due to counting statistics.
c) Measured with D_2^+; all other measurements were performed with D^+.

The values for the (100) and (111) surfaces, which are included in Table I, are the preliminary results reported before [6]. Work is in progress to repeat these measurements under various surface conditions of the nickel target.

Discussion

At present no theoretical calculations of the electron spin polarization at surfaces of magnetized nickel single crystals are available. On the other hand there exists evidence that the captured electrons are linked to the bulk electron states rather than to surface or surface impurity states. By reducing the vacuum in the target chamber from $4 \cdot 10^{-9}$ Torr to about $5 \cdot 10^{-6}$ Torr, the accelerator working pressure, the apparent polarization remains almost constant at first and then drops gradually until after a few hours of continued irradiation almost zero polarization is reached. Then an optically visible surface contamination, presumably predominantly carbon, has built up in the beam spot. We estimate the thickness of the contamination layer to be ∿500 Å and conclude that not until such a thickness is reached are the electrons involved in the neutralization of the deuteron beam

Table II

Electron spin polarization of nickel single crystal surfaces.

Reflection Plane	Predominant Spin Orientation Experimental[a] Sign P_{el} (hkl)	Beam Direction	Polarization Observed %	Predominant Spin Orientation Theory Sign P_{el} (hkl)						
				Ref: [10]	[11]	[12]	[13]	[14]	[15]	[16]
(110)	+	[110]	96.1±2.8 ⎫							
(110)	+	[1$\bar{1}$2]	23±2b)c) ⎬	+	+	+	+	+	+	+
(100)	+	[110]	19±2b)	+	+	+	+	+	+	+
(111)	+	[1$\bar{2}$1]	12±2b) ⎫							
(111)	+	[110]	10±2b) ⎬	+	+	+	+	+	+	+
(120)	−	[$\bar{2}$10]	16±1b)	−	−	−	−	−	−	−

a), b), c): See Table I

provided predominantly from the surface material. Cleaning of the surface restores the former polarization completely.

Therefore, we proceed by attempting to link the present data to the volume spin polarization calculations, which have become available recently [10-16]. The next difficulty which we encounter at this stage is the ignorance about any dependence of the electron capture process on the relative momentum between the electron and the moving deuteron. Suitable data for an experimental investigation of this matter are in progress but are not yet available. Therefore, we shall ignore for the time being any such dependence on the relative momentum and put forward three postulates on the basis of which the experimental data are compared with the theoretical volume spin polarization calculations.

The deuterons in grazing incidence interact only with the tail of the electron density at the reflecting nickel surface (distance of closest approach ~ 1 Å in present work). This leads to the first postulate:

   (i) Only electrons in the highest populated energy states, in particular electrons at the Fermi surface, are captured.

The second postulate is then consequently:

   (ii) The degree of polarization is given by the relative fraction of the density of states with spin up and spin down in these high energy states.

Finally:

   (iii) The polarization is determined by electrons with k-vector normal to the reflecting surface (hkl).

With these assumptions we have extracted from band structure calculations the spin polarization for the various (hkl) surfaces used in the experiments. In Table II these data are shown. As far as the present experiments are concerned the agreement in magnitude of polarization is unsatisfactory, except for the (110) case, however perfect but perhaps fortuitous with regard to the sign of polarization. Further experimental and theoretical work is needed to fully comprehend the relation between electron k-vector and electron pick-up so that this spin spectroscopy technique can be used in checking band structure calculations of ferromagnetica or as an analytical tool per se.

Acknowledgments

We thank Mr. O. Scherber for assistance in preparing the nickel samples. One of the authors (R.S.) is grateful to Dr. A. Merlini and Dr. W. Schuele, Euratom, Ispra for their hospitality while completing this manuscript. The research has received financial support from the Bundesministerium für Bildung und Wissenschaft, FRG.

References

*Permanent address: Sektion Physik der Universität München, Munich 40, FRG.

[1] W. Brandt and R. Sizmann, Phys. Lett. 37A, 115 (1971).

[2] E. K. Zavoiskii, Soviet Phys. JETP 5, 378 (1957).

[3] M. Kaminsky, Phys. Rev. Lett. 23, 819 (1969).

[4] L. C. Feldman, D. W. Mingway and J. P. F. Sellschop, Radiation Effects 13, 145 (1972).

[5] M. Kaminsky, ANL-Report No. 7971, p. 178 (unpublished), U.S. Patent 3. 700. 899.

[6] C. Rau and R. Sizmann, Phys. Lett. 43A, 317 (1973).

[7] W. Haeberli, Ann. Rev. Nu. Sci. 17, 373 (1967).

[8] G. Clausnitzer, Nucl. Instrum. Methods 23, 309 (1963).

[9] W. J. McG. Tegart in The Electrolytic and Chemical Polishing of Metals, Pergamon Press 1959.

[10] L. Hodges, H. Ehrenreich and N. D. Lang, Phys. Rev. 152, 505 (1966).

[11] J. W. D. Connolly, Phys. Rev. 159, 415 (1967).

[12] E. I. Zornberg, Phys. Rev. B1, 244 (1970).

[13] B. A. Politzer, Diss. Univ. Pennsylvania (1971).

[14] J. Langlinais and J. Callaway, Phys. Rev. B5, 124 (1972).

[15] G. Meister, Diss. Univ. Munich (1971).

[16] J. Callaway and C. S. Wang, Phys. Rev. B7, 1096 (1973).

ELECTRON PICK UP BY PROTONS EMERGING FROM SOLID SURFACES

W. BRANDT

Department of Physics, New York University
New York, New York, U.S.A.

and

R. SIZMANN

Sektion Physik, University of Munich
F.R.G.

ABSTRACT

From experimental evidence it follows that
the neutral fraction ϕ_0 of protons emerging from
a solid surface with energy <1 MeV depends on the
proton velocity v_1 but hardly on target atomic
number Z_2.

It is the purpose of this paper to present
simple statistical approximations for the electron
pick up processes in a wide range of proton energies.
When v_1 is comparable to the Fermi velocity but
small compared to the electron velocities in the ion
cores, $\sim Z_2^{2/3} v_0$, electron capture occurs in the tail
of the electron distribution at the target surface
($v_0 = 2.2 \cdot 10^8$ cm/s). Under these conditions ϕ_0
is nearly the same for all surfaces and depends only
on v_1. When v_1 approaches $Z_2^{2/3} v_0$, electron capture
is influenced primarily by the interaction of the
proton with the cores of the target atoms, but the
relevant cross sections depend on Z_2 only as $Z_2^{1/3}$,
and, thus ϕ_0 depends only slightly on Z_2. When v_1
exceeds $Z_2^{2/3} v_0$ and thus has exhausted the electron
velocity space proffered to the proton beam by the
target, double scattering processes are required for
electron capture; ϕ_0 then depends on v_1 and strongly
on Z_2, to order Z_2^3.

Details of the electron capture and loss
processes require further elucidation, but in the

light of the simplicity of the model underlying
the considerations, theory and experiments are
found to be in satisfactory agreement over the
projectile energy range from E $\sim$20 keV to E $\sim$0.6
GeV, where ϕ_0 ranges from $\sim$0.6 to $\sim$10^{-12}.

EXPERIMENTAL NEUTRAL CHARGE FRACTIONS IN PROTON BEAMS EMERGING FROM SOLIDS*

A. CHATEAU-THIERRY and A. GLADIEUX

Institut National des Sciences et Techniques Nucléaires/Saclay
B.P. N° 6 91190 Gif-sur-Yvette, France

ABSTRACT

Experimental neutral fractions of proton
beams emerging from foils of C, Al, Cr, Ni, and
Au with energies in the range of 0.15 to 1.8 MeV
are reported. In this range the neutral fraction
varies from about 10% to about 0.01%, the high Z
material producing slightly higher neutralization
than the low Z foils.

Introduction

When protons emerge from a solid target, they can capture an
electron thus forming a neutral beam fraction $\Phi_0 = n_0/(n_0 + n_+)$,
where n_0 and n_+ are the numbers of projectiles which emerge as H
atoms and protons, respectively. Here we report on first measurements of Φ_0 for protons emerging from thin targets of C, Al, Cr,
Ni and Au in an energy range between 0.15 and 1.8 MeV. The results
are in support of a statistical approach presented in a previous
paper by Brandt and Sizmann [1].

Experimental

Figure 1 shows the experimental arrangement. Protons and
neutral hydrogen atoms emerge from the foil and are electrostatic-
ally separated through a 7 degrees deflection. They are detected
simultaneously thereby eliminating the influence of intensity
fluctuations in the incident beam; the protons by a CsI(Tl) or
plastic scintillator coupled to a photomultiplier, the neutrals by
a Si(Li) detector kept at the temperature of liquid nitrogen. The
energy resolution was better than 5 keV at 1 MeV. In comparing
the effect of various targets, the entrance energy of the protons

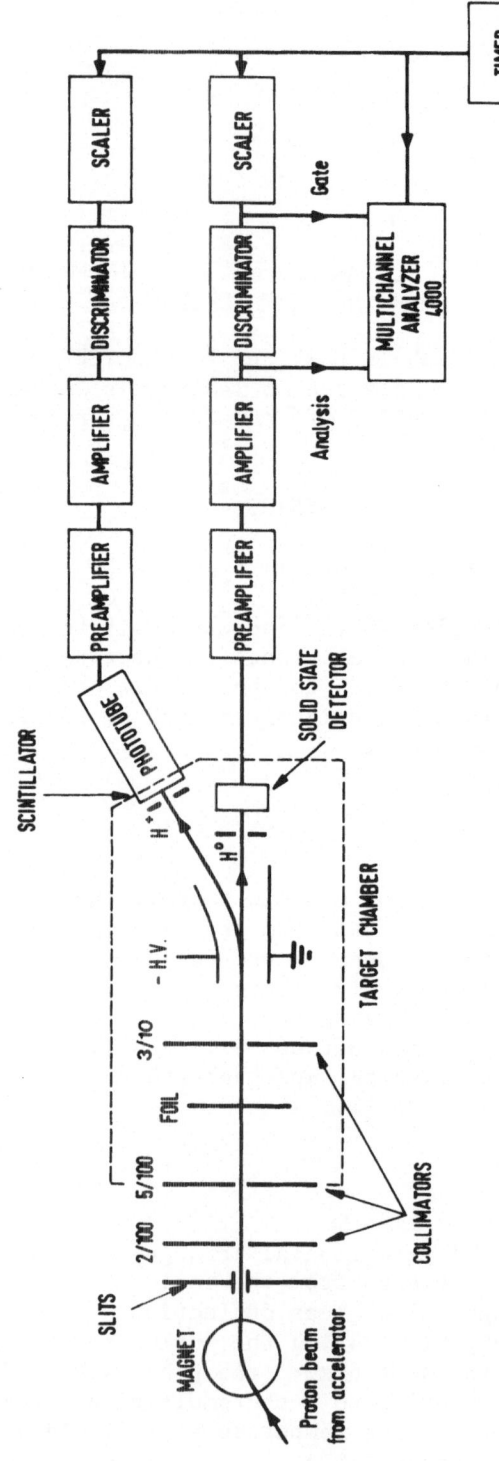

Fig. 1. Experimental set-up.

was always adjusted so to give the same exit energy for every
target. The collimators and slits in the various beam lines are
such that the divergence of the entrance beam is less than 0.015
degree, the angular apertures of the detected neutrals and protons
emerging from the foil are 0.32 degree. The energy of the proton
beam incident on the foils is calibrated with the resonance at
991.87 keV of the reaction $^{27}Al(p,\gamma)^{28}Si$.

The foils are mounted on a target holder with four holes, each
6 mm in diameter. Two are covered with a 10 $\mu g/cm^2$ carbon foil as
an internal standard and the other is left open for beam calibration.
The foils used are carbon (5, 10, 20 and 40 $\mu g/cm^2$), aluminum
(12 and 20 $\mu g/cm^2$), gold (200 $\mu g/cm^2$) which were separated from
their support. Other foils were prepared by evaporation on 10 $\mu g/cm^2$
carbon foils; nickel (9 $\mu g/cm^2$), chromium (7 $\mu g/cm^2$), gold
(19 $\mu g/cm^2$). No systematic trends in Φ_0 could be detected depending on these various preparation techniques. The composite
metal-carbon foils were placed in the target holder with the metal
layer downstream.

It is also possible to tilt the target holder relative to the
beam direction through an angle θ, from $\theta = 0$ to $\theta = 60$ degrees.
Thus, the dependence of the neutral beam fraction Φ_0 on the exit
angle of the beam, θ, can be investigated.

Results

Figure 2 illustrates the dependence of Φ_0 for $\theta = 0$ on the
exit energy E(MeV), found in the present experiments. For comparison with theoretical considerations, a double-log plot is used
with Φ_0 and $1 + v_1^2$ as the variables. Here v_1 is the particle
velocity in units of $v_0 = e^2/\hbar = 2.2 \times 10^8$ cm/s. The relation
between v_1 and the kinetic energy E_p of the protons is $v_1^2 = 40.2 E_p$
(MeV). We have included the experimental data of Kaminsky [2] and
Berkner et al. [3]. The locus of all these points can be viewed
as defining an experimental curve that ranges over four orders of
magnitude in Φ_0 from ~ 1 to 10^{-4} and over nearly three orders of
magnitude in energy. For all foils Φ_0 is a rapidly decreasing
function of energy. The experimental slope d log Φ_0/d log$(1 + v_1^2)$
in Figure 2 is -2.7. No distinctive dependence of Φ_0 on foil
thickness was found.

The dependence on target material, represented by its atomic
number Z_2, is almost zero at low velocities, except for a fine
structure, but increases with increasing energy, viz. Figure 3.
Here, the ratio Φ_0 (Au)/Φ_0 (C) increases from 1 to about 2.5 at
high energies which is consistent with $\{Z_2 (Au)/Z_2 (C)\}^{1/3} = 2.4$,
the theoretical limit [1]. This dependence, however, is still a
relatively small perturbation of Φ_0, since Φ_0 itself drops over
orders of magnitude in the energy range under consideration.

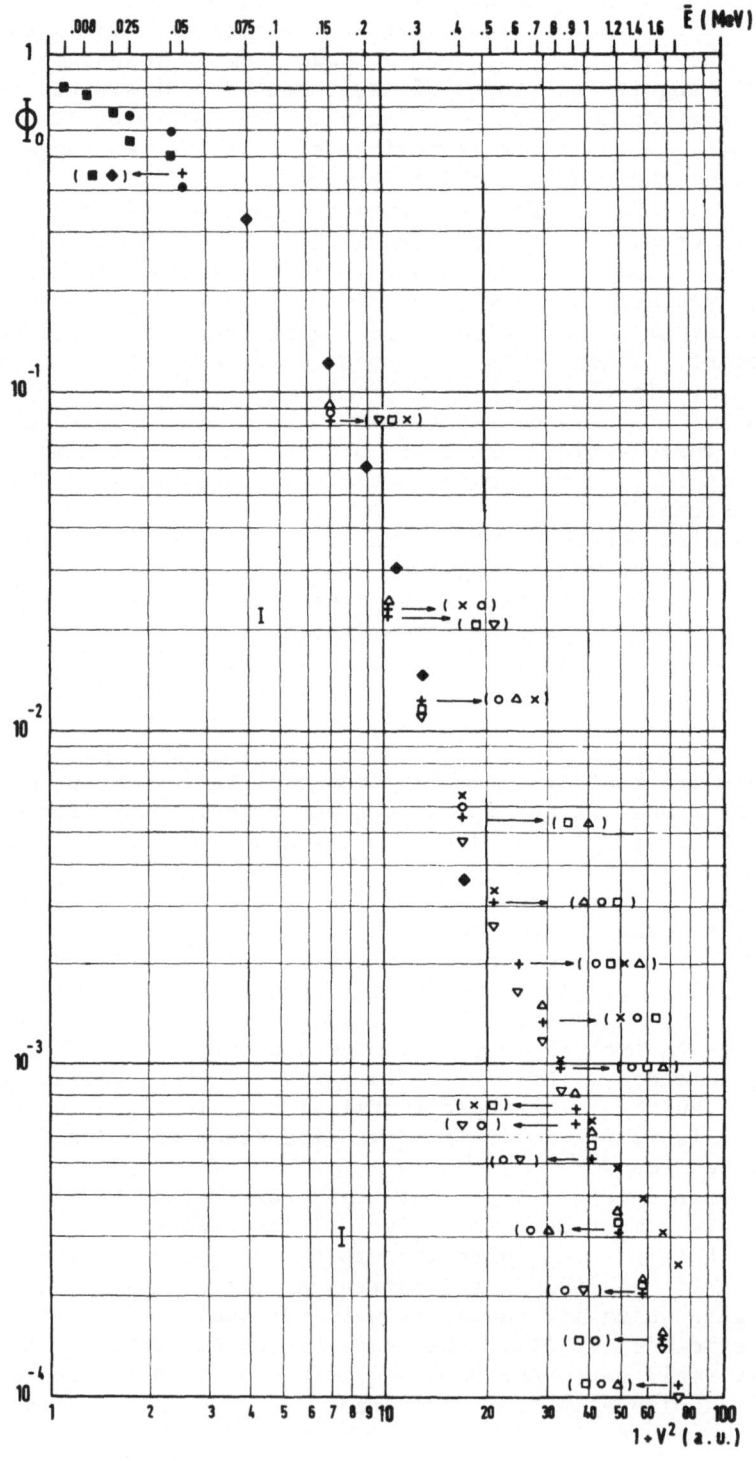

Fig. 2. Neutral beam fraction Φ_0 as a function of $1 + v_1{}^2$, where v_1 is the velocity of the emerging particles. This work: C (∇), Al ($\square$), Cr (o), Ni (Δ), Au (X) with incident protons; Ref. 2: Ni ($\blacklozenge$) with incident deuterons; Ref. 3: C ($\blacksquare$), Au ($\bullet$) with incident deuterons.

The ratios Φ_0 (Al)/Φ_0 (C) and Φ_0 (Ni)/Φ_0 (C) are always slightly larger than 1 but their dependence on energy is not clearly resolved within the experimental accuracy.

Figure 4 shows the dependence of Φ_0 on the tilt angle θ. Plotted are the ratios Φ_0 (θ)/Φ_0 (0) for aluminium and gold targets at a constant 1.4 MeV exit energy of the protons. The neutral fraction decreases with increasing obliquity of the beam direction towards the foil surface. Qualitatively, this is in agreement with the theoretical expectation; electron loss processes dominate over electron capture in the tail of the electron overspill at the surface, a large θ lengthens the path of a neutral hydrogen atom in this surface electron distribution. The variation of Φ_0 with θ increases with Z_2 as shown in Fig. 4. Quantitatively, however, change of Φ_0 with θ exceeds the theoretical predictions. This discrepancy needs further clarification.

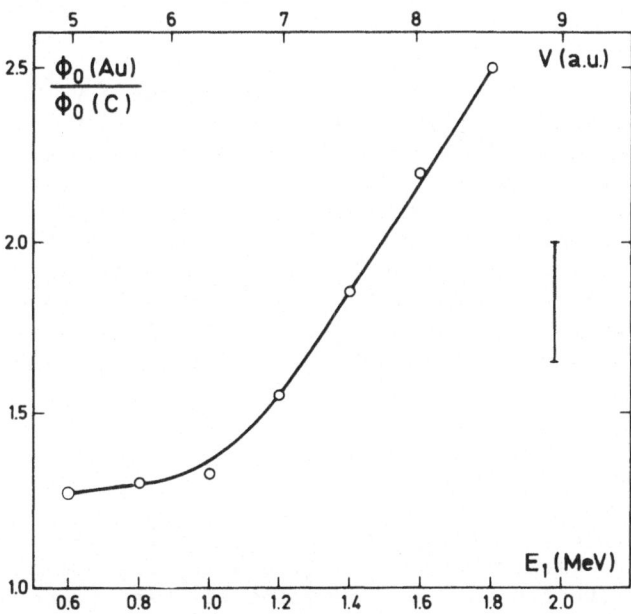

Fig. 3. Ratio of Φ_0 (Au)/Φ_0 (C) as a function of proton energy.

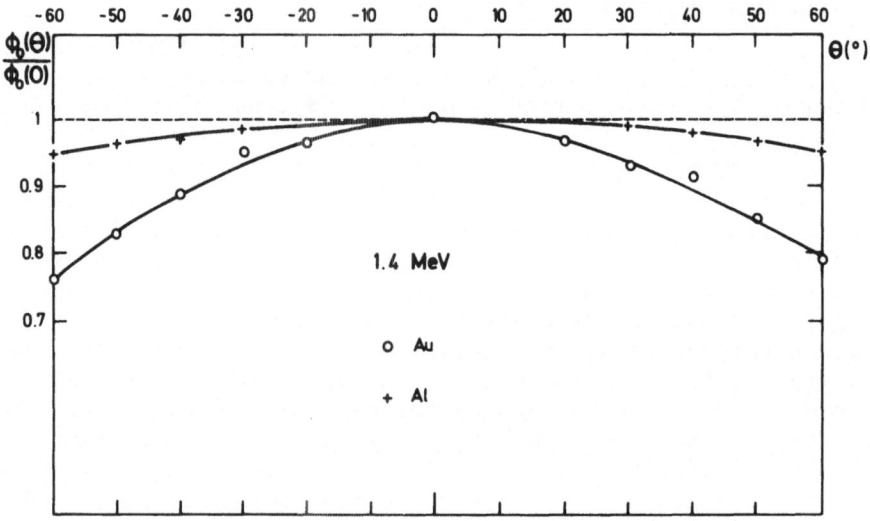

Fig. 4.   Ratio $\Phi_O (\theta)/\Phi_O (0)$ as a function of the exit angle θ for aluminium and gold foil at 1.4 MeV.

Conclusions

We have demonstrated that in the regime where v_1 is larger than v_O the neutral fraction Φ_O is essentially independent of the material and the foil thickness. Its absolute value decreases strongly with increasing v_1, from about 10% at $v_1 = 2.5\ v_O$ to about 10^{-4} at $v_1 = 8.5\ v_O$. At the high velocities, Φ_O for gold compared with carbon Φ_O exhibits a Z_2 dependence which, however, is a small variation compared with its strong absolute energy dependence. Finally, we found that Φ_O varies with the angle between foil surface and beam direction which amounts to a decrease of about 20% (with gold, less with aluminium) for an oblique angle of 60 degrees.

Acknowledgments

The authors are particularly indebted to Professor W. Brandt and Professor R. Sizmann for many stimulating discussions. We greatly appreciate the valuable experimental assistance of R. Penard. We sould also like to thank R. Ripon for providing electronic facilities.

References

*Work supported by the French Atomic Energy Commission.

[1] W. Brandt and R. Sizmann, "Electron Pick Up by Protons
    Emerging from Solid Surfaces," this conference.

[2] M. Kaminsky, Bull. Am. Phys. Soc. 14, 846 (1969);
    ANL Report 7620, p. 181, 1969 (unpublished).

[3] K. H. Berkner, I. Bornstein, R. V. Pyle and J. W. Stearns,
    Phys. Rev. A6, 278 (1972).

[4] We are grateful to J. Remillieux for making the accelerator
    facilities at the University of Lyon (Institut de Physique
    Nucléaire) available to us to check the Φ_0 (θ) dependence.

CHARGED FRACTION OF 5 keV TO 150 keV HYDROGEN ATOMS AFTER EMERGENCE FROM DIFFERENT METAL SURFACES

by

R. BEHRISCH, W. ECKSTEIN, P. MEISCHNER,
B. M. U. SCHERZER and H. VERBEEK
Max-Planck-Institut für Plasmaphysik
EURATOM Association D-8046 Garching, Germany

ABSTRACT

The charged fraction of hydrogen atoms backscattered from Be, V, Cu, Nb, Mo and Ta surfaces has been measured for energies between 5 keV and 150 keV and a wide range of angles of emergency. Hydrogen particles with energies above 20 keV are counted and energy analysed by a surface barrier detector. Charged particles are separated from the neutrals by means of electrical deflection plates between target and detector. Neutrals with energies below 20 keV are partly ionized in a calibrated gas stripping cell. They are energy analysed in a subsequent electrostatic spectrometer and counted by a channeltron multiplier. The backscattered ions were recorded with no gas in the stripping cell. Only small differences are found for the charged fraction for different materials as long as the surface is covered by a layer of adsorbed impurities. There is, however, for most materials a change in the charged fraction due to annealing the target. For emergence energies above ~ 40 keV it is lower than for unannealed targets. An observed dependence of the charged fraction on the angle of emergence was generally just slightly above the experimental error.

The measured results are compared with
theoretical curves of Zaidins, Trubnikov et al.,
and Brandt and Sizmann. The best agreement is
found with values given by Zaidins. The
peculiarities observed at annealed surfaces
are not predicted by theory.

Introduction

The charge state of energetic atoms after leaving the surface
of a solid has been investigated since a long time with increasing
effort [1,2,3]. It is of considerable interest in beam foil spectroscopy [4], surface analysis by ion backscattering [3,5] and the
investigation of electron states in the surface region [6,7]. The
results for hydrogen and helium atoms are also of interest for the
problem of recycling in high temperature plasma experiments, as
only the neutrals can move back into the plasma [8].

In several papers it has been assumed that the charge state of
an atom after leaving a surface is the same [9,11] as the charge
state while penetrating the solid. It is determined by the cross
sections for electron loss and capture of the atom in the solid
[12,13]. However, other theories assume that the final charge state
is reached mainly in leaving the surface, i.e., in passing the decreasing electron density in the near surface region [6,14,15].
This means that the charge state is determined by the last surface
layer and really clean surfaces should be used in the measurement.

In this work the charged fraction of backscattered hydrogen
after leaving the surface of different materials has been measured
and is compared with the results of others and with theoretical calculations [6,10,14,15,16].

Experiments

The measurements have been performed by backscattering of protons from solid surfaces. For a given incident energy the energy
distribution of the backscattered particles covers the whole energy
range from zero up to almost the primary energy [17,18,19]. Except
for the highest energy in the spectra the particles are backscattered
from a depth of more than $\sim$10 Å. Therefore, the charge state of the
backscattered particles is expected to be the same as in transmission experiments, where the atoms enter a thin solid foil on one
side and leave it on the other side.

Two different experimental set ups, shown schematically in
Fig. 1, have been used. One covers the energy range from 5 to 20
keV, the other one the energy range from 20 to 150 keV. In both
experiments magnetically analyzed well collimated proton beams
[20,21] are incident on the target. The backscattered particles
are analyzed according to their energy and their charge.

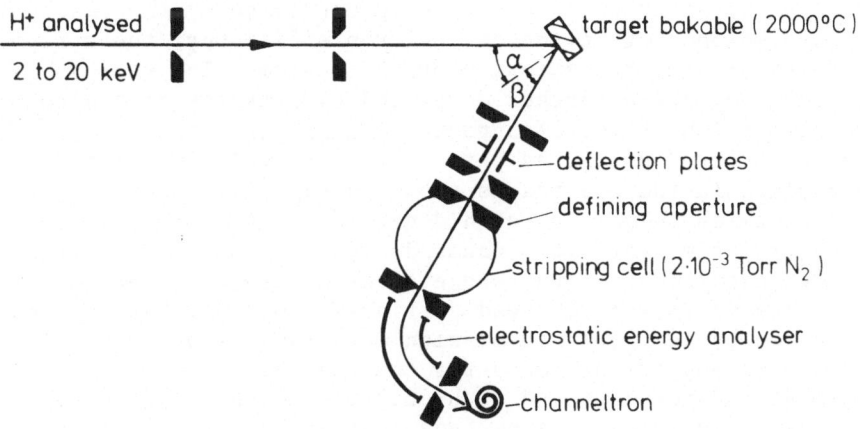

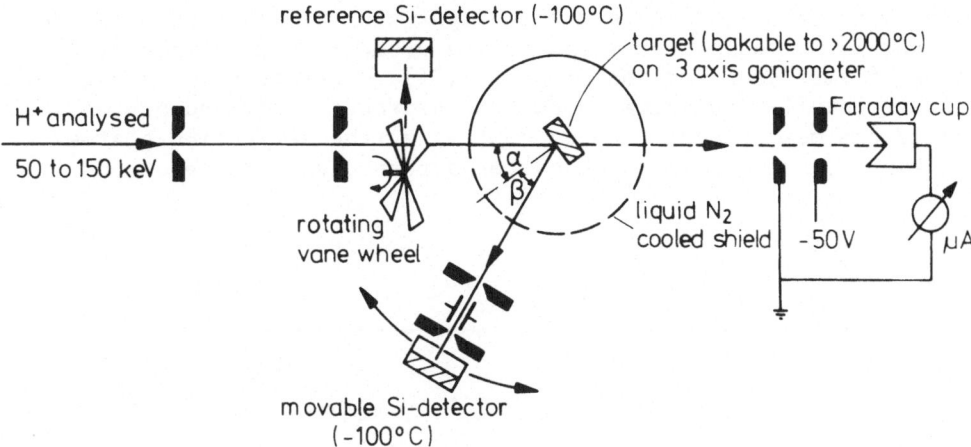

Fig. 1. Schematic of the experimental setups.

For energies below 20 keV a gas stripping cell followed by an electrostatic energy analyser was used.[†] This was mounted at a scattering angle of 135°. The positively charged particles have been measured with no voltage applied to the deflection plates and without gas in the stripping cell (see Fig. 1, top). In order to detect the neutrals the charged particles were removed by the deflecting plates and the gas cell was filled with $2 \cdot 10^{-3}$ Torr dry nitrogen. Both for the neutrals and the charged particles the acceptance angle as determined by the front aperture of the gas cell was 10^{-6} sr.

[†]This particle detector has been constructed and calibrated by Dr. J. P. Girard and Dr. F. P. G. Valckx, Ass. EURATOM, C.E.A. Fontenay aux Roses.

The measured energy distributions have been corrected by the calibration curves, which contain the channeltron efficiency, the stripping probability and the transmission of the system. The energy
spectra obtained have been further corrected by a factor of $\frac{1}{E}$ account-
ing for the dispersion of the electrostatic analyzer. Thus, the
number of backscattered particles per energy interval (in arbitrary
units) are plotted in the spectra as shown in Fig. 2a. Charged particles with energies below 1.4 keV did not reach the detector because of the stray field of a few Gauss due to the ion pump magnets.
Therefore, the measurements have been evaluated only for energies
above 5 keV. The negatively charged particles could not be detected.
However, as shown by earlier transmission measurements on thin films
[2,16] they become appreciable only for energies below 5 keV.

Hydrogen atoms backscattered with energies above 20 keV were
counted and energy analyzed by a surface barrier detector with an
energy resolution better than 3 keV [23]. Electrical deflection
plates were installed between the target and the detector. As the
detector is moveable in the vacuum measurements can be made at all
angles of emergence and incidence [19,24]. Surface barrier detectors are equally sensitive to neutral hydrogen atoms and protons.
Thus in one measurement with no voltage at the deflection plates
(see Fig. 1, bottom) all backscattered particles were recorded.

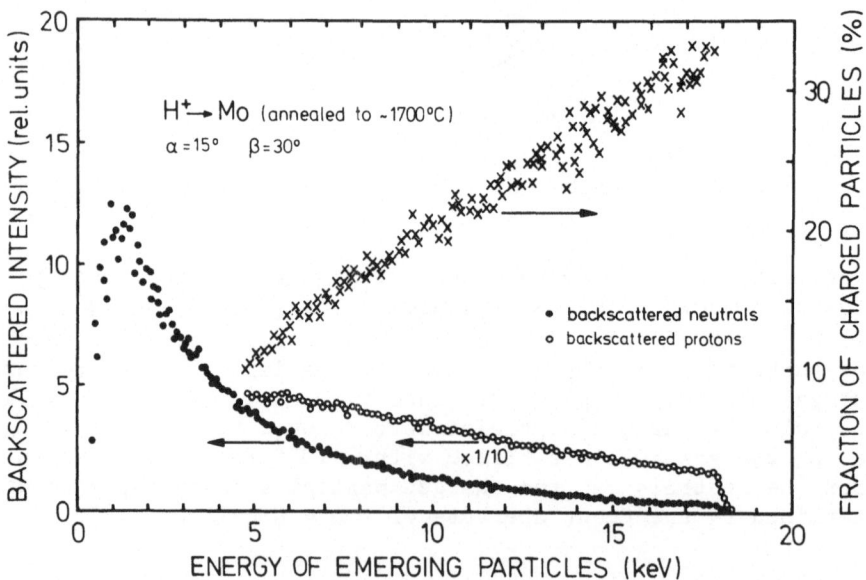

Fig. 2a. Energy spectra of the backscattered neutrals and protons
for 18 keV protons incident at an angle of 15° relative to the normal on a molybdenum surface.

In a second measurement with the same primary ion dose the charged
particles were deflected and only the neutrals were counted. A
typical spectrum is shown in Fig. 2b. The energy distribution of
all backscattered particles has further been used for a Rutherford
backscattering analysis of surface and bulk contaminants [25].
This was especially valuable in the case of Be.

     The data have been recorded on a multichannel analyzer (DIDAC
800) and evaluated on the IBM 360/91 computer. The charged fraction,
i.e., the number of particles backscattered positively or negatively
charged divided by the total number of backscattered particles has
been calculated for every energy interval corresponding to one channel. As there were generally some statistical fluctuations in the
points the curves have been smoothed by averaging over 5 to 10
channels at every point. The targets used for the experiments
were polycrystalline high purity materials of Be, V, Cu, Nb, Mo,
and Ta. They have been mechanically polished before being mounted
in the target chamber. The fraction of charged particles was
measured on the mechanically polished targets as they were built in
and after annealing the target in situ in several steps up close to
the melting point. For the low energy accelerator the base pressure

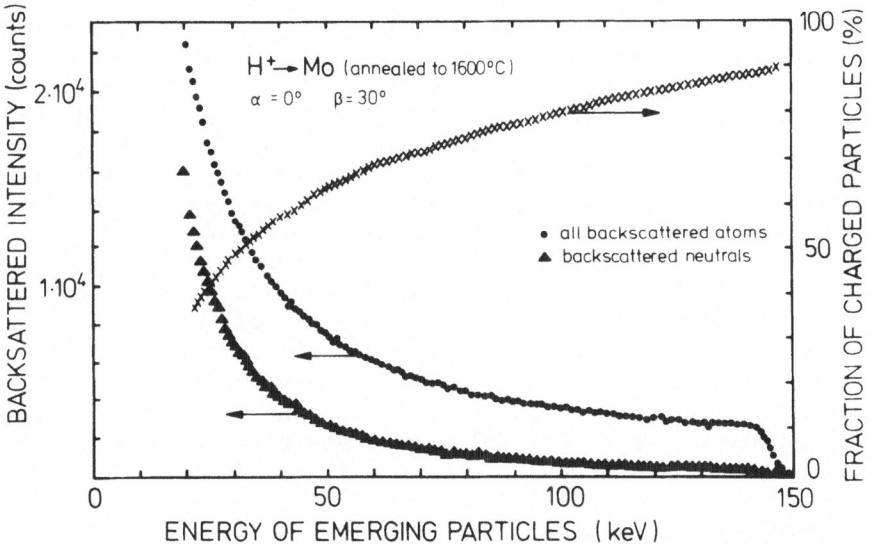

Fig. 2b. Energy spectra of all backscattered particles and of the
backscattered neutrals for 150 keV protons at normal incidence on
a molybdenum surface. In both figures the fraction of charged
particles is also plotted as calculated from the two distributions.

in the target chamber was 10^{-9} Torr. During the measurements of the neutrals the pressure rose up to 2×10^{-7} Torr N_2 due to the leakage from the stripping cell. For the high energy accelerator the residual gas in the target chamber was 10^{-7} Torr Hg from the Hg diffusion pumps as well as CO and H_2O both in the 10^{-9} range. The target was surrounded by a copper shield at liquid nitrogen temperature, to improve the vacuum conditions around it. No mercury could be detected on the target surface as was concluded from the backscattering spectra.

Results

In the energy range investigated the fraction of charged hydrogen atoms after leaving the solid surface increases with emerging energy from about 10% to about 90%. No dependence on the angle of incidence of the primary ions has been observed. This general feature holds for all materials investigated. The differences measured for different materials and different angles of emergence and annealing conditions are generally small and partly just above the uncertainties of the measurements.

Beryllium

The results for Be are shown in Fig. 3. After annealing, the charged fraction increased at energies above 50 keV, while at energies below 50 keV the fraction of charged particles decreased. This effect has already been observed by Phillips [16] in a transmission experiment. His absolute values agree well with those obtained by us after annealing to 1000°C and 1100°C. Rutherford backscattering surface analysis [25] showed that the mechanically polished Be was covered with a beryllium oxide layer containing about $2 \cdot 10^{17}$ oxygen atoms/cm². This could not be removed by 4 keV Ar bombardment up to doses of $3 \cdot 10^{16}$/cm², but after bombardment implanted Ar was found. During the annealing process the oxygen started diffusing into the bulk material at $\sim$900°C. At 1100°C platinum from the Pt-PtRh thermocouple diffused all over the target. However, this had no influence on the fraction of charged particles.
The angle of emergence showed only a small effect on the fraction of charged particles. For energies above $\sim$50 keV the charged fraction decreased continuously with increasing emergence angle (about 5% absolute for 30° to 84°). For energies below 50 keV no angular dependence was found.

Vanadium

The results of the measurements on vanadium are plotted in Fig. 4. The difference in the fraction of charged particles for the mechanically polished surface and the well annealed surface

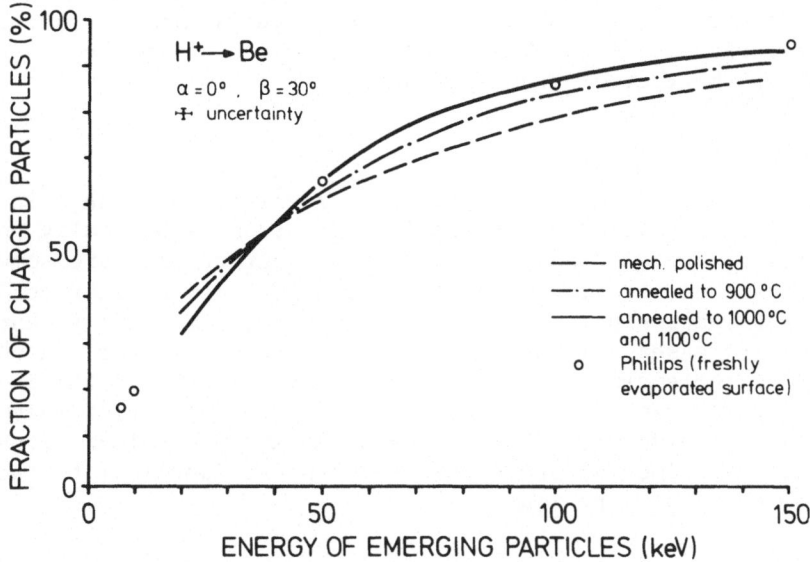

Fig. 3. Fraction of charged particles backscattered from a Be target, mechanically polished and annealed in situ to 900, 1000 and 1100°C. The angle of emergence was β = 30° relative to the normal. The values measured by Phillips [16] are also included.

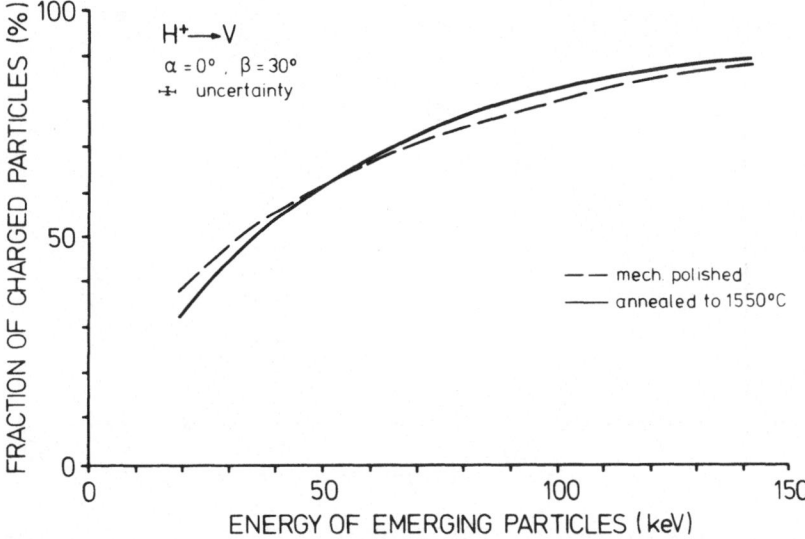

Fig. 4. Fraction of charged particles backscattered from vanadium, mechanically polished and annealed in situ to 1550°C. The angle of emergence was β = 30°.

observed is similar to that on Be. The fraction of charged particles did not change with angle of emergence within the experimental accuracy of 2-3%. The vanadium surface is expected to be
clean after the thermal treatment applied.

Copper

Figure 5 shows the results for Cu. There was no change in
the fraction of charged particles in going from mechanically polished to well annealed surfaces. However, our results are somewhat
higher than those of T. Buck et al. measured under similar conditions. For increasing angles of emergence the fraction of charged
particles increased slightly. In order to check whether the charged
fraction of the emerging atoms depends on the energy of the incident
ions and on the use of atomic or molecular ions, a set of measurements has been made with 75 keV H^+ and 150 keV H_2^+ primary ions.
Within the experimental errors the results are identical to those
with 150 keV H^+ primary ions.

Niobium

The measurements on niobium surfaces have been performed on
single crystals annealed to 2200°C. A surface analysis from the

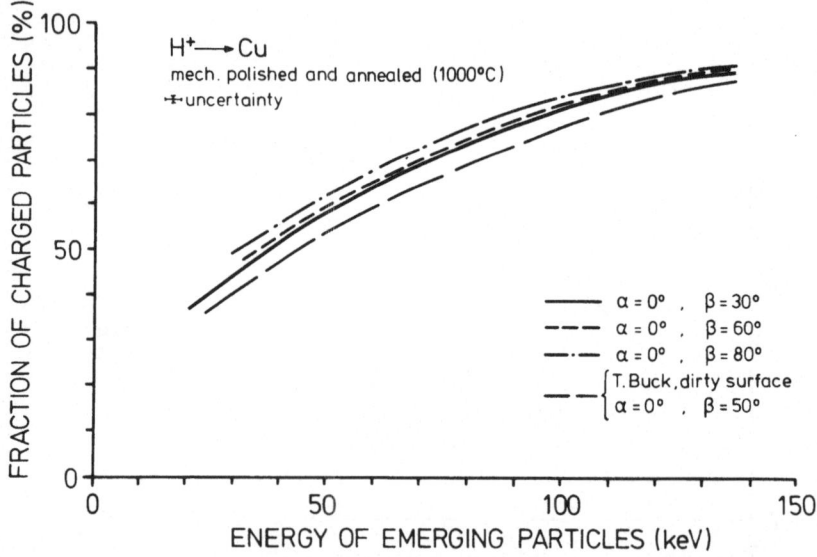

Fig. 5. Fraction of charged particles backscattered from copper
for angles of emergence β between 30 and 80°. There was no
difference between the results at the mechanically polished and the
annealed (1000°C) surface. The values measured by Buck [3] et al.
are also included.

energy spectrum of backscattered protons in double alignment
showed that the surface was well ordered and the oxygen content
below 1/10 of a monolayer [26]. The fraction of charged particles
has been measured for emergence parallel to close packed directions
(blocking) and well off close packed directions (random). Figure 6
shows the results for random incidence and emergence together with
the results of K. H. Berkner et al. [2]. Within the experimental
errors no dependence on the emergence angle could be found. For
blocking conditions, i.e., emergence in a close packed direction
the measured fraction of charged particles was lower by 10 to 20%;
however, statistics were generally poor. This is contrary to what
has been observed in transmission experiments at single crystal
foils. Here the charged fraction was found to be higher for channeling conditions, i.e., transmission in close packed directions
compared to random directions [27,28]. This can be explained by
the difference of the ion trajectories in both cases. For blocking
conditions, only those particles can emerge in a close packed direction which have made close collisions with the last atoms of a
terminated lattice row. In transmission experiments, however, the
particles transmitted in close packed directions always stayed far
away from any lattice row.

Molybdenum

The charge states measured on the mechanically polished and
on the annealed target are shown in Fig. 7. They show a similar

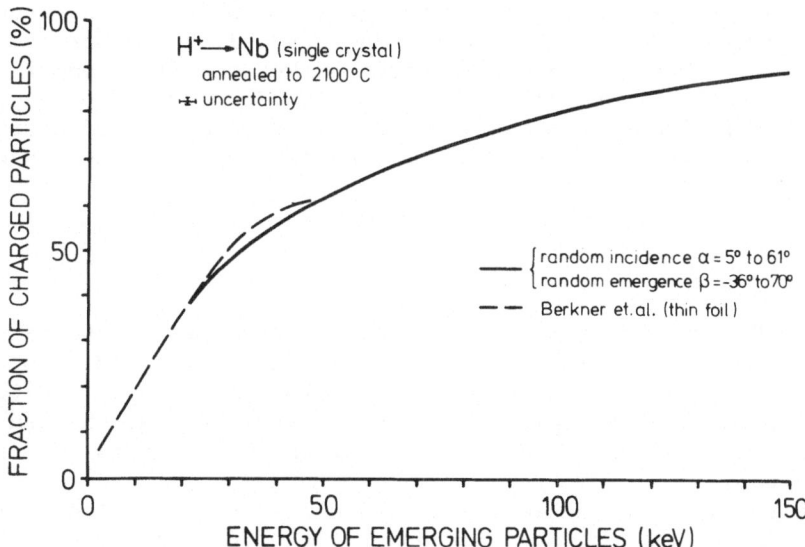

Fig. 6. Fraction of charged particles backscattered from a niobium
single crystal annealed to 2200°C for different angles of emergence
together with the results of Berkner et al. [2].

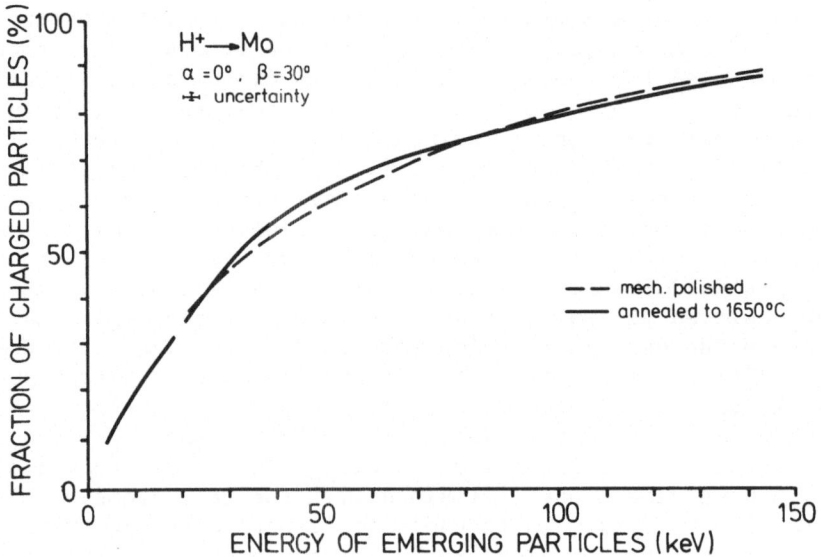

Fig. 7. Fraction of charged particles backscattered from a molybdenum surface, mechanically polished and annealed to 1650°C. The measurements at energies below 20 keV have been performed with the low energy accelerator.

tendency as observed on Be and V, however, even more complex. The fraction of charged particles at energies below 20 keV has been measured only on Mo, the results match very well those for energies above 20 keV. While at energies above 50 keV the fraction of charged particles increased about 3-4% with increasing angle of emergence relative to the surface normal, the fraction of charged particles decreased slightly with increasing angle of emergence β for energies below 20 keV.

Tantalum

Fig. 8 shows the fraction of charged particles for tantalum. The surface was mechanically polished but the target could not be annealed, i.e. the surface was covered with $\sim$ 30 Å Ta_2O_5. With increasing angle of emergence a slight increase of the charged fraction was found which was, however, within the limits of experimental error.

The results for the well annealed surfaces are collected in Fig. 9. Though the energy dependence of the fraction of charged particles is different for the different materials all curves except the one for Be coincide within $\sim$5%. The values measured at the mechanically polished surfaces fall into the same range. At about 40 keV the fraction of charged particles is 50%, decreasing to lower energies and increasing to higher energies.

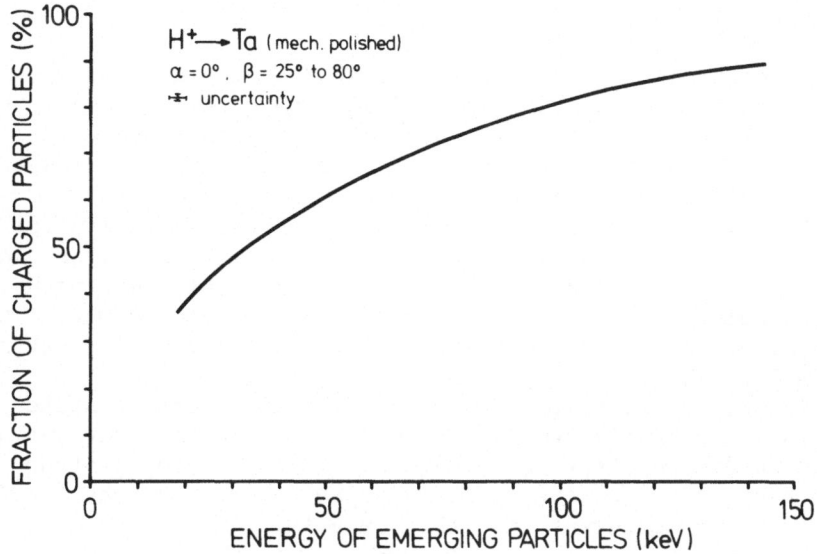

Fig. 8. Fraction of charged particles backscattered from a tantalum surface covered with ~30 Å Ta_2O_5.

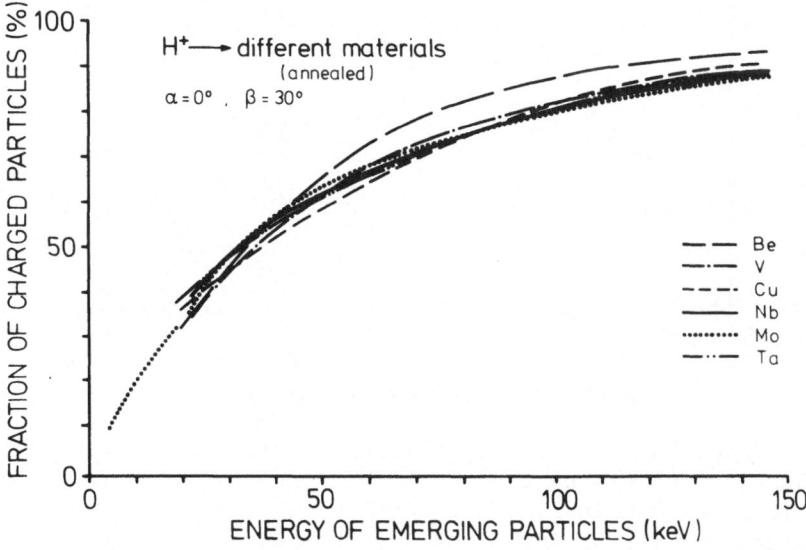

Fig. 9. Fraction of charged particles for Be, V, Cu, Nb, Mo and Ta for an angle of emergence of 30°. All targets except Ta have been well annealed in situ.

Comparison with Theory

The theoretical calculations published on the charge state of hydrogen atoms after leaving a solid surface give only semi-quantitative results. Fig. 10 shows the different theoretical curves [6,9,10,14,15] together with our experimental results and with those of others [2,3,9,16,27,29].

C. S. Zaidins [10] has calculated the charged fraction for hydrogen atoms after leaving the surface from the model of electron loss and pickup in the solid. No details are published and nothing is said about the dependence on angle of emergence. However, his results agree well with the experimental results of this work as shown in Fig. 10.

Yavlinski et al. [14,15] have calculated the charge states from the probability of electron pickup of the ion in the near surface region. He distinguishes two velocity ranges for the protons v_p with respect to v_0, the velocity of an electron at the Fermi-surface, giving different results. He gets:

$$v_p \ll v_0 : \frac{N^+}{N^+ + N^0} = \exp -\sqrt{\frac{E^*}{E_p}}$$

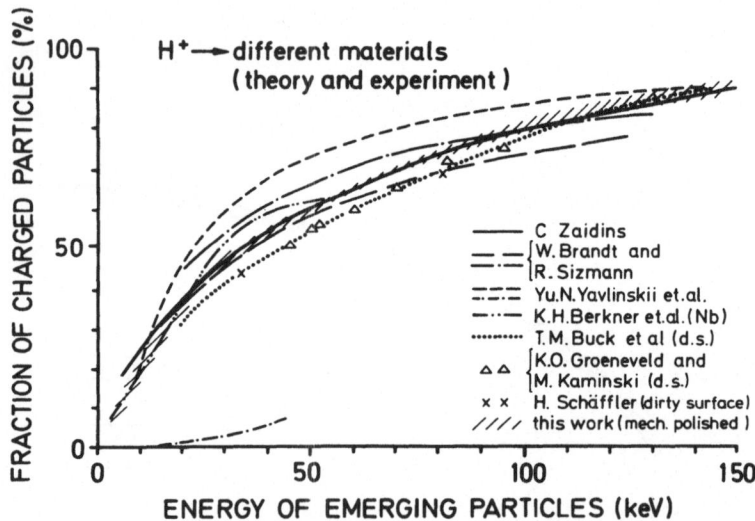

Fig. 10. Fraction of charged particles backscattered from different materials as predicted by the theory of Zaidins [10], Brandt and Sizmann [6], and Yavlinski and Trubnikov [25], together with the experimental results of this work and different other groups.

and
$$v_p \gg v_o : \frac{N^+}{N^+ + N^o} = \exp - \frac{E^{**}}{E_p}$$

with
$$E^* = 136 \left(\frac{n_o}{10^{22}}\right)^{1/3} [\text{keV}] \quad \text{and} \quad E^{**} = 6.5 \left(\frac{n_o}{10^{22}}\right)^{1/2} \text{keV}$$

The measurements reported here lie in the range $1 < v_p/v_o < 10$. The values from the formula (1) or (2) are introduced in Fig. 10, but they deviate largely from experimental results.

Recently Brandt and Sizmann [6,28] calculated the energy dependence of the fraction of charged hydrogen atoms at energies of about 10 keV to 100 keV. They get:

$$1 - \frac{1}{1 + 0.7 \dfrac{E_p [\text{keV}]}{25}} \lesssim \frac{N^+}{N^+ + N^o} \lesssim 1 - \frac{1}{1 + \dfrac{E_p [\text{keV}]}{25}} \tag{3}$$

The results of formula (3) are also introduced in Fig. 10. Though formula (3) gives again values of the right order of magnitude the general form and energy dependence of the measured curves is not predicted by this theory. However, Brandt and Sizmann further conclude that the charge state should be nearly independent of the angle of emergence from the solid. This has been well confirmed by the experimental results.

All theories give the correct order of magnitude of the experiments, the special features, i.e. the change of the charged fraction in going from mechanically polished to well annealed, i.e. reasonable clean, surfaces are not predicted.

Summary

Backscattering seems to be a valuable straightforward method to investigate the charge states of ions emerging from a solid surface. Though small differences are observed for different materials, the measured curves lie very close together. The last surface layer has a major influence on the charged fraction. There is a considerable difference in the fraction of charged particles between dirty and annealed surfaces; the latter give generally higher values for energies above 40 keV and lower values for energies below 40 keV. The fraction of charged particles is 50% near $\sim$ 40 keV. The special features, i.e. the dependence on surface conditions observed are not predicted by any theory.

Acknowledgments

The authors would like to thank Professor Brandt and Professor Sizmann for discussing the theoretical aspects of ion surface neutralization. The skillful technical assistance by R. Hippele, S. Schrapel, and H. Wacker is kindly acknowledged. The drawings have been made by C. Drewes and the manuscript was typed by I. Wunderlich. We thank Dr. Girard, Dr. Valckx, and Dr. Eckhardt for their kind collaboration in handling the neutral particle detector.

Literature

[1] S. K. Allison, Rev. Mod. Phys. $\underline{30}$, 1137 (1958).
[2] K. H. Berkner, I. Bornstein, R. V. Pyle, J. W. Stearns, Phys. Rev. A $\underline{6}$, 278 (1972).
[3] T. M. Buck, G. H. Wheatley, L. C. Feldman, Surf. Sci. $\underline{35}$, 345 (1973).
[4] See: Nucl. Instr. Meth. $\underline{110}$, 1-522 (1973).
[5] E. S. Mashkova and V. A. Molchanov, Rad. Eff. $\underline{16}$, 143-187 (1972).
[6] W. Brandt and R. Sizmann, Phys. Lett. $\underline{37A}$, 115 (1971).
[7] C. Rau and R. Sizmann, Phys. Lett. $\underline{43A}$, 317 (1973).
[8] R. Behrisch, W. Heiland, 6th Symp. on Fusion Technology, Aachen, 1970.
[9] T. Hall, Phys. Rev. $\underline{79}$, 504 (1950).
[10] C. S. Zaidins, Ph. D. Thesis, Appendix 1, California Institute of Technology 1967; see also: J. B. Marion and F. C. Young, Nucl. Radiation Analysis, North Holland 1968, p. 36.
[11] M. E. Ebel, Phys. Rev. Lett. $\underline{24}$, 1395 (1970).
[12] N. V. Federenko, Sov. Phys. Techn. Phys. $\underline{15}$, 1947 (1971).
[13] H. Tawara and A. Russek, Rev. Mod. Phys. $\underline{45}$, 178 (1973).
[14] B. A. Trubnikov and Yu. N. Yavlinski, Sov. Phys. JETP $\underline{25}$ 1089 (1967).
[15] Yu. N. Yavlinski, B. A. Trubnikov, V. F. Elesin, Bull. Acad. Sci., USSR Phys. Sov. $\underline{30}$, 1996 (1968).
[16] J. A. Phillips, Phys. Rev. $\underline{97}$, 404 (1955).
[17] S. Rubin, Nucl. Instr. Meth. $\underline{5}$, 177 (1959).
[18] E. Bøgh, Can. J. Phys. $\underline{46}$, 653 (1968).
[19] R. Behrisch, Thesis, Techn. Univ. of Munich (1968)
[20] W. Eckstein and H. Verbeek, IPP Report 9/7, 1972 and Vacuum $\underline{23}$, 159 (1973).
[21] R. Behrisch, Vak. Techn. $\underline{10}$, 250 (1967).
[22] A. Egidi, R. Marconero, G. Pizella, Rev. Sci. Instr. $\underline{40}$, 88 (1969).
[23] H. Schmidl, IPP Report 9/3 (1971).
[24] B. M. U. Scherzer, Thesis, Techn. Univ. of Munich (1969).
[25] See for example: M. A. Nicolet, J. W. Mayer, I. V. Mitchell, Science $\underline{177}$, 481 (1972).
[26] R. Behrisch, B. M. U. Scherzer, H. Schulze, Rad. Eff. $\underline{13}$, 33 (1972).

[27] K. O. Groeneveld and M. Kaminsky, Bull. Am. Phys. Soc. <u>14</u>, 1246 (1969) and private communication.
[28] R. Sizmann, private communication.
[29] H. Schäffler, Thesis, Technical University of Munich (1973).

CHARGE NEUTRALIZATION OF MEDIUM ENERGY H AND ^{4}He IONS BACKSCATTERED FROM SOLID SURFACES, EFFECTS OF SURFACE CLEANING

T. M. BUCK, L. C. FELDMAN, and G. H. WHEATLEY
Bell Laboratories
Murray Hill, New Jersey   07974, U.S.A.

ABSTRACT

Experiments on charge states of H and ^{4}He backscattered from silicon and gold surfaces in the energy range 25-190 keV have been extended to investigate surface impurity and surface cleaning effects. The average charge state of ^{4}He scattered from iodine or gold impurities on silicon surfaces was close to that of ^{4}He scattered from silicon at the same energy. Argon bombardment cleaning of a gold surface caused substantial changes in neutralization for both ^{4}He and H, similar to results of Phillips in the case of hydrogen, and suggestive of more than one neutralization mechanism. The results are consistent with a neutralization step occurring outside the surface as the particle leaves. It is pointed out, however, that at low energies, 0.5-5 keV for ^{4}He, preferential neutralization of ions which penetrate beyond the surface has been inferred from electrostatic analyzer data.

Introduction

Charge neutralization of light ions as they are backscattered from a solid surface is important in electrostatic energy analysis at medium energies (50-200 keV) and, especially, at low energies (0.5-5 keV). Measurements of ion-fractions in the medium-energy range and observations of ESA data in the low energy region have raised some questions regarding the neutralization mechanisms which will be pointed out in this paper.

At medium energies the ion-fractions of H and ^{4}He scattered from "practical" surfaces (not cleaned in ultra-high vacuum) appear to be smooth functions of the exit velocity with little or no dependence on penetration depth [1]. A small influence of the target material was observed; He$^+$ ions were neutralized more efficiently at a gold surface than at a silicon surface. Also, ^{4}He scattered from unidentified heavy impurity atoms on a silicon surface ($\sim 5 \times 10^{13}/cm^2$) exhibited the behavior typical for a silicon surface.

In this paper another gold-silicon comparison is reported, together with results for ^{4}He scattered from trace surface impurities of iodine and gold on silicon. The medium energy measurements for H and ^{4}He have also been made on gold surfaces cleaned by argon bombardment in UHV and sensitivity to surface cleanliness is apparent. The medium energy results we have observed thus far are consistent with a neutralization process occurring as the particle leaves the surface rather than inside. Yavlinskii, et al. [2] and Brandt and Sizmann [3] have pointed out that Debye-type shielding by electrons should prevent bound states for hydrogen traveling through a metal.

However, at low energies, ≤ 5 keV for ^{4}He, there has been evidence in ESA data for additional neutralization processes occurring inside the solid. In fact, one of the most interesting features of low energy ion scattering for surface analysis, i.e., the resolution of peaks for light elements on heavy atom substrates without ion channeling, has been attributed [4,5,6] to the more complete neutralization of noble gas ions which penetrate beyond the first one or two atom layers, thereby reducing the background yield from deeper substrate atoms. This apparent difference between charge neutralization behavior at medium and low energies has not yet been resolved.

Experimental Procedure

The experimental system and procedure for medium energy charge exchange experiments have been described previously [1,7]. In each experiment two or more backscattered spectra were obtained alternately (using the same incident dose): A) Ions plus neutrals; B) neutrals alone, by deflecting ions out of the scattered beam. The energies of the scattered particles were measured by a silicon surface barrier detector in conjunction with a multichannel pulse height analyzer. Letting A and B represent the respective yields in counts per channel, the ion fraction corresponding to a channel number or energy is then given by $y^+/y = (A-B)/A$.

Target samples were prepared as follows: <u>polycrystalline gold</u>; 99.8% pure, rolled and annealed to give crystallite size of 0.1-0.2 mm; rinsed in trichlorethylene and acetone; electropolished in a saturated solution of NaCl in H_2O(80 parts) and HCl(20 parts); rinsed in deionized water and methanol. A gold sample prepared this way was mounted in the chamber and experiments on a "practical" surface were carried out without further cleaning, at pressures of 5×10^{-9} to 2×10^{-8} torr in the differentially pumped chamber. Experiments on "clean" surfaces were performed on the same sample after heating to 800°C and argon ion bombardment (460 eV, $3 \times 10^{16}/cm^2$). <u>Silicon</u>; single crystal, 10 ohm-cm n-type with (111) surface Syton polished, lightly etched in HNO_3-HF and rinsed in deionized water. One Si slice was given an iodine-methanol [8] treatment before mounting in the vacuum chamber in order to provide a spectrum with an iodine impurity peak. Another slice was etched, soaked in HF, and dipped in a very dilute solution of KAuCN to produce a small gold peak in the spectrum.

Experimental Results

   ^{4}He scattered from practical surfaces

   Figure 1 shows curves of y^+/y vs energy for ^{4}He scattered from practical surfaces of gold and silicon, and also silicon contaminated with iodine ($\sim 2 \times 10^{13}/cm^2$) or gold ($\sim 1.4 \times 10^{13}/cm^2$). Points to be made about these data are: a) The curve for ^{4}He scattered from

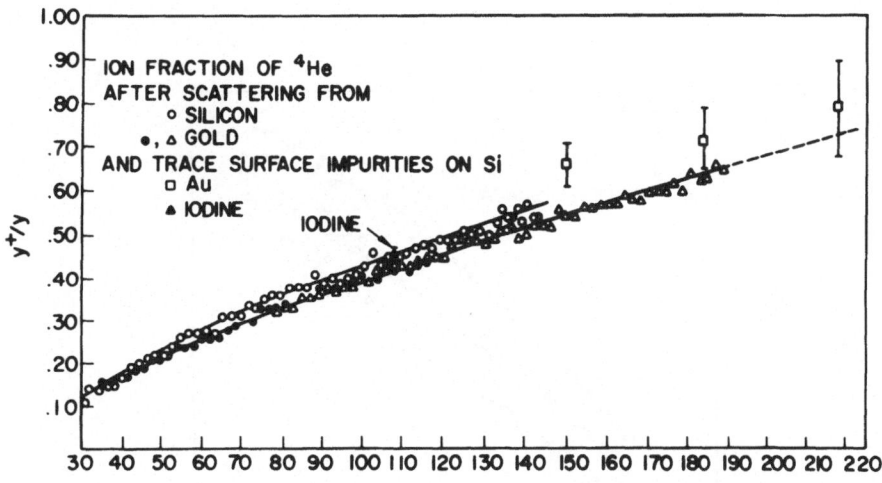

Fig. 1. Curves of ion-fraction vs scattered energy for ^{4}He$^+$ incident on single crystal silicon in channeling direction (110) at 240 keV, and on polycrystalline gold at 120 and 200 keV. Practical surfaces. ▲ and □ are points derived from iodine and gold as impurities on silicon. Scattering angle 130° for these and other spectra.

silicon lies slightly higher than the one for [4]He scattered from
gold, as was found previously [1]. b) Points for two different
incident energies on solid gold, 120 and 200 keV, fall on the same
curve, showing no effect of penetration depth, as was also observed
before. c) The point at 108 keV for [4]He scattered from iodine on a
silicon surface falls on the silicon curve, again showing no special
effect of a target atom known to be very close to the surface, and
suggesting that the ion fraction is determined by the substrate
surface environment rather than the particular identity of an isolated target atom on the surface. d) The points for gold on silicon
are definitely higher than the curve for solid gold and are closer
to an extrapolation of the silicon curve, again suggesting the influence of the substrate surface environment.

[4]He and H scattered from clean gold surface

     We use the word "clean" in a relative sense here. The argon
bombardment treatment described above should have been sufficient
to remove adsorbed gases, although no independent check, e.g., by
Auger electron spectroscopy, was made. Pressure in the chamber
after bombardment was about 2×10^{-8} torr and the "clean" scattering
experiments were done quickly after bombardment. In Fig. 2 the
ion-fraction curve for [4]He scattered from the cleaned surface is
seen to be substantially lower than the curve for a practical surface, while the curve for 12 hours exposure at 2×10^{-8} Torr lies

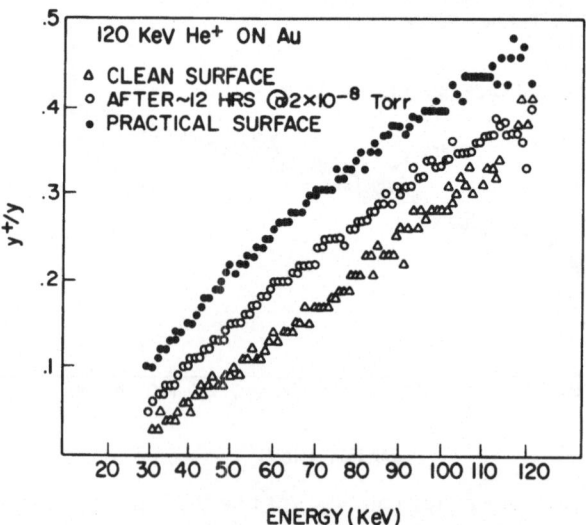

Fig. 2. Ion-fraction vs scattered energy for [4]He scattered from
polycrystalline gold: before bombardment cleaning (●), after heating to 800°C and argon bombardment cleaning (Δ), and 12 hours after
cleaning (o).

in between. Figure 3 shows data for hydrogen corresponding to similar surface conditions. The behavior is more complicated: at the
low energy end the "clean" surface curve lies substantially lower
than the "practical" surface curve but it rises steeply and crosses
over the practical curve at about 50 keV, then crosses again at
90 keV.   After the surface had been exposed to a background pressure of 2×10^{-8} torr for 12 hours, the y^{+}/y curve had a shape more
like that of the practical surface curve, but was lower. The
intricate behavior of the clean surface curve happens to match
remarkably well with some data of Phillips [9] published in 1955
as shown in Fig. 4. Phillips' data was obtained on hydrogen beams
transmitted through thin Al foils and scattered through 90°. His
practical, or "dirty," surface condition was that of the Al foil
after long standing in the chamber. The clean surface condition
was achieved by evaporating a fresh layer of gold shortly before
the scattering experiment.

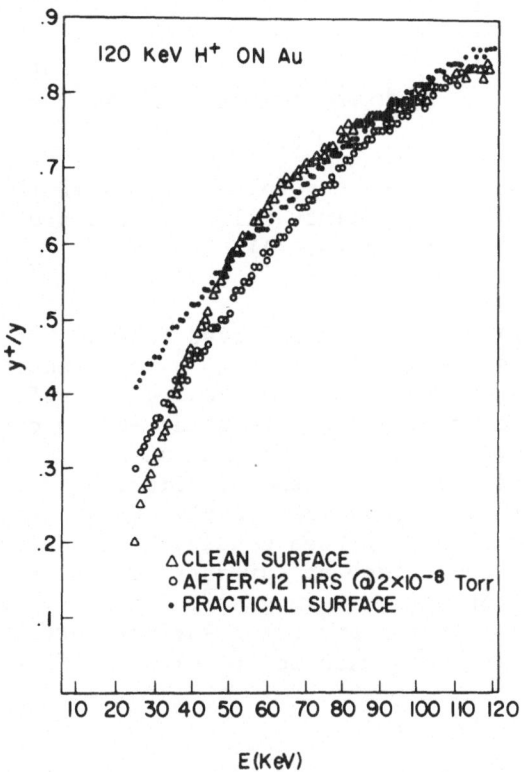

Fig. 3. Ion-fraction vs scattered energy for H scattered from
polycrystalline gold: before bombardment cleaning (●), soon after
argon bombardment (Δ), and 12 hours after cleaning (o).

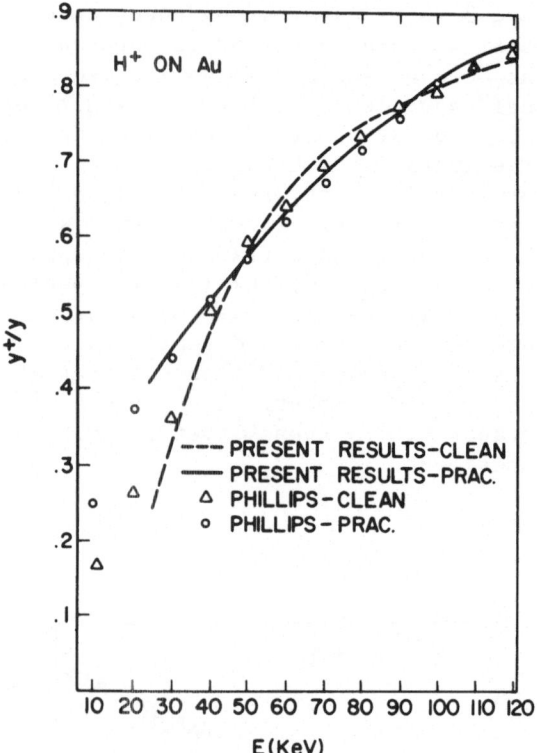

Fig. 4. Comparison of present results for H with those of Phillips [9] for H after traversing "dirty" Al foil (o), and Al foil with fresh gold layer on exit surface (Δ).

Since particle velocity is believed to have the dominant influence on charge state we show in Fig. 5 ion fraction vs velocity for both ^{4}He and H on clean and practical gold surfaces. The ^{4}He curves join the H curves in the same order—clean curve lower than practical. As previously found there is some difference in slope between the ^{4}He and H curve for the practical surface.

Since there is a considerable supply of data on charge exchange in gases and since solid surfaces usually have adsorbed gas layers we have included a comparison of our data with some equilibrium charge state data for He beams in gases, O and Ar, (Fig. 6). Both the shape and height of the practical surface curve are more characteristic of the gas data than is the clean surface data.

Discussion and Summary

The charge state data for ^{4}He and H scattered from solid surfaces at medium energies illustrate several points, all of which support a mechanism [2,3] of ion neutralization occuring outside the surface as the particle leaves:

1.   There is little or no evidence of a depth effect, i.e.,
of deeper penetrating ions being more completely neutralized. Thus,
as has been found before, curves of ion-fraction vs backscattered
energy coincide, regardless of incident energy in the medium range.
Furthermore, ion fractions of particles scattered from impurity
atoms (I or Au), known to be deposited on the surface, fall on or
close to the substrate curve, suggesting that the neutralization is
dominated by the electron distribution of the substrate surface.

2.   There is a small but significant influence on charge
state by the target material for practical surfaces. Solid gold
neutralizes ^{4}He more efficiently than does a silicon surface. But

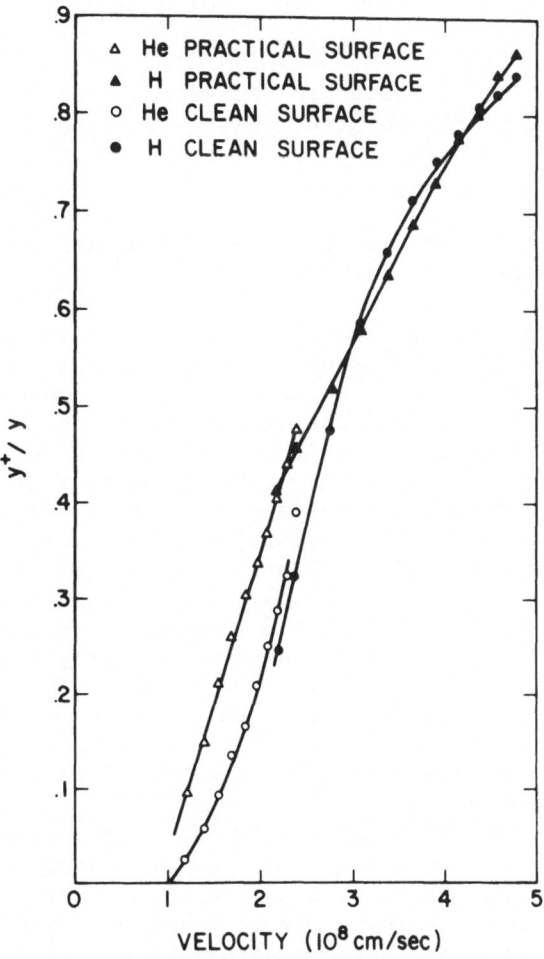

Fig. 5. Ion-fraction vs particle velocity for ^{4}He and H after
backscattering from practical and clean gold surfaces.

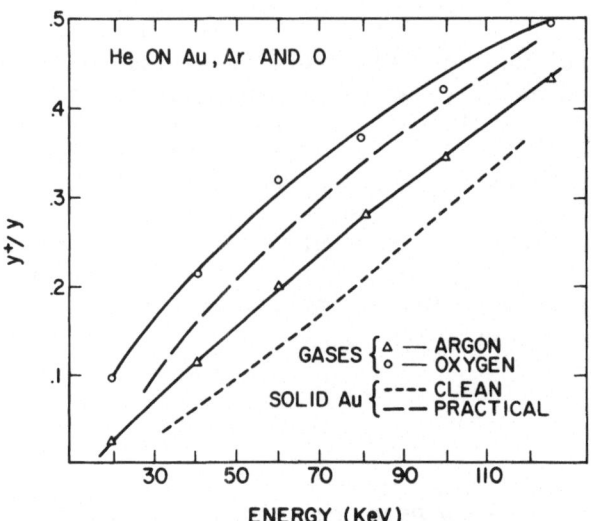

Fig. 6. Ion fraction vs scattered ^{4}He velocity after scattering from gold (present work) and after passing through gases (Barnett and Stier [10]).

ion fractions of ^{4}He scattered from gold atoms deposited on silicon, at very low concentration, lie closer to the silicon curve than to the gold curve.

3. Argon bombardment cleaning of gold produces substantial reduction in ^{4}He ion fractions over the energy range 30-120 keV, and 12 hour aging of the surface at 2×10^{-8} torr causes the ion fraction curve to rise toward the practical surface curve. This behavior suggests charge neutralization occurring outside the surface by a tunneling mechanism [2] which would be inhibited by a surface film. The hydrogen results (25-120 keV) which correspond to higher scattered velocities exhibit a more complicated behavior, with the clean curve starting out below the practical curve at 25 keV, then crossing over at 50 keV, and recrossing at 90 keV, in a manner very similar to Phillips' [10] results. This seems to indicate one or more additional mechanisms competing with the tunneling mechanism at higher velocity.

4. Although the evidence cited above points strongly to neutralization outside the surface for medium energies, we would like to point out that certain low-energy electrostatic analyzer data seem to suggest neutralization occurring inside. An example is given in Fig. 7 which shows ESA spectra for ^{4}He scattered from polycrystalline gold, for 4 different incident energies [6]. For 25 keV incident energy there is no surface peak detected and the yield falls off slowly at lower energy because of analyzer resolution and somewhat greater charge neutralization. As incident energy is reduced there is a very rapid decrease in the low energy yield and a very pronounced surface peak develops. Either the scattering

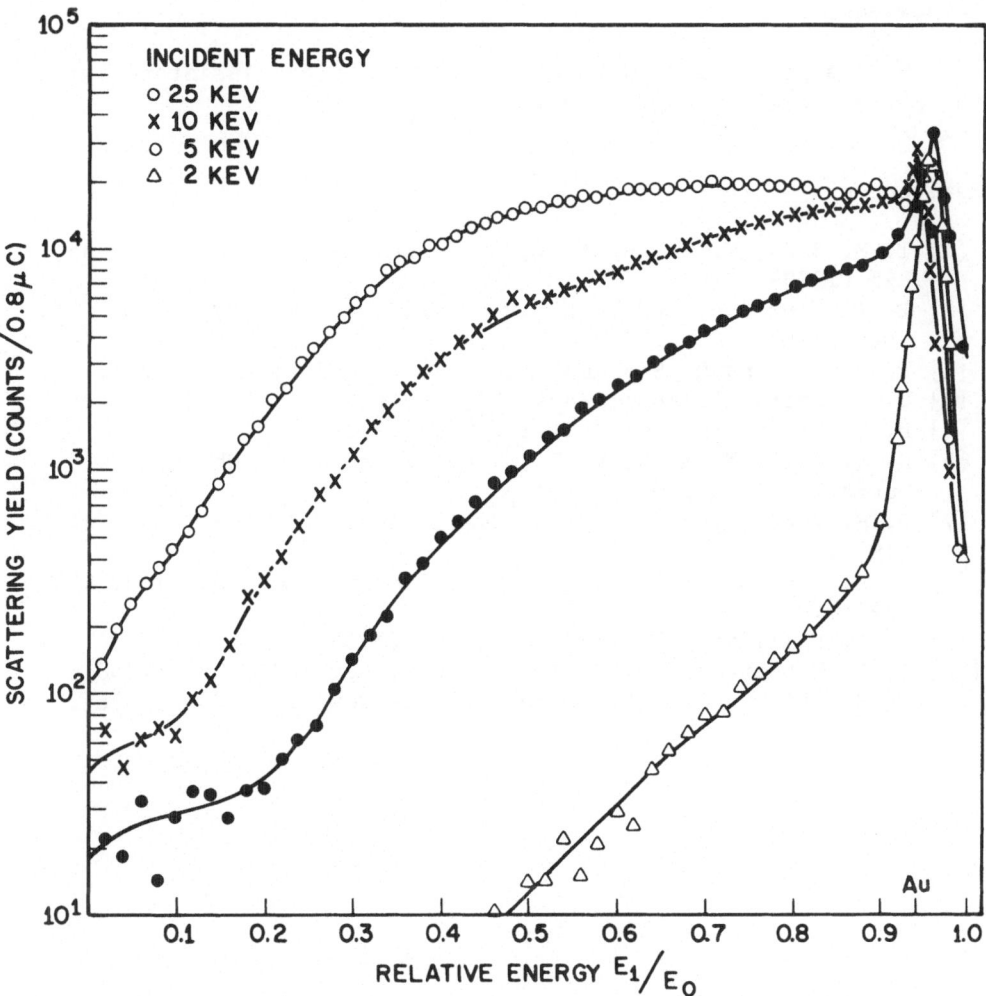

Fig. 7. Backscattered energy spectra for ^{4}He incident at 4 different energies. Measured by electrostatic analyzer, 120° scattering
angle. Illustrating the drastic reduction in yield below surface
peak for low incident energy (Ref. 6).

yield from beneath the surface has decreased dramatically or, as
has been supposed, there is more efficient neutralization of ions
which penetrate beyond the surface. This is consistent with the
possibility that bound states can exist in the solid for lower
velocity particles.

Acknowledgment

We are grateful to J. Morrison for technical assistance with the surface cleaning operation.

References

[1]   T. M. Buck, G. H. Wheatley, L. C. Feldman, Surface Sci. <u>35</u>, 345 (1973).
[2]   B. A. Trubnikov and Y. N. Yavlinskii, Soviet Phys., JETP <u>25</u>, 1089 (1967).
[3]   W. Brandt and R. Sizmann, Phys. Letters <u>37A</u>, 115 (1971).
[4]   D. P. Smith, J. Appl. Phys. <u>38</u>, 340 (1967).
[5]   D. P. Smith, Surface Sci. <u>25</u>, 171 (1971).
[6]   D. J. Ball, T. M. Buck, D. MacNair, and G. H. Wheatley, Surface Sci. <u>30</u>, 69 (1972).
[7]   G. H. Wheatley and C. W. Caldwell, Rev. Sci. Inst. <u>44</u>, 744 (1973).
[8]   R. Lieberman and D. L. Klein, J. Electrochem. Soc. <u>113</u>, 956 (1966).
[9]   J. A. Phillips, Phys. Rev. <u>97</u>, 404 (1955).
[10] C. F. Barnett and P. M. Stier, Phys. Rev. <u>109</u>, 385 (1958); P. M. Stier and C. F. Barnett, Phys. Rev. <u>103</u>, 896 (1956).

SECTION V
X-RAYS

SOLID EFFECTS ON INNER SHELL IONIZATION

F. W. SARIS

F.O.M.-Institute for Atomic and Molecular Physics
Kruislaan 407, Amsterdam/Wgm.
The Netherlands

ABSTRACT

X-ray yields from ion bombardment of solid
or gaseous targets differ markedly, due to the
steady state excitation of the projectile when
penetrating a solid target, and due to recoil
effects. The observed influence of outer-shell
ionization on inner-shell excitation yields information on the state of energetic ions moving through
solids. It is observed that the equilibrium fraction of ions which have one or more inner-shell
vacancies can be as large as 10-100%. This leads
necessarily to collision-induced deexcitation
processes one of which is the emission of MO x rays.
These charge state effects are reviewed along with
an illustration of the importance of recoil effects
as well as a discussion of possible future investigations.

Introduction

It is not only for historical reasons that solid rather than
gaseous targets are so often used in ion-induced x-ray studies.
By making use of solid targets and particularly by comparing with
results from gaseous targets much is to be learned about solids and
particle-solid interactions. For instance: a detailed analysis of
x-ray spectra should in principle reveal information on the band
structure and related chemical and screening effects. It was specu-
lated recently that also plasmon excitation could be observed.
Looking at the projectile x rays might give information on the steady
state excitation of projectiles penetrating solids. Using monocrystalline material one can envisage coherent inner-shell excitation

of channeling projectiles. Then one does not really need much
imagination to think up x-ray lasing. Finally, one would like to
make use of the large cross sections and low bremsstrahlung back-
grounds to use ion-induced x rays as an analytical tool in trace
element analysis of solids.

Before putting ion-induced x rays to use, quite a few fundamental atomic collision problems have to be solved. True, for
light ions (protons and α particles) the Merzbacher formula [27]
can be used to deduce cross sections out of thick target yields.
But heavier ions introduce recoil and charge state effects. New
collision induced excitation and deexcitation processes appear
because the time interval between two successive collisions of the
projectile in a solid is often short compared with the relaxation
of its atomic shells.

Here we will first review steady state excitation effects on
characteristic x-ray production in solids and its significance for
ion-solid interaction studies. Then we will proceed discussing
molecular x rays. Finally, the importance of recoil effects will
be illustrated. Since these discussions are all in the context
of the molecular model for heavy-ion-atom collisions this model
will be introduced briefly.

Molecular Model for Heavy-Ion-Atom Collisions [1,2]

In the molecular model the atomic wave functions of the separated atoms are supposed to change quasi-adiabatically into molecular wave functions during the collision. At very small internuclear
distances these molecular states change into the atomic states of
the united atom. Fano and Lichten and later Barat and Lichten
have formulated the correlation rules, which prescribe the correlation of the separated and united atom states via the molecular
orbitals (see references listed in [1] and [2]). In Fig. 1 a so-
called correlation diagram is shown of which two things are worth
noting. First, due to the Pauli exclusion principle some electronic
levels have to be promoted, i.e., the principle quantum number is
increased when going from infinity to zero internuclear distance
(for instance $1S_B \rightarrow 2p$ and $2P_B \rightarrow 4f$). Secondly, at various internuclear distances crossings occur of MO's. At the crossings elec-
trons may get trapped in vacant MO's, which implies that a vacancy
is formed in a lower orbital when the particles separate. Thus
inner-shell ionization occurs, provided the internuclear distance
during the collision becomes small enough to reach a crossing (or
give sufficient coupling) of two MO's and provided the outer orbital
has a vacancy.

Steady State Excitation of Heavy Ions Penetrating Solids

The excitation or ionization state of heavy projectiles prior
to a collision is of crucial importance for inner-shell ionization

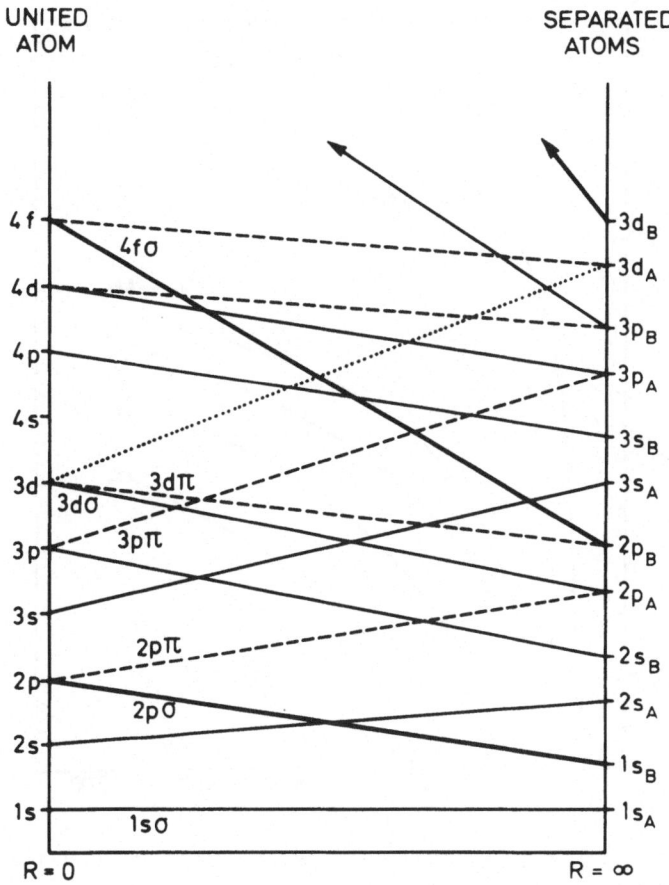

Fig. 1. Molecular orbital correlation diagram for an asymmetric system.

processes [28]. This is clearly observed for neon projectiles. K-shell ionization occurs via rotational coupling between the 2pπ and 2pσ MO's provided there is a vacancy in the 2pπ MO [25]. This orbital is correlated with the 2p orbital of the projectile which is the outer shell. Fig. 2 shows that the cross section for K x-ray production in Ne-Ne collisions increases indeed by a factor 2 if the projectile charge state is increased from 1+ to 2+.

It is well known that the state of a fast projectile is changed upon entering a solid target due to electronic excitation and ionization. The distribution of excited and ionized states obtained is determined by the projectile type and velocity and to a lesser extent by the kind of solid material. Neon projectiles of keV energy are probably ionized in the 2p shell which has an interesting consequence for neon bombardment of Mg, Al, Si, P. According to the correlation diagram of Fig. 1 Ne K electrons can only be

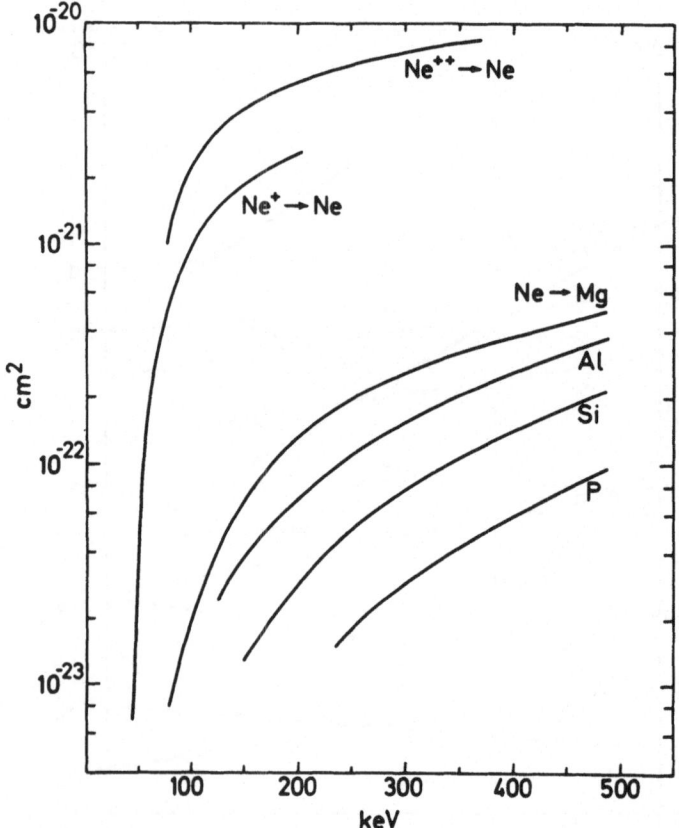

Fig. 2. Neon K x-ray production cross sections versus the energy
of neon ions incident on solid targets of the elements Mg, Al, Si,
P, from ref. 29. For comparison are also shown the K x-ray production
cross section in Ne+-Ne and Ne++-Ne collisions, from ref. 36.

excited if there is a vacancy in the 2p shell of the target atom,
which is not the case. However, the outer-shell ionization of the
Ne projectile causes a swapping of the correlation. Because of
the difference in screening in the ionized neon, its electron
binding energies increase, and can become larger than the 2p binding
energy of Mg, Al, Si and P successively. Neon charge states of 2,
3, 4 and 5+ are needed respectively, to cause swapping of the Ne-2p
and target 2p shells. Then the Ne-2p shell is correlated with
the 2pπ MO and thus the K-shell excitation channel via 2pπ-2pσ
rotational coupling is open. Cross sections are shown in Fig. 2.
Similar cases have been reported in the literature which at first
sight seemed to be in contradiction with the predictions of the
molecular model but which are attributable to these typical solid
effects [29].

The proof of this effect is of course in the comparison with gas targets [37]. This has been done for I bombardment of Se and Br_2 or Kr targets of identical thickness [30]. Fig. 3 shows that the I L x-ray line when bombarding the Se foil is almost an order of magnitude larger than when the beam is passing through Br_2 or Kr gas. It is interesting that the reverse is true for the I M x ray. This could be due to the fact that at 48 MeV, I is stripped of its N and O electrons and therefore no bound electrons are available for transitions to M-shell vacancies.

Similar effects have been noticed by Schnopper et al. [31] for 125 and 140 MeV Br, where only a small additional number of Br L_α x rays are produced in thick targets compared to thin targets, whereas the thick to thin ratio is large for the target lines and for Br K_α. This suggests that only K-shell and L-shell electrons (a total of 10 electrons) remain bound to the 140 MeV Br projectile traversing a solid target. It is interesting to note that the mean number of electrons that are ionized by a solid stripper for 140 MeV Br is ∿24 out of a total of 35 [35]. This clearly illustrates some of the potentialities of ion-induced x rays in studying the penetration of ions through matter.

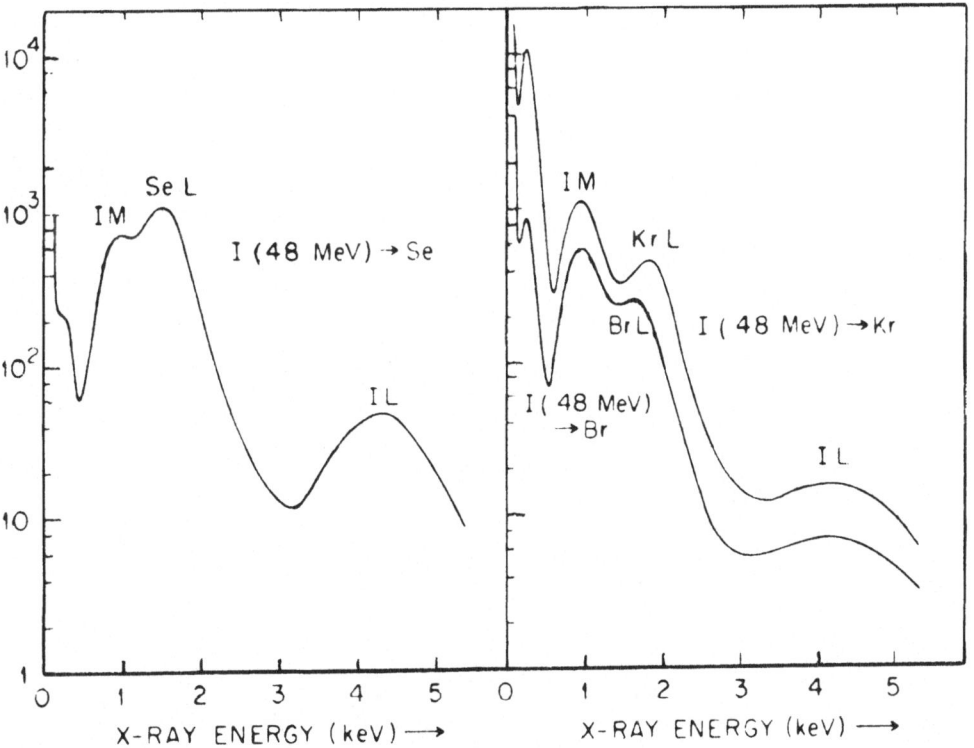

Fig. 3. X-ray spectra excited by impact of 48 MeV I ions on Se, Br and Kr targets, from ref. 30.

  If outer-shell ionization is already so influential that it
changes the MO correlation as we have seen above, then it is to be
expected that inner-shell ionization of projectiles may cause even
more dramatic swapping effects. Fig. 4 shows a detailed Ar L x-ray
spectrum obtained with a grazing incidence monochromator viewing
a Ti target bombarded by 35 and 100 keV Ar ions [32]. Four peaks
are identified and listed below.

X-Ray Energy [eV]	Radiative Transition
220	$3s \rightarrow 2p$ in neutral Ar
235	$3s \rightarrow 2p$ in Ar^{4+} single 2p vacancy
258	$3s \rightarrow 2p$ in Ar^{4+} double 2p vacancy
280	$3d, 4s \rightarrow 2p$ in Ar^{4+} single 2p vacancy

Three peaks appear to originate from 4 times ionized argon. The
220 eV peak is probably from argon target atoms which have been
implanted during prolonged bombardment of the Ti target.

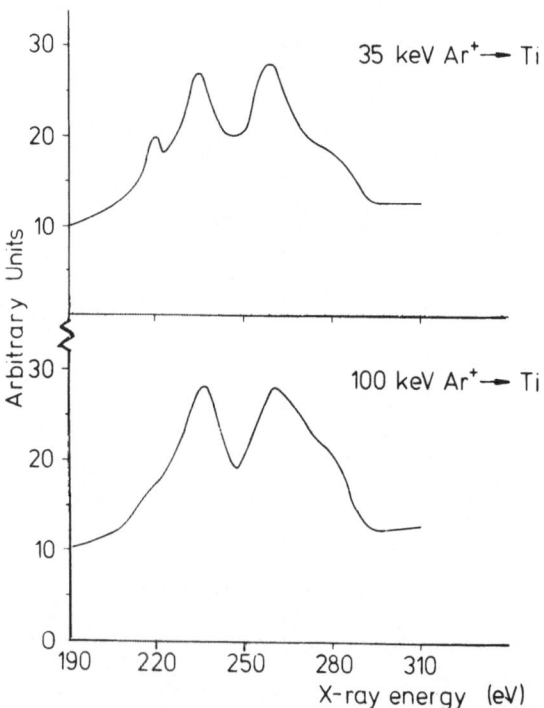

Fig. 4. Ar L x-ray spectrum recorded by a grating incidence monochromator when 35 keV and 100 keV Ar ions are incident on Ti, from
ref. 32.

A double L-shell vacancy in the Ar projectile may be produced in a single Ar-Ti collision [33] because the Ar-2p orbital is promoted via the 4 fσ MO, and many crossings occur. On the other hand in the Ti solid also double collisions may be responsible [5]. Given an Ar L vacancy production cross section of 2×10^{-17} cm^2 in Ti-Ar collisions at 100 keV the mean free path for L-shell ionization of an Ar projectile in Ti is of the order of 100 Å. For various defect configurations the inner-shell vacancy lifetime can become longer than 10^{-14} sec. [34] and therefore the extinction distance (the ion velocity times the mean lifetime) can become longer than 100 Å, thus giving a large probability for the production of a second inner-shell vacancy before the first decays. Therefore, the use of solid targets introduces ambiguities in determining the number of inner-shell vacancies per collision.

Moreover, double L-shell vacancies in the Ar projectile will cause very dramatic changes in the Ar-Ti correlation diagram. For neutral atoms the Ar-2p and Ti-2p orbital energies are separated by ∿200 eV. If a double vacancy in the Ar-2p shell is present along with some ionization of the M shell then the Ar-2p and Ti-2p orbitals are mixed and an interchange of MO's will occur allowing a correlation of the Ar-2p orbital with the united atom Br-2p orbital via the 2pπ MO. Then the excitation channel for the Ar K shell is opened via 2pσ-2pπ rotational coupling. Ar K x rays are observed indeed, see Fig. 5. The broad feature at lower x-ray energies is the Ar-Ti MO x ray to which we will focus our attention later.

First, some additional remarks on the study of steady state excitation of projectiles in solids via x-ray spectra. Extinction distances for inner-shell vacancies can be several hundred Å long. Since the mean free path for collisions involving outer shells is only 2-3 Å there will be more than 30 of such collisions before the inner-shell vacancy decays. Therefore, the x-ray spectrum that is radiated does in principle reflect the equilibrium state of ionization/excitation of the projectile outer shell. Yet to determine these equilibrium charge states from x-ray spectra with some accuracy is very difficult if not impossible. As is pointed out by Bhalla [34] at this conference, the x-ray energy differences from different defect configurations are frequently too small to be observed, yet their fluorescence yields differ widely. Therefore, we conclude that it is not possible as yet to obtain detailed quantitative information concerning the electronic configuration of many electron carrying ions in solids. On the other hand enough interesting features are observed in x-ray spectra to allow speculations on steady-state excitation of ions traversing solids which go beyond the existing models of Bohr-Lindhard or Betz -Grodzins, see ref. 35.

MO X Rays

The long excitation distances for projectile inner-shell vacancies have still another consequence which was only appreciated

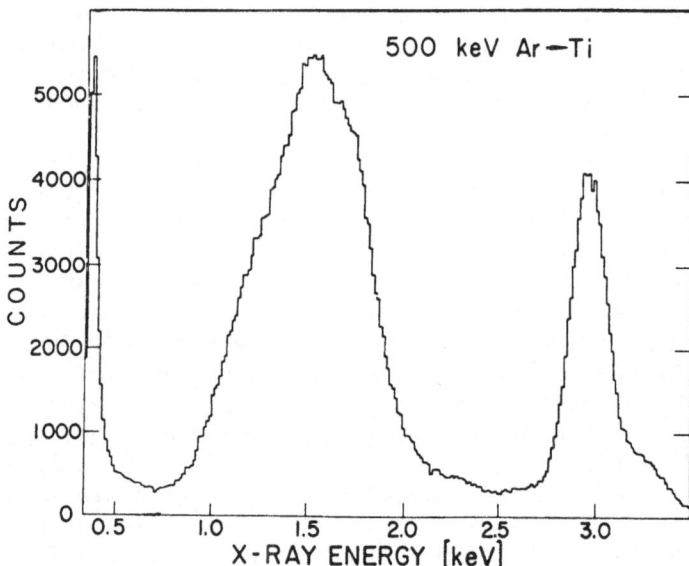

Fig. 5. X-ray spectrum recorded when 500 keV Ar ions are incident
on Ti. A broad x-ray distribution is observed along with the
Ar-K$_\alpha$ line (2.96 keV) from ref. 5.

very recently. After an inner-shell vacancy is produced, in a subsequent collision it can be brought into deeper MO's where it may
decay to give rise to x rays not characteristic for either of the
two colliding atoms but for the quasi-molecule formed. These
"MO x rays" were first observed from collisions of Ar ions with Ar
atoms previously implanted in a solid silicon target [3]. The
equilibrium fraction of Ar projectiles that have an L-shell vacancy
when traversing a silicon target at 100 keV is at least 10%. Then
in Ar-Ar collisions this 2p vacancy may follow the 2pπ MO which
is correlated with the united atom (Kr) 2p shell. It is interesting to note that the 2p-vacancy lifetime in Ar is 4 x 10^{-15} sec
whereas in Kr it is 5 x 10^{-16} sec. So as the collision proceeds
the 2pπ vacancy lifetime decreases by an order of magnitude to
become of the same order as the collision time of $\sim10^{-16}$ sec. In
addition the fluorescence yield changes over two orders of magnitude from $\sim10^{-4}$ to $\sim10^{-2}$, making the possibility of observing a
radiative decay in the quasi-molecular states very likely. Such a
radiative decay will give rise to a broad x-ray band because the
vacancy can decay at all internuclear distances from infinity to
the distance of closest approach. Over this region, R, the energy
difference, ΔE, between, for instance, the 3dπ and 2pπ MO's, varies
from Ar-L (220 eV) to Kr-L (1.64 keV). It is indeed in this x-ray
region where we first observed a broad band up to an x-ray energy
of 1.5 keV (see Fig. 6), when bombarding Ar implanted Si with 100
keV Ar ions [3]. The high energy side of this band shifts to

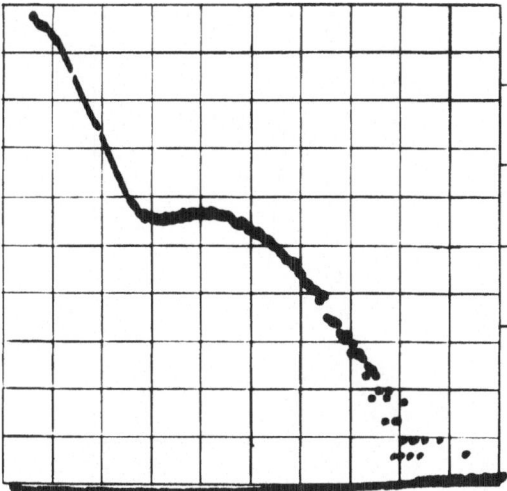

Fig. 6. X-ray spectrum obtained with a Si(Li) detector and mylar window observing the broad x-ray band next to Ar-L x rays emitted during 100 keV Ar bombardment of Ar saturated silicon.

higher x-ray energies as the beam energy is increased, for ΔE increases with smaller R. The observation of the endpoint of the band near 1.64 keV (Kr-L) is obscured by the appearance of Si-K characteristic x rays (1.74 keV). The intensity of the so-called MO x-ray band was strongly beam dose dependent until the Si target was saturated with Ar at a dose of $\sim 3 \times 10^{17}$ cm^{-2} . Similar bands were observed for Ar bombardment of C, Al and Fe [3,4] as well as solid Ar [4]. We simulated a solid argon target using KCl [5]; potassium is only 1 atomic number higher and chlorine is 1 atomic number lower than argon. The resulting spectra, see Fig. 7, clearly show the shift of the band when the beam energy is increased from 100 to 250 keV whereas hardly any shift is observed if the beam energy is increased further to 400 keV. The background at higher energies, however, does change along with the appearance of Cl-K x rays above 250 keV impact energy. Apparently this energy is sufficient to transfer the 2pπ vacancy via rotational coupling into the 2pσ MO. If the vacancy can decay in the 2pπ, it will also in the 2pσ MO, which probably accounts for the observed background at higher x-ray energies.

MO x rays have also been investigated for asymmetric systems; Ar on unimplanted Ti [5], low dose Ar on Si, Al [4,8] very similar bands, but different in x-ray energy, were detected. These studies have not merely been restricted to MO-L x rays but M x rays [6] and K x rays [7,8,9] have also been reported.

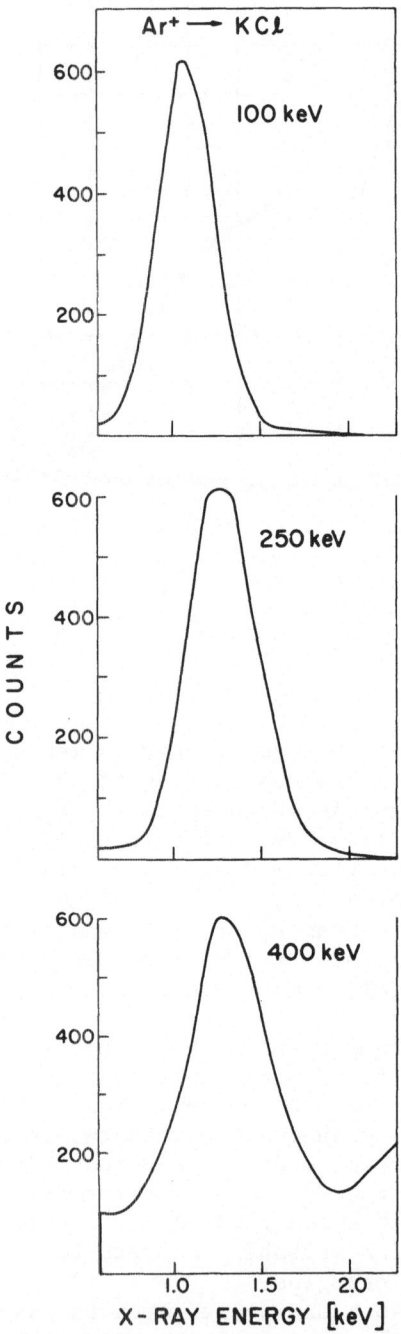

Fig. 7. X-ray spectra for 100, 250, and 400 keV argon ions incident on KCl. The intensities are normalized in order to compare
the shapes of the x-ray bands, which are attenuated on the low
energy side by Be filters.

Solid targets were used in order to achieve, in a double collision, the necessary inner-shell vacancy during the second encounter. Recently we have found strong evidence that this collision
condition can be met also by the use of a molecular gas target
[21,22], thus eliminating complicating solid effects. Double
scattering in one molecule can produce an inner-shell vacancy in
the projectile during the first collision which will persist until
the next collision, thus opening the possibility of observing MO
x rays from the radiative filling of the vacancy during the second
collision. Fig. 8 shows x-ray spectra obtained during Ar-Ar, Ar-HCl
and Ar-Cl$_2$ collisions. In the Ar-HCl and Ar-Ar cases only L x rays
are observed. In the Ar-Cl$_2$ case also Cl-K x rays are detected
as well as MO x rays. For a detailed description of the experiments
one is referred to ref. 22. Clearly this opens some interesting
new ways of experimental investigation, not in the least the search
for MO-Auger electrons.

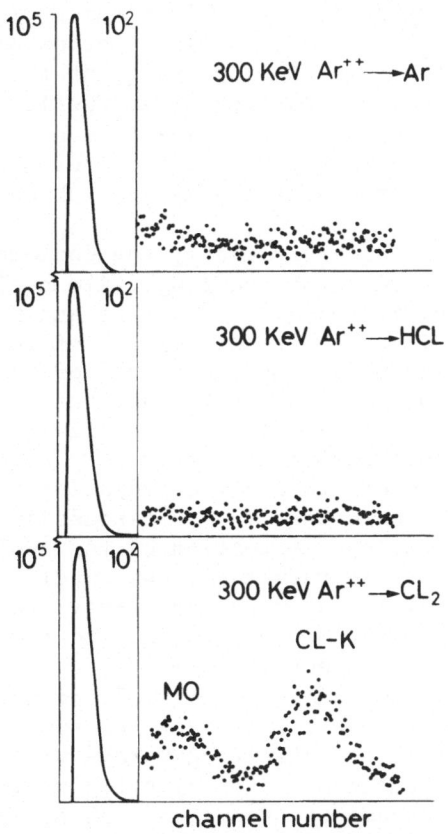

Fig. 8. X-ray spectra obtained from 300 keV Ar^{++} bombardment of Ar,
HCl and Cl$_2$. The spectra have been normalized at the L x-ray
intensity.

Ion-Induced Continuum X-Rays

Rather than giving an extensive review of published data on MO
x rays, I will proceed with a discussion of various effects that can
give rise to continuum x rays in ion-induced x-ray spectra, since experimentally we have to distinguish such x rays from the MO x rays.
The following phenomena can be listed:

        1) secondary electron induced bremsstrahlung
        2) inverse and inner bremsstrahlung
        3) nuclear bremsstrahlung
        4) nuclear γ rays
        5) collision broadening
        6) decay processes involving few electron systems
        7) radiative Auger transition and plasmon excitation.

1) Secondary electrons can be produced by a direct energy
transfer of beam ions to continuum electrons, but the maximum
energy transfer is limited to $E_e = (4m/M) \times E_i$, where E_i is the ion
energy, M its mass, m the electron mass and E_e the energy it
acquires. Much more energy can be transferred to electrons bound
to the target inner shells. Here the maximum E_e is [10]: $E_e =
(4m/M)E_i + 4\sqrt{(m/M)E_iE_b}$. Here E_b is the electron binding energy.
Clearly, since the effective binding energy of the innermost electrons can be very large, one can expect energetic electrons to
traverse the target which will give rise to a bremsstrahlung background. In addition one should note that inner-shell ionization
generally initiates Auger electron emission, which may cause a
bremsstrahlung background also.

2) The bremsstrahlung generated by the Coulomb deflection of a
target electron in the field of the projectile nucleus has been
recognized recently by Schnopper et al. [26] and Kienle et al. [11].
Schnopper considers "inverse bremsstrahlung" of an unbound electron
at rest in the field of a fast moving proton or oxygen ion. Kienle
et al. take into account the momentum distribution of electrons
ionized out of inner shells and therefore call it "inner bremsstrahlung". Since both processes require extremely high projectile
energies (>10 MeV) in order to produce radiation in the x-ray region
beyond 1 keV, we can ignore them in our discussion.

3) Radiation due to the deflection of the incident particle
in the Coulomb field of the target nucleus (nuclear bremsstrahlung)
is particularly significant at very high x-ray energies, but is
vanishingly small for heavy projectiles due to the factor
$(Z_1/A_1 - Z_2/A_2)$ in the cross section [12].

4) For very small impact parameter collisions at relatively
high beam energies Coulomb excitation of the nucleus or nuclear
reactions are possible. The resulting γ rays will cause a broad
continuum background in the detector directly or via Compton effects.

5) The decay of a vacancy produced in an ion moving through a
solid lattice and suffering multiple collisions, should be treated
as a case of collision (pressure) broadening, analogous to the decay
of an atom under bombardment by other atoms in a gaseous environment.

Briggs [16] has shown that the Lorentz tail of characteristic lines
due to collision broadening can be very pronounced but it is smaller
in intensity than the MO x-ray continuum, whose shape differs from
a Lorentzian line shape also.

   6) A large variety of effects is related to the production of
few electron systems during stripping at high beam energy. We
mention: radiative electron capture [11,13], two photon decay of
metastable states [15], and the many unresolved lines at the series
limits for high quantum numbers in the Lyman or Balmer series.

   7) Finally, there are radiative-Auger transitions [23] and in
solid targets the related plasmon excitations [17]. These effects
as well as bremsstrahlung from Auger electrons always come at the
low x-ray energy side of a characteristic line. However, we observe
MO x rays long before excitation of a deeper shell gives rise to
an additional characteristic line.

   Obviously it is not trivial to identify the cause for an
observed continuum x-ray spectrum, in particular not at relatively
high beam energies. However, at the beam energies below $\sim$1 MeV
one can neglect processes 2, 3, 4, 5, 6, 7. The production of fast
secondary electrons is not very likely either, because at these
beam energies the collisions are still rather adiabatic and the
molecular model is quite applicable which tells us that electrons
are promoted into some higher empty states perhaps in the continuum,
but electrons with excessive energies are not produced. Because
of its larger mass, the heavier projectile can transfer less energy
to electrons than protons of equal energy.

   In order to get a feeling for the importance of ion-induced
continuum x rays, especially bremsstrahlung from fast electrons,
we have bombarded various thick targets with 2 MeV helium and
nitrogen ions [18]. As stated above, at equal beam energy, helium
ions are more likely to produce fast electrons and subsequent
bremsstrahlung than nitrogen ions. Still the comparison of Figure 9
clearly shows that continuum x-ray bands are considerably more
intense in the nitrogen induced x-ray spectra. For the arguments
listed above, which are discussed extensively in ref. 18, we attribute these nitrogen induced bands to MO x-ray emission. However,
from a correlation diagram of the N-Ti system, it appears that the
deepest orbital that might be vacant to give rise to MO x rays is
the $3d\sigma$ MO and radiative decay will result in too low an x-ray
energy. Even for vacancies in the $2p\pi$ MO the energy would not be
sufficient to explain the observed x-ray band up to 3 keV x-ray
energies. The same holds for the heavier targets, Ni and Ge.

   Taulbjerg and Sigmund [19] have demonstrated clearly the
importance of recoil effects in the production of characteristic
x rays in solid targets. Therefore, recoil effects should be
equally important in the production of MO x rays from solid targets.
The proposed mechanism is the following: N $\rightarrow$ Ti $\rightarrow$ Ti$\sim\rightarrow$ MO x rays.
L vacancies that are likely to be present in the recoiling Ti atoms
can during Ti-Ti collisions follow the $2p\pi$ MO and there give rise
to L-shell MO x rays. This will lead to a broad x-ray band, depending on the beam energy used, extending to the united atom, Ru,

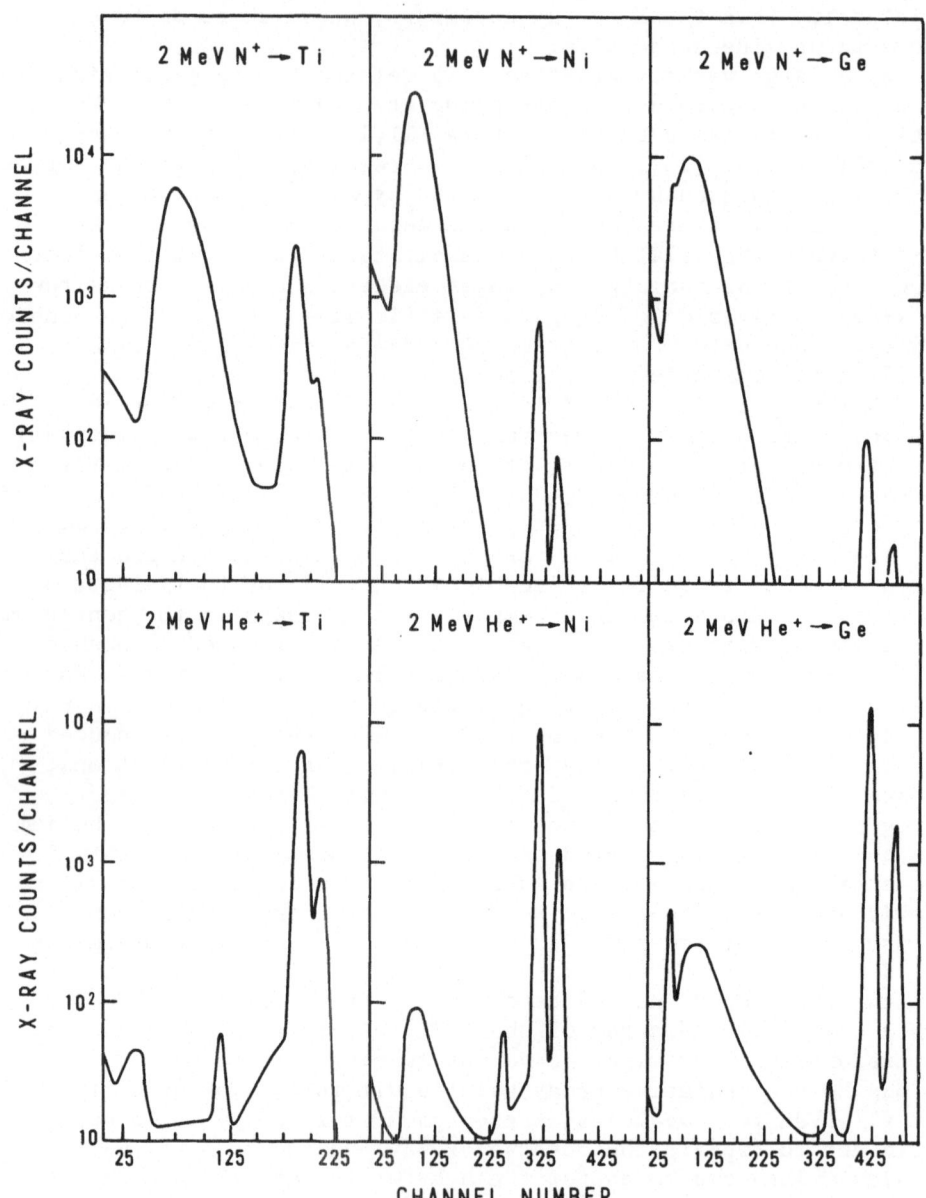

Fig. 9. X-ray spectra for 2 MeV N and He bombardment of Ti, Ni and Ge. Note that the spectra are heavily attenuated at the low energy x-ray side due to the use of a 150 μ Be absorber.

L-shell, x-ray energies of 2.5-3.0 keV. If the recoiling Ti atoms
have sufficient energy to reach in Ti-Ti collisions such small
internuclear distances that the united atom L shell is formed,
then vacancies in the 2pπ MO will have some probability to be transferred into the 2pσ MO via rotational coupling of the orbitals.
In other words, if the Ti-Ti MO x-ray band reaches energies as
high as Ru-L then one should expect Ti-K x rays to appear in the
spectrum also. In order to check this we bombarded 0.3-2.0 MeV N
on Ti with 100 keV intervals, see Fig. 10. For 300 keV beam energy
one observes a broad x-ray band only (Ti-L x rays are absorbed in
the 150μ Be absorber used to avoid pulse pile up). As the beam

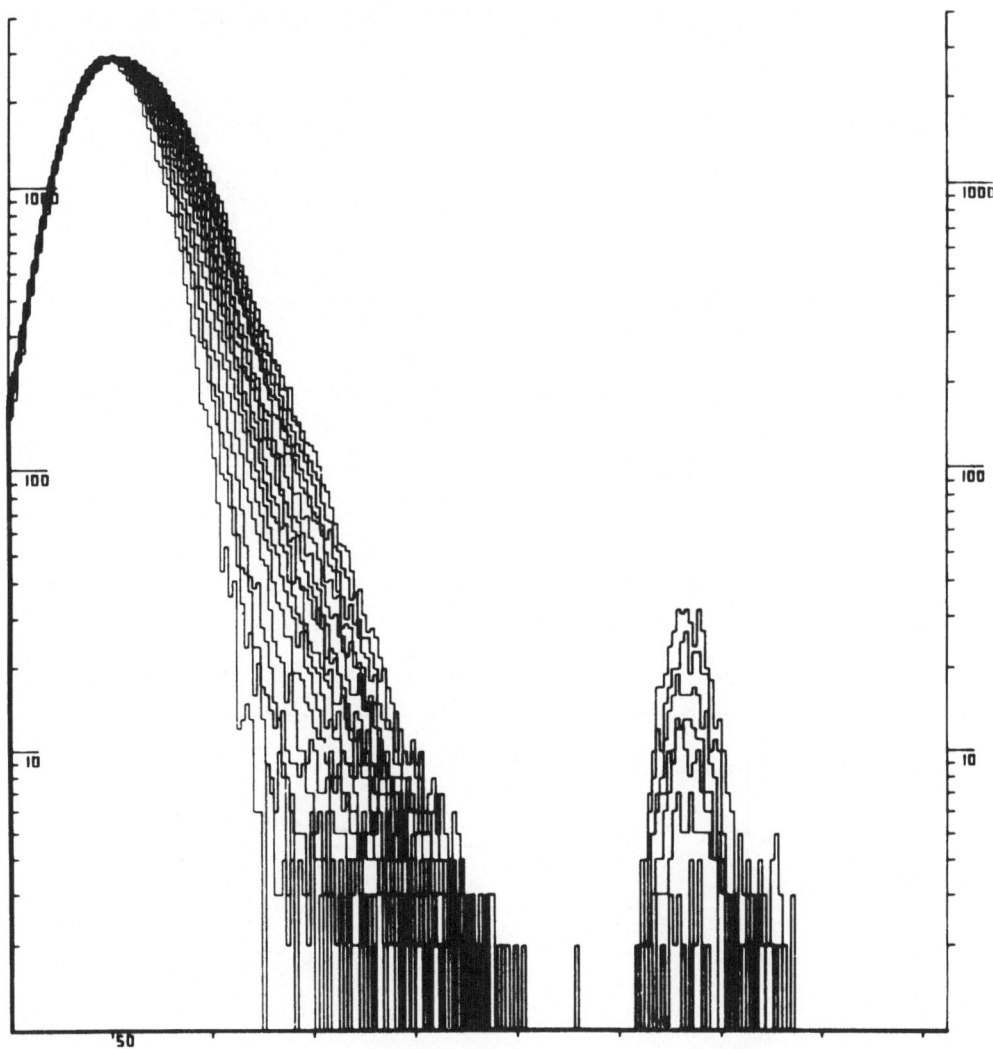

Fig. 10. X-ray spectra for 0.3 - 2.0 MeV N bombardment of Ti. The
spectra are all normalized in channel 50.

energy is increased the high energy side of the band shifts further
out until 800 keV when this shift becomes less pronounced and Ti-K
x rays appear.   In Fig. 11 the end-point energies of the x-ray band
have been plotted as function of beam energy.   At the turnover
point near 800 keV (when an x-ray energy of 3.0 keV is reached) the
excitation function for Ti-K sets in clearly.   It is interesting
to note that the maximum energy transfer of N to Ti is $0.7\ E_O$.
Thus at 800 keV N the maximum energy a Ti recoil can have is 560
keV.   This is close to the threshold energy (of $\sim$600 keV) for Ti-K
ionization via rotational coupling using the calculated cross section of Briggs and Macek [25] scaled to Z=22 (Ti).   Of course, one
should also take into account Coulomb ionization by the fast N
impact.   A simple comparison with the heavier projectile Ne shows

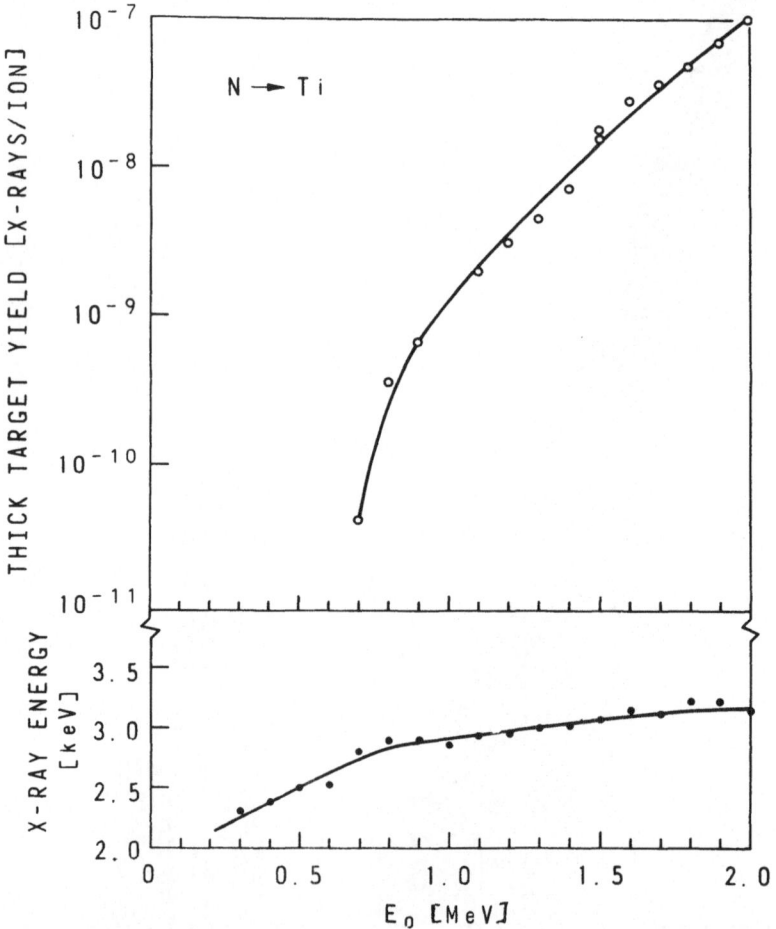

Fig. 11.   Bottom part shows the x-ray energy of the endpoints of
the bands of Fig. 5 as function of N beam energy.   Top part shows
the Ti-K thick target yield as function of N beam energy.

some interesting effects, see Fig. 12. For Ti and V the Ne induced cross section is larger than for N bombardment. This gives strong evidence for the recoil effect since one can transfer more energy with the heavier Ne than with the lighter N projectile. Yet, the Coulomb excitation cross section is not negligible and apparently taking over for heavier targets Co, Ni, Cu, Ge where the N induced x-ray yield is found to be higher than the Ne induced yield; for a further discussion of these results see ref. 18, 20.

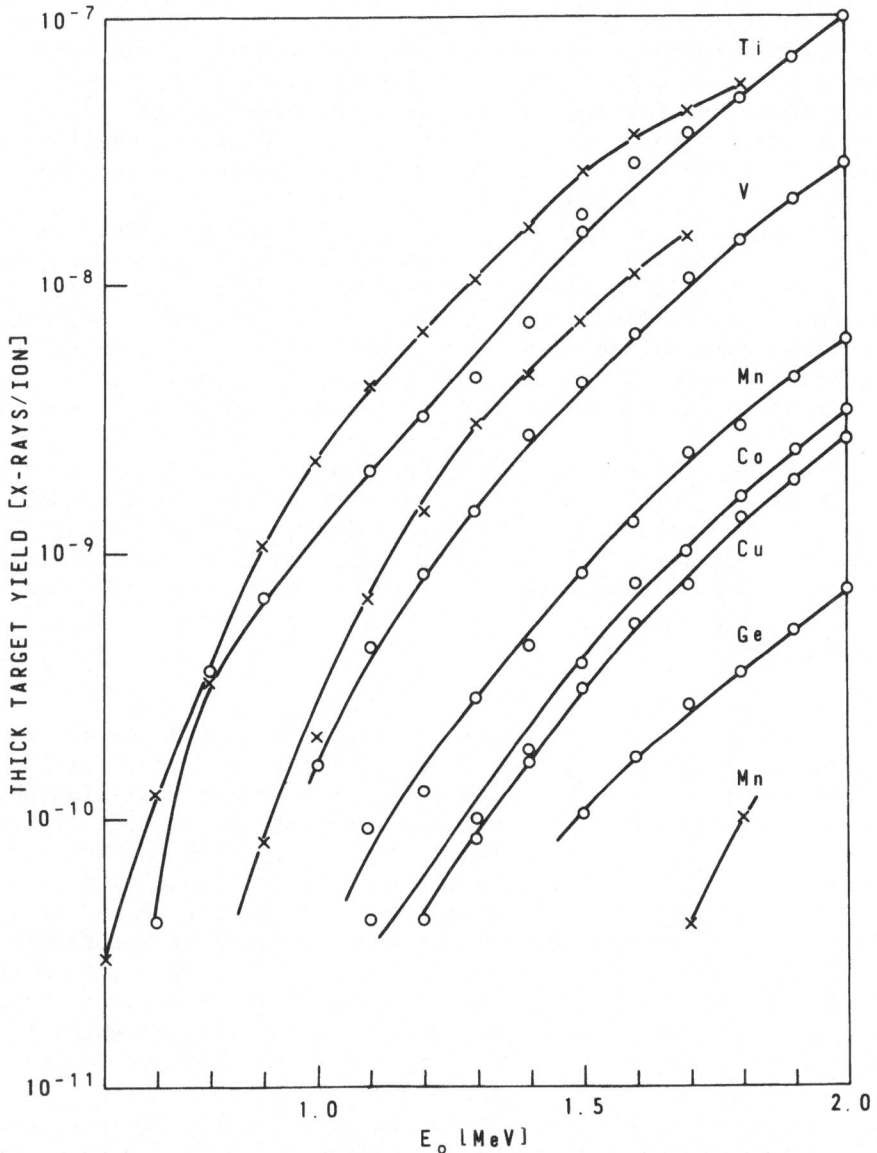

Fig. 12. Thick target yields for N (o) and Ne (x) bombardment of Ti, V, Mn, Co, Cu, Ge as function of beam energy.

Line Shapes

In an attempt to deduce more information on MO's we have analyzed our data further, assuming that the highest energy radiated
in the MO x-ray band comes from a radiative transition at the classical turning point, or distance of closest approach in a head-on
collision of the most energetic recoiling Ti atom produced (compare also ref. 7). The energy of this Ti atom is known to be 0.7
x E_0. Using a screened Coulomb potential one can convert this energy
into a distance of closest approach during a head-on Ti-Ti collision. Then the end-point energies of the observed MO x-ray band
can be plotted versus this distance, see Fig. 13. In the same
figure we have sketched in the correlation diagram for the Ti-Ti
system. Since internuclear distances considered are an order of
magnitude smaller than the radius of the united atom M shell it is
suggested that the changes in observed x-ray energy reflect the
change in shape of the lower level, the 2pπ MO.
    The flaw in this analysis is determining the endpoint energies.
From Fig. 10 it is clear that the x-ray band falls off exponentially,
thus any endpoint is meaningless since with a little more counts in
the maximum of this figure the endpoint will shift to higher energies.
    A close examination of all MO x-ray bands reported so far shows
a very similar exponential shape at the high energy side. We attribute this to collision broadening effects on the MO x-ray line
shape.
    A recoiling Ti atom at 2 MeV energy has a velocity of ∿3 x 10^{16}
Å/s. The time it will spend within an internuclear distance of
∿0.02 Å with the target Ti is equal to ∿1.5 x 10^{-18} sec. The
Heisenberg uncertainty principle then tells us that the uncertainty
with which one can measure the MO x-ray energy is ∿500 eV. Compared with a 3.0 keV x-ray energy this is a considerable line
broadening effect.

    In conclusion it is to be expected that with the present pace
of experimental and theoretical investigations, MO x rays will
soon have developed into a separate field of atomic collision physics. This development is undoubtedly inspired by the work of Greiner
and coworkers [24]. They predict that for $Z_{cr} \geq 169$ the 1s level
will "dive" into the negative continuum. If a 1s vacancy is brought
into a collision system whose united atom is heavier than Z_{cr}, it is
predicted to lead to spontaneous positron creation at internuclear
distances large compared to nuclear dimensions. However, from the
above discussions on the molecular model it will be clear that it
is very unlikely that U ions get ionized in the K-shell while traversing a U target even at a few hundred MeV beam energy, for no vacancies are present in higher orbitals like the L and M shell. Therefore, we suggest to make use of the recoil effect in Coulomb ionization of U by fast light projectiles. During this small impact parameter

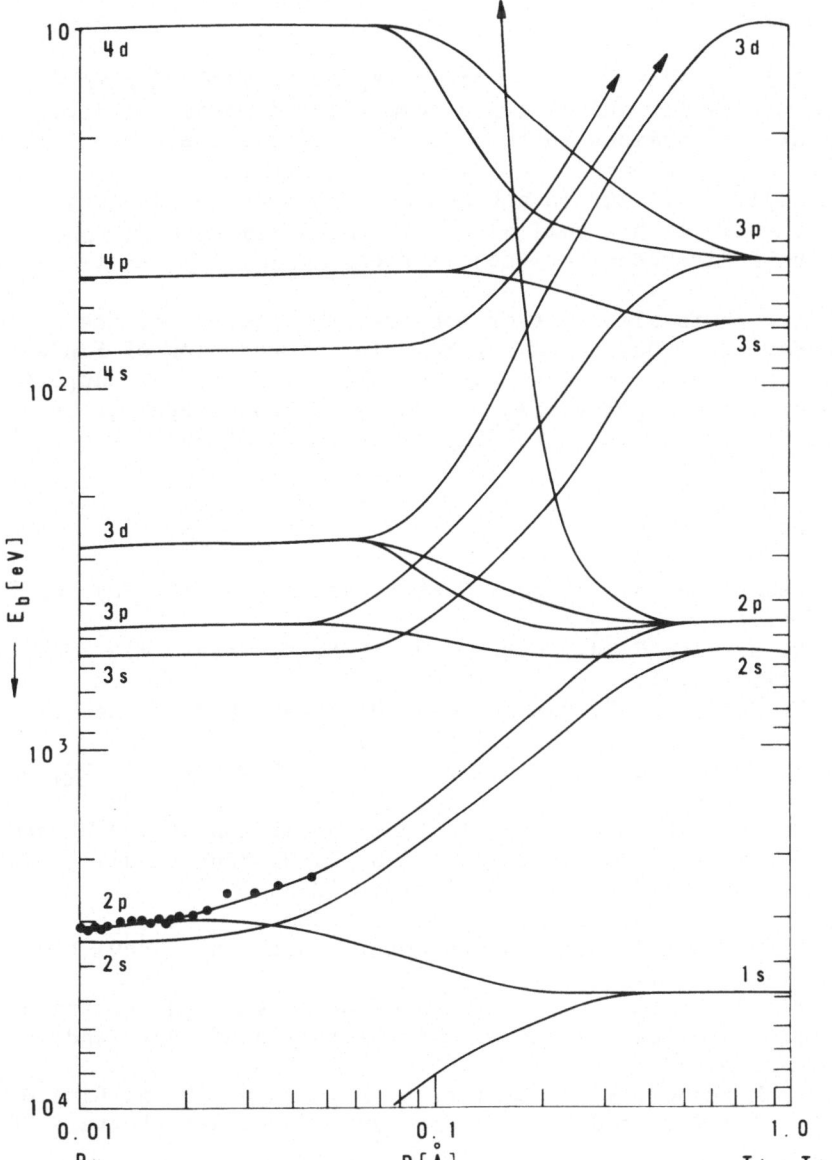

Fig. 13. X-ray energies of the endpoints of Fig. 6 now plotted versus
distance of closest approach in a head-on collision. Sketched in is
the Ti-Ti molecular energy level diagram.

collision the kinetic energy transfer can be large enough to create
U atoms with a K-shell vacancy recoiling through the U target. If
the recoil energy is sufficient to reach the critical internuclear
distance in subsequent U-U collisions one might find Greiner's
positrons, as well as MO K x rays of Z=184!

Acknowledgments

In this "preview" of a rapidly expanding field I have made use of many results and concepts communicated to me in preprints and private discussions in the past two years. Therefore I feel indebted to the authors listed in the references.

In addition I want to acknowledge the warm hospitality of I. V. Mitchell and his colleagues, of the Solid State Science Branch at the Chalk River Nuclear Laboratories, extended to me during all of 1972.

This work is also part of the research program of the "Stichting voor Fundamenteel Onderzoek der Materie" (Foundation of Fundamental Research on Matter) and was made possible by financial support of the "Nederlandse Organisatie voor Zuiver Wetenschappelijk Onderzoek" (Netherlands Organization for the Advancement of Pure Research).

References

[1] Q. Kessel and B. Fastrup, Case Studies in Atomic Physics 3, 137 (1973).

[2] J. D. Garcia, R. J. Fortner and F. M. Kavanagh, Rev. Mod. Phys. 45, 111 (1973).

[3] F. W. Saris, W. F. van der Weg, H. Tawara and R. Laubert, Phys. Rev. Letters 28, 717 (1972).

[4] J. R. MacDonald and M. D. Brown, Phys. Rev. Letters 29, 4 (1972).

[5] F. W. Saris, I. V. Mitchell, D. C. Santry, J. A. Davies and R. Laubert, Proc. Int. Conf. Inner Shell Ionization Phenomena and Future Applications, edited by R. W. Fink et al. (U.S. AEC Report No. Conf. 720669 (1973)) p. 1255.

[6] P. Mokler and J. Stein, ref. 5, p. 1283.

[7] J. R. MacDonald, M. D. Brown and T. Chio, Phys. Rev. Letters 30, 471 (1973).

[8] J. Bissinger and L. Feldman this conference and private communications; J. Cairns and A. Marwick this conference and private communications.

[9] W. E. Meyerhof, T. K. Saylor, S. M. Lazarus, W. A. Little, B. B. Triplett and L. F. Chase, Jr., Phys. Rev. Letters 30, 1279 (1973).

[10] E. Rudd and J. Macek, Case Studies in Atomic Physics 3, 46 (1973).

[11] P. Kienle, M. Kleber, B. Povh, R. M. Diamond, F. S. Stephens, E. H. Grosse, M. R. Maier and D. Proetel, Lawrence Berkeley Lab. Report LBL-1961 and private communication.

[12] K. Adler, A. Bohr, T. Huus, B. Mottelson and A. Winter, Rev. Mod. Phys. 28, 432 (1956); F. Folkmann, private communication.

[13] H. Schnopper, H. Betz, J. P. Delvaille, K. Kalata, A. R. Sohval, K. W. Jones and H. E. Wegner, Phys. Rev. Letters 29, 898 (1972).

[14] J. R. MacDonald, P. Richard, C. L. Cocke, M. Brown and I. A. Sellin, Phys. Rev. Letters 31, 684 (1973).

[15] R. Marrus, R. W. Schmieder, Phys. Rev. $\underline{A5}$, 1160 (1972).

[16] J. Briggs, private communication and this conference.

[17] H. Oona, J. D. Garcia, VIII ICPEAC, Belgrade 1973, p. 723.

[18] F. W. Saris, I. V. Mitchell and J. F. Chemin, J. Phys. B,
     Atomic and Molecular Physics, to be published.

[19] K. Taulbjerg and P. Sigmund, Phys. Rev. $\underline{A5}$, 1285 (1972).

[20] P. Sigmund and F. W. Saris, to be published.

[21] C. Foster and F. W. Saris, VIII ICPEAC, Belgrade 1973, p. 716.

[22] F. W. Saris, C. Foster, A. Langenberg and J. v. Eck, J. Phys. B,
     Atomic and Molecular Physics, to be published.

[23] R.E. la Villa, ref. 5, p. 509 and references in there.

[24] B. Müller, H. Peitz, J. Rafelski and W. Greiner, Phys. Rev.
     Letters $\underline{28}$, 1235 (1972).

[25] J. S. Briggs and J. Macek, J. Phys. B: Atom. Molec. Phys. $\underline{5}$,
     579 (1972), $\underline{6}$, 982 (1973).

[26] H. Schnopper, to be published.

[27] E. Merzbacher and H. W. Lewis, 1958, Handbuch der Physik
     (Springer-Verlag, Berlin) $\underline{34}$, 166.

[28] F. W. Saris and D. J. Bierman, Phys. Letters $\underline{35A}$, 199 (1971).

[29] K. Taulbjerg, B. Fastrup and E. Laegsgaard, Phys. Rev.,to be
     published.

[30] H. O. Lutz, H. J. Stein, S. Datz and C. D. Moak, Phys. Rev.
     Letters $\underline{28}$, 8 (1972).

[31] H. Schnopper, A. R. Sohval, H. Betz, J. Delvaille, K. Kalata,
     K. Jones and H. Wegner, ref. 5, p. 1348.

[32] H. Tawara and F. W. Saris, unpublished.

[33] F. W. Saris, Physica $\underline{52}$, 290 (1971).

[34] C. P. Bhalla, Proceedings of this conference.

[35] H. Betz, Rev. of Mod. Phys. $\underline{44}$, 465 (1972).

[36] H. Tawara, C. Foster and F. J. de Heer, Physics Letters $\underline{43A}$,
     266 (1973).

[37] It was found recently that the cross sections for Ne K x-ray
     production in the gas collisions $Ne^+ \to SiH_4$ and $Si^+ \to$ Ne were
     equal to those measured in solid collisions $Ne^+ \to Si$. See
     C. Foster, T. Hoogkamer and F. W. Saris, J. Phys. B, Atomic &
     Molecular Physics, to be published.

QUASI-MOLECULAR APPROACH TO THE THEORY OF
ION-ATOM COLLISIONS

J. S. BRIGGS
Theoretical Physics Division
A.E.R.E. Harwell
Didcot, Berkshire, England

ABSTRACT

The usefulness of recent calculations of the
Hartree-Fock potential energy, electronic wavefunctions and orbital energies of a diatomic ion-atom
collision pair in the description of elastic
scattering and the excitation and radiative de-
excitation of inner-shells is described. Good
agreement with experimental data on K-shell
vacancy production and x-ray emission indicates
that the molecular promotion model of Fano and
Lichten can be put on a quantitative basis.

Introduction

A treatment of the interaction of heavy ions moving through
solids with velocities of the order of 1 atomic unit (energy $\sim$
25 keV/a.m.u.), requires a knowledge of the cross-sections (both
elastic and inelastic) for the individual ion-atom collisions. At
these velocities, which are comparable with outer-shell electron
orbital velocities, a detailed treatment of the outer-shell processes is prohibitively difficult, particularly as regards the
influence of the binding of the outer electrons in the solid. However, the inner-shell electrons are moving rapidly compared with
the rate of change of the internuclear distance, which enables one
to recognise, as did Fano and Lichten [1], that states of these
electrons during the collision should be well-described by the
electronic states which a molecule formed from the collision partners would possess could it be stabilized at each internuclear
distance. Transition of electrons between such stationary states
is caused by the perturbing influence of the nuclear motion and

gives rise to inelastic scattering. The importance of the Fano-
Lichten model is that it demonstrates that, when an electron occupies a molecular orbital (MO) during a collision, its binding
energy may be diminished greatly from that in the unperturbed
atomic orbital and hence an inelastic transition to an outer
unoccupied MO is more easily made. Thus this phenomenon of electron promotion explains the enhanced cross-section for inner-shell
vacancy production at low velocities compared with that expected
from a consideration of the <u>atomic</u> binding energies.

Fortunately, the approximate solution of the time-independent
Schrödinger equation for a stationary molecule is a problem which
can be formulated in terms of a solution of the Hartree-Fock equations and computer codes exist which can solve these equations to
arbitrary accuracy, subject to boundary conditions of time and
expense. In this paper, recent progress in the quantitative application of the Fano-Lichten model to the treatment of the excitation
and de-excitation of the inner-shells of ions moving through solids
will be described. A further product of the Hartree-Fock method is
an accurate adiabatic potential energy as a function of internuclear
distance, which could be used to describe elastic scattering.

The Hartree-Fock Potential Energy

The Hartree-Fock method [2] generates a set of one-electron
orbital wavefunctions and energies for a given internuclear separation via a variational principle for the total energy of the molecular system. In practice, an approximate solution of the Hartree-
Fock equations is sought in which the individual one-electron wavefunctions are expanded as a linear combination of a finite set of
analytic basis functions of Gaussian or Slater type. The minimisa-
tion procedure on the total energy then provides an optimum set of
coefficients of the expansion. Large computer codes for the solution
of the Hartree-Fock equations have been developed and used extensively in calculations on stable molecules (i.e., internuclear distances
$\sim 1 - 3$ Å). Recently, calculations have been made for certain
di-atomic systems over the whole range of internuclear distance
from the united atom limit out to the completely separated atom
limit. This has been done either as a means of studying the MO
correlation diagram itself [3] or in exploring its application to
the Fano-Lichten model of heavy-ion-atom inelastic scattering
[4,5,6,7].

From the Hartree-Fock calculation of the total energy U(R) at
a particular internuclear distance R the potential energy may be
obtained from

$$V(R) = U(R) - U(\infty) \tag{1}$$

This potential energy is an adiabatic potential in that the Hartree-
Fock calculation of the molecular ground state places electrons in

the lowest energy orbitals available. This, in common with most
simple approximations of the interatomic potential, makes no
allowance for inelastic transitions and their effect on elastic
scattering. However, it should provide an accurate representation
of the truly elastic potential. Although at velocities in the
order of 1 a.u. the probability of strict elastic scattering is
low, the outer electrons provide only a small part of the total
energy in regions of close approach. The repulsive part of the
Hartree-Fock potentials for the diatomic systems N-N and NeO are
shown in Fig. 1 together with simple approximations to the internuclear potential. In both cases the exponentially screened
Coulomb potential is adequate out to distances greater than the K-
shell radii of the collision partners. At larger distances the
Hartree-Fock potential differs significantly from both the
Thomas-Fermi potential and the screened Coulomb potential, but in
both cases lies between the two. It should be noted that departures
of the approximate potentials from the Hartree-Fock adiabatic
potential are significant even up to where the potential has a
strength of a few keV. These results suggest the need for a systematic study of the departure of the adiabatic potential from the
commonly used simple approximations.

Excitation of Inner-shells

     The Hartree-Fock calculation provides a set of one-electron
energy levels which constitute the correlation diagram between the
atomic levels of the united atom and those of the separated atoms.
An example of this is the correlation diagram of the N-N system
calculated by Briggs and Hayns [6] and shown in Fig. 2. Diagrams
of this type, estimated rather than calculated and with avoided
crossings ignored, were first used by Lichten and co-workers [1,8]
to discuss electron promotion in ion-atom collisions and since that
time have found increasing use in the qualitative description of
inelastic scattering.
     In the case of a simple electronic transition Briggs and Macek
[9] have shown that an inelastic scattering calculation based on
an expansion of the time-dependent wavefunction in terms of the
one-electron adiabatic functions gives good agreement with existing
experimental data. This transition is of an electron in the $1\sigma_u$
orbital (Fig. 2) to the $1\pi_u$ orbital in a symmetric (homonuclear)
di-atomic collision. These orbitals are degenerate at the united
atom 2p level. The transition between them requires a prior vacancy
in the $1\pi_u$ orbital which correlates with the separated atom 2p
level. The coupling between the two orbitals is effected by the
rotation of the internuclear axis and leads to a vacancy in the
K-shell after the collision. This transition is exactly that which
leads to excitation of the 2p level in proton-hydrogen collisions
at velocities much less than 1 a.u. [10]. Indeed, from a detailed
calculation of N-N collisions, Briggs and Macek [9] have shown

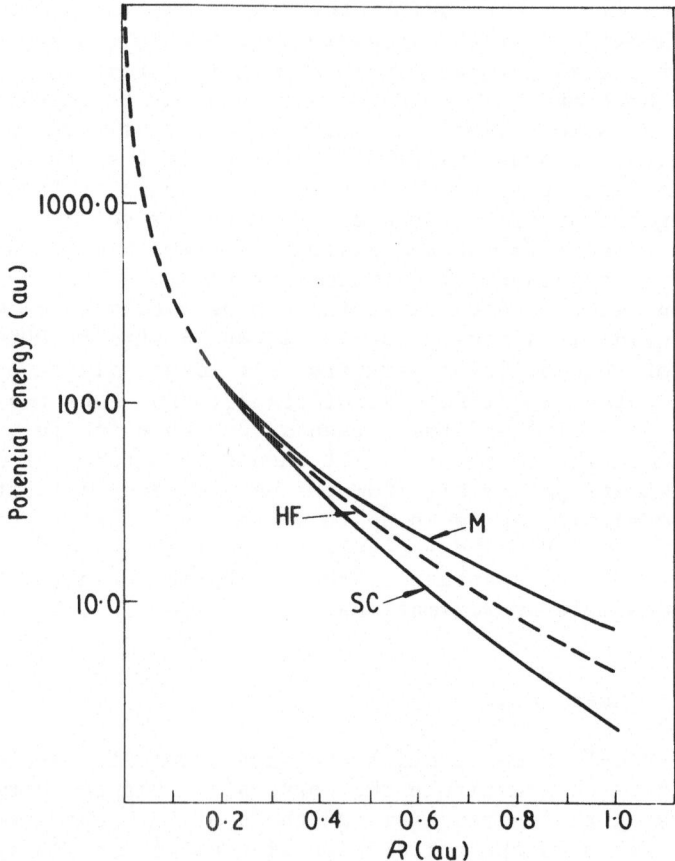

Fig. 1a. The potential energy of N_2: M Moliere approximation to
to Thomas-Fermi; HF Hartree-Fock; SC exponentially screened Coulomb.

that it is possible to scale the cross-section for $1\sigma_u - 1\pi_u$ transitions in the scattering of deuterium ions from deuterium atoms
to the cross-section for the same process in any symmetric collision
(the change to mass ratio of 1/2 is necessary to scale the internuclear motion). The scaling law is

$$\sigma(E) = \frac{N}{6} \frac{1}{Z_s^2} \sigma D (E_D) \tag{2}$$

where

$$E = M Z_s^2 E_D .$$

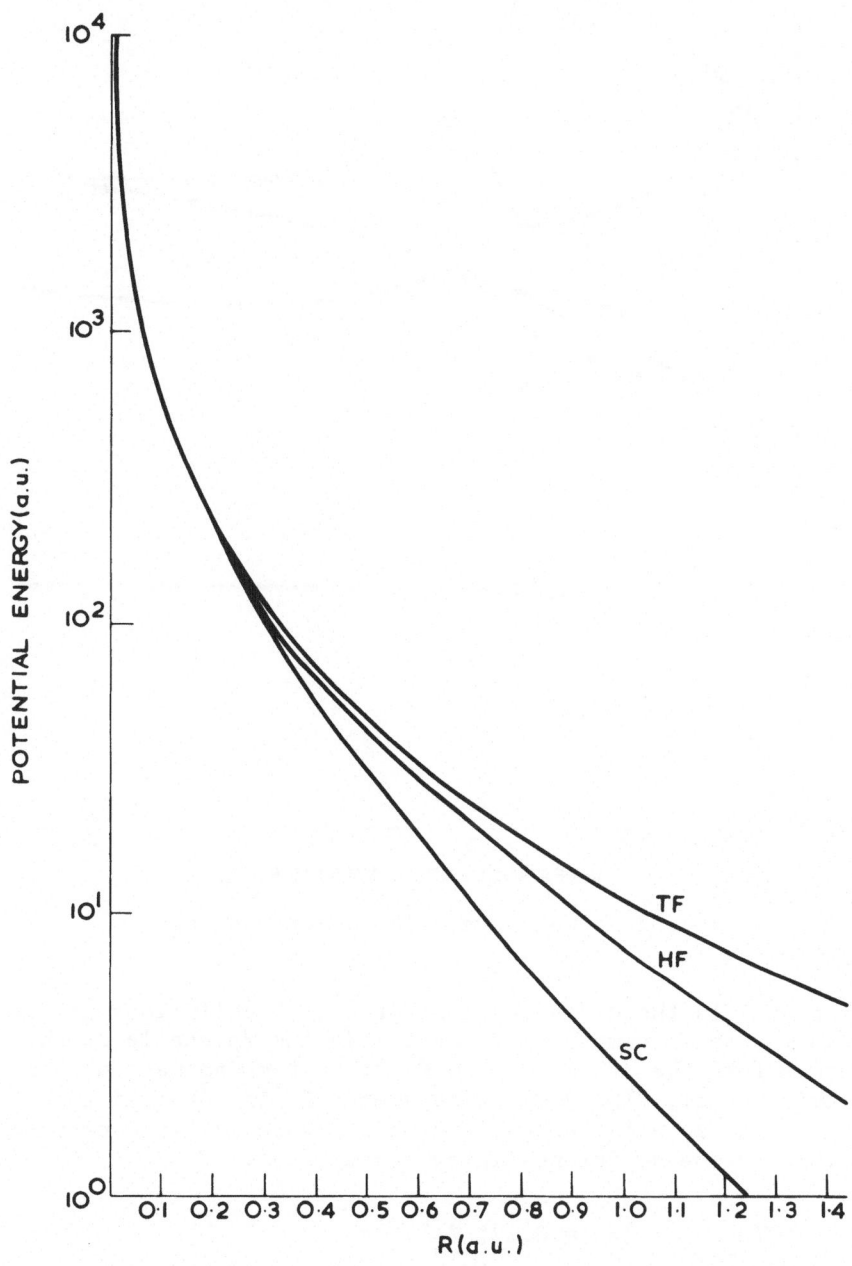

Fig. 1b. The potential energy of NeO: TF Moliere approximation to Thomas-Fermi; HF Hartree-Fock; SC exponentially screened Coulomb.

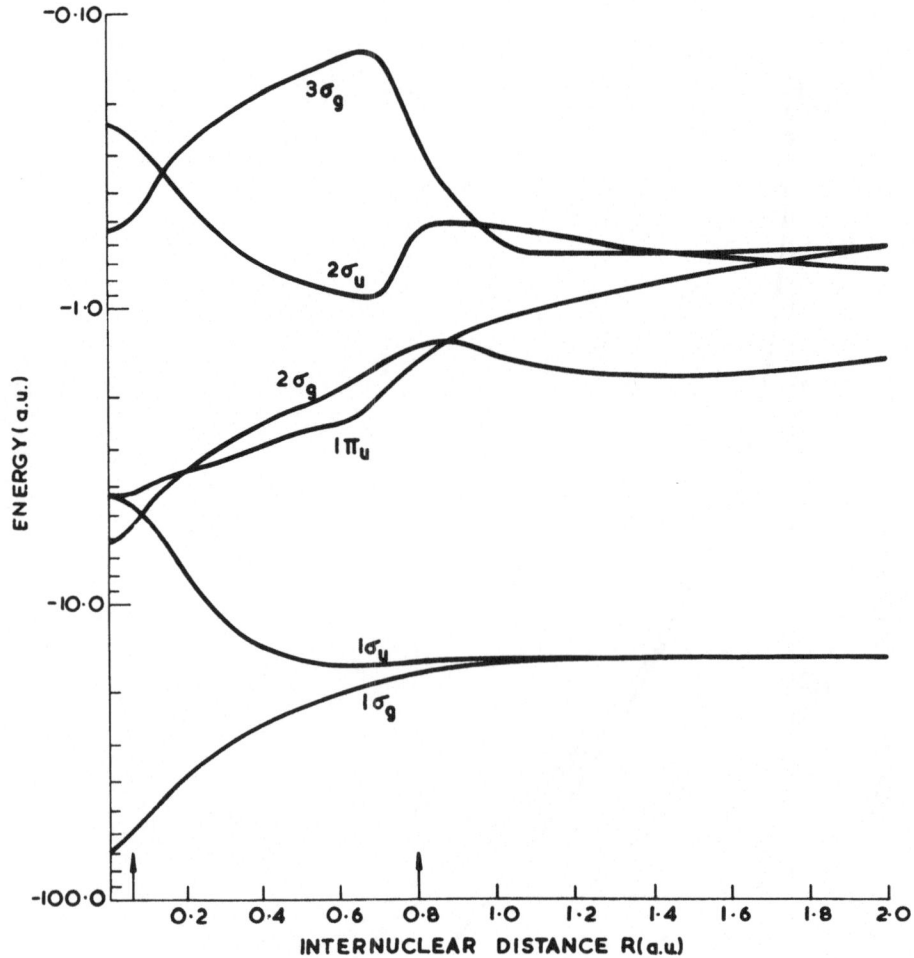

Fig. 2.   The correlation diagram of N_2.

Here σ_D (E_D) is the cross-section for D^+ – D collisions at energy E_D, N is the total number of vacancies in the 2p shells of the partners before the collision and M is the reduced mass number of the atomic nuclei. The scaling parameter Z_s is obtained from the scaling of the 2p – 1s energy level difference in the separated atoms to the same difference in hydrogen, i.e.,

$$\varepsilon(2p) - \varepsilon(1s) = 0.375 \ Z_s^2 \ \text{a.u.} \tag{3}$$

The parameter Z_s always lies between Z and Z–1, where Z is the nuclear charge. The scaling law requires a Coulomb potential to be a good description of the nuclear potential, which is clearly only valid at very short distances. However, the $1\sigma_u$ – $1\pi_u$ coupling is important only inside the K-shell so that the scaling is reasonable so long as the impact energy is above the threshold region for this process. A comparison of scaled cross-sections and experimental data is given in Fig. 3.

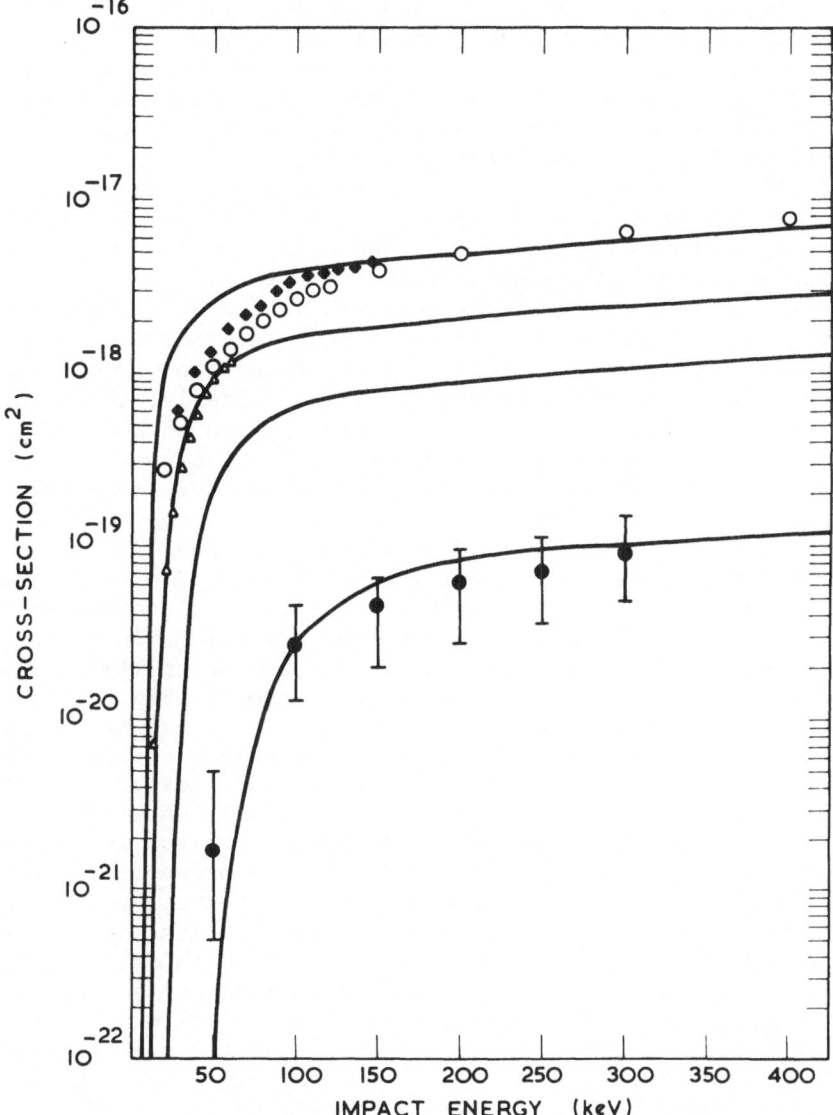

Fig. 3. The continuous curves are the calculated K-shell excitation cross sections for, in order of increasing threshold energy, $C^+ - C$, $N^+ - N$, $O^+ - O$, $Ne^+ - Ne$. The experimental points are: ● $Ne^+ - Ne$, Auger electron, Ref. 11; △ $N^+ - N_2$, Auger, Ref. 12; ○ $C^+ - C$, thick target x ray, Ref. 13; ◆ $C^+ - C$ (molecular), x ray, Ref. 14.

Radiative De-excitation of Inner Shells

The observation of characteristic x rays from the decay of
inner-shell vacancies has been used widely in the study of atomic
collisions in solids, both as a means of studying inelastic collisions [15] and also as a diagnostic for channeling and lattice
location [16] and detection of surface species [17,18]. This usage
leans heavily on the assumption that all lines observed are characteristic of the emitting element so that the resolution of proportional counters or solid state detectors is sufficient to identify
the emitter. However, recently Saris et al. [19] observed a broad
x-ray line in the Ar^+ bombardment of aluminium which was characteristic of neither argon nor aluminium. These authors suggested that
the band may be due to the emission of x rays between MO's during
the short time of the collision. Subsequently, similar bands have
been observed in other ion-solid collisions [20,21,22]. It tran-
spires that the true shape of the emission is a continuous shoulder
on the high-energy side of a soft x-ray peak and that its appearance
as a discrete peak is due to the drastic attenuation of soft x-ray
components by the counter window [20,22,23]. Clearly, it is of
interest to investigate the conditions under which such non-characteristic x rays might be observed and to seek to corroborate or
otherwise the identification of this emission as due to decay of
vacancies in MO's.

MacDonald et al. [20] have observed a continuous distribution
of x-ray emission on the high-energy side of the C-K line in the
C^+ bombardment of solid carbon. This emission is distorted into a
broad line with a peak around 1 keV by the beryllium window of the
detector used. The correlation diagram of the C-C collision system
is similar in all respects to that for N-N shown in Fig. 2. The
calculated MO energies of N-N will be used to discuss the emission
spectrum of a first-row ion moving through a solid lattice of its
atoms at a velocity in the region of 1.0 a.u. and less. Even at
these low velocities, an incident ion will readily obtain a K-shell
vacancy by promotion of an electron on the $1\sigma_u$ orbital as considered
in the previous section (the threshold for this process is v ∿
0.2 a.u. in nitrogen). A nitrogen atom with a K-shell vacancy will
travel ∿200 Å at a velocity of 1 a.u. during the lifetime of the
vacancy and therefore will make many collisions with other target
atoms. At these velocities the K-shell electrons can adjust adiabatically during the collision and in a close collision the K-shell
vacancy may occupy the $1\sigma_u$ or the $1\sigma_g$ orbital with equal probability
(see Fig. 2). On the $1\sigma_g$ orbital the vacancy can decay with an
energy which varies between the 2p-1s transition energy in nitrogen
and the 2p-1s energy in the united atom, in this case silicon. The
yield of x-rays from a decay when the vacancy occupies this MO will
be estimated.

This problem is essentially one of collision broadening, where
the frequency and intensity of emitted photons varies as a result

of the perturbing effect of other atoms with which the emitter is
in collision. Such emission has been widely studied in optical
spectroscopy [24] but in the case of x rays and keV impact energies
the shifts are much more dramatic.

The probability of decay of a vacancy during a time interval
dt during which its energy varies from E to E + dE is simply dt/τ_x,
where τ_x is the instantaneous radiative lifetime. Then the probability of decay per unit energy interval can be written as

$$P(E) = \frac{\frac{dR}{dE}}{v_R \tau_x} \qquad (4)$$

where v_R is the radial velocity along the incident ion trajectory
(for heavy ion-atom collisions at these velocities a classical
description of the internuclear motion is adequate). The cross-
section for emission at a particular energy E_0 corresponding to a
distance of approach R_0 is obtained by integrating P(E) over all
impact parameters less than that for which R_0 constitutes the distance of closest approach. This step may be performed analytically.
The yield of photons of energy E_0 is then obtained by multiplying
by the density n of target atoms and the total distance $v \tau$ moved
in the total lifetime τ of the vacancy. (Note that this estimate
then neglects any effects of slowing down during this lifetime:
these should be small since $v \tau$ is much less than the total range).
The yield of photons of energy E_0 is given by

$$Y(E_0) = \frac{n \tau 4 \pi R_0^2}{\tau_x} \cdot \left(1 - \frac{U(R_0)}{T}\right)^{1/2} \cdot \left.\frac{dR}{dE}\right|_{R_0}, \qquad (5)$$

where $U(R_0)$ is the internuclear potential and T is the incident
kinetic energy. The yield then cuts off at an energy corresponding
to the distance of closest approach in a head-on collision.

The spectral distribution of emitted radiation is more correctly described by a Fourier transform of the time-dependent decay
probability. From such a treatment it may be shown [25] that the
expression (5) arises if the approximation is made that the photon
energy is identical with the energy difference between the decaying
levels as indeed is implicit in the discussion leading to eq. (5).
This approximation is essentially the assumption that at each internuclear distance the decaying levels are stationary. Clearly the
transience of these levels does impart a smearing to the distribution (5). A direct Fourier transform [25] shows that this is
particularly noticeable at the highest energies where the sharp
cut-off is broadened by ~100-200 eV. However, its effect in other
regions is small and the more direct and simple expression of
eq. (5) will be used.

In deciding whether emission which is characteristic of the transient MO levels will be distinguishable it is necessary to estimate the contribution of the broadening of the central "atomic" line in energy regions far removed from the atomic 2p-1s energy. This contribution is difficult to estimate reliably as the distinction between what is "molecular" emission and what is "atomic" emission is clearly arbitrary. However, the region of internuclear distance where large changes in radiative lifetime and orbital energy are initiated is well defined. The atomic emission can then be treated as that of an oscillator at a fixed frequency whose emission is interrupted after a finite interval by close collision with a lattice atom. The emission spectrum is then of Lorentzian form with a width Γ which is inversely proportional to the average time between collisions,

$$f(E) = \frac{\Gamma/\pi}{(E - E_o)^2 + \Gamma^2} \tag{6}$$

The total yield of this emission per K-vacancy created is obtained by multiplying by the atomic K-vacancy fluorescence yield (the total MO emission is a small fraction of the total "atomic" emission and can be neglected in computing the "atomic" yield).

The computation of the molecular yield (5) requires a knowledge of the $1\pi_u = 1\sigma_g$ orbital energy difference (this is the dominant mode of decay of $1\sigma_g$ vacancies), the corresponding radiative rate and the internuclear potential. The orbital energies and the potential energy are provided directly by the Hartree-Fock calculation. The radiative rate was computed from the orbital wavefunctions by evaluation of the electric dipole transition matrix element between the $1\pi_u$ and $1\sigma_g$ orbitals.

The two contributions to the x-ray yield at energies removed from the atomic line are shown in Fig. 4. The "molecular" emission forms a distinct and visible shoulder on the high energy side of the nitrogen-K line extending up to energies near to the united atom K x-ray energy. This result supports the identification of emission in this region in the similar $C^+ - C$ system [20] as being due to emission from the MO levels during the collision. It also makes plausible the identification of emission around 1 keV in the Ar^+ bombardment of some solid targets [19,22] as due to a similar "molecular" emission, although in this case the radiative decay is to a π orbital which correlates the 2p shell of Ar with the united atom 2p shell. Further, it has been shown [23] that when the transmission function of typical detector windows is folded into the emission spectrum of Fig. 4 the result is a broad peak with a position and width in qualitative agreement with the observed emission in C^+ bombardment of solid carbon [20].

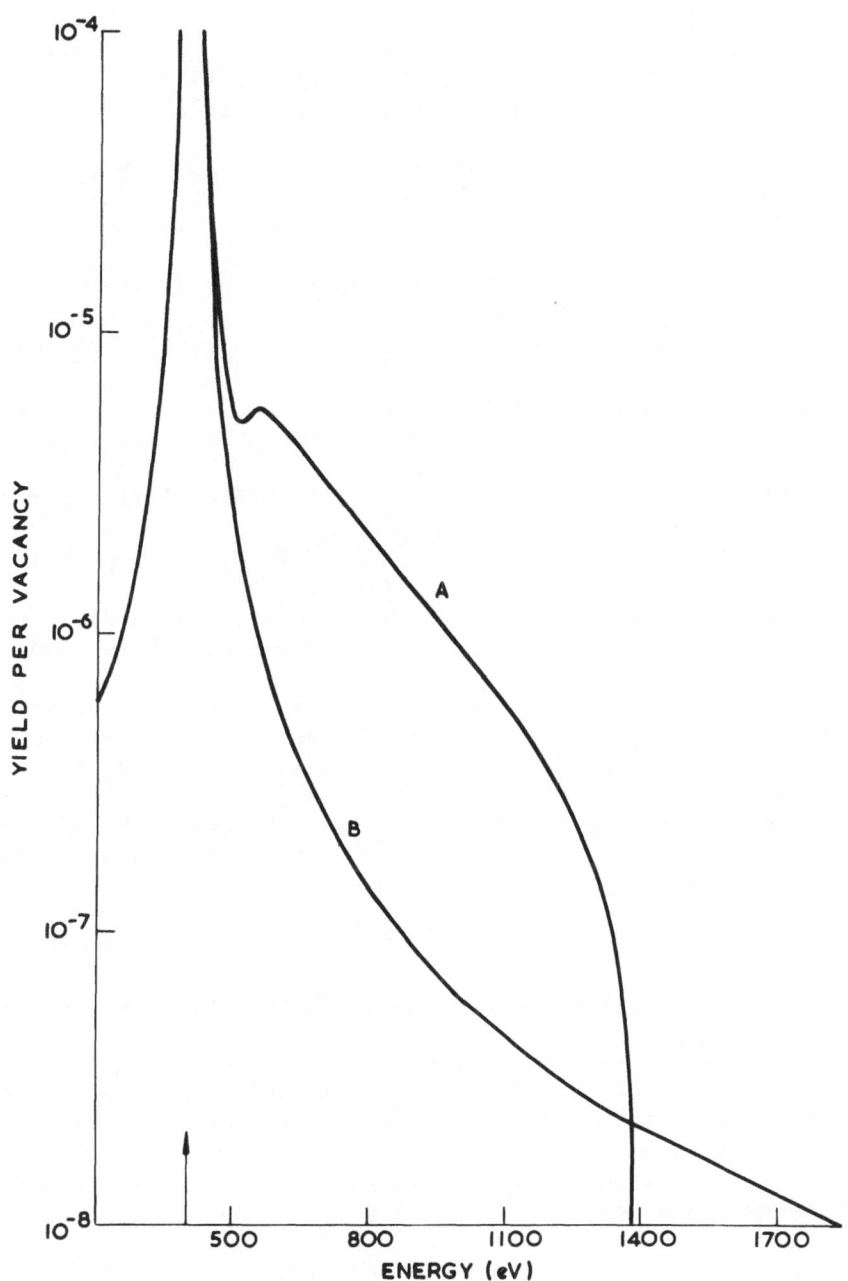

Fig. 4. The emission spectrum of a K-excited N atom moving through solid nitrogen. A is the yield from MO emission, B is the atomic line centered on a position marked by the arrow.

References

[1] U. Fano and W. Lichten, Phys. Rev. Lett. 14, 627 (1965).
[2] C. C. J. Roothaan, Revs. Mod. Phys. 23, 69 (1951).
[3] R. S. Mulliken, Chem. Phys. Lett. 14, 137 (1972).
[4] F. P. Larkins, J. Phys. B5, 571 (1972).
[5] E. W. Thulstrup and H. Johansen, Phys. Rev. A6, 206 (1972)
[6] J. S. Briggs and M. R. Hayns, J. Phys. B6, 514 (1973).
[7] B. Sidis, Proc. VII ICPEAC, 1971 (Amsterdam: North Holland) p. 120.
[8] W. Lichten, Phys. Rev. 164, 131 (1967); M. Barat and W. Lichten, Phys. Rev. A6, 211 (1972).
[9] J. S. Briggs and J. Macek, J. Phys. B5, 579 (1972); B6, 982 (1973).
[10] D. R. Bates and D. A. Williams, Proc. Phys. Soc. 83, 425 (1964).
[11] R. K. Cacak, Q. C. Kessel and M. E. Rudd, Phys. Rev. A2, 1327 (1970).
[12] B. Fastrup and G. A. Larsen, Proc. VII ICPEAC, 1971 (Amsterdam: N. Holland) p. 392.
[13] R. J. Fortner, B. P. Curry, R. C. Der, T. M. Kavanagh and J. M. Khan, Phys. Rev. 185, 164 (1969).
[14] H. Tawara and F. J. deHeer, Physica (to be published).
[15] J. D. Garcia, R. J. Fortner and T. M. Kavanagh, Revs. Mod. Phys. 45, 111 (1973).
[16] J. A. Cairns, A. D. Marwick, R. S. Nelson and J. S. Briggs, Rad. Effects 12, 7 (1972); J. A. Cairns, R. S. Nelson and J. S. Briggs in Ion Implantation in Semi-conductors, Ed., I. Ruge and J. Grand (Springer: Berlin) 1972, p.299.
[17] J. A. Cairns in Inner Shell Ionization Phenomena and Future Applications, Vol. II, Eds., R. W. Fink et al. (1973) (USAEC).
[18] L. C. Feldman, J. M. Poate, F. Ermanis and B. Schwarts (to be published).
[19] F. W. Saris, W. F. van der Weg, H. Tawara and R. Laubert, Phys. Rev. Lett. 30, 471 (1973).
[20] J. R. MacDonald, M. D. Brown and T. Chiao, Phys. Rev. Lett. 30, 471 (1973).
[21] F. W. Saris in Inner Shell Ionisation Phenomena and Future Applications, Vol. II, Eds., R. W. Fink et al. (1973) (USAEC).
[22] G. Bissinger and L. C. Feldman, Phys. Rev. A8, 1624 (1973).
[23] J. S. Briggs, J. Phys. B7, 47 (1974).
[24] H. R. Griem, Plasma Spectroscopy (McGraw-Hill: New York) 1965.
[25] J. Macek and J. S. Briggs (to be published).

AN INVESTIGATION OF THE PROCESSES INVOLVED IN THE PRODUCTION OF NON-CHARACTERISTIC X RAYS DURING ION BOMBARDMENT OF SOLID TARGETS

J. A. CAIRNS, A. D. MARWICK, P. J. CHANDLER,
and J. S. BRIGGS
AERE Harwell, Didcot, Berkshire, England

ABSTRACT

This work can be subdivided into two main areas. First, it examines the role played by the projectile in generating both characteristic and non-characteristic x rays during argon ion bombardment of silicon and silicon carbide. This involves an investigation of the effect of projectile build-up in these targets, and some high resolution spectral examination of the ArL x rays. It emerges that Ar → Si collisions are mainly responsible for the non-characteristic x rays; the dominant mechanism at the energy investigated is a double scattering process, although recoils also play a role in the Ar → Si system. Second, the differential cross-sections for non-characteristic x ray produced during C → C collisions have been measured over the energy range 40 → 280 keV. This work demonstrates a gradual levelling off of these cross sections and also highlights the appearance of x rays having energy greater than the united atom limit, arising from a collision broadening effect.

Introduction

A recent interesting phenomenon, observed during collisions between energetic ions and solid targets, has been the production of x rays which cannot be ascribed to either the target or the projectile, and which therefore may be described as non-characteristic. Fig. 1 shows the production of such x rays during argon

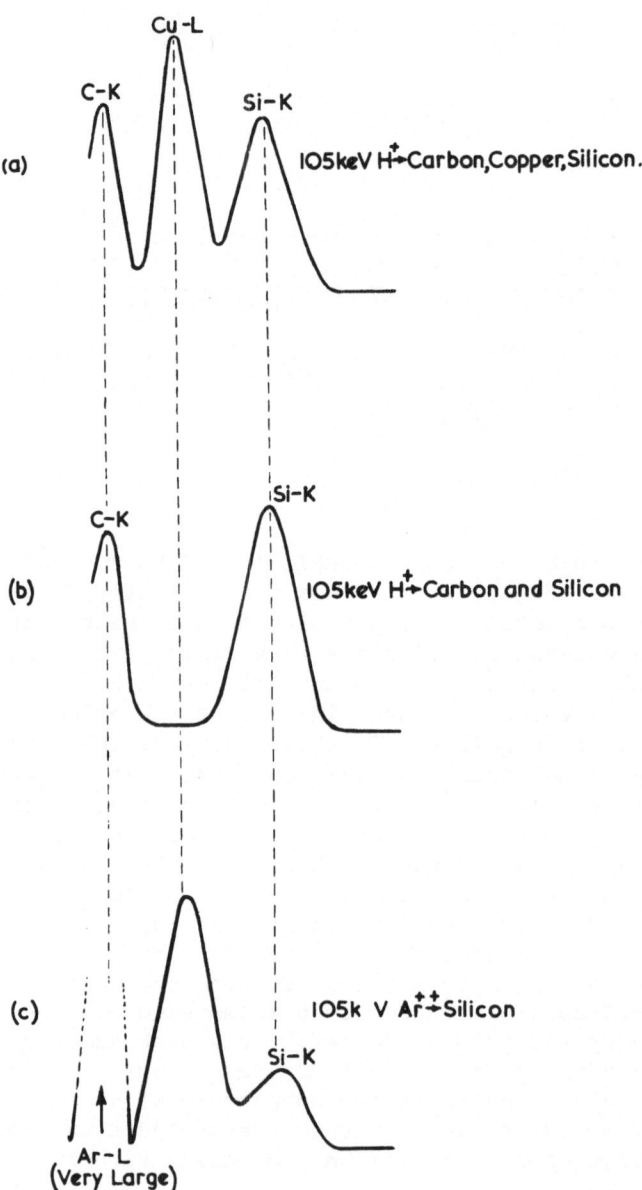

Fig. 1. To illustrate the production of non-characteristic
x rays by 105 kV Ar++ bombardment of solid silicon. X rays
from proton irradiation of carbon, copper, and silicon are
shown for comparison (spectra taken with proportional
counter [11]).

bombardment of silicon, while Fig. 2 shows those produced by argon
bombardment of aluminium. Each figure also displays for comparison

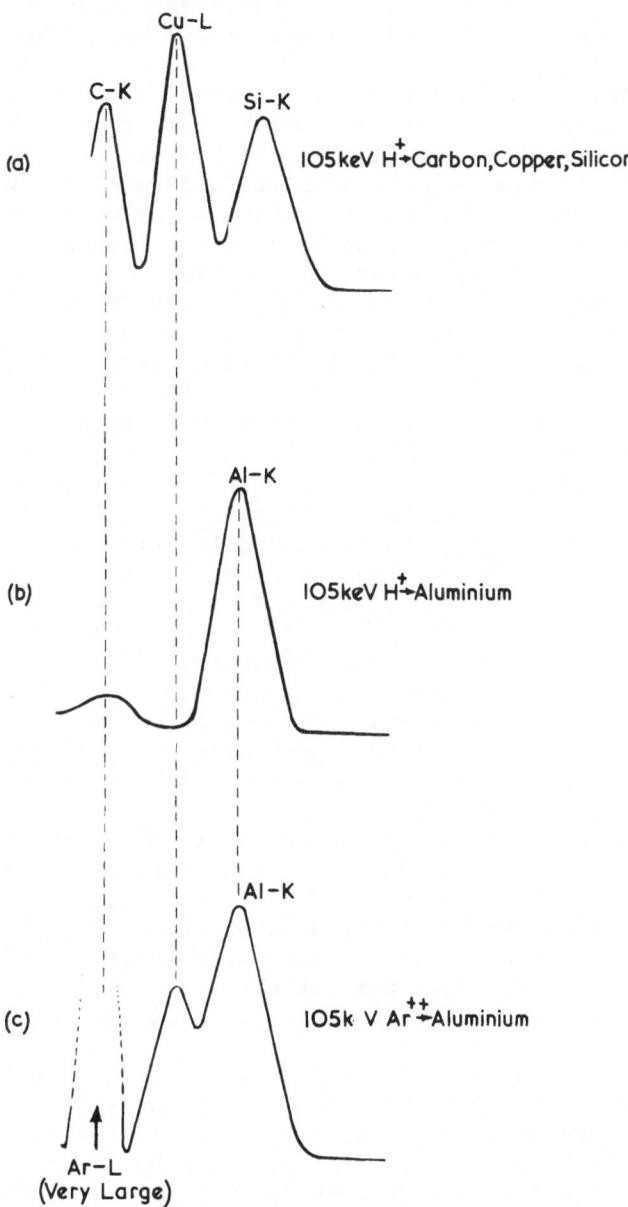

Fig. 2. To illustrate the production of non-characteristic
x rays by 105 kV Ar++ bombardment of aluminium. X rays from
proton irradiation of carbon, copper, and silicon are shown
for comparison (spectra taken with proportional counter [11]).

some characteristic lines produced by proton bombardment of solid targets, which illustrate that the non-characteristic band apparently peaks at ∿ the same energy as CuL, i.e. ∿ 1 keV. These figures, which were taken with a gas flow proportional counter, illustrate further that the non-characteristic x-ray band can be of considerable magnitude, particularly in the Ar → Si case where, at the projectile energy shown, it dominates the SiK yield.

An explanation for the production of these x rays was proposed by Saris et al.[1]: they considered them to arise during the lifetime of the pseudomolecule which is formed during collisons between incoming argon ions and argon atoms which had previously been implanted into the solid silicon. Since then, other observations [2-5] have increased the catalogue of examples, but the Ar → Si system still remains one of the most attractive for study because of the relatively large yields of non-characteristic x rays produced (within a certain projectile energy range).

If we use the Fano-Lichten model [6-8] to predict the probability of SiK x-ray production, it becomes apparent that this is in itself unlikely, since the SiK electron requires to be promoted into an argon 2p level which is normally already full. However, Macek et al.[9] were able to demonstrate that the appropriate 2p vacancy could be created in a prior Ar → Si collision, and exist sufficiently long for a second collision to take place. During this second collision, either SiK or non-characteristic x rays may be produced; hence we see the interconnection between them. (Note incidentally that this picture would predict that the non-characteristic x rays should appear as a very broad band, covering the energy range from SiK down to ArL; the peaking shown in Figs. 1 and 2 at ∿ 1 keV is now generally accepted [2,5] to arise from preferential absorption of the softer x-ray components by the detector window).

There is also an alternative process by which SiK x-rays may be produced [10]. This can arise when a silicon atom, recoiling from a collision with an argon projectile, collides with another silicon atom. Since the recoiling silicon can carry a 2p vacancy, it is apparent that SiK x rays can also be produced in these Si → Si collisions. We shall demonstrate that our results are consistent with a fraction of the observed yields being due to such recoils.

Thus, in the present work, we propose to look at all three x-ray bands produced during argon bombardment of silicon. These x rays will be studied as a function of argon projectile dose, and their behaviour compared to that obtained by using a complementary target, namely, silicon carbide. Since one of the above pictures [9] of the production of SiK and non-characteristic x rays has invoked the prior electronic excitation of the argon projectile, we shall present spectral measurements of the argon L x rays, recorded during argon bombardment of both silicon and silicon carbide.

In addition, we present some recent measurements of a
quantitative study of non-characteristic x-ray production during
carbon → carbon collisions. This system, which has the attraction of relative simplicity compared to the above, shows an
interesting behaviour exhibited by the differential cross-section
as a function of projectile energy.

Experimental

Three different instruments were employed to measure the
x-rays.
  (i) A gas flow proportional counter [11] was used to measure
      the yields of the three x-ray bands emitted during argon
      bombardment of silicon and silicon carbide.
 (ii) A grating spectrometer [12] containing a new high
      resolution miniature proportional counter [13] was
      employed to resolve the argon L spectra generated during
      these collisions.
(iii) A high resolution solid state detector system, employing
      pulsed optical feed back electronics [14] was used to
      measure the non-characteristic x rays emitted during
      C → C collisions. The detector was housed behind an
      interchangeable window composed of 6 μm mylar or 1 μm
      polypropylene, and was able to resolve soft x rays,
      including carbon K. Fig. 3 shows an example of its

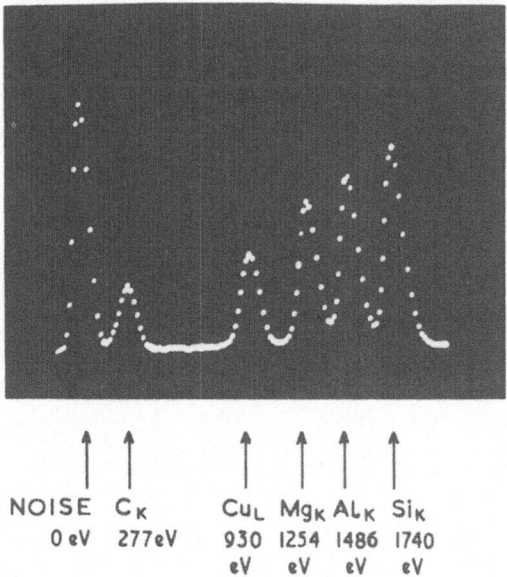

Fig. 3. An example of the favourable soft x-ray resolution exhibited by the solid state detector used in this work.

resolution.   Its favourable performance, as compared
to that of a gas flow proportional counter, has been
discussed in another publication [15].

Results and Discussion

The effects of projectile build-up in targets

Fig. 4 shows the yields, measured with a proportional counter,
of the three sets of x rays generated by argon bombardment of
silicon (Fig. 4a) and silicon carbide (Fig. 4b), as a function of
argon dose.   In their original study, Saris et al.[1] noted that

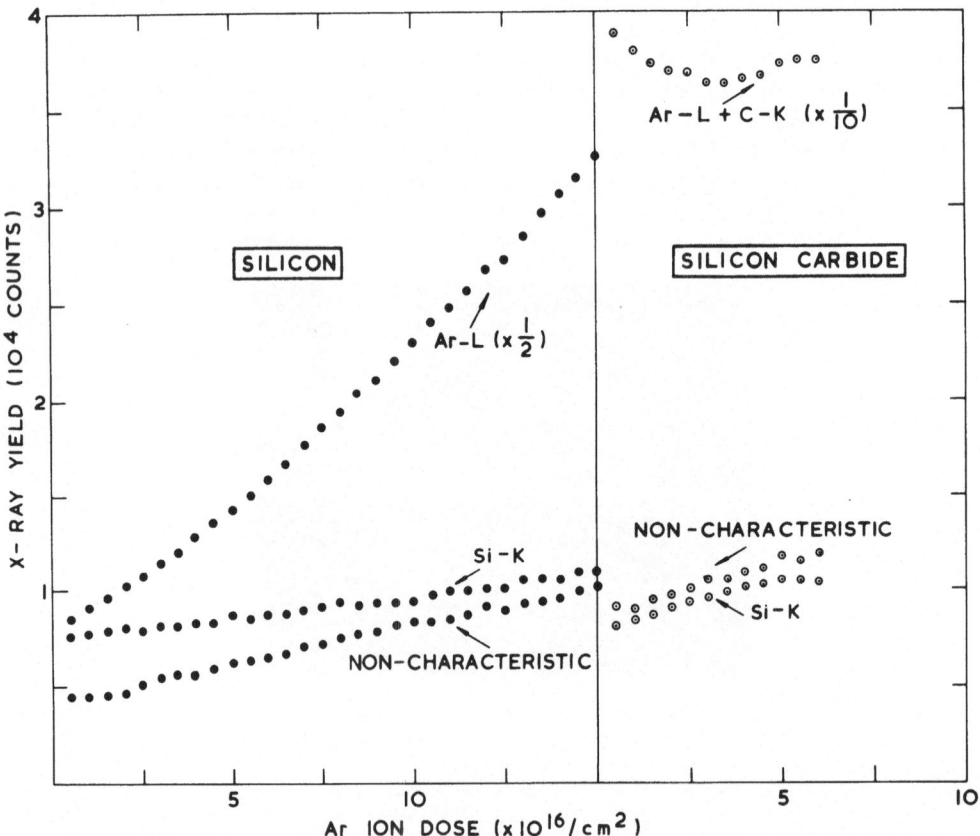

Fig. 4. To illustrate the build-up of x-ray yields emitted during
270 keV Ar^{++} bombardment of silicon (referred to as Fig. 4a) and
silicon carbide (referred to as Fig. 4b).

the non-characteristic x-ray yield increased as a function of projectile dose. This led them to postulate that these x-rays were due to Ar → Ar collisions. However, we see from Fig. 4 that not only do the ArL and non-characteristic x-ray yields build up, but so also do the SiK x-ray yields. This can be understood when we realize that the build up of argon in the target increases the probability of argon 2p vacancy production from Ar → Ar collisions. Then, according to the double scattering mechanism outlined earlier, this in turn should increase the probability of production of both non-characteristic x rays and SiK x rays. However, we may then ask why all three sets of x rays do not build up at the same rate. Fig. 4a shows that the rate of ArL build up is ∿ twice that of the non-characteristic x rays. This can be understood by noting that ArL x-rays can arise from both argon projectiles and previously-implanted argon, but only projectiles with a 2p vacancy are likely to contribute towards the production of non-characteristic x rays, because it can be shown that an argon projectile is ∿ 100 times more likely to acquire a 2p vacancy than to cause another argon to recoil with energy greater than one quarter of the incoming projectile energy. So far as SiK is concerned, the slow build up rate, which is only one quarter of that of the non-characteristic x-ray yield, is rather instructive. It leads us to conclude that ∿ 75% of the SiK yield is due to recoils. This is in good agreement with the work of Taulbjerg et al.[11].

It is also worth noting that all three x-ray yields build up linearly. This is to be expected if they depend linearly on argon concentration. Thus the linear increase in the non-characteristic x-ray yield with argon dose favours the interpretation that they arise from Ar → Si collisions, rather than from Ar → Ar collisions, which would cause the yield to build up quadratically.

These considerations would appear to be substantiated by the Ar → SiC results shown in Fig. 4b. In this system, the production of 2p vacancies in the incoming argon projectile is expected to be enhanced by the presence of carbon atoms. In fact we see that the yields of both non-characteristic x rays and SiK x rays are higher than in the Ar → Si system. The yield of ArL is complicated by the simultaneous generation of carbon K x rays, as these two x rays cannot be separated by the proportional counter. However, the high resolution measurements described below prove that the ArL yield is also higher in the Ar → SiC system than in the Ar → Si system. Fig. 4b also shows that, as in the Ar → Si system, the yields of all three x rays builds up linearly with argon dose, again indicating that the non-characteristic x rays are most likely due to Ar → Si collisions. One further interesting feature of the dose dependence shown in Fig. 4b is that now the yields of non-characteristic x rays and SiK x rays build up at the same rate. This would imply that now recoil effects play a negligible role in the production of the SiK x rays, and indeed this is to be expected, for the presence of abundant carbon in the target means a lower likelihood of a recoiling silicon making an efficient collision with another lattice silicon atom.

High resolution examination of the ArL x-ray band

A further insight into these mechanisms is obtained by examining the degree of excitation of the argon projectile during the collisions. This was done by using the grating spectrometer referred to earlier [12]. Fig. 5a shows the argon L spectra generated during Ar → Si bombardment at a projectile energy of 270 keV, before and after argon build up in the target. Fig. 5b shows the same situation when a silicon carbide target was used.

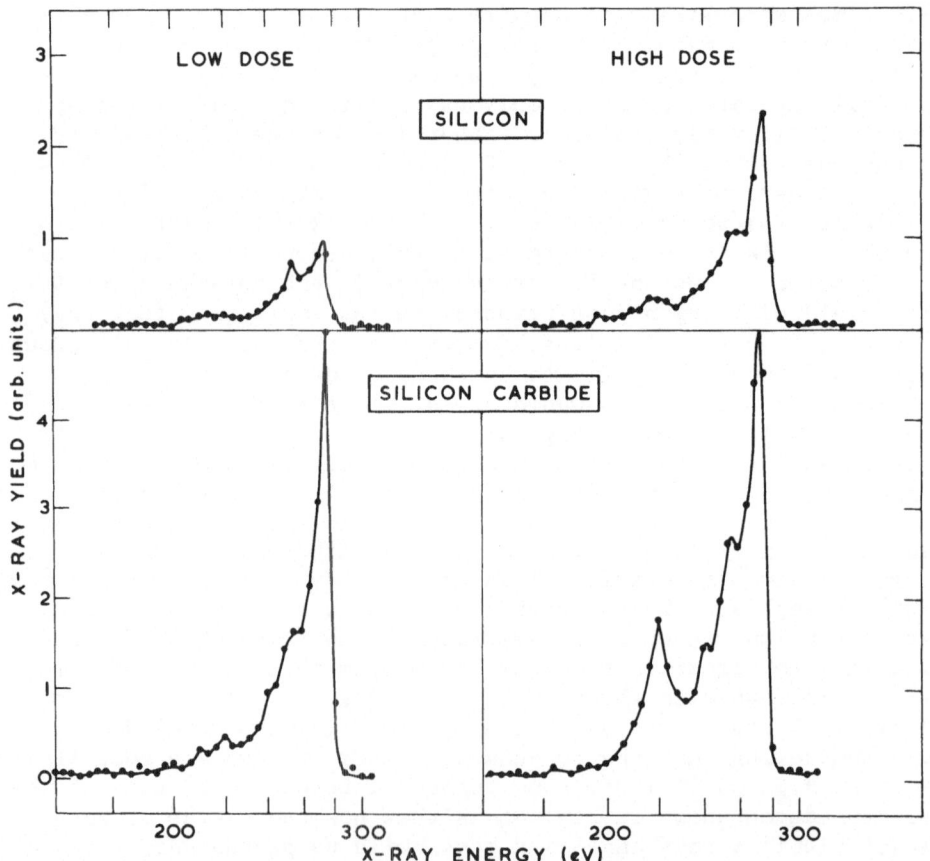

Fig. 5. Soft x-ray spectra, covering the energy range 200–300 eV, emitted during 270 keV Ar++ bombardment of silicon (referred to as Fig. 5a) and silicon carbide (referred to as Fig. 5b).

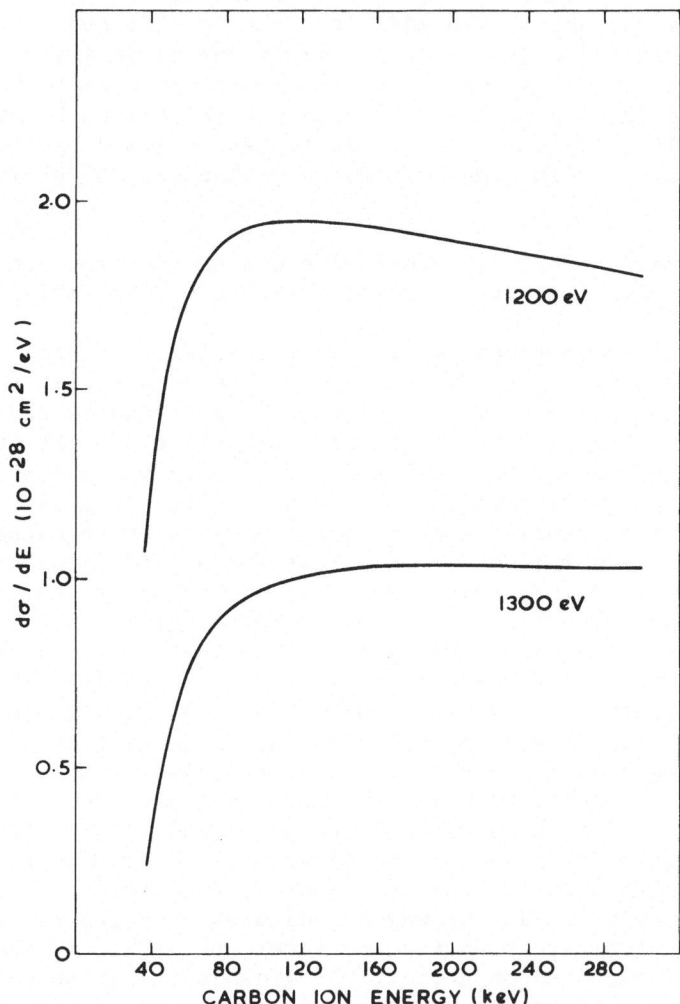

Fig. 6. Differential cross section of non-characteristic x rays
generated by bombardment of solid carbon target by carbon ions
of energy 40 - 280 keV.

We note that, as anticipated, the ArL yield at 220 eV is
higher in the Ar → SiC case, and although building up with projectile dose, is dominated by the CK line at 277 eV, which of
course does not build up. In addition to the normal line at 220
eV, both spectra contain an additional satellite at ∿ 250 eV,
which most likely arises from argon ions having one 2p vacancy and
several outer vacancies. In addition, the Ar → Si spectra show
another line at ∿ 280 eV, which is likely due to an argon ion
having two 2p vacancies plus some additional outer vacancies.
This line should be produced with greatest efficiency by Ar → Ar
collisions [16], and this would seem to be confirmed by the fact
that it builds up with dose more quickly than the 250 eV line.

Measurement of the differential cross-section for non-characteristic x rays produced during C → C collisions

The results described so far are consistent with the idea of
non-characteristic x rays arising from a pseudomolecular state
produced during ion - atom collisions, as advocated by Saris
et al.[1]. We have sought to clarify the role played by the projectile in these processes. Therefore, it is now appropriate to
look at the non-characteristic x rays themselves in a little more
detail; for this purpose we chose the C → C system, because this
is free from the complication of simultaneous production of higher
energy characteristic x rays. In addition, there was the availability of some recent calculations performed by Briggs [17] on
the similar N → N system. We used the new solid state detector
system, described in the experimental section, to measure the
non-characteristic x-rays yields over the projectile energy range
40-280 keV. From these, the differential cross section was calculated, using a method similar to that employed by Macek et al.[9].
In the course of this, we measured the CK x-ray production cross
section and found good agreement with published values [18]. Since
the non-characteristic x-ray band is very broad, two regions were
examined, centered at 1200 eV and 1300 eV, respectively. The
results are shown in Fig. 6, which indicates that the cross sections
rise rapidly at first, then tend to level off. This is the behaviour to be expected on geometrical grounds [17], for an x ray
of a particular energy can be produced only at or near to a given
internuclear separation, and therefore once the energy necessary
to achieve this has been surpassed (i.e. the threshold energy)
then the x-ray cross section will be proportional to the time spent
by the projectile within the critical region.
    The appearance of x rays at 1300 eV (and above) is in itself
significant, because the maximum energy to be expected from a
simple consideration of the united atom limit is 1250 eV (MgK).
The production of x rays beyond this limit is likely to arise because of the rapid change of molecular energy levels with

internuclear distance, so that x rays within a particular energy range are emitted in a time which is short compared to the appropriate radiative life-time.

An implication of this broadening, which can be as large as several hundred eV, is that it imposes a limit on the accuracy with which one can map out molecular energy levels, as had been attempted by MacDonald [5].

References

[1] F. W. Saris, W. F. van der Weg, H. Tawara, Phys. Rev. Lett. 28, 717 (1972).

[2] F. W. Saris, I. V. Mitchell, D. C. Santry, J. A. Davies, R. Laubert, Proc. International Conference on Inner Shell Ionization Phenomena, Atlanta (USAEC Conf.-720404 (4 vols.)) 1972.

[3] P. H. Mokler, H. J. Stein, P. Armbruster, Phys. Rev. Lett. 29, 827 (1972).

[4] J. R. MacDonald, M. D. Brown, Phys. Rev. Lett. 29 (1972).

[5] J. R. MacDonald, M. D. Brown, T. Chiao, Phys. Rev. Lett. 30, 471 (1973).

[6] U. Fano and W. Lichten, Phys. Rev. Lett. 14, 628 (1965).

[7] W. Lichten, Phys. Rev. 164, 131 (1967).

[8] M. Barat, W. Lichten, Phys. Rev. A 6, 211 (1972).

[9] J. Macek, J. A. Cairns, J. S. Briggs, Phys. Rev. Lett. 28, 1298 (1972).

[10] K. Taulbjerg, B. Fastrup, E. Laegsgaard, Phys. Rev. 8, 1814 (1973).

[11] J. A. Cairns, C. L. Desborough, D. F. Holloway, Nucl. Instr. Meth. 88, 239 (1970).

[12] M. Steadman, J. A. Cairns, A. D. Marwick (in preparation).

[13] J. A. Cairns, D. F. Holloway, G. F. Snelling, Nucl. Instr. Meth. 111, 419 (1973).

[14] K. Kandiah, Nucl. Instr. Meth. 95, 289 (1971).

[15] J. A. Cairns, A. D. Marwick, I. V. Mitchell, presented at the International Conference on Ion Beam Surface Layer Analysis (IBM Research, N.Y., June 1973); Thin Solid Films 19, 91 (1973).

[16] B. Fastrup, G. Hermann, K. J. Smith, Phys. Rev. A 3, 1591 (1971).

[17] J. S. Briggs, J. Phys. B 7, 47 (1974).

[18] J. M. Khan, D. L. Potter, R. D. Worley, Phys. Rev. A 139, 1735 (1965).

OBSERVATION OF THE UNITED CARBON-CARBON ATOM K-SHELL X-RAY BAND FOR INCIDENT CARBON ION ENERGIES OF 30-2500 keV*

ROMAN LAUBERT [†]

Department of Physics, New York University
New York, New York 10003, U.S.A

ABSTRACT

The centroid energy and x-ray cross section for the non-characteristic carbon-carbon x-ray band produced by 30-2500 keV carbon ions incident on thick solid carbon targets are reported. We test the double collision assumption for the production of these non-characteristic x rays and find that, for 250 keV carbon ions, the experimental results are consistent with this assumption.

Introduction

The observation of x rays, in atom-atom collisions, which cannot be associated with the characteristic radiation of either atom was first reported [1] just one year ago. Since that time other investigators [2-5] have substantiated this observation. In this paper we report the observation of non-characteristic x rays when 30-2500 keV carbon ions are incident on thick solid carbon targets.

It has been suggested [1] that these non-characteristic x rays result from a radiative decay of an inner shell vacancy, in this case a carbon K-shell vacancy, during an atomic collision. Since the velocity of the colliding particles is slow, the atomic wave functions of the projectile-target combination are able to adjust to the imposed perturbation and form molecular orbitals [6]. The resulting electron energy level diagram of the system would be that of the isolated atoms at large internuclear separation and that of the united atom at very small internuclear separation. If a carbon K-shell vacancy brought into the collision decays during the collision then the emitted radiation can vary in energy, depending on the internuclear separation at the instant of de-excitation, from an energy associated with the carbon K-shell (0.277 keV) to an energy associated with the united atom, in this case a Mg atom ($\sim$1.25 keV). The requirement that an inner shell vacancy be brought into the collision has led to the assumption that two collisions are required to produce the observed x-ray band.

Experiments were performed that test the double collision assumption and we find that, for 250 keV carbon ions, two collisions are required to produce these non-characteristic x rays. We also find that the energy of the emitted x rays can exceed the combined atom limit, which in this case is 1.25 keV. The cross section for observing non-characteristic x rays greater in energy than $\sim$700 eV is reported for the single and double collision.

Experimental Results and Discussion

The experimental arrangement is shown schematically in Fig. 1. A mass and energy analyzed $^{12}_{6}$C ion beam in the energy range of 30–2500 keV impinges on a solid carbon target. We used the New York University accelerator [7] to obtain carbon ions in the energy range of 30–300 keV and the Brookhaven 3 MeV Van de Graaff accelerator for ion energies of 400–2500 keV. The target is heated to 150°C to prevent the buildup of hydrocarbons and electrically biased with respect to the target chamber to obtain a proper current measurement. The emitted x rays are registered and energy analyzed by a Si(Li) x-ray detector which has an energy resolution, as

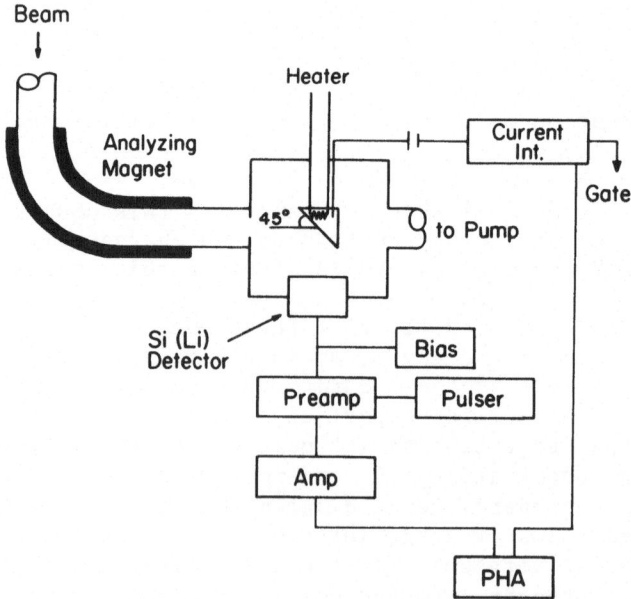

Fig. 1. Schematic of the experimental arrangement.

measured by the full width at half maximum (FWHM), of 192 eV determined by proton bombardment of Al and Si targets. The "window"
between the target and the x-ray detector was made of beryllium and
two different thicknesses were used; 25 µm and 8 µm. The target
was a thick vitreous carbon of 5 N metal purity which was first
abraded to remove the glasslike surface layer and then washed in
acid and distilled water. The cleanliness of the target was checked
by low energy proton bombardment and no x rays were observed in the
x-ray region of interest.

When carbon ions are incident on the carbon target the observed
x-ray distribution is shown in Fig. 2. This is a linear and logarithmic picture of the observed x-ray distribution using the 8 µm
Be window for 1 MeV carbon ion and a total dose of 4 µC. In all
cases the count rate was kept below 1 K cts/sec to minimize dead-
time and pulse pileup. The distribution extends from ∿500 eV to
1.5 keV and has a centroid at ∿1 keV, so that we can characterize
the distribution by a centroid energy, FWHM and the total number
of x rays observed in a specified x-ray energy interval.

Alternatively one can deconvolute the observed x-ray energy
distribution by the instrument function of the detector and the
transmission function of the beryllium window to obtain the actual
energy distribution of the emitted x rays. Since we were unsuccess-
ful in this endeavor, all of the results presented here will utilize
the simple data reduction procedure. It has been shown [5,8] that
the energy distribution of the emitted x rays extends from the

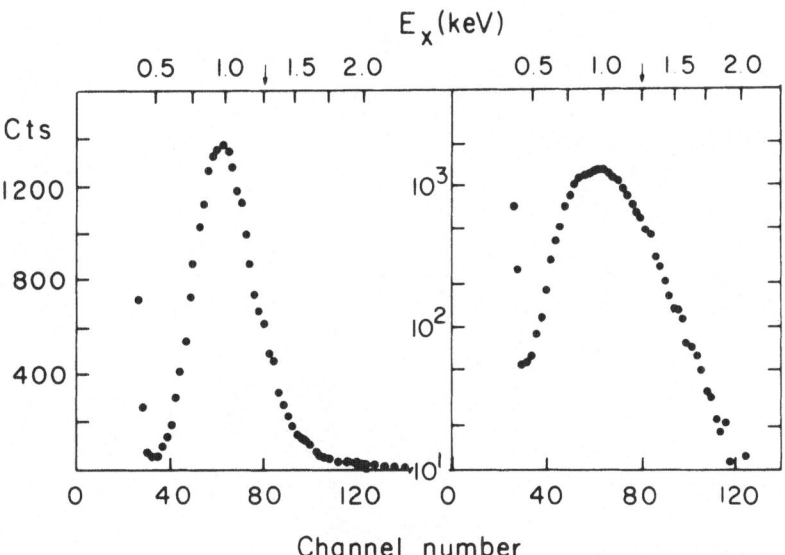

Fig. 2. A linear and logarithmic picture of the C–C x-ray band as
viewed by a Si(Li) detector with an 8 µm Be window for 1 MeV carbon
ions and a total dose of 4 µC. The corresponding x-ray energy is
marked across the top of the figure. The arrow marks the location
of the combined atom K_α x ray.

carbon K x-ray energy to some unspecified upper limit and that the
Si(Li) detector allows us to observe the higher energy x rays con-
voluted by the instrument and transmission functions.

The centroid energy, in keV, of the observed C-C x-ray band,
using the 8 μm Be window, as a function of the square root of the
incident ion energy is shown in Fig. 3. The velocity of the incident projectiles in atomic units is indicated across the top of
the figure. As is evident from the figure the centroid energy,
after an initial rise, increases linearly with increasing projectile
velocity. This increase amounts to ∿85 eV for a change of 1 MeV
in projectile energy. The horizontal bars, in Fig. 3, are the
observed FWHM which increase from ∿400 eV to ∿700 eV. This latter
figure corresponds to ∿4 times the instrumental resolution. Hence
x rays are emitted which have an energy greater than the combined
atom K_α x ray of 1.25 keV confirming the results of Macdonald et
al.[4]. This reflects the existence of L-shell vacancies [4] in
these highly distorted "atoms" and possibly, as suggested earlier
[9], is due to collision broadening. The results for the 25 μm Be
window show similar trends except that the centroid energy is
∿110 eV higher and the FWHM is reduced somewhat (∿20%). With the
25 μm Be window the centroid energy of the x-ray band can exceed
the combined atom limit for incident carbon ion energies greater
than 1.6 MeV.

We determine the x-ray yield by integrating the spectrum over
an appropriate energy region and correcting for the transmission

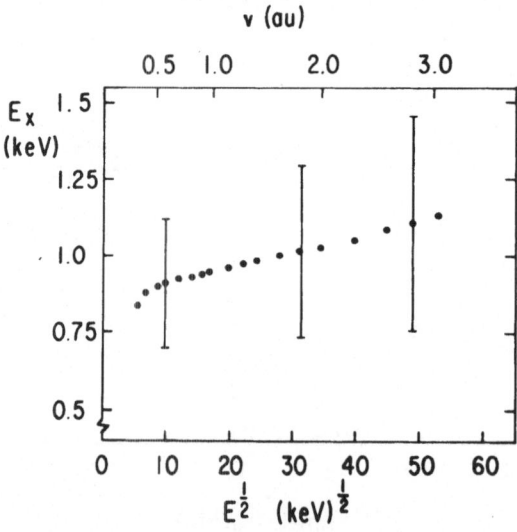

Fig. 3. The centroid energy of the C-C x-ray band, using the 8 μm
Be window as a function of the square root of the incident ion
energy. The velocity of the incident projectiles, in atomic units,
is indicated across the top of the figure.

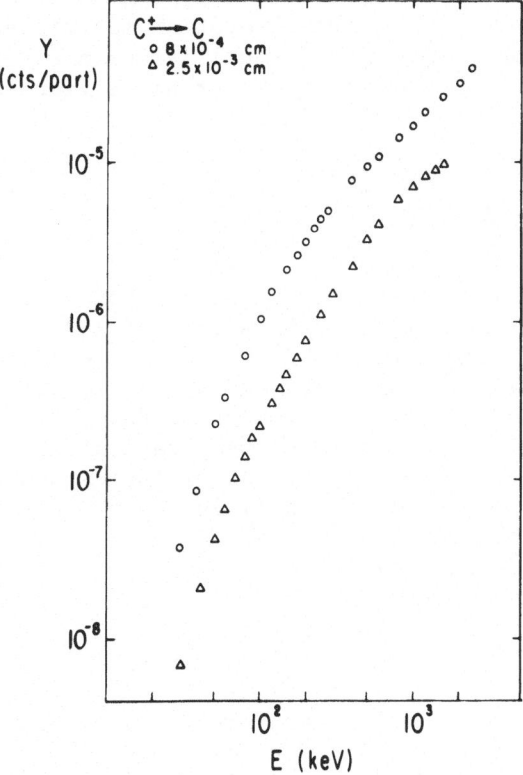

Fig. 4. The C-C x-ray band yield per incident projectile as a function of the incident ion energy for the 25 μm and 8 μm Be window.

of the Be window at the centroid energy. The results are shown in Fig. 4. Note that the absolute yield obtained with the 8 μm window is greater than the yield obtained with the 25 μm window. This indicates that not all of the emitted x rays are counted and that the cross section determined from this data will be an underestimate of the true cross section.

If single collisions are the dominant mode of generating the observed x-ray band, then the x-ray cross section, $\sigma_x(E_1)$, can be determined from the data of Fig. 4 by the equation

$$\sigma_x(E_1) = \frac{1}{n} \left[\frac{dy}{dE} \Big|_{E=E_1} S(E_1) + \mu Y(E_1) \right] , \qquad (1)$$

where E_1 is the incident ion energy, n is the number density of target atoms, $\frac{dy}{dE} \Big|_{E=E_1}$ is the slope of the x-ray excitation function

Y(E) evaluated at $E=E_1$, $S(E_1)$ is the stopping power of the material for the incident ions, and μ the absorption coefficient of the target for the emitted x rays. If two collisions are required to produce the observed x-ray band then the cross section can be determined from [10]

$$\sigma_x^*(E_1) = \frac{\sigma_x(E_1)}{nv[\sigma_A(C)\tau_A + \sigma_x(C)\tau_x]} \quad , \qquad (2)$$

where $\sigma_x(E_1)$ is given by Eq. 1, v is the velocity of the incident projectile, $\sigma_A(C)$ and $\sigma_x(C)$ are the Auger and x-ray cross sections of carbon for producing a carbon K-shell Auger electron or K-shell x ray in carbon-carbon collisions, and τ_A and τ_x are the respective lifetimes. The term in the denominator of Eq. 2 reflects the fraction of the beam that is in the desired state of having a K-shell vacancy. Since the fluorescence yield of the carbon K-shell, ω_k, is much less than unity, Eq. 2 simplifies to

$$\sigma_x^*(E_1) = \frac{\sigma_x(E_1)}{nv\sigma_x(C)[\frac{\tau_A}{\omega} + \tau_x]} \quad . \qquad (3)$$

Using the carbon K-shell x-ray cross section, $\sigma_x(C)$, as determined by Fortner et al. [11], we find that more than 100% of the beam is in the desired state of having a K-shell vacancy [12]. To circumvent this problem we arbitrarily neglected the Auger term contribution in determining the x-ray cross section.

The resultant cross sections for single and double collisions is shown in Fig. 5 for the data collected using the 8 μm Be window. For single collisions the cross section increases rapidly until ∿100 keV and then levels off with increasing bombarding energy. In this energy region the values of the stopping power employed in Eq. 1 become of paramount importance. In compiling Fig. 5 we used the electronic stopping power as tabulated by Northcliff [13] and nuclear stopping powers calculated according to Lindhard [14]. If we would have used the same stopping power of carbon for carbon ions as employed by Fortner et al. [11] in their carbon K-shell ionization studies, then the single collision cross section would exhibit a maximum at ∿800 keV. Using Eq. (3) for determining the cross section for double collisions we find that the cross section exhibits a maximum at ∿50 keV incident ion energy and subsequently decreases monotonically. Note that this cross section is approximately a factor 10^2 to 10^3 lower than the carbon K-shell x-ray cross section for carbon ions.

To ascertain that the observed x-ray band results from two collisions we attempted to detect these x rays in a gaseous target. In this case the mean time between collisions would be increased by about a factor of 10^9 and hence practically all of the K-shell

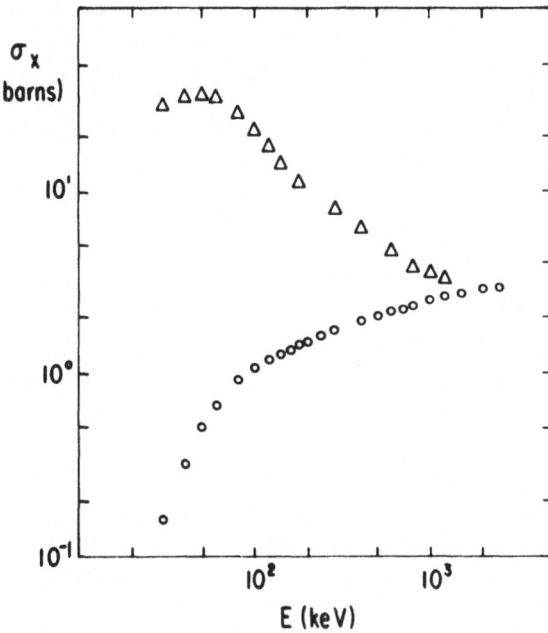

Fig. 5. The x-ray cross section in barns of the C-C x-ray band as a function of the incident ion energy. The bottom curve is obtained by assuming single collisions and using Eq. (1), while the top curve is for the double collision case using Eq. (3).

vacancies would decay before another collision occurred. If the x-ray band can be produced in a single collision then the cross section would be the same in a gaseous or solid target. Using 250 keV carbon ions incident on acetylene we were not able to observe any combined atom x rays and the upper limit for this cross section was determined to be $5 \times 10^{-26} cm^2$, or approximately two orders of magnitude lower than what is observed for a solid target. To check experimentally whether we could observe such a cross section we repeated the experiment using 200 keV helium ions incident on an argon gas target. The corresponding argon K-shell x rays were observed where the x-ray cross section was determined to be $\sim 7 \times 10^{-27} cm^2$. Hence, the upper limit for the C-C x-ray band cross section in a gaseous target is realistic.

Conclusion

     The centroid energy of the x-ray band observed in C-C collisions, after an initial steep rise, increases linearly with the velocity of the projectile. X rays greater in energy than the combined atom limit are observed. Assuming that a C K-shell vacancy has to be brought into the collision to produce the x-ray band, the

cross section for observing x rays >700 eV is ∿35 barns at an
incident ion energy of 50 keV and decreases monotonically to ∿3.5
barns for an incident ion energy of 1 MeV. Noting the differences
in the cross sections between a gaseous and solid targets we conclude that for carbon ions up to 250 keV two collisions are required to produce the observed x-ray band.

Acknowledgments

I would like to thank Dr. W. Brandt and Dr. A. Schwarzschild
for many helpful discussions and Mr. C. Peterson for his assistance
in performing the experiments.

References

*Work supported by the U. S. Atomic Energy Commission.
†Present address: Department of Physics, University of
Tennessee, Knoxville, Tennessee 37916 and Oak Ridge National
Laboratory, Oak Ridge, Tennessee 37830.

[1] F. W. Saris, W. F. van der Weg, H. Tawara, and R. Laubert,
Phys. Rev. Lett. 28, 717 (1972).

[2] J. R. Macdonald and M. D. Brown, Phys. Rev. Lett. 29, 4 (1972).

[3] F. W. Saris, I. V. Mitchell, D. C. Santry, J. A. Davies, and
R. Laubert, Proc. of Intl. Conf. on Inner Shell Ionization
Phenomena, Atlanta, Georgia, 1972, Eds., R. W. Fink, J. T. Manson,
I. M. Palms, and R. V. Rao (USAEC, Oak Ridge, Tenn., 1973).
p. 1255 ff; J. A. Cairns, P. J. Chandler, and A. D. Marwick,
this publication.

[4] J. R. Macdonald, M. D. Brown, and T. Chiao, Phys. Rev. Lett.
30, 471 (1973).

[5] G. Bissinger and L. C. Feldman, Phys. Rev. A8, 1624 (1973).

[6] W. Lichten, Phys. Rev. 164, 131 (1967); M. Barat and W. Lichten,
Phys. Rev. A6, 211 (1972).

[7] R. Laubert and N. Wotherspoon, IEEE Trans. Nucl. Sci. 12, 285
(1965).

[8] F. W. Saris, private communication.

[9] J. S. Briggs, this publication.

[10] J. Macek, J. A. Cairns, and J. S. Briggs, Phys. Rev. Lett. 28,
1298 (1972).

[11] R. J. Fortner, B. P. Curry, R. C. Der, T. M. Kavanagh, and
J. M. Khan, Phys. Rev. 185, 164 (1969).

[12] Comment by R. J. Fortner: Eq. 3 is valid only if the denominator
<<1; when the denominator in Eq. 3 is ∿1 a more rigorous deriva-
tion is required.

[13] L. C. Northcliffe and R. F. Schilling, Nucl. Data 7, 233 (1970).

[14] J. Lindhard, V. Nielsen, M. Scharff, and P. V. Thomsen, Kgl.
Danske Videnskab. Selskab, Mat.-Fys. Medd. 33, No. 10 (1963).

CROSS SECTIONS FOR THE PRODUCTION OF X RAYS FROM HEAVY ION COLLISIONS AT MeV ENERGIES

F. BELL and H.-D. BETZ
Sektion Physik, Universität München
D8000 München
Amalienstr. 54/III, Germany

ABSTRACT

We have systematically investigated x rays which are emitted in heavy ion collisions. Absolute cross sections, σ, are reported for the production of Cl-K x rays in collisions of 30- to 120-MeV chlorine ions with various thin solid targets ranging from $Z_T=6$ (carbon) to $Z_T=92$ (uranium). Surprisingly pronounced oscillations are found in the dependence of σ on Z_T; in contrast to general expectations these oscillations (i) are very sharp, (ii) tend to increase with increasing energy of the projectile, and (iii) can not, at present, be satisfactorily explained by Coulomb excitation or molecular orbital considerations. Furthermore, we present some detailed results on the newly discovered phenomenon of radiative electron capture (REC); the mechanism to produce REC, line position and width of REC are discussed.

Introduction

The Tandem van de Graaff facilities at Brookhaven National Laboratory and at the University of Munich have been used to accelerate beams of p, O, F, S, Cl, Br, I, and U to energies between 6 and 156 MeV. These ions were directed onto many thin and thick solid targets ranging from beryllium to uranium, and the induced x rays were observed with a high-resolution Si(Li) detector in the x-ray energy range from 1 to 30 keV. Some results from this work have already been published; the first observations of REC [1-5], selected projectile and target x-ray production cross sections [2], and measurements and interpretations of x-ray line shifts [6]. Publication of most of our further results is in preparation [7].

X-ray spectra have been measured at 90° with respect to the beam direction by means of a Si(Li) detector with a resolution of approximately 250 eV FWHM at 5.9 keV. The targets had thicknesses between ∿5 and 150 μg/cm^2 and were oriented at 45° with respect to the beam direction. In order to obtain absolute cross sections, (i) Rutherford scattering has been measured with two Si particle detectors, placed at different angles, and (ii) the transmitted beam current has been integrated and target thicknesses have been measured. A least squares fitting routine has been employed to extract line intensities from the accumulated spectra. Particular problems in this analysis are due to considerable line broadenings, non-gaussian shapes, and high-energy tails; these effects are important when it is attempted to obtain cross sections of better than ∿10%. Only x-ray production cross sections have been evaluated, and no attempt has been made to obtain experimental vacancy production cross sections via fluorescence yields.

Results and Discussion – Total Cross Sections

In the present paper, we concentrate on the production cross section, σ, of chlorine K x rays observed in collisions of 30-, 60-, and 120-MeV chlorine ions with various thin solid targets with Z_T=6-92. Fig. 1 shows the resulting dependence $\sigma(Z_T)$. The cross section exhibits an oscillatory structure with peak-valley ratios of almost an order of magnitude. It is not surprising that oscillations appear at all; this effect is known from several earlier investigations [8-10] and has been attributed to level matching and, more recently, to molecular orbital effects. Our data adds some new aspects which appear to be remarkable for reasons examined below.

The collisions studied here are far from being adiabatic. Following the definition given by Brandt [11], we determine an adiabaticity parameter, ξ_K, from the ratio of the mean electron revolution time of a chlorine K electron and the time it takes the chlorine K shell to pass by a target atom. The resulting values are ξ_K=0.7, 1.0, and 1.4 for 30-, 60-, and 120-MeV chlorine, respectively. It has been believed in the past that when ξ_K approaches and exceeds unity, Coulomb excitation becomes the dominant excitation mechanism and, consequently, no pronounced structures should be found in $\sigma(Z_T)$. Our data, however, indicate that even in exceedingly fast collisions the dominating excitation mechanism need not be simple Coulomb excitation.

Let us examine to what extent our findings can be explained by the molecular orbital mechanism, provided that we disregard the fundamental problem of non-adiabaticity. For $Z_T < 17$, we must reckon with almost pure Coulomb excitation. When Z_T is near 17, both K-level matching occurs and the exit channel for promoted Cl-1s electrons is open. When we assume a most probable charge of 12$^+$ for the chlorine ion, the 2pπ orbital ends up in the practically

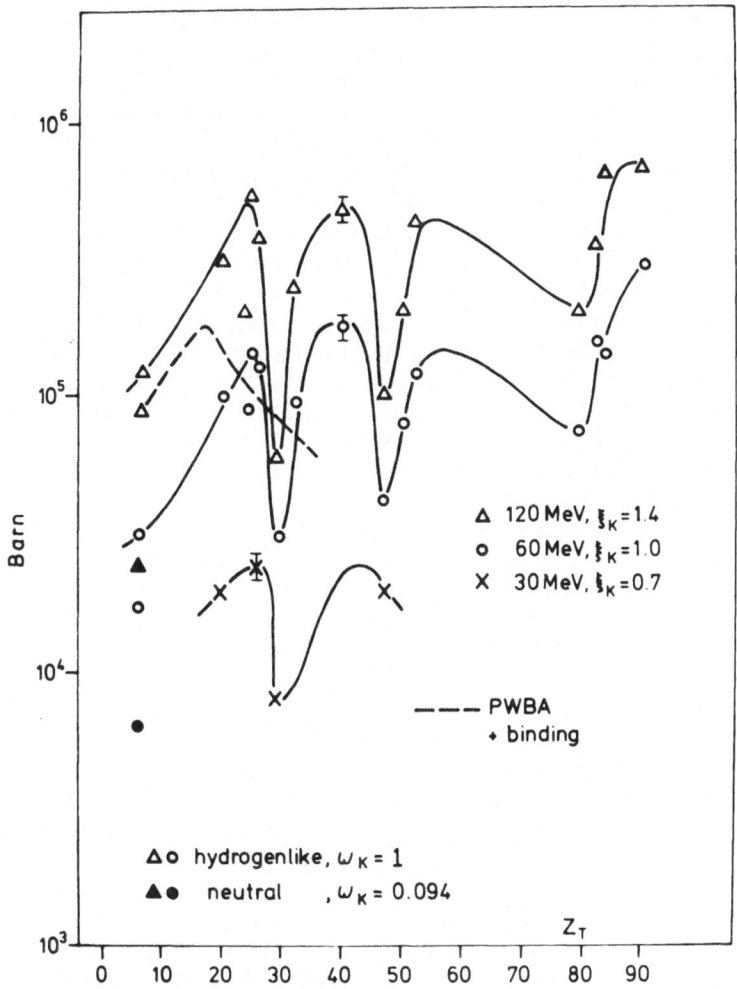

Fig. 1. Production cross sections for chlorine K x rays, as a function of the atomic number of the target atoms. The unconnected symbols $(0, ●, \triangle, ▲)$ for $Z_T=6$ and the dotted line represents theoretical estimates (see text).

empty 2p states of Cl after separation of the colliding partners, i.e., Cl–1s electrons will be promoted to the Cl–2p shell. For this reason, large cross sections are expected––and are indeed observed. This promotion mechanism can remain effective until the exit channel closes. When Z_T is increased, the 2p state of Cl^{12+} finally swaps with the 2p state of the target atom near $Z_T=25$. From here on, one would expect a sharp drop in σ for still increasing Z_T. This reduction of σ is also observed (Fig. 1). Recent measurements by Kubo et al. for Ni and Br ions show a similar

trend [12]. Following this line of argumentation, another major
increase of σ is not expected until Z_T reaches 46, where the exit
channel opens via 3dσ - 3dπ coupling to the 2p state of chlorine.
However, our experimental results show that σ increases much
earlier, namely near Z_T=32. It is no longer possible to explain
the experimental structures σ(Z_T) for Z_T > 29 on the basis of the
above promotion model, which would predict the second rise above
$Z_T \gtrsim$ 46. This latter channel would close near Z_T=53 when the
target 3d state swaps with the Cl-2p state. A third rise would be
expected above Z_T=89 where 4fσ-4fπ coupling correlates Cl-1s to
Cl-2p. It is obvious from Fig. 1 that the data can not be reconciled with this simple MO model. At present, it is not clear in
what direction refinements could be attempted. It should also be
pointed out that matching and swapping of 2p states in the above
discussion must be viewed with some suspicion; for these states,
the adiabaticity parameter is an order of magnitude larger than
ξ_K and one can not easily understand how these MO states could
develop during the fast collision.

    As regards the increase of the amplitude of oscillation with
increasing beam energy, we note that the average degree of stripping
of the chlorine ions increases with projectile energy. This leads
to a more effective removal of L electrons and, thus, to more
effective opening of exit channels for the chlorine 1s promotion.

    We have argued above that Coulomb excitation is the dominant
mechanism for excitation in chlorine collisions with Z_T < 17. For
the case of a carbon target, we have estimated the cross section on
the basis of the plane wave Born approximation [11]. In this calculation, binding effects due to the target atom potential have been
taken into account, but polarization effects of the initial state
have been disregarded. Two extreme cases have been considered:
(i) a neutral Cl atom with fluorescence yield 0.094 (full symbols
in Fig. 1), and (ii) heliumlike chlorine ion with a fluorescence
yield of unity (open symbols in Fig. 1). The resulting x-ray
production cross section from (ii) exceeds the one from (i) by a
factor of 3.7 (2.6) at 120 MeV (60 MeV). This difference results
partly from the enhanced 1s binding energy in heliumlike Cl ions
compared to neutral Cl atoms. Fig. 1 shows that experimental
results at 120 and 60 MeV exceed the theoretical estimates from (ii)
by less than a factor of 2. It must be kept in mind that it is not
quite clear to what extent the L shell is emptied in Cl ions inside
solids; even when the average total charge is less than 15^+ after
emergence from the foil, it is possible that many L electrons are
in excited states while the ions traverse the target [13]. It is
also known that the fluorescence yield drops quickly when the L
shell is filled. These are part of the reasons for which more
sophisticated theoretical estimates are difficult to carry out.
Nevertheless, the approximate agreement indicates that Coulomb
excitation is certainly of importance in this domain. The dotted
line in Fig. 1 illustrates a projected calculation with higher
values of Z_T for heliumlike chlorine at 120 MeV. The decrease of

the dotted curve for $Z_T > 17$ occurs because the strongly increased
effective binding of the Cl-1s electron during a close collision
due to the Coulomb field of the target overwhelms the Z_T^2 dependence
of the PWBA cross section. Although such a detailed calculation
stretches Brandt's theory [11] too far and should not be taken too
seriously, it indicates that simple Coulomb excitation becomes
less important when Z_T is increased to values above 17.

Radiative Electron Capture (REC)

Among important spectral features which often interfere with
more characteristic x rays we discuss briefly the phenomenon of
REC. Figs. 2 and 3 illustrate the occurrence in x-ray spectra.
In accordance with our previous publications [1-5], we define REC
as a dipole transition for which the initial state is in the resting
target and the final state is in the moving projectile, i.e., the
moving ion captures a target electron and emits the liberated energy
by means of a single photon. In the case of capture of free
(unbound) target electrons, REC is identical with the inverse photo-
effect. On this basis, the peak energy, E_P, of REC can be easily
approximated;

$$E_P \simeq \frac{m}{2} v^2 + E_B \tag{1}$$

where m is the electron mass, v the ion velocity, and E_B the binding
energy of the captured electron in the ion. Fig. 4 shows experimental results for E_P obtained from collisions of 35- to 110-MeV
sulphur ions with carbon and aluminium targets. The straight line
reflects a best fit when only data above 55 MeV are considered.
It is obvious that this fit is in good agreement with Eq. (1).
The slope is correct, but there is a slight offset to lower x-ray
energies, $\Delta \simeq 100$ eV (150 eV) for C (Al). Thus, the main features
of REC, E_P and shift of E_P with projectile energy, are well explained. But it must be realized that deviations from the fit in
Fig. 4 must occur for both lower and higher beam energies. A
value $E_B = 3.25$ keV, i.e., the ionization energy of a 1s electron
in sulphur, is correct only for S^{14+}. At energies much higher
than 100 MeV, S^{16+} will become the dominant charge state and E_B
will rise to 3.5 keV. At energies below ~ 50 MeV, the L shell will
be filled more effectively and, thus, the increased screening will
reduce E_B to values below 3.25 keV. The latter trend is visible
in Fig. 4. Though REC occurs most obviously when ions are highly
stripped, we point out that this is not at all a necessary condition
for observing REC. Fig. 3 illustrates that REC is still present
for 16 MeV S ions where the average charge of S is no more than
$\sim 9+$ [13]. A precise value of E_B is difficult to calculate because
the state of excitation of S during REC is not known. It is
interesting to mention that in the low-energy limit, when collisions

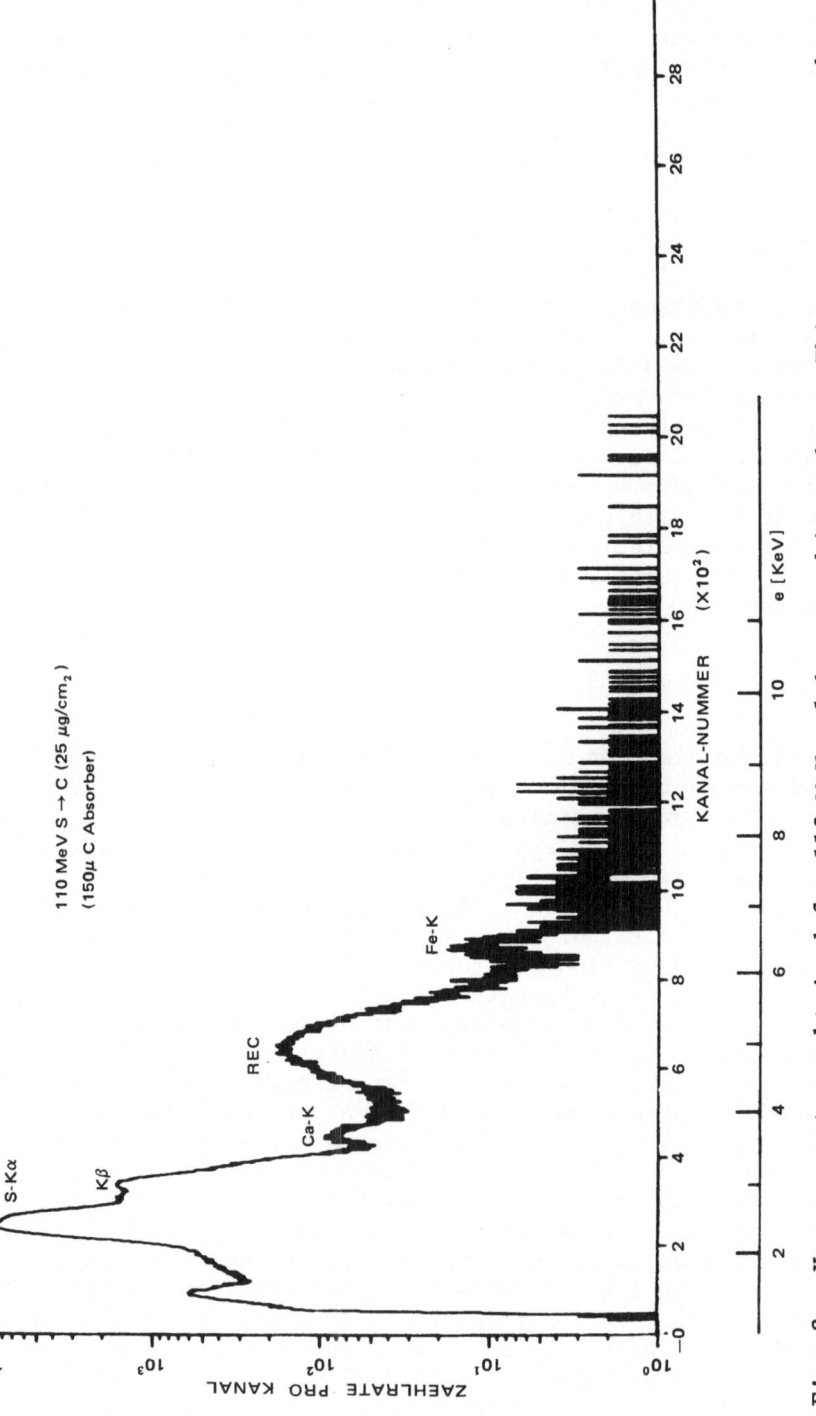

Fig. 2. X-ray spectrum obtained for 110-MeV sulphur on thin carbon. This spectrum was taken with a Si(Li) detector with resolution of ∼ 200 eV. Small impurities of Ca and Fe are present, but are well separated from the REC peak.

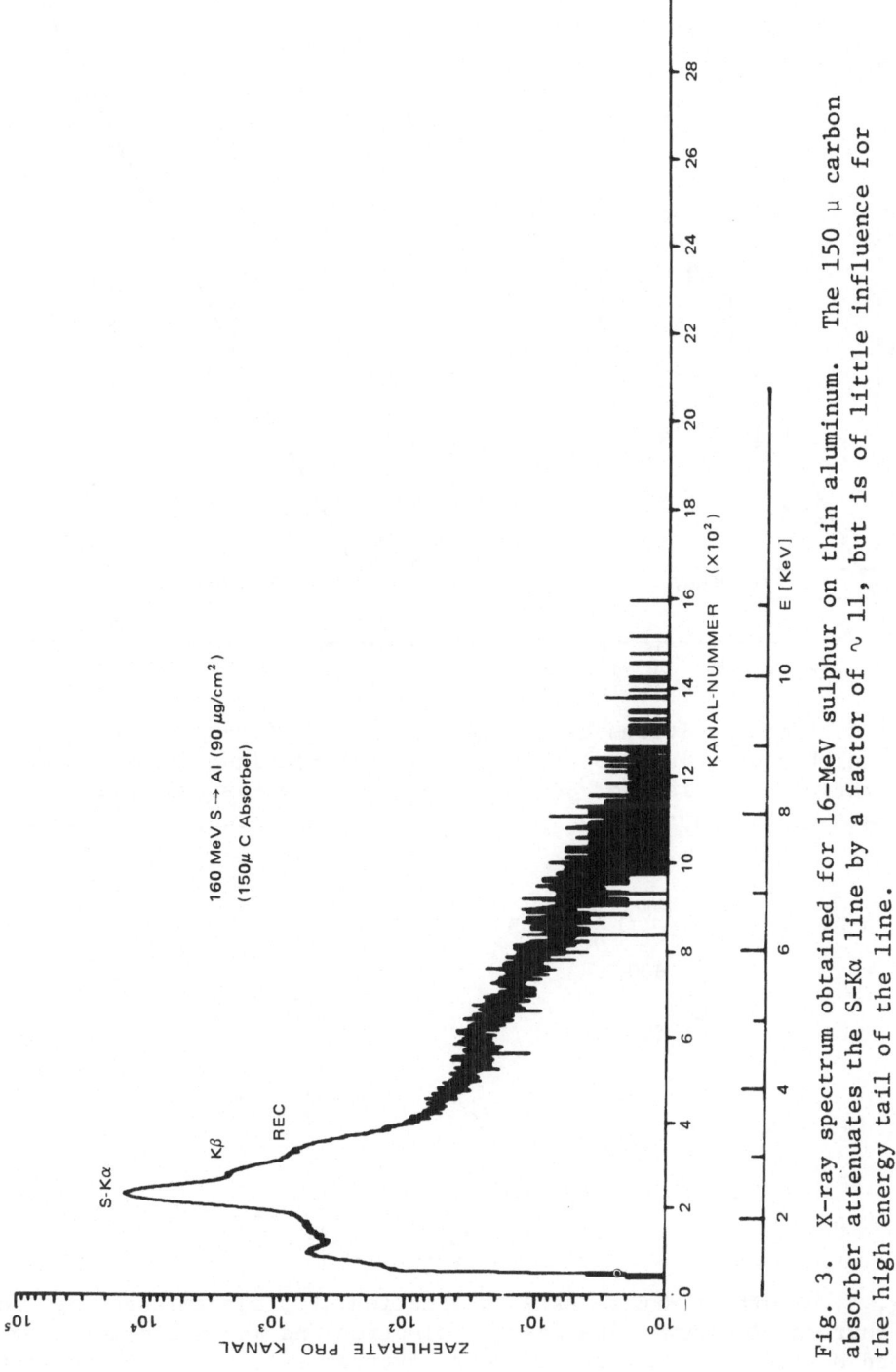

Fig. 3. X-ray spectrum obtained for 16-MeV sulphur on thin aluminum. The 150 μ carbon absorber attenuates the S-Kα line by a factor of ∿ 11, but is of little influence for the high energy tail of the line.

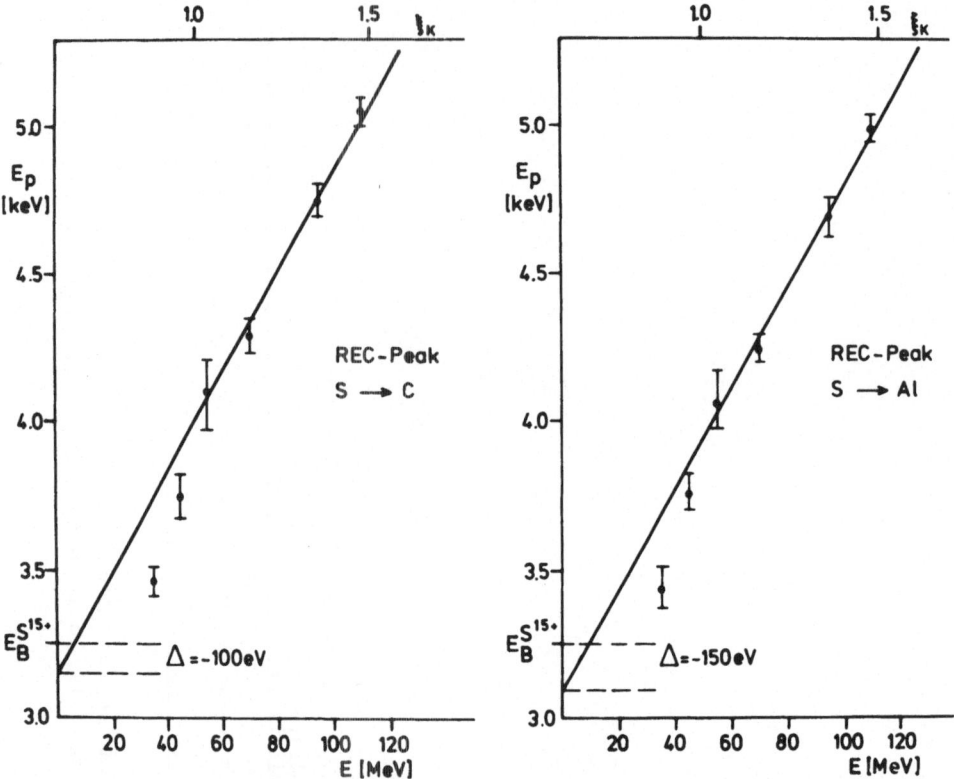

Fig. 4. Peak positions of REC as a function of projectile energy
for sulphur on thin carbon (left) and thin aluminum (right). The
ionization potential for S-14+ is indicated by E_B=3.25 keV. This
is part of the photon energy emitted when REC occurs for S-15+.

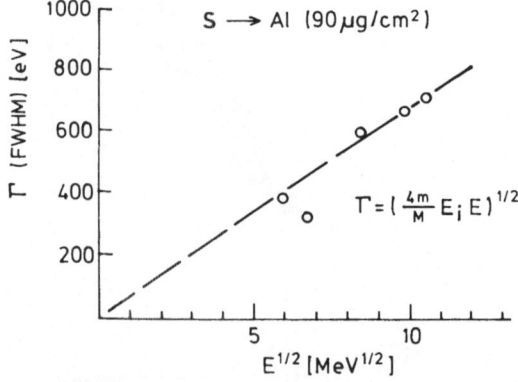

Fig. 5. Width of REC peaks as a function of projectile energy, for
sulphur on aluminum. The mass of electron and ion are denoted by
m and M, respectively, and E_i is an average effective binding energy
of the captured electrons. The dotted line is a fit to the data
and yields E_i=80 eV.

become extremely adiabatic, REC should become what one often calls
molecular-orbital x rays.

The distribution width, Γ, of REC for S on Al is shown in
Fig. 5. There is no doubt that the observed values of Γ can not
be explained by REC of unbound target electrons. It is necessary
to take into account capture of bound target electrons. Our calculations show that already the outermost bound target electrons
contribute most to Γ. Thus, Γ reflects the momentum distribution
of bound target electrons. An average effective binding energy
can be extracted from Fig. 5, $E_i \simeq 80$ eV. Detailed calculations
are under way.

Finally, we comment the line tails which are visible, for
example, in Fig. 3. When x-ray spectra are measured with care,
one observes these tails on the high-energy side of any line for
any collision partners at any bombarding energy. At present, we
can not attribute these tails to REC processes, and there is no
obvious reason to connect these tails, which show no cut-off what-
soever, with so-called MO x rays. Further analysis of these aspects
of our data will be published separately.

References

[1] H.-D. Betz and H. W. Schnopper, Communications of the European
    Conference on Nuclear Physics, Compt. Rend. Proc. II, 118 (1972).
[2] H. W. Schnopper, A. R. Sohval, H.-D. Betz, J. P. Delvaille,
    K. Kalata, K. W. Jones, and H. E. Wegner, Proceedings of the
    Conference on Inner Shell Ionization Phenomena and Future
    Applications, ed. by R. W. Fink, Atlanta, USAEC Technical
    Information CONF-720404, 1348 (1972).
[3] H.-D. Betz, Proceedings of the Heavy Ion Summer Study, ed. by
    S. T. Thornton, Oak Ridge, USAEC CONF-720669, 524 (1972).
[4] H. W. Schnopper, H.-D. Betz, J. P. Delvaille, K. Kalata,
    A. R. Sohval, K. W. Jones, and H. E. Wegner, Phys. Rev. Letters
    29, 898 (1972).
[5] H.-D. Betz, Proceedings of the Conference on Particular Prob-
    lems in Heavy Ion Research, ed. by P. Armbruster and B. Fricke,
    GSI Darmstadt, 1973 (to be published).
[6] H.-D. Betz, J. P. Delvaille, K. Kalata, H. W. Schnopper, A. R.
    Sohval, K. W. Jones, and H. E. Wegner, Proceedings of the
    Conference on Inner Shell Ionization Phenomena and Future
    Applications, ed. by R. W. Fink, Atlanta, USAEC CONF-720404,
    1374 (1972).
[7] H.-D. Betz and H. W. Schnopper (to be published).
[8] H. J. Specht, Z. Physik 185, 301 (1965).
[9] F. W. Saris, Physica 52, 290 (1971).
[10] T. M. Kavanagh, M. E. Cunningham, R. C. Der, R. J. Fortner,
     J. M. Khan, E. J. Zaharis, and J. D. Garcia, Phys. Rev. Letters
     25, 1473 (1970).

[11] G. Basbas, W. Brandt, and R. Laubert, Phys. Rev. A $\underline{7}$, 983 (1973).

[12] H. Kubo, F. C. Jundt, and K. H. Purser, Phys. Rev. Letters $\underline{31}$, 674 (1973).

[13] H.-D. Betz, Rev. Mod. Phys. $\underline{44}$, 465 (1972).

DE-EXCITATION OF SULPHUR L-SHELL VACANCIES PRODUCED IN ION-ATOM COLLISIONS IN SOLIDS*

C. P. BHALLA

FOM-Institute for Atomic and Molecular Physics
Amsterdam, The Netherlands
and
Kansas State University, Department of Physics
Manhattan, Kansas, U. S. A.

ABSTRACT

The inherent difficulties in obtaining relative populations of electronic defect configurations (and charge states) are presented for a 100 keV sulphur projectile traversing through a solid. It is concluded that this type of information cannot be deduced, as yet, from x-ray spectra of sulphur projectiles. Theoretical transition energies, transition rates and fluorescence yields are presented for sulphur with single 2p and multiple M vacancies.

Introduction

Several investigators have summarized [1-6] measurements involving heavy ion-atom collisions. A recent review of the subject with particular emphasis on differences observed with the solid and gas targets have been given by Datz [7]. Fortner et al. [8] have more recently measured the x-ray and Auger-electron spectra of projectiles of bombarding energies $\simeq$ 100 keV in solids.

These are general features in heavy ion-atom collisions which depend on the bombarding energies, for example

    (a) experimental observations of the charge state distribution [9],

    (b) increase in x-ray transition energies,

    (c) decrease in Auger electron energies [10-12],

    (d) the appearance of both satellite [13] and hypersatellite [14], K_α lines at bombarding energies $\simeq$ 1 MeV/amu; the satellite Auger electrons [11],

and
       (e) significant deviations of experimental fluorescence yields
          [1,10,11] from the normal values.
These measurements imply the creation of multiple vacancies in
addition to a vacancy in the inner shell.

    There are very few detailed calculations of transition energies,
and the x-ray and Auger rates which can be used in the analyses of
the data involving heavy ion-atom collision. Bhalla and Walters [15]
calculated the dependence of ω_{2p} for argon as a function of the
number of electrons in the valence shell. Bhalla and Hein [16],
and Bhalla, Folland and Hein [17] report on extensive calculations
for neon. More recently calculations for the K shell of argon [18]
and aluminum [19] have been presented.

    We discuss in this paper the inherent difficulties in obtaining
quantitative information concerning equilibrium charge state distributions (and electronic defect configurations) of a porjectile
(sulphur ion) traversing through a solid from x-ray spectra. The
theoretical transition energies and rates for variously ionized
sulphur are also presented to facilitate the discussion.

Theoretical Model

    In this section, we use atomic units ($e = m_e = \hbar = 1$). The
units for energy and inverse time are 2 Ry (27.2 eV) and 4.134 x
10^{16} sec^{-1} = 27.212 eV/$\hbar$ respectively.

Transistion energies

    The x-ray transition energies and the Auger electron energies
were calculated for the initial configuration $(1s)^2(2s)^2(2p)^5(3s)^r(3p)^q$
of sulphur by computing the appropriate differences of the total
energy of the atom in the initial state and in the final state.

$$E_X(3s \rightarrow 2p) = E_T(1s^2 2s^2 2p^5 3s^r 3p^q) - E_T(1s^2 2s^2 2p^6 3s^{r-1} 3p^q)$$

$$E_A(2p\text{-}3s\text{-}3s) = E_T(1s^2 2s^2 2p^5 3s^r 3p^q) - E_T(1s^2 2s^2 2p^6 3p^q)$$

$$E_A(2p\text{-}3s\text{-}3p) = E_T(1s^2 2s^2 2p^5 3s^r 3p^q) - E_T(1s^2 2s^2 2p^6 3s^{r-1} 3p^{q-1})$$

$$E_A(2p\text{-}3p\text{-}3p) = E_T(1s^2 2s^2 2p^5 3s^r 3p^q) - E_T(1s^2 2s^2 2p^6 3s^r 3p^{q-2})$$

The total energies, E_T, are defined by Slater [20] in terms of I's,
average kinetic energy and average electron-nucleus interaction
energy and the electrostatic integrals F^k and G^k.

    The results of these calculations are presented in Table 1.
The wave functions unsed in the calculations are described below.

Table 1

Total energies in Rydbergs of various defect electronic configurations $[2p^\ell, 3s^m, 3p^t]$ of sulphur.[a]

Configuration			E_T
ℓ	m	t	
1	0	0	−782.2549
1	0	1	−780.5042
1	0	2	−777.8213
1	0	3	−774.1196
1	0	4	−769.3318
1	1	0	−779.5085
1	1	1	−776.8650
1	1	2	−773.2163
1	1	3	−768.4888
0	1	0	−793.2395
1	1	0	−791.6968
2	1	0	−789.2827
3	1	0	−785.9199
4	1	0	−781.5183
0	2	0	−790.6323
1	2	0	−788.2444
2	2	0	−784.9175
3	2	0	−780.5635

[a]   t = 4−q and m = 2−r in Table 2.

Wave functions

The one-electron bound-state functions [21] were taken of the form $P_{n\ell}(r)\, Y_{\ell m}(r)/r$. The central field potential is then defined as sum of two terms

$$V'(r) = V_d(r) + V_x(r) \tag{1}$$

where the direct term is

$$V_d = -\frac{Z}{r} + \int \frac{\rho(r')}{|\vec{r} - \vec{r}'|}\, d^3r' \,, \tag{2}$$

and the exchange term

$$V_x = -2\ \left[\frac{3\rho}{8\pi}\right]^{1/3} (1 + \tanh G) \,. \tag{3}$$

The density ρ is defined as

$$4\pi r^2 \rho(r) = \sum_{n\ell} N(n,\ell) \; P^2_{n\ell}(r). \tag{4}$$

The second term in eq. (3) includes inhomogeneity correction factor,

$$G = \beta \; [2 \; (\nabla\rho/\rho)^2 - 3 \; \nabla^2\rho/\rho]/\rho^{2/3} \tag{5}$$

The parameter $\beta = 0.0028$, consistent with the optimum choice suggested by Herman and Schwarz [22], is used in all caculations. The central field potential is obtained by appending $V'(r)$, as defined above, a "Latter tail" correction [21].

$$V(r) = \begin{cases} V'(r) & , \; -V'(r) > (Z-Z_o+1)/r \\[2ex] -(Z-Z_o+1)/r & , \; -V'(r) \leq (Z-Z_o+1)/r \end{cases}$$

where Z is the atomic number and Z_o is the total number of electrons.

The modified program of Herman and Skilman [21] with the inclusion of this exchange contribution was used in all the calculations of bound-state wave functions. The calculations of continuum state wave functions are described elsewhere [23].

Transition rates

We denote by ϕ_i and ϕ_f the properly antisymmetrized many electron wave functions for the initial state and the final state respectively. The standard perturbation theory gives for the <u>total</u> Auger transition rate, Γ_A

$$\Gamma_A = 2\pi \; \sum \; | \; < \phi_f \; (1,2,\ldots N) \; | \; \sum_{i>j} \frac{1}{r_{ij}} \; | \; \phi_i \; (1,2,\ldots N) \; > \; |^2 \tag{6}$$

The density of final states is unity when the continuum-state wave functions are normalized in the energy scale, as is done in this work. The symbol Σ denotes the average and the sum over the initial and the final states respectively.

It should be noted that Γ_A in eq. (6) can be expressed in terms of the average group rates, T_A. For two electrons participating in the Auger process, $n_1\ell_1$ and $n_2\ell_2$, with the filling of the vacancy in $n_3\ell_3$ shell and the Auger electron, $E\ell_4$, in the continuum state,

$$T_A(n_3\ell_3 - n_1\ell_1 n_2\ell_2) = 2\pi \; N_{12} \; \Sigma \; |M|^2 \tag{7}$$

with

$$M \equiv \; < \phi \; (n_3\ell_3, E\ell_4) \; \left| \frac{1}{r_{12}} \right| \; \phi \; (n_1\ell_1, n_2\ell_2) \; > \tag{8}$$

The two-electron antisymmetrized wave functions are represented by ϕ in eq. (8). The calculation of the weighting factor, N_{12}, requires an additional assumption for the case when there is a single vacancy in $n_3\ell_3$ shell and when either (or both) shells $n_1\ell_1$ and $n_2\ell_2$ are not full. With the assumption of statistical population of the multiplet states arising from a defect electron configuration, the weighting factor is given by eq. (9):

$$N_{12} = \begin{cases} \dfrac{N_1 N_2}{(4\ell_1+2)(4\ell_2+2)} & \text{for inequivalent electrons} \\[3ex] \dfrac{N_1(N_1-1)}{(4\ell_1+2)(4\ell_1+1)} & \text{for equivalent electrons} \end{cases} \tag{9}$$

N_1 and N_2 are the occupation numbers of electrons in the $n_1\ell_1$ and the $n_2\ell_2$ shells. Equation (7), therefore, refpresents the Auger group $(n_3\ell_3-n_1\ell_1 n_2\ell_2)$ averaged over the statistical population of the multiplet states which can be formed from a particular electronic defect configuration with one vacancy in the $n_3\ell_3$ shell and any arbitrary number of electrons missing from other shells. These considerations are important since the Auger and x-ray rates are different for each multiplet state [24]. In a heavy ion-atom collision the population of the multiplet states is more likely to be statistical. However, this may not be the case for other types of collisions [29].

Similarly the <u>average</u> x-ray rate in atomic units for a single vacancy is given by

$$T_x(n_1\ell_1 \rightarrow n_3\ell_3) = \left(\frac{N_1}{4\ell_1+2}\right) \cdot \frac{4}{3} k^3 \; \frac{\ell_>}{(2\ell_3+1)} \left| \int_0^\infty P_{n_1\ell_1} P_{n_3\ell_3} \; r \middle| dr \right.^2 \tag{10}$$

where N_1 is the number of electrons in a shell from which an active electron makes the transition to fill the vacancy in the $n_3\ell_3$ shell. The quantity $\ell_>$ represents the larger value of ℓ_3 and ℓ_1. Equation (10) is the expression for electric dipole transitions.

Fluorescence yields

The fluorescence yield for the 2p shell, ω_{2p}, is defined as

$$\omega_{2p} = \frac{\Gamma_x}{\Gamma_x + \Gamma_A} \tag{11}$$

The total x-ray and Auger rates are denoted by Γ_x and Γ_A respectively.
It should be noted that when <u>average</u> quantities are used in eq. (11)
then ω_{2p} represents the fluorescence yield averaged over the multiplet states with a statistical population (see [28]).

Numerical Results

The transition energy of the (3s → 2p) x-ray was calculated
by the adiabatic procedure, which was described in the section,
Theoretical Model - Transitions energies. Several states of ionization were considered for sulphur. Table 2 contains a comparison
of our calculations (HFS) with those of Scofield, who had used the
relativistic Hartree-Fock (HF) model. The agreement between the
two calculations is remarkable. We have also listed other transition
energies, which were not calculated with the relativistic HF model.

Table 2

The transition energies (3s → 2p) calculated with the non-relativistic
HFS adiabatic model and those with the relativistic HF model.[a]

Initial state $(1s)^2(2s)^2(2p)^5(3s)^r(3p)^q$		Non-relativistic HFS	Relativistic HF
r	q	eV	eV
2	4	149.5	148.8
1	4	151.3	–
2	3	152.3	–
1	3	154.8	–
2	2	155.9	156.0
1	2	159.2	160.0
2	1	160.6	160.9
1	1	164.3	164.2
2	0	165.8	166.5
1	0	–	169.9

[a] Calculations performed by J. H. Scofield, as reported by Fortner
et al.

Table 3

Auger electron and x-ray energies in eV, group and total Auger rates and x-ray rates in atomic units and fluorescence yields for the 2p shell for variously ionized sulphur.

Identification	Defect Configuration				
	$[2p]$	$[2p,3p]$	$[2p,3p^2]$	$[2p,3p^3]$	$[2p,3p^4]$
$E_A(2p-3s-3s)$	114.0	105.3	96.55	87.67	78.67
$E_A(2p-3s-3p)$	128.5	119.4	110.2	100.7	–
$E_A(2p-3p-3p)$	140.6	131.6	122.5	–	–
$E_X(2p-3s)$	149.5	152.3	155.9	160.6	165.8
$10^5 \times T_A(2p-3s-3s)$	4.39	5.04	5.80	6.67	7.78
$10^4 \times T_A(2p-3s-3p)$	7.45	6.53	4.98	2.79	–
$10^4 \times T_A(2p-3p-3p$	13.4	8.37	3.35	–	–
$10^3 \times$ Total Auger rate	2.16	1.54	0.891	0.346	0.0778
$10^7 \times$ Total x-ray rate	4.09	4.45	4.89	5.42	6.07
$10^4 \times \omega_{2p}$ [28]	1.89	2.89	5.49	15.6	77.4

Table 4

Auger electron and x-ray energies in eV, group and total Auger rates and x-ray rates in atomic units and fluorescence yields for the 2p shell for variously ionized sulphur.

Identification	Defect Configuration			
	$[2p,3s]$	$[2p,3s,3p]$	$[2p,3s,3p^2]$	$[2p,3s,2p^3]$
$E_A(2p-3s-3p)$	118.9	109.6	99.96	90.06
$E_A(2p-3p-3p)$	133.0	123.2	113.0	–
$E_X)2p-3s)$	151.3	154.8	159.2	164.3
$10^4 \times T_A(2p-3s-3p)$	4.47	3.28	2.84	1.59
$10^4 \times T_A(2p-3p-3p)$	17.2	10.3	3.97	–
$10^4 \times$ Total Auger rate	21.7	14.1	6.81	1.59
$10^7 \times T_X(2p-3s)$	2.20	2.40	2.65	2.95
$10^4 \times \omega_{2p}$ [28]	1.01	1.70	3.89	18.2

The Auger electron energies, Auger group rates, x-ray rates and ω_{2p} are given in Table 3 for variously ionized sulphur in the valence shell. The largest Auger group rate, $T_A(2p-3p-3p)$, depends sensitively on the number of electrons in the 3p shell through the weighting factor defined in eq. (9). Consequently the changes in the total Auger rate and, therefore, in ω_{2p} are very much pronounced as a function of the number of electrons missing from the 3p shell.

Table 4 contains theoretical results for the defect configurations $[2p,3s,3p^t]$ of sulphur for t=0 to 3.

Discussion

It is extremely difficult to distinguish between the configurations: $1s^2 \, 2s^2 \, 2p^5 \, 3s^1 \, 3p^q$ and $1s^2 \, 2s^2 \, 2p^5 \, 3s^2 \, 3p^{q-1}$ from measurements of transition energies either for x-rays or for Auger electrons of sulphur when q > 1. This follows since the differences in transition energies for the above mentioned configurations are very small, typically $\simeq$ 1 eV (see tables 3 and 4). The relative Auger intensities, shown in Fig. 1 and 2, could distinguish between these configurations. Considering even the fact there will be Doppler broadening of Auger lines for the projectile ($\simeq$ 0.8 to 2 eV for 100 keV sulphur ion with typical experimental conditions), the gas experiments [25] appear to be fruitful. Similarly experimental measurements of fluorescence yield [12] for a gas target can dis tinguish between the defect configurations with a vacancy in the

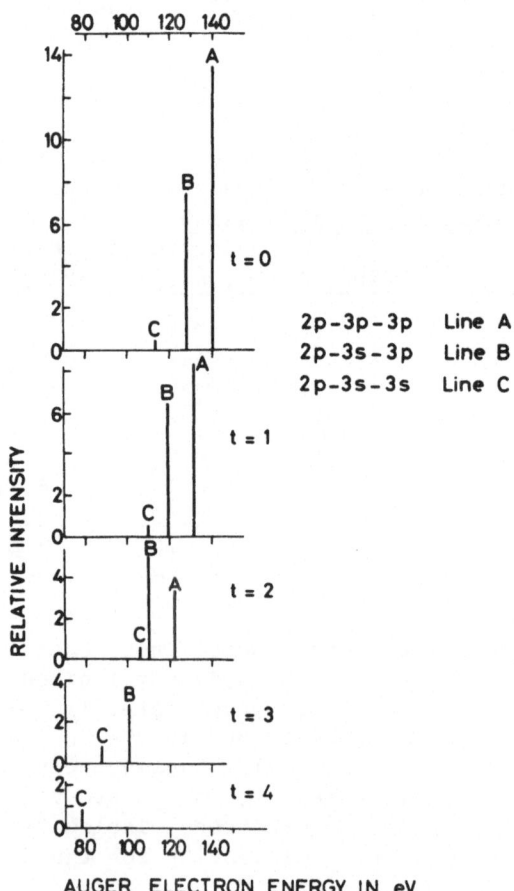

Fig. 1. Theoretical Auger spectra for defect configurations $[2p, 3p^t]$ of sulphur.

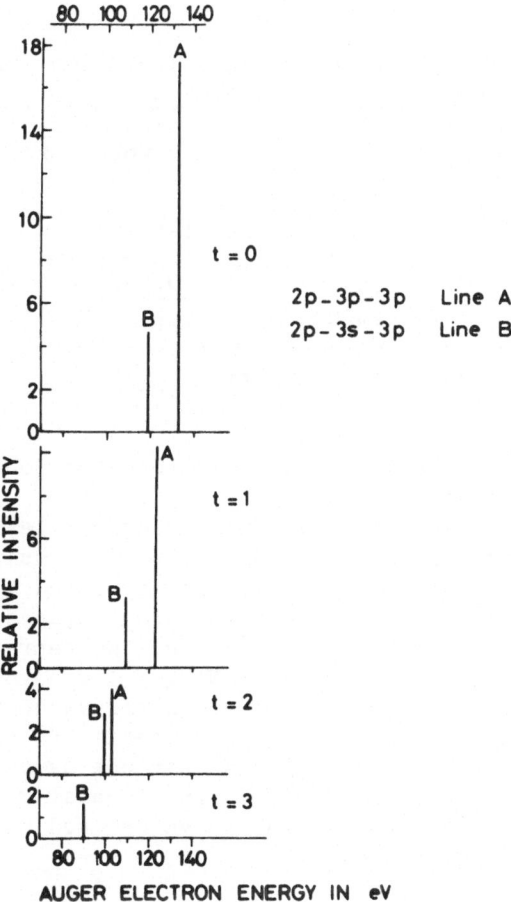

Fig. 2. Theoretical Auger spectra for defect configurations
[2p,3s,3p^t] of sulphur.

2p shell and the same number of electrons in the M shell (see tables
3 and 4). For a solid target the Auger electron spectra and experimental fluorescence yield will require formidable corrections to
ascertain defect configurations of the projectile.

It is instructive to investigate whether charge states, reflected by x-ray spectra of the projectile in a solid, are related
to equilibrium charge states of the ion. From the calculated
transition rates, presented in this paper, one can deduce the
extinction distances, L_e. Table 5 contains values of L_e for a
100 keV sulphur ion with different electronic defect configurations.
(It should be noted that the multiplet states, which will have
different L_e values, have been assumed to have a statistical population for a defect configuration.) Since in all cases the extinction distance is at lease one order of magnitude larger than the
mean free path for outer shell ionization, the ions attain an

Table 5

Extinction distance[a] of vacancies in sulphur ions of 100 keV for various defect electronic configurations.

Defect Configuration	L_e (Angstrom)	Defect Configuration	L_e (Angstrom)
$[2p]$	87		
$[2p,3p]$	122	$[2p,3s]$	86
$[2p,3p^2]$	211	$[2p,3s,3p]$	134
$[2p,3p^3]$	545	$[2p,3s,3p^2]$	276
$[2p,3p^4]$	2420	$[2p,3s,3p^3]$	1183

[a] L_e = velocity of ion x mean life time.

equilibrium charge state distribution before the de-exciation of the inner-shell vacancy occurs.

Now we discuss why the x-ray spectra cannot lead to a unique determination of equilibrium charge states of a (sulphur) projectile traversing through a solid. We have already noted that one cannot distinguish between the defect configurations, $[2p,3s,3p^{i-1}]$ and $[2p,3p^i]$ from experimental x-ray energies. We represent this x-ray intensity by $I(2p,M^i)$. We denote by $N(2p,M^i)$ the equilibrium fraction of sulphur ions with single 2p- and i-M shell vacancies. The probabilities of defect configurations with single 2p- and i-M shell vacancies are represented by α_i and Δ_i for one 3s vacancy and double 3s vacancy states respectively. The experimental x-ray intensities, $I(2p,M^i)$, are given in terms of the above defined quantities and fluorescence yields as follows:

$$I(2p,M^0) = \{\omega(2p,M^0) \qquad\qquad\qquad\qquad\qquad\} N(2p,M^0)$$

$$I(2p,M^1) = \{\omega(2p,3s)\,\alpha_1 + \omega(2p,3p)(1-\alpha_1) \qquad\qquad\} N(2p,M^1)$$

$$I(2p,M^2) = \{\omega(2p,3s,3p)\,\alpha_2 + \omega(2p,3p^2)(1-\alpha_2-\Delta_2)\} N(2p,M^2)$$

$$I(2p,M^3) = \{\omega(2p,3s,3p^2)\,\alpha_3 + \omega(2p,3p^3)(1-\alpha_3-\Delta_3)\} N(2p,M^3)$$

$$I(2p,M^4) = \{\omega(2p,3s,3p^3)\,\alpha_4 + \omega(2p,3p^4)(1-\alpha_4-\Delta_4)\} N(2p,M^4)$$

$$I(2p,M^5) = \{\omega(2p,3s,3p^4)(1-\Delta_5) \qquad\qquad\qquad\} N(2p,M^5)$$

$$(12)$$

The number of unknown parameters in the above equations are 14 consisting of six, equilibrium fractions (one would like to determine), and eight other parameters which represent relative populations of specific defect configurations. There are six possible experimental quantities, $I(2p,M^i)$. The fluorescence yields, which appear in eqs. 12, are significantly different for configurations with the same number of M-shell vacancies (see Tables 3 and 4). Furthermore, defect configurations with double 3s vacancies complicate the analysis because these have no (3s → 2p) x-ray transition. At any rate, there is not way to determine uniquely 14 parameters from 6 equations. Therefore, it is not possible to deduce equilibrium charge state fractions of a sulphur projectile traversing through a solid from x-ray spectra. We have ignored the possible excited states in our considerations since firstly more parameters will have to be introduced and secondly electrons in excited states have negligible effects on x-ray energies.

The approximations to reduce the number of parameters, which can be useful in the gas experiments or when the emphasis is on the target x rays [19] rather than the projectile, cannot be used for the solid targets because of the samll value of the mean free path as compared to extinction distances [26-27]. The average number of hard collisions in one mean life time of inner-shell vacancy is given by the ratio of L_e and the mean free path for hard collisions, which can be estimated to be ≃ 20 Å in solids [27]. It is clear from table 5 that sulphur ions with single 2p and multiple M vacancies are involved in many hard collisions before de-excitation of the 2p vacancy occurs. Furthermore, the number of soft collisions will be larger by an order of magnitude as compared to hard collisions. In all these collisions of the projectile with an inner shell vacancy present, there are several features (e.g., de-excitation and transfer of vacancies, excitation, ionization and rearrangement of electrons) about which little is known either experimentally or theoretically. The data concerning these effects are essential in the interpretation of experimental x-ray spectra.

It appears to us, therefore, that it is not possible as yet to obtain detailed quantitative information concerning either the charge state or the distribution of active electrons, e.g., 3s shell versus 3p shell of sulphur ions, for a low Z projectile with energies ≃ 100 keV traversing in a solid target.

Acknowledgments

It is a pleasure to thank Dr. F. W. Saris for several stimulating discussions. The warm hospitality at the FOM-Institute of ATomic and Molecular Physics is gratefully acknowledged.

This work is sponsored by F.O.M. with financial support by Z.W.O. (Netherlands Organization for the Advancement of Pure Research).

References

*Work supported, in part, by the U.S. Army Research Office, Durham, N. C. at Kansas State University

[1] F. W. Saris in Physics of Electronic and Atomic Collisions
     VII ICPEAC, (North-Holland Publishing Company, Amsterdam,
     1972) 181. This paper contains an excellent review of the
     subject.

[2] Inner-Shell Ionization Phenomena and Future Applications,
     edited by R. W. Fink, S. T. Manson, I. M. Palms and P. Venugopala
     Rao, U. S. ATomic Energy Commission Report No. CONF-720404 (1973).
     See invited papers by W. Brandt, T. M. Kavanagh, E. Merzbacher,
     D. I. Nagle, P. Richard and F. W. Saris. There are also many
     contributed papers in this report.

[3] Heavy-Ion Summer Study, edited by S. T. Thornton, U. S. Atomic
     Energy Commission Report No. CONF-720669 (1973). See papers
     by P. Richard and E. Merzbacher.

[4] W. Brandt in Atomic Physics 3, edited by S. S. Smith, G. K.
     Walters and L. M. Volsky (Plenum Press, 1973) 155.

[5] J. D. Garcia, R. J. Fortner and T. M. Kavanagh, Rev. Mod. Phys.
     45, 111 (1973).

[6] Q. C. Kessel and B. Fastrup, Case Studies in Atomic Physics 3,
     137 (1973)(North-Holland Publishing Company, 1973).

[7] Sheldon Datz, invited talk, American Physical Society Meeting,
     Washington, D. C. (1973).

[8] R. J. Fortner, T. M. Kavanagh and R. C. Der, Bull. Am. Phys.
     Soc. 18, 662 (1973).

[9] Hans-Dieter Betz, Rev. Mod. Physics 44, 465 (1972). Also see
     Ref. [3].

[10] M. E. Rudd in Ref. [1].

[11] N. Stolterfoht, Eighth Int. Conf. on the Physics of Electronic
     and Atomic Collisions - VIII ICPEAC Abstracts (1973).

[12] D. Burch, W. B. Ingalls, J. S. Risley and R. H. Heffner, Phys.
     Rev. Lett. 29, 1719 (1973).

[13] A. R. Knudson, D. J. Nagel, P. G. Burkhalter and K. L. Dunning,
     Phys. Rev. Lett. 26, 1149. Also see Ref. [2].

[14] P. Richard, W. Hodge and C. Fred Moore, Phys. Rev. Lett. 29,
     393 (1972).

[15] C. P. Bhalla and D. L.Walters, Ref. [2].

[16] C. P. Bhalla and M. Hein, Phys. Rev. Lett. 30, 39 (1973.

[17] C. P. Bhalla, N. O. Folland and M. Hein, Phys. Rev. A 8, 649
     (1973).

[18] C. P. Bhalla, Phys. Lett. A (in press)(1973); Phys. Rev. A 8,
     2877 (1973).

[19] F. Hopkins, D. O. Elliott, C. P. Bhalla and P. Richard, Phys.
     Rev. A 8, 2952 (1973).

[20] J. C. Slater, Quantum Theory of Atomic Structure, Vol. I, II
     (McGraw-Hill Book Company, New York 1960).

[21] F. Herman and S. Skillmann, Atomic Structure Calculations
     (Prentice Hall, Inc., Englewood Cliffs, N. J. 1963).

[22] F. Herman and K. Schwarz in Computational Solid State Physics
     (Plenum Press, New York, 1972).

[23] D. L. Walters and C. P. Bhalla, Phys. Rev. A 3, 1919 (1971).

[24] C. P. Bhalla, Physics Lett. 46A, 185 (1973).

[25] P. Dahl et., Electronic and Atomic Collisions, VIII ICPEAC Abstracts (1973) 708.

[26] F. W. Saris, W. F. van der Weg, H. Tawara and R. Laubert, Phys. Rev. Lett. 28, 717 (1972); see F. W. Saris et al. in Ref. [2].

[27] F. W. Saris, Proceedings of this Conference (1973).

[28] The definition given in eq. (11) is not correct. The fluorescence yield, averaged over the spectroscopic terms designated by L and S with a statistical population (2L + 1)(2S + 1), is as follows:

$$<\omega> = \frac{\Sigma(2L + 1)(2S + 1)\omega(L,S)}{\Sigma(2L + 1)(2S + 1)}$$

The multiplet fluorescence yield, $\omega(L,S)$, is the ratio of the total x-ray rate and the total decay rate for the particular state specified by L and S. The numerical results of $<\omega>$ differ significantly from the ω_{2p}-values in tables 4 and 5. These considerations do not change the values of average extinction distances and the general arguments and conclusions presented here.

[29] C. P. Bhalla, D. L. Matthews and C. F. Moore, Phys. Lett. 46A, 336 (1973).

ION EXCITATION OF AL K X-RAY SPECTRA

A. R. KNUDSON, P. G. BURKHALTER and D. J. NAGEL
Naval Research Laboratory
Washington, D.C.   20375, U.S.A.

ABSTRACT

High resolution measurements of ion-excited
Al K spectra were extended to include Al projectile spectra and Al target spectra excited by
heavier projectiles (Ar,Ni). Spectra from 3.0-MeV
Al projectiles showed that the number of L vacancies
present at the time of K vacancy decay is determined
largely by collisions subsequent to production of
the K vacancy. Heavy projectiles on Al targets
produce Al spectra which appear to be the result
of two ionization mechanisms operating simultaneously. Molecular orbital formation may enter into
the interpretation of satellite structure in both
Al projectile and Al target spectra.

Introduction

Of all the ion-excited spectra measured with high resolution,
the Al K spectra have been the most thoroughly studied [1-7].
These data provide a useful basis on which one may attempt to con-
struct an understanding of the multiple ionization which occurs so
frequently with ion excitation. Attention will be focused here on
L-shell ionization accompanying K-shell ionization. X-ray emission
has been observed from vacancy configurations KL^N (N = 0 .... 6).
Studies of Al ionization mechanisms can be grouped by the binding
energies of the levels of the collision partner relative to the
Al K and L shells as shown in Figure 1. Binding energies of neutral
atoms are plotted here for illustration, even though projectile
energy levels will be more tightly bound due to electron stripping.
Previous studies have indicated that multiple vacancy production by He ions is well accounted for on the basis of multiple,
independent Coulomb ionizations with no evidence, at present, for
cooperative many electron effects [1]. Measurements of spectra

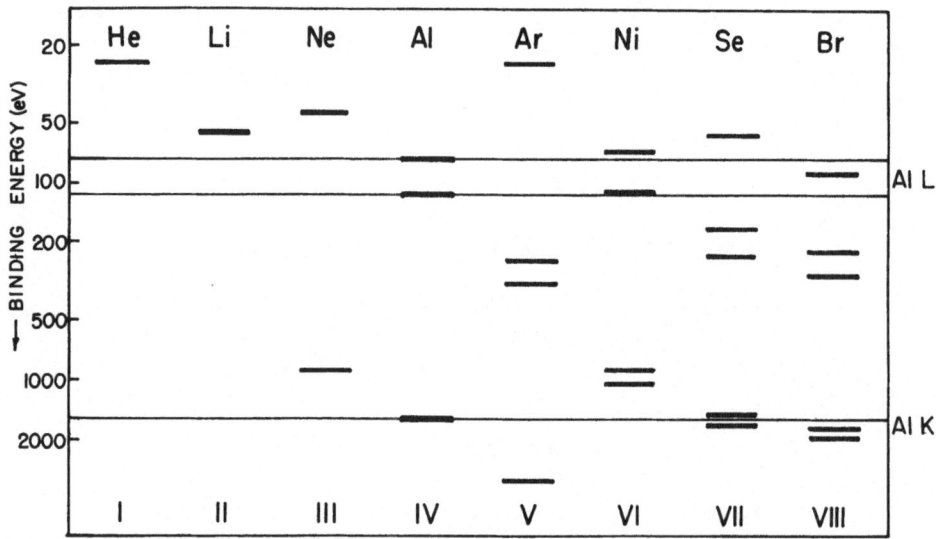

Fig. 1. Binding energies for electron shells in various elements compared with the K- and L-shell binding energies of Al. Many of these elements have shells whose binding energies approximately match that of an electron shell in Al.

excited by heavier ions, such as N [2,3] and Ne [4], have shown high probabilities for multiple ionization. The relative intensity distribution $I(KL^N)/\Sigma_N I(KL^N)$ can be fit fairly well with a binomial distribution [8], whose dependence on ion energy follows that of σ_L qualitatively. These heavy-ion observations again point to a mechanism involving simultaneous but independent electron ejection events.

In this work we extend to higher Z projectiles, the investigation of high resolution Al target spectra. In addition we have measured spectra and yields for Al K x rays emitted by Al projectiles incident on solid targets in order to provide a more complete picture of multiple ionization.

Experimental

X-ray yields have been measured with a Si(Li) detector for 3.0 and 4.0 MeV Al ions incident on thick targets. High resolution measurements were performed with a flat crystal Bragg spectrometer and an EDDT crystal. The targets used in this work were LiF, NaF, Mg, Al, Si, KCl, Ti, Ni, and Ge. Effects due to implantation of Al into the target were seen with C targets, but were not observed for any of these targets.

Results

Al projectiles

Figure 2 shows the K x-ray yield as a function of target Z as measured with a Si(Li) detector. The yield has a maximum for Z between 13 and 14 where the target and projectile K-shell binding energies match and rises again at larger Z as the projectile K-shell and target L-shell binding energies approach equality. From the peak value around Z = 13 the yield decreases sharply by a factor of 50 at the lowest Z for which data are presented and falls

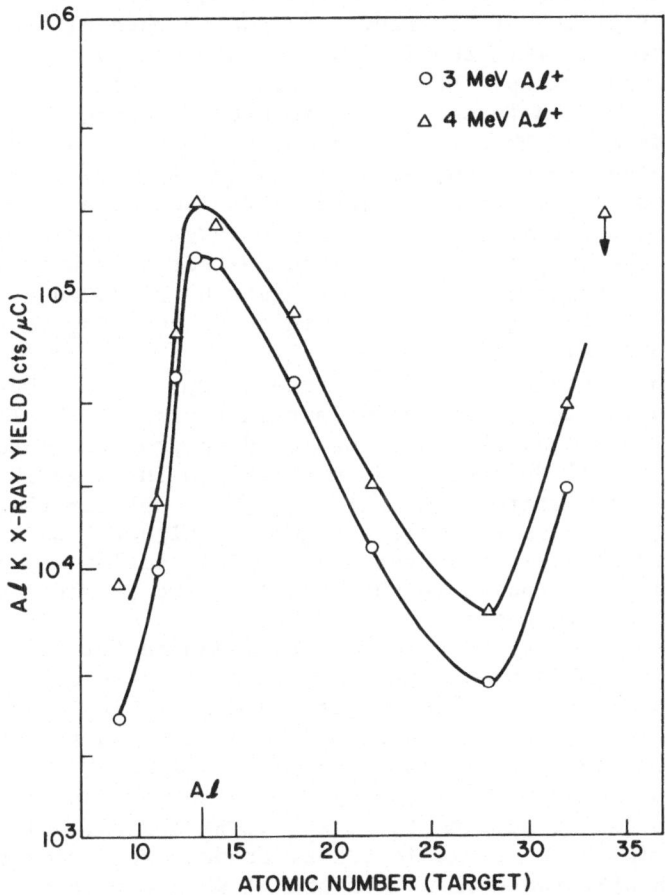

Fig. 2. Al projectile x-ray yield for an Al beam incident on thick targets. The yield for an Al target has been divided by two before plotting. The point at Z = 34 on the 4 MeV curve represents the combined yield of Al K and Se L x rays. The yield from the LiF target is plotted at Z = 9, since most of the yield is due to the fluorine.

by a factor of 40 in the valley at Z = 28(Ni). These data suggest
that a molecular orbital (MO) promotion mechanism [9] is mainly
responsible for production of the 1s vacancy in the Al projectile.
The velocity of a 3 MeV Al ion is about 1/5 the velocity of the
Al 1s electron and is within range of applicability of the MO model.
This ratio of projectile velocity to 1s electron velocity is slightly smaller than that corresponding to the 80 MeV Br beam used by
Kubo, Jundt, and Purser [10] who measured the Br Kα yield as a
function of target Z. However, their peak occurred for target Z
equal to projectile Z plus four or five, whereas in the present
data the peak occurs for target Z equal to projectile Z plus 1/2.

The appropriate conditions for excitation of a 1s electron by
an MO mechanism would appear to be present only for Al and Si. As
indicated in Figure 3, for targets of lower Z than Al, the Al 1s
level is correlated with the 1s level of the united atom and no
promotion occurs, except through coupling to higher orbitals. An
Al charge state of +3 is sufficient to cause swapping of the Al
and Si 2p levels, so that the Al 1s and 2p levels are correlated
with the 2p level in the united atom via the 2pσ amd 2pπ MO's and
Al 1s electrons may be promoted into Al 2p vacancies. For targets
heavier than Si, the Al 1s level and the target 2p level are correlated, but the target 2p level is not expected to have any vacancies into which projectile electrons might be promoted. However,
higher charge states of Al could again cause swapping of projectile
and target 2p levels [11]. For example, for charge state +6 the Al
2p level is more tightly bound than the Cl 2p level. Then the
projectile 1s and 2p levels would correlate with the 2p level in
the united atom and Al 1s electrons could be promoted into the
vacancies in the Al 2p shell. So the tail on the high charge state
side of the charge state distribution may account for the more
gradual decline of the yield as one goes to higher Z targets. A
possible alternative is that the Coulomb interaction may produce
the necessary 2p vacancies in the target as the projectile and
target atoms come together.

High-resolution Al Kα spectra were measured for 3.0 MeV Al
projectile ions incident on the same set of targets used for the
yield data in Figure 2. These spectra are shown in Figure 4. Each
spectrum consists of the $K\alpha_{1,2}$ line at the left and five satellite
lines, each due to an additional L vacancy, although in most of the
spectra one or more of these components is so weak as to be undis-
cernible. It is interesting to note that Si and Ge, which occupy
the same column in the periodic table, produce such similar spectra.
Starting with LiF, the envelope of the Al Kα spectrum moves to the
left as the target Z is increased. This left shift corresponds to
a decreasing number of L-shell vacancies present at the time of
K x-ray emission. At Z = 14(Si) the maximum leftward displacement
is observed. Z = 13(Al) is a special case which will be discussed
later. The envelope then moves to the right with the maximum
shift occurring around Z = 22(Ti). The information contained in

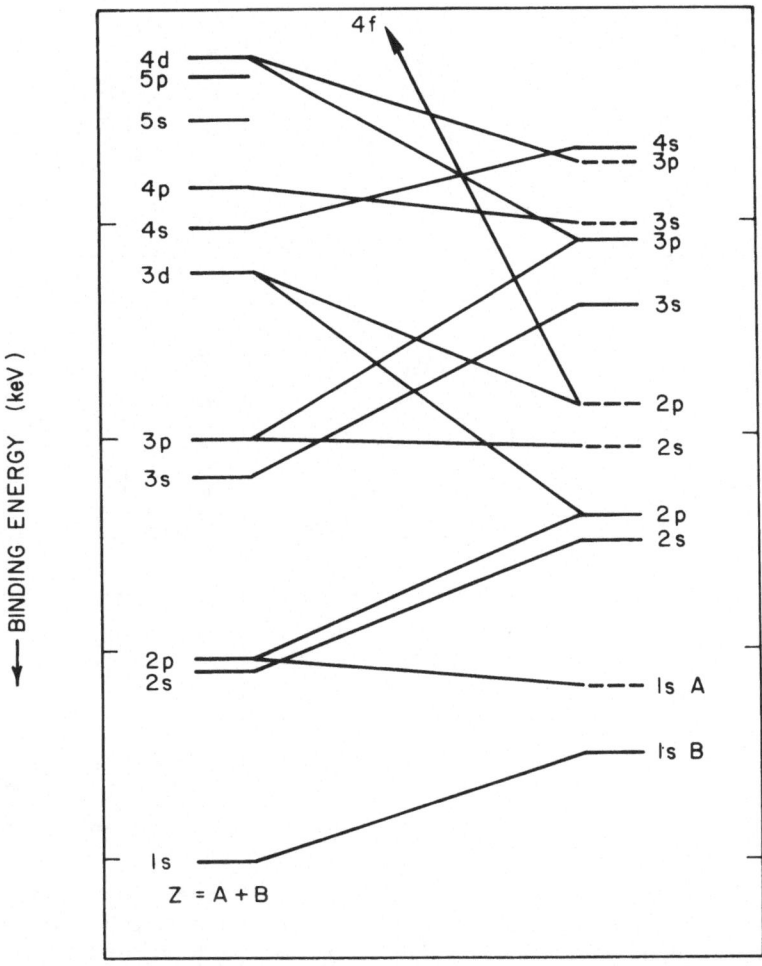

Fig. 3. MO correlation diagram [9] used to discuss the collision
systems considered in the text. Atom A represents the lower Z
element of the Al-collision partner pair while atom B is the higher
Z element.

these spectra is displayed in more quantitative form in Figure 5,
which shows the relative intensities of the components as a function
of target Z. The maximum leftward shift, as indicated by the maximum in the KL^0 curve and the minimum in the KL^5 curve, occurs at
about the same Z as the peak of the K x-ray yield as shown in
Figure 2. However, the minimum in the KL^0 curve and the maximum
in the KL^5 curve occur well before the Z value corresponding to
the minimum in the K x-ray yield.

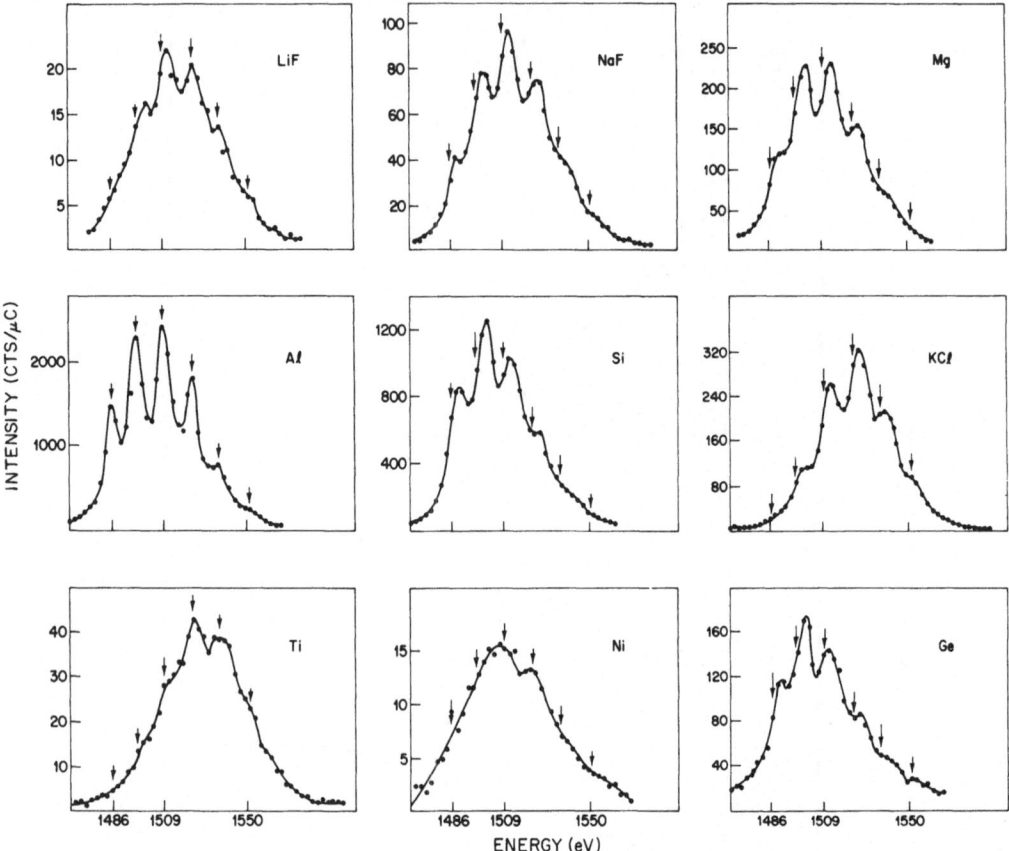

Fig. 4. Al Kα spectra for 3.0 MeV Al ions incident on a variety
of targets. The arrows indicate the energies at which the peaks in
the Al target spectrum occur. The leftmost peak around 1486 eV
is the Kα$_{1,2}$ transition.

    The Al spectrum is of special interest since it is a super-
position of spectra emitted by target and projectile ions. The
target spectrum reflects the distribution of L-shell vacancies
produced simultaneously with the K-shell ionization. Because of
prior L-shell vacancies, the projectile very probably has a different distribution of L-shell vacancies than the target immediately
after K-shell ionization. This distribution is modified by electron
capture and loss processes prior to filling of the K-shell vacancy.
The effect of x-ray emission from Al target atoms is to shift the
envelope of the Kα spectrum to the right as compared to spectra
from Mg and Si. In Figure 5 the Al data points, with the exception
of KL^{0}, depart noticeably from smooth curves drawn through the
other data points. The contribution from the target atoms also

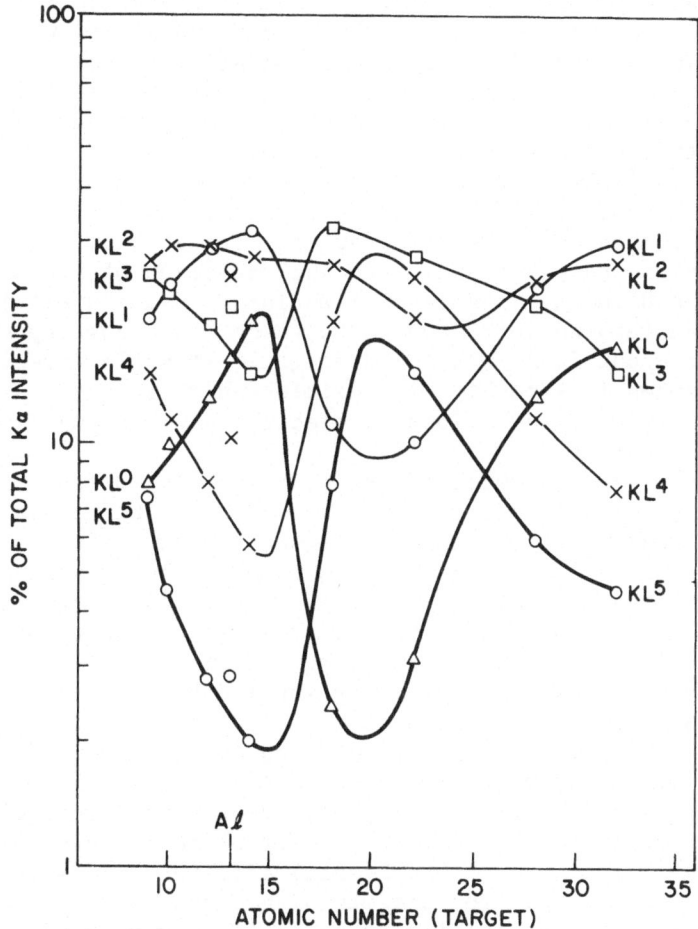

Fig. 5. Fractional intensity of components in Al Kα spectra from
Al projectiles as a function of target Z. Each component is
labelled by the initial state vacancy configuration.

causes each peak in the spectrum to be shifted to lower energy due
to the presence of three M-shell électrons on each target atom.
These electrons have been stripped from the projectile ions. The
peaks in the Al target spectrum are also sharper than those in any
of the other spectra, presumably because the x rays emitted by
target ions have less Doppler broadening.

The changing intensities of the various components in these
spectra, as shown in Figure 5, are the result of a complex inter-
play between the vacancy configurations produced in the Al-target
atom collision, which ionizes the Al K-shell, and subsequent collisions in which the Al ion captures and loses L- and M-shell
electrons. A 3 MeV Al ion travels about 90 Å in the life-time of

a K-shell vacancy, when no L-shell vacancies are present. This distance will be even greater in the case of an Al ion with several L-shell vacancies since the Auger and radiative transition rates will be reduced. According to calculations by Bhalla [12] the 1s-hole lifetime will approximately double for three L-shell vacancies. Electron capture and loss cross sections are of the order of 10^{-15} to 10^{-16} cm^2 for most charge states of interest here. The data of Ryding et al.[13] indicate that 1 MeV Al$^+$ ions in nitrogen reach charge equilibrium after traversing about 10^{16} atoms/cm^2. Other data also indicate that, with the exception of light target gases such as H and He, charge equilibrium is usually attained after passage through 10^{16} atoms/cm^2, which would correspond to a thickness of about 30 Å in a solid target. It would appear then that the changes in satellite intensities in these spectra are determined mainly by variations of electron capture and loss cross sections with target Z.

A maximum in the number of L-shell vacancies present at the time of x-ray emission occurs for Z around 21. This could be caused either by a maximum in the loss cross section or a minimum in the capture cross section for L-shell electrons. Electron loss cross sections measured in gases by Dmitriev et al. [14] show a peak around Z = 18 which may be related to the present observations.

The possibility of using an MO mechanism to explain the changes in electron loss and capture implied by the data of Figures 4 and 5 is deserving of consideration. The velocity of a 3 MeV Al ion is about 90% of the velocity of a 2p electron bound to Al, so electron promotion effects are expected to be small for L-shell electrons. However, Kubo et al. [10] have reported such effects in the yield of Br Lα x rays with 110 MeV Br ions where the ratio of projectile velocity to bound L-shell electron velocity is similar to that for 3.0 MeV Al. For targets with atomic number less than 13, promotion of target 1s electrons can fill in some of the 2p vacancies as may be seen in Figure 3. Also, the Al 2p level is correlated with the target 2p level in the 3d level of the united atom, so that target 2p electrons can be demoted into the Al 2p level. Whe the Al 2p level becomes less tightly bound than the target 2p level as one goes to higher Z targets, promotion of target 1s electrons into the Al 2p level ceases, corresponding to a decreased probability of electron capture. Furthermore, the 4fσ exit channel becomes available to Al 2p electrons, producing an increased probability of electron loss. The left shift of the envelope for Z greater than 22 may be due to filling of the target 3d level with possible demotion of these electrons into the Al 2p level. An MO mechanism, then, provides at least a plausible explanation of the effects seen in the Al projectile spectra.

Al target

Data were also obtained for several cases shown in Figure 1 in which Al was the target instead of the projectile. These spectra

are shown in Figures 6, 7, and 8 along with the corresponding Al
projectile spectra. The case of Ne on Al and Al on NaF is shown
in Figure 6. 3 MeV Al on NaF has the same relative velocity as
2.25 MeV Ne on Al. There is reasonable agreement between the
spectrum obtained by interpolating between the 1.5 and 3.0 MeV
Ne-excited spectra and the Al on NaF spectrum. This indicates
that essentially the same L-shell vacancy distributions are obtained
at the time of K x-ray emission for both Ne on Al and Al on NaF.
These spectra are of the type which can be fit by using a binomial
distribution to calculate the intensities of the components.

     Spectra for Ar and Ni on Al are shown in Figures 7 and 8 to-
gether with the inverse systems. These heavy projectiles produce
spectra unlike those excited by lighter projectiles. With Ar and
Ni excitation, the intensity distribution is no longer in accord
with a binomial distribution. The $K\alpha_{1,2}$ line and the first satellite are enhanced and, in the case of Ar excitation, the intensity
falls off more slowly than usual as one goes to the higher energy
satellites. The possibility was considered that these spectra are
the superposition of a spectrum excited directly by the incident
beam and a spectrum generated by collisions of recoiling Al ions
with other Al atoms, with the latter contributing mainly to the
parent line and the first satellite or two. However, this explanation is not tenable; the integrated intensity in the Ni-excited
spectrum is about 1/20 of the integrated intensity in the Ar excited
spectrum, but the cross section for recoil excitation should be
larger for Ni projectiles than for Ar. Hence, even if the entire
Ni-excited spectrum is attributable to Al recoil excitation, this
process could contribute only a few percent to the intensity of
the Ar-excited spectrum. To aid in interpreting these spectra, data

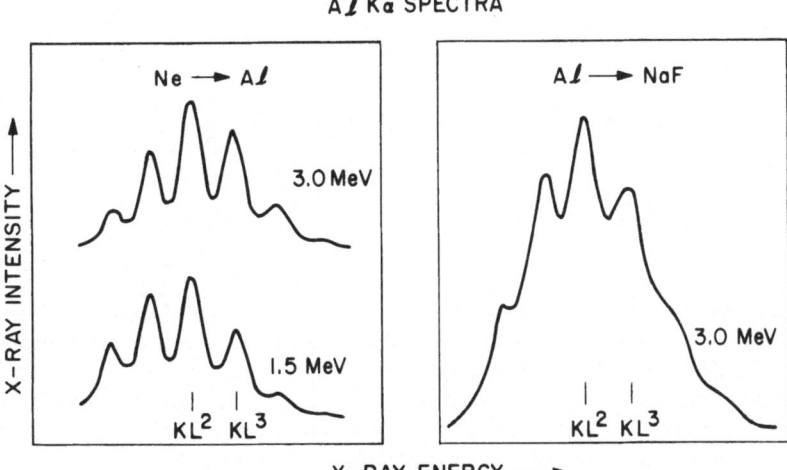

Fig. 6. Al Kα spectra for Ne on Al and Al on NaF.

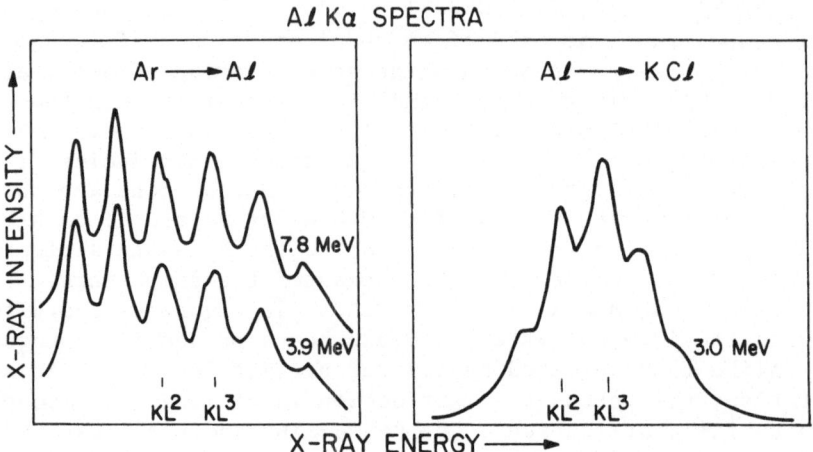

Fig. 7. Al Kα spectra for Ar on Al and Al on KCl.

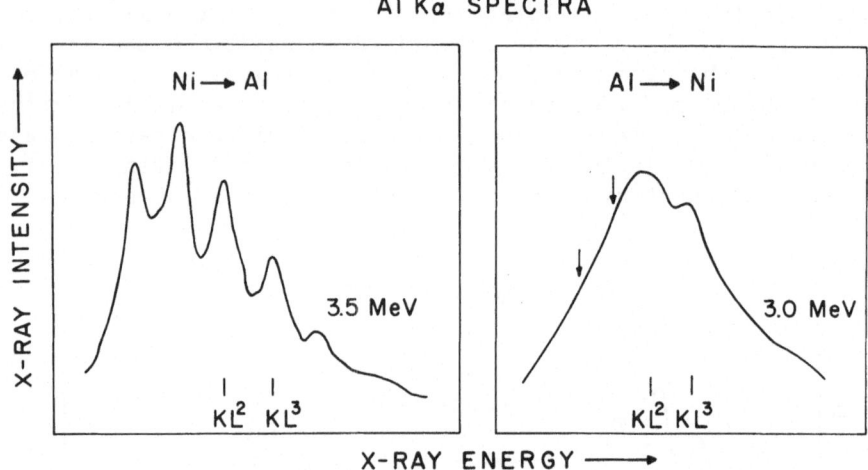

Fig. 8. Al Kα spectra for Ni on Al and Al on Ni.

were obtained for Ar projectiles on Mg and Si as shown in Figure 9. The spectra for Mg and Al targets are comparable, but the enhancement of the KL^0 and KL^1 components decreases substantially for Si. The explanation for these spectra is uncertain, but the distribution of L-shell vacancies appears to be due to a superposition of two mechanisms, such as Coulomb excitation and possibly electron promotion.

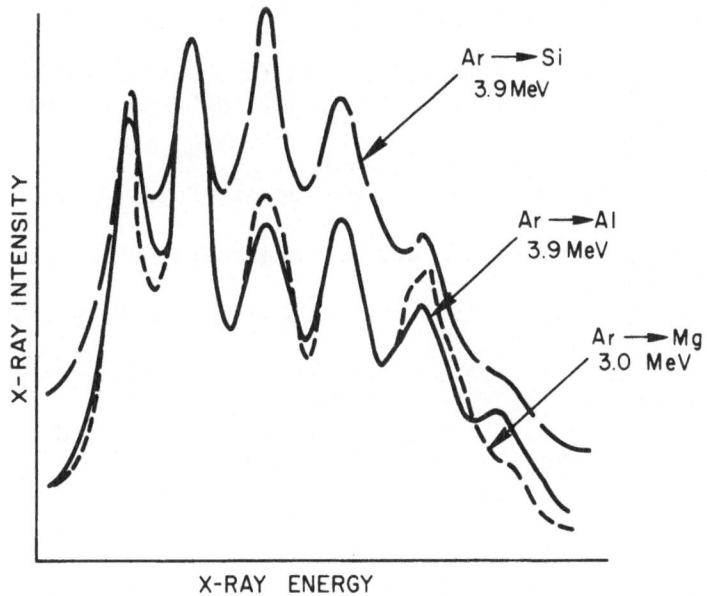

Fig. 9. Target Kα spectra for Ar on Mg, Al, and Si.

Summary

        Excitation of K-shell vacancies in Al projectiles is explain-
able by an MO mechanism. It has been shown that subsequent electron
loss and capture may have an important modifying effect on the distribution of L-shell vacancies produced in K-ionizing collisions,
when projectile x-ray spectra are measured. Variations of electron
loss and capture cross sections with target Z have been discussed
in terms of an MO model. In the case of target x rays produced by
collision with heavy projectiles, the spectra may result from a
superposition of mechanisms for producing L-shell vacancies.

References

[1] A. R. Knudson, P. G. Burkhalter and D. J. Nagel, Proc. of Int.
    Conf. on Inner Shell Ionization and Future Applications, R. W.
    Fink, Ed., AEC Report CONF-720404, 3, 1675 (1973).
[2] A. R. Knudson et al., Phys. Rev. Ltrs. 26, 1149 (1971).
[3] F. C. Jundt, H. Kubo, K. H. Purser, A. R. Knudson, and D. J.
    Nagel, to be published.
[4] A. R. Knudson, D. J. Nagel, P. G. Burkhalter, Phys. Ltrs. 42A,
    69 (1972).
[5] P. Richard et al., Bull., APS 18, 560 (1973).
[6] P. G. Burkhalter, A. R. Knudson, D. J. Nagel and K. L. Dunning,
    Phys. Rev. A6, 2093 (1972).

[7] P. Richard, C. F. Moore, and D. K. Olsen, Phys. Ltrs. $\underline{43A}$, 519 (1973).

[8] D. Burch in Proc. of Int. Conf. on Inner Shell Ionization and Future Applications, R. W. Fink, ed., AEC Report CONF-720404, $\underline{2}$, 1464 (1973).

[9] M. Barat and W. Lichten, Phys. Rev. $\underline{A6}$, 211 (1972).

[10] H. Kubo, F. C. Jundt, and K. H. Purser, Phys. Rev. Lett. $\underline{31}$, 674 (1973).

[11] K. Taulberg, B. Fastrup, and E. Laegsgaard, Phys. Rev. $\underline{A8}$, 1814 (1973).

[12] F. Hopkins, D. O. Elliott, C. P. Bhalla, and P. Richard, Phys. Rev. $\underline{A8}$, 2952 (1973).

[13] G. Ryding, A. Wittkower, G. Nussbaum, A. Saxman, and P. H. Rose, Phys. Rev. $\underline{A2}$, 1382 (1970).

[14] S. S. Dmitriev, V. S. Nikolaev, L. N. Fateeva, and Ya. A. Teplova, Sov. Phys. JETP $\underline{15}$, 11 (1962).

X-RAY EMISSION FOLLOWING ION BEAM AND PLASMA EXCITATIONS

DAVID J. NAGEL
Naval Research Laboratory
Washington, D. C. 20375, U.S.A.

ABSTRACT

The hallmark of much recent work in x-ray spectroscopy is the study of emission from high ionization states. Such states can be produced by heavy-ion impact or in very hot plasmas. Ion excitation is a flexible way to produce spectra similar to and intermediate between those resulting from electron beam and plasma excitation. Ion excited x-ray spectra are useful in atomic and collision physics research. High temperature plasmas occurring in fusion power devices emit intense x radiation which provides a useful diagnostic of conditions within such plasmas. X-ray spectra from ion impact and hot plasmas are compared in this paper. They are shown to be complementary in some cases and very similar in others.

Introduction

Until a few years ago, most x-ray spectroscopy research was concerned with single or double ionization of atoms. Recently, very-energetic heavy ion impacts were shown to produce higher multiple ionization in core levels. High-resolution x-ray spectra have proved useful for study of the mechanisms by which heavy ions eject several core electrons in a single collision. Similar measurements of x-ray emission from highly ionized projectiles during or after transit of a target give information on projectile charge states and the decay rates of highly excited states. Concurrent with such ion beam research, several sources of hot plasmas have been developed further and used. Laser and plasma focus, vacuum spark, exploding wire and other devices produce plasmas consisting of highly stripped atoms with temperatures of about 10^7°K. Such high temperatures result in strong x-ray emission near 1 keV,

making x-ray spectroscopy a prime diagnostic tool for ultra-hot
plasmas. Comparison of ion beam and plasma production of highly
ionized states, and the x-ray spectra arising from their decay, is
discussed in this paper.

A brief summary of methods of multiple vacancy production and
the resulting states which can be produced in electron beam, ion
beam and hot plasma experiments follows:

Method of Excitation	Species	Electron Vacancies	
		Core	Outer
Electron Beam	Target Atom	Few	Some
Light Ion Beam	Target Atom	More	Some
Heavy Ion Beam	Target Atom	Many	If Recoiling
Heavy Ion Beam	Projectile	Many	Many
Plasma	Ion in Plasma	Few	Many

This table shows that ion excitation provides a bridge between
conventional electron beam excitation, which produces relatively
little ionization, and plasma excitation, which leads to highly
stripped atoms. That is, light ions such as protons produce ionization and spectra similar to those with electron-beam excitation,
and heavy projectiles stripped in a beam foil are similar in their
vacancy states and spectra to ions stripped by energetic electrons
in a hot plasma. Emission following electron and light ion impact
consists of familiar x-ray (core electron) spectra in contrast to
emission from stripped projectiles and plasmas which is high-energy
"optical" (outer electron) radiation.

The table also emphasizes the complementary nature of some
ion-excitation and plasma experiments. Copious ionization can result
in both cases but there is an important difference. Target ions
struck by a heavy projectile tend to be ionized in core levels, with
weakly bound outer electrons perturbed but not really separated
from the target atom due to low outer electron ionization cross
sections and the high electron density in solids. In contrast to
this, electrons are preferentially stripped off ions in a plasma
from the outer orbitals to the core orbitals; i.e., the core tends
to remain intact. The x-ray spectra from these two cases reflect
the complementary nature of heavy ion and plasma excitation, as
will be illustrated later.

Ion and plasma spectra can also be very much alike. In the
case of projectile (beam foil) spectra, the x-ray emission is composed primarily of Rydberg series very similar to those from ions
in plasmas. Then the spectra are directly comparable, as will also
be illustrated.

We will examine the excitation processes, energies of the
resulting vacancy states and the x-ray spectra emitted during their

decay; both ion and plasma excitation are discussed in the following sections.

Vacancy State Production

    Multiple vacancy production in a target atom tends to be a
single step excitation process while stripping of projectiles in
targets and ions in plasmas involve multiple ion-electron collisions. We note in passing that single collision processes are
very rapid ($\sim 10^{-15}$ sec) compared to multiple-step processes such
as electron loss and pickup by projectiles and collisional stripping of atoms in a plasma.
    Several mechanisms are now known to cause vacancy production
in a single projectile-target collisions [1]. These include simple
Coulomb processes applicable to light-ion impact and very high-
energy heavy-ion bombardment. Pauli excitation processes are important in collisions where the orbital electronic energy of the
collision partners matches for at least a pair of shells. The
Pauli mechanism leads to the formation of molecular orbitals for
slow collisions and to a statistical ionization picture at high
collision energies. These single-step mechanisms tend to produce
multiple core orbital vacancies in the collision partners. They
do not occur in plasmas.
    The stripping of outer electrons from projectiles in a target
and ions in a plasma results from successive collisions with electrons [2]. But there are significant differences in excitations:

Situation	Ions	Nearby Electrons	Charge State Depends on
Projectiles in Target	High Velocity	Cold Bonding Electrons	Ion Velocity
Ions in Plasma	Relatively Slow	Hot Free Electrons	Plasma Temperature

The point here is simply that the relative velocity between the
ion and background electrons is the important factor. Figure 1
illustrates the dependence of charge state on velocity for an ion
moving in a solid. The greater the projectile velocity, the more
frequent are the collisions with ambient electrons. Figure 2
shows the variation in charge state as a function of average energy
per ion, which is related to plasma temperature as shown. The
greater the temperature, the greater the energy of the plasma
electrons in their collisions with the ions.

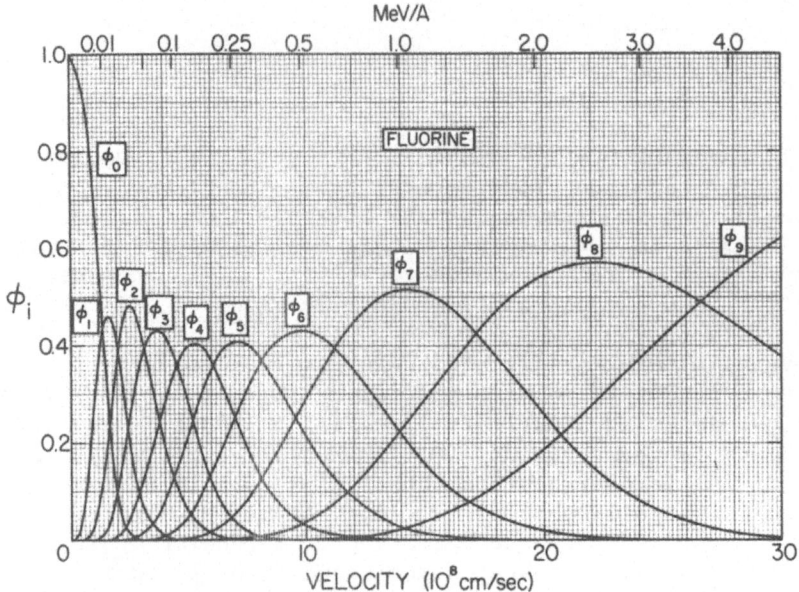

Fig. 1. Distribution of charge states for fluorine ions in matter
[3]. ϕ_i, the fraction of ions with i electrons removed, depends on
the ion velocity.

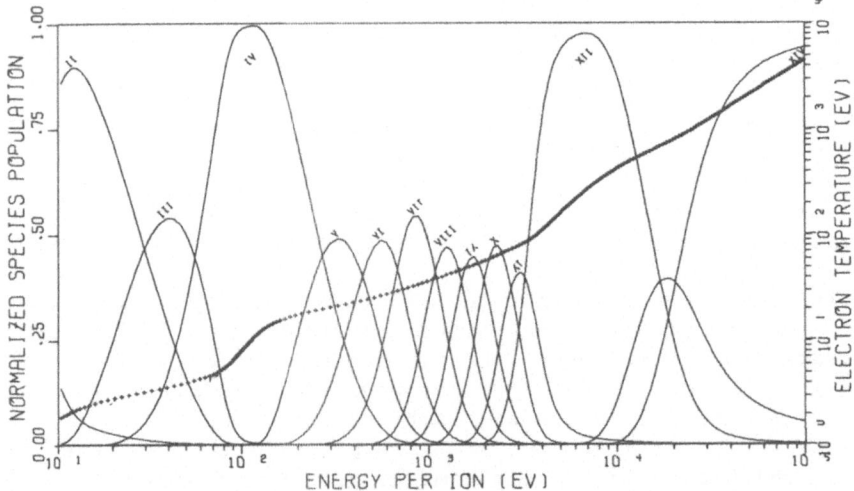

Fig. 2. Distribution of charge states for aluminum ions in a hot,
dense plasma [4] (density = 1×10^{19} ions/cm^3). The fraction of ions
with a particular charge state (e.g., Al IV = Al^{3+}) depends on the
average energy per ion which increases monotonically with the
plasma temperature.

Vacancy State Energies

The particular states produced by a collision or in a plasma depends on the number and arrangement of electron vacancies. Single ion collisions produce up to about 10 vacancies while it is not uncommon to have over 20 outer electrons missing in an x-ray emitting plasma. The energies of vacancy states are of interest since x rays emitted during their decay have energies equal to the difference in the excitation energies of the states involved in the transition. Figure 3 gives energies for some of the lower ionization states in

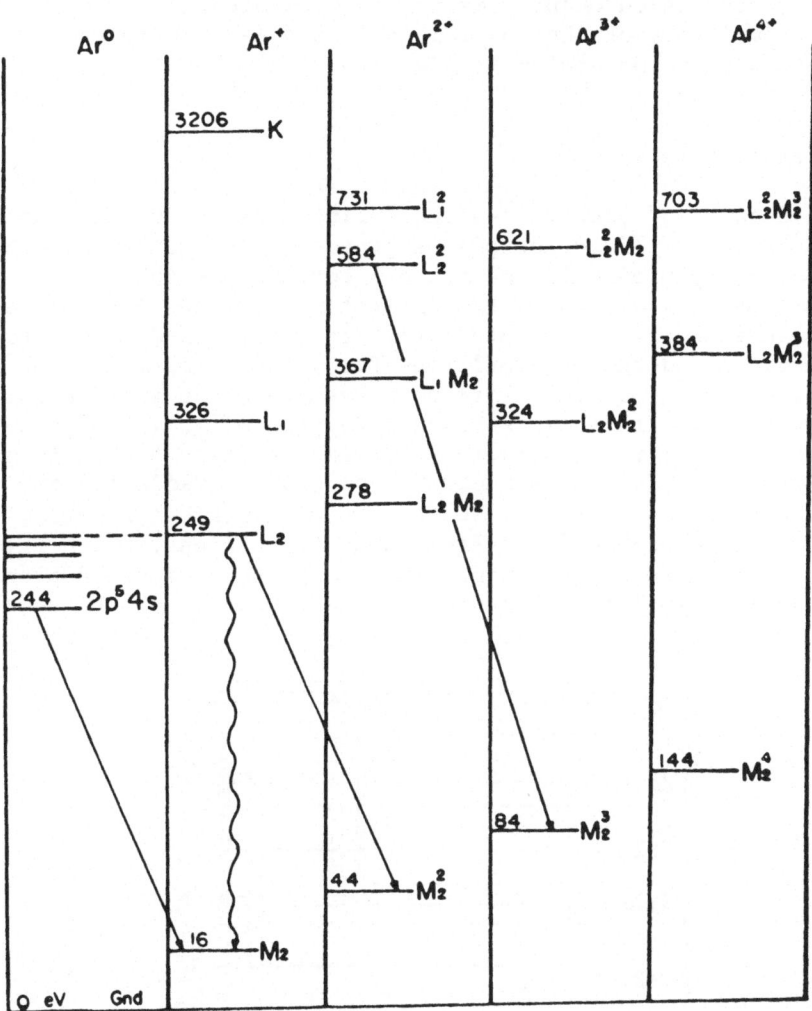

Fig. 3. Excitation energy diagram for argon [5]. L_1 indicates, for example, a missing L_1(2s) electron. An x-ray transition is shown by wavy lines, while Auger (electron-emitting) transitions are given by solid lines.

Ar. X-ray energies are equal to differences between such excitation energies for ions with the same charge state.

No multiplet structure of the vacancy state levels is shown in Figure 3. But, in fact, interaction between electrons in open shells produces multiplet splitting. The term level diagrams for ion and plasma excitation can be very similar since two holes produced in a filled shell by ion-bombardment interact just as do two electrons outside of closed shells in the plasma case. Figure 4 shows the levels produced by the indicated double vacancy arrangements. This diagram is used to interpret multiple-vacancy structure in ion-excited x-ray spectra. Figure 5 gives the levels for double electron states which occur commonly in low-atomic-number plasmas. Note the similarity of the initial 1s-2s and 1s-2p hole states in Figure 4 and electron states in Figure 5.

Vacancy State Decays

The details of spectra emitted during decay of vacancy states produced by ion collision or plasma processes depend on the excitation and decay probabilities (rates) for the possible vacancy distributions in addition to the excitation energies just discussed. Both absolute and relative intensities are determined by excitation cross sections and the probabilities for vacancy rearrangement and radiative decay. Full calculation of even relative x-ray spectra is very complex. It requires production probability, energy level and decay probability computations for each vacancy state and decay path. Transport effects on incident and emitted radiation are usually complicated. In this section we will exhibit spectra, which include bound→bound and free→bound transitions, which illustrate the complementary and comparative aspects of ion and plasma excited x-ray emission. No thorough interpretation of the spectra is given here.

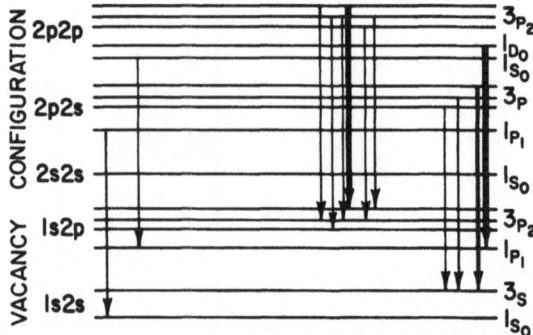

Fig. 4. Multiplet diagram for initial 1s2s and 1s2p states in elements such as Al [6]. Here excitation energy increases downward. Arrow widths indicate relative intensities for equal production of 1s2s and 1s2p vacancy states.

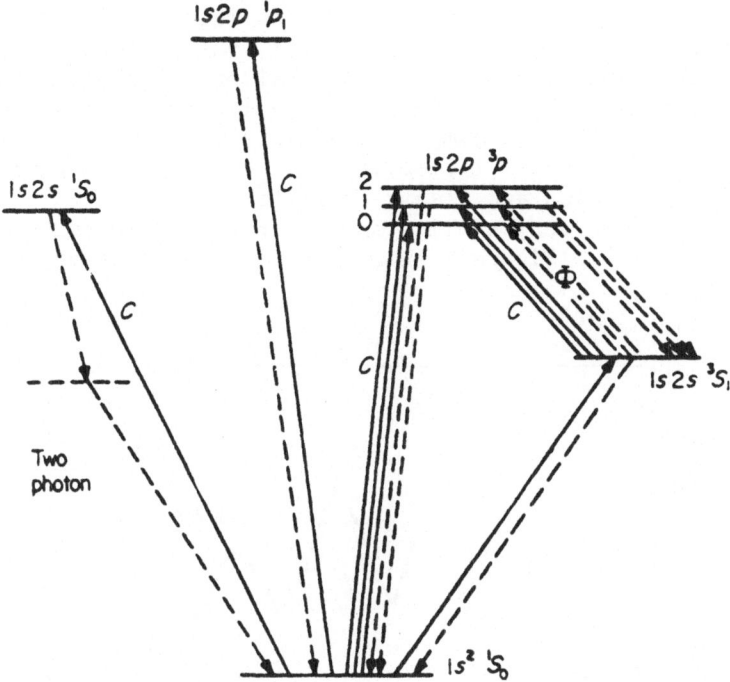

Fig. 5. Multiplet diagram for He-like ions [7]. Collisional (c) and photo (φ) excitation are indicated, with radiative decays shown by downward, dashed arrows.

The complementary nature of electron, light-ion, and heavy-ion excitation is shown in Figure 6. Satellites, which occur at high energies due to the loss of screening associated with additional vacancies, appear in all spectra but with very different intensities. Incident electrons produce one or two extra (L-shell) holes by the shakeoff process. Light ions such as protons or alpha particles have a higher probability of ejecting L-shell electrons, but not as great as that for heavy projectiles such as nitrogen ions. The relation of heavy-ion and plasma excitation is illustrated in Figure 7. The vertical marks indicate the Al peak energies obtained with 5 MeV N^+ projectiles, as in Figure 6. The spectrum of Figure 7 was emitted by the plasma produced at the focus of a high-power laser. It consists of three types of radiation: (a) Rydberg series (np → ls, n = 2.3..) in one- and two-electron ions, (b) satellites S which occur at low energies due to the additional screening introduced by additional bound electrons, and (c) the He-like free-to-bound continuum. The 2p → ls transition energy varies systematically with increasing stripping. Calculations [11] show that M electron loss has little effect on the energy of this transition. L-electron loss, as in the ion-impact work, causes greater energy shifts, which increase with the degree of L-shell ionization. Removal of a K shell electron (in going from He to H-like) involves the greatest change in screening and the largest jump in the transition energy. Taken together, Figures 6 and 7 illustrate the complementary nature of emission following ion and plasma excitation.

We now examine very similar spectra obtained in ion beam and plasma experiments. Figure 8 shows the fluorine K spectra obtained

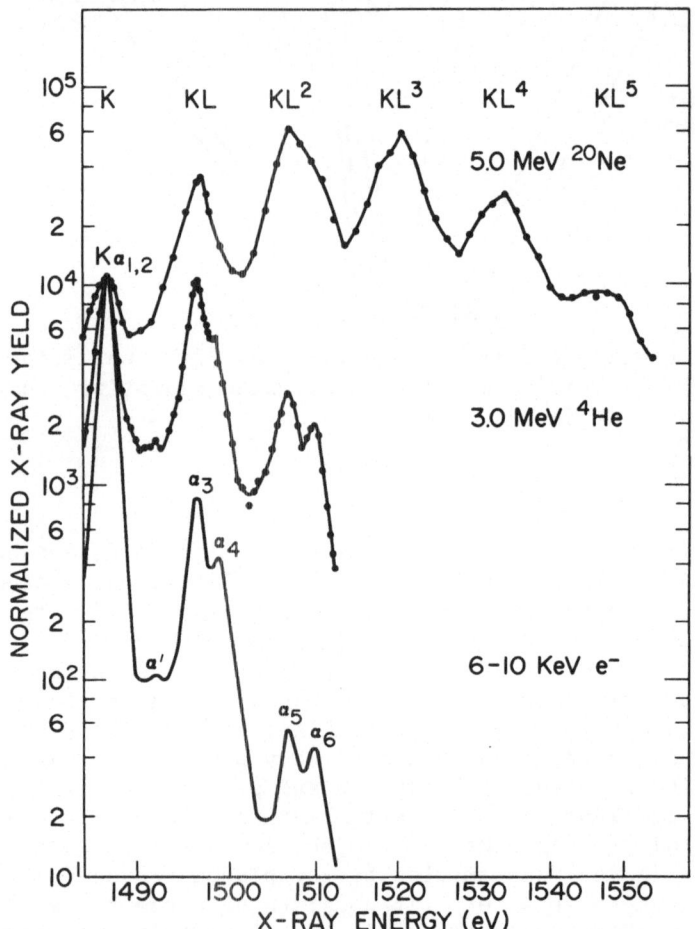

Fig. 6. Al K α x-ray spectra from electron, light ion (^{4}He) and heavy ion (^{20}Ne) bombardment [8]. K_L^N indicates one K-shell and N L-shell electron vacancies in the initial state. The $K\alpha_{1,2}$ (single K hole) lines are normalized to exhibit increasing multiple L-shell ionization. The $\alpha'\alpha_3\alpha_4$ structure is due to the transitions indicated in Figure 4 broadened by the K hole lifetime width.

with a crystal spectrometer by bombarding a thin (20 μg/cm²) carbon
foil with a 22.2 MeV F beam. Spectra were measured at various
distances behind the foil in order to obtain decay rate information.
The spectrum obtained at the foil contains Rydberg-series lines
from short-lived states. Figure 9 shows F spectra from laser-
produced plasmas which are very similar to that in the bottom of
Figure 8. The high intensity available from the plasma permitted
use of a high-resolution grating spectrograph. Many more lines as
well as the free-bound continua are obtained from the plasma compared to the ion beam. Hence, while ion-excitation allows determination of decay times, plasma excitation is useful for accurate

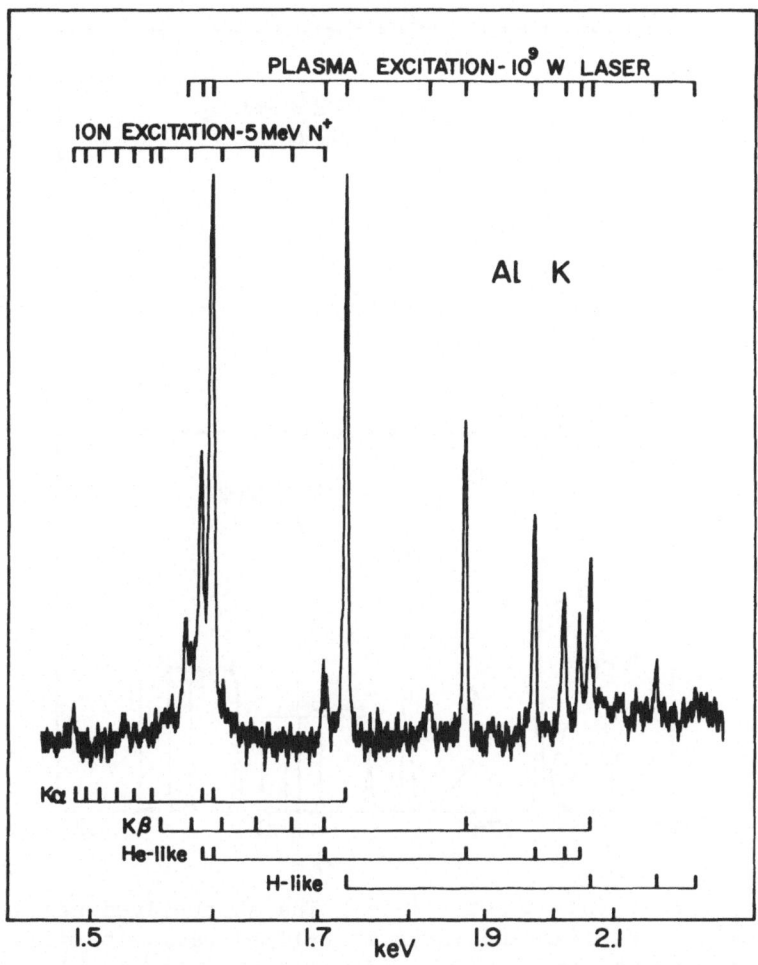

Fig. 7. Al K optical spectrum from a laser-produced plasma [9].
X-ray line and satellite energies produced with ion excitation [10]
are indicated. These are intermediate between the normal $K\alpha_{1,2}$
and the Rydberg series from the plasma.

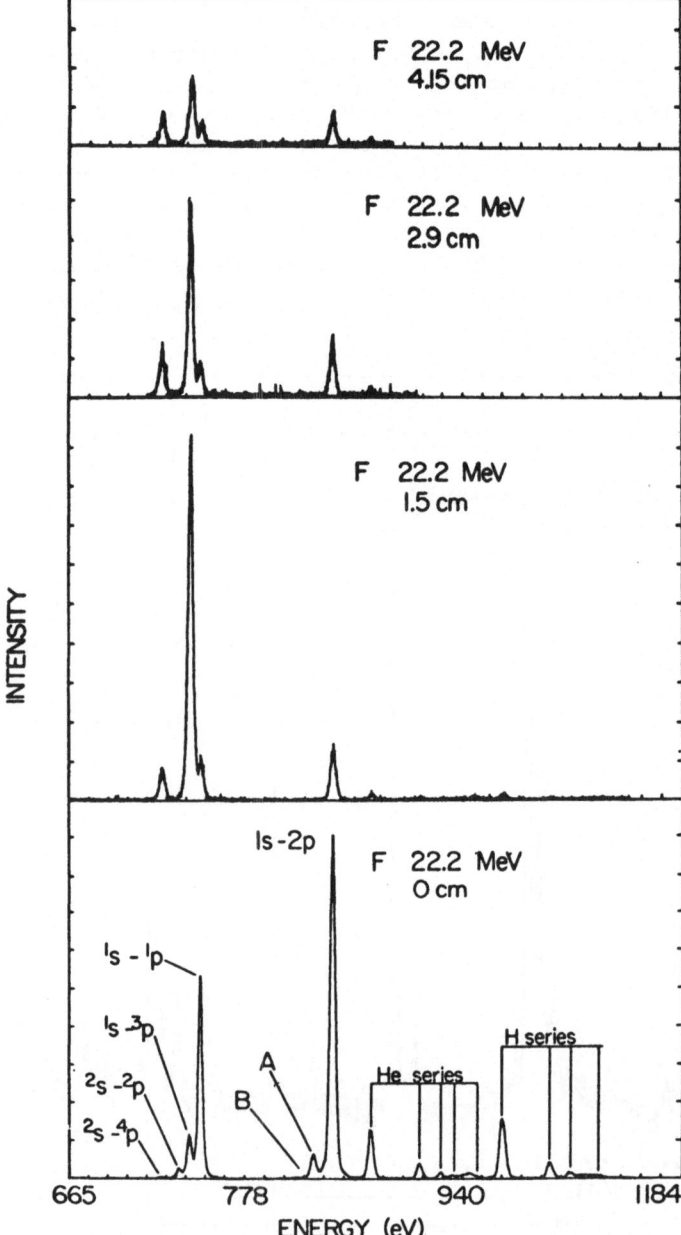

Fig. 8. F K beam foil spectra [12]. The short-lived resonance
(e.g., $^1S-^1P$ in He-like and 1s-2p in H-like) radiation dies off
rapidly compared to the $^1S-^3P$ He-like intercombination line. The
He-like multiplet structure for the 2p → 1s transition is that
shown in Fig. 5.

wavelength determination. But intensities are more easily measured
(electronically) in the beam foil work compared to photographic
recording. No detailed comparison of relative intensities of
Rydberg series from ion and plasma excitation has been made yet.
Such a comparison may illuminate excitation and cascade effects in
the ion beam data. Details of the laser plasma spectrum in Figure
9 have been used to extract information on the emitting plasma
temperature, density and equilibrium conditions.

The complementary (Figures 6 and 7) and directly comparable
(Figures 8 and 9) aspects of ion and plasma excitation were just
discussed in terms of bound→bound transitions. But free→bound
transitions also occur in spectra from both modes of excitation.
Transitions of free electrons to bound levels in a plasma, as in
Figure 6, have long been known. The emitted continuum radiation
occurs at higher energies due to the electron kinetic energy.
Recently it was found that radiative capture of bonding electrons
to vacant levels in a moving projectile can also occur [14]. Since
bonding electrons in a material have many of the characteristics
of a free electron gas, radiative electron capture by a projectile
is similar to a free-to-bound or recombination process in a plasma.
But, there are two differences. In the ion beam case, (a) the
radiation does not form in a wide continuum due to the finite
energy width of the valence-electron band and (b) the ion motion
rather than the electron velocity supplies the photon energy in
excess of the core-level binding energy.

The discussion so far in this section has concentrated on the
major features of x-ray spectra from ion-beam and plasma experiments.
Similarities and differences in the fine structure of spectra from
these sources can also be examined. For example, velocity (Doppler)

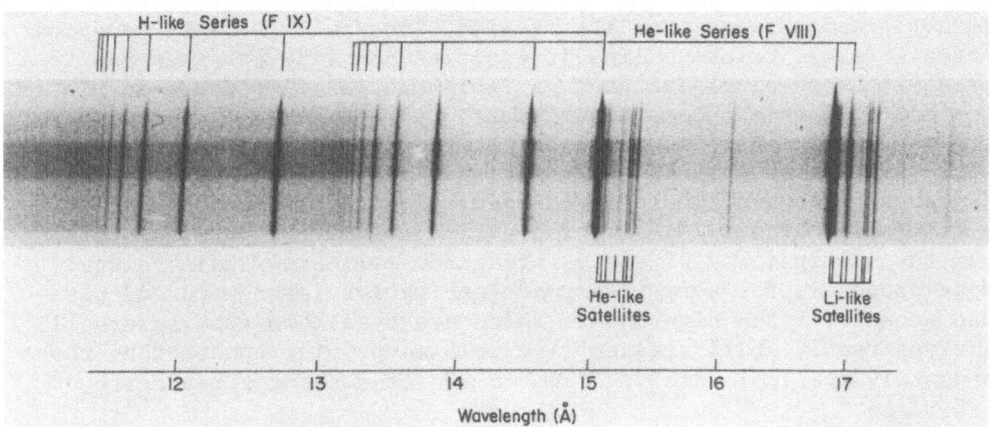

Fig. 9. Fluorine lines from laser produced plasmas [13]. (Entire
width-2 shots, central band-4 shots). The energy-wavelength scale
is the reverse of that in Fig. 8 so that the structure on the right
side here corresponds to that on the left side in Fig. 8.

line broadening is well-known to plasma spectroscopists. Recently, it was also found in ion-excited spectra [15]. But, again the similarity is limited. In the ion-beam work, (a) multiplet structure rather than the Doppler broadening determines the spectral width and (b) projectiles and recoiling target atoms have preferred directions, unlike the thermalized ions in high-temperature plasmas. Other influences on line-profiles, such as electric (Stark) and magnetic (Zeeman) field effects, are well-known in plasma spectra. But they have not been found yet in ion-excited x-ray spectra.

Discussion

K x-ray spectra from relatively low atomic number elements were used to compare ion and plasma excitation in this paper. They were found to be either complementary or much alike, depending on whether thick or thin (beam foil) targets were used in the ion experiments. Similar situations have been demonstrated for L and M x-ray spectra from higher atomic number elements. However, both ion and plasma-generated x-ray spectra are increasingly complex and less useful in going from K to L to M shell spectra.

Laser-plasma spectra were used in this paper, but emission from other hot plasmas could have been invoked as illustrations. Spectra from 1 and 2 electron species of elements beyond argon are now available from plasma-focus, vacuum spark, and exploding-wire plasmas as well as those produced at a laser focus. Such plasmas figure prominently in research aimed at fusion power generation. High resolution x-ray spectroscopy is one of the major tools in such work.

Ion excitation experiments are commonly done with either solid ($\sim 10^{22}$ at./cm^3) or gas ($\lesssim 10^{19}$ at./cm^3) targets. Ionic densities in hot plasmas range from 10^{17} to 10^{21} ions/cm. Density-dependent effects (e.g., relative line intensities and line profiles due to pressure broadening) are thus variable over a wider range in plasmas compared to ion beam experiments.

A final point of comparison between ion and plasma excitation of x-rays. Both methods of producing vacancies have been proposed as means to obtain the inverted population necessary to make an x-ray laser. Traveling-wave ion-excitation may be possible [16] but the multiplicity of states produced implies low gain values. Most proposals for x-ray laser action involve laser-produced plasmas because of the high powers which are available from lasers [17]. But apparently still greater laser-plasma pumping powers than those presently available are required to overcome short x-ray state lifetimes.

In summary, high-ionization states produced in ion-beam and hot-plasma work and the spectra resulting from their decay can be either complementary or similar. Ion-excitation produces spectra ranging from ordinary x-ray (inner shell) to optical (outer electron) emission. It bridges the gap between conventional and plasma excitation. X-ray optical spectra from plasmas give useful information on high temperature plasmas produced in fusion research.

Acknowledgments

The author is privileged to collaborate with A. R. Knudson, P. G. Burkhalter and F. C. Jundt on ion-excitation of x rays, and G. A. Doschek, C. M. Dozier, U. Feldman, B. M. Klein and R. R. Whitlock on plasma x-ray spectroscopy. The receipt of preprints from R. W. Boyd and P. Richard and comments on the manuscript by L. S. Birks are greatly appreciated.

References

[1] R. W. Fink et al. (eds), Proc. Int. Conf. on Inner Shell Ionization Phenomena and Related Phenomena (Apr. 72), AEC Conf-720404 (1973).

[2] R. C. Elton in Methods of Experimental Physics, V. 9, Part A, Academic Press, New York (1970) p. 115.

[3] J. B. Marion and F. C. Young, Nuclear Reaction Analysis, North-Holland, New York (1968) p. 34.

[4] R. W. Boyd, M. S. Thesis, Air University (1972).

[5] M. E. Rudd in Phys. of El. and At. Coll., VII ICPEAC, 1971, North Holland (1972) p. 107.

[6] H. Hartmann and D. Hendel, Th. Chim. Acta (Berlin) $\underline{15}$, 303 (1969).

[7] A. H. Gabriel and C. Jordan, Mon. Not. R. Astr. Soc. $\underline{145}$, 241 (1969).

[8] A. R. Knudson, P. G. Burkhalter and D. J. Nagel in reference 1, p. 1675.

[9] D. J. Nagel et al., to be published.

[10] A. R. Knudson, D. J. Nagel, P. G. Burkhalter and K. L. Dunning, Phys. Rev. Letts. $\underline{26}$, 1149 (1971).

[11] L. L. House, Astro. J. Suppl. $\underline{18}$, 21 (1969).

[12] R. L. Kauffman et al., J. Phys. B. $\underline{6}$, 2197 (1973).

[13] U. Feldman et al, Astro J. $\underline{187}$, 417 (1974).

[14] H. W. Schnopper et al., Phys. Rev. Letts. $\underline{29}$, 898 (1972).

[15] P. G. Burkhalter, A. R. Knudson and D. J. Nagel, Phys. Rev. A, $\underline{7}$, 1936 (1973).

[16] R. A. McCorkle, Phys. Rev. Ltrs., $\underline{29}$, 982 (1972).

[17] M. A. Duguay in reference 1, p. 2350.

APPLICATION OF A CAUCHOIS SPECTROMETER TO MEASUREMENT OF ION–EXCITED Ni Kα SATELLITE SPECTRA

F. C. JUNDT*
Nuclear Structure Research Laboratory
University of Rochester
Rochester, N. Y., U.S.A.

and

D. J. NAGEL
Naval Research Laboratory
Washington, D. C. 20375, U.S.A.

ABSTRACT

A transmission crystal (Cauchois) spectrometer was used for the first time for high resolution measurements of ion-excited x rays. Ni Kα satellite spectra produced in collisions of 61 MeV Ni ions with a Ni target showed that the simply-constructed instrument is useful for high-energy atomic collision studies. Dependence of measured and calculated satellite energies on atomic number was examined. Kα satellite peak spacings computed from Hartree–Fock–Slater theory were found to be significantly smaller than experiment for elements Ca–Ni. The measured Ni Kα spectrum contains structure not found in spectra from lighter atoms.

Introduction

The mechanisms by which light and heavy ion projectiles eject electrons from target atoms have received intensive study in recent years. During this time, a variety of experimental methods have been brought to bear, including charge state and energy loss measurements, and electron and x-ray spectroscopy [1,2]. In particular, x-ray emission has been measured with increasing resolution, as shown in Figure 1 [3-5]. The early energy-dispersion x-ray measurements were made with proportional counters. These and Si(Li) detector work yielded a great deal of ionization cross section

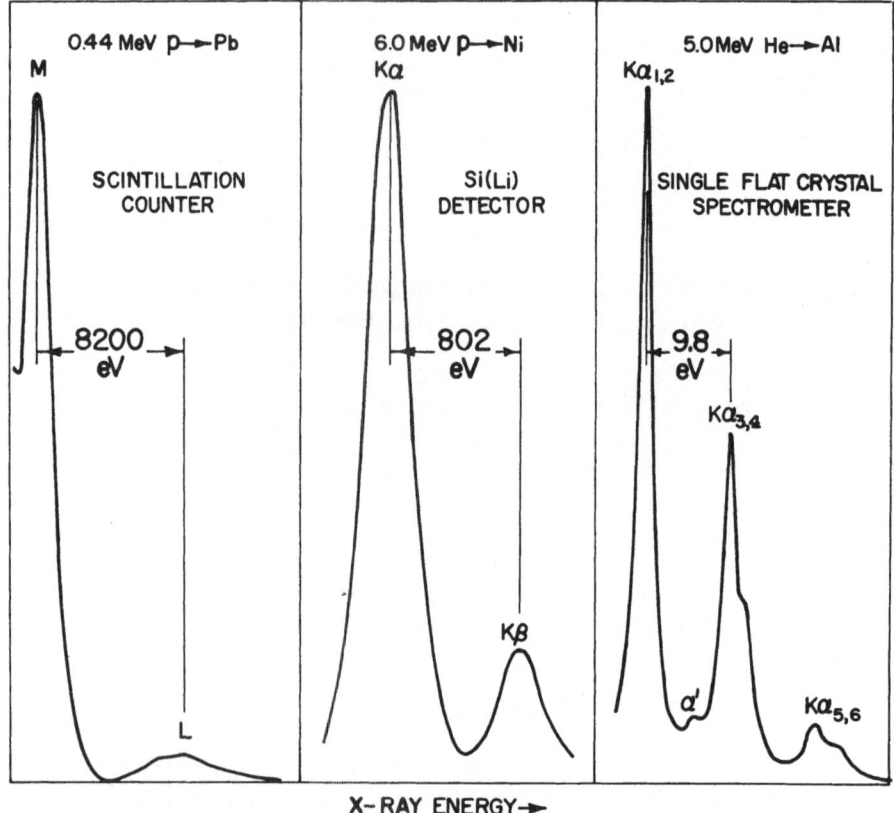

Fig. 1.   Spectra representative of resolution of ion-excited x rays
in 1962 (left [3], 1969 (center [4], and 1972 (right [5].

data.   Then, observations of multiple ionization under heavy ion
bombardment made with Si(Li) detectors provoked many high-resolution
wavelength-dispersive spectrometer measurements of x-ray spectra.
Initially,flat-single-crystal x-ray spectrometers were used.   While
these simple devices are still very useful, more efficient curved-
single-crystal spectrometers are employed for spectral measurements
where low intensity is a problem.   Both reflection and transmission
curved-crystal geometries can be used.   The transmission, or Cauchios
[6], spectrometer is well-suited for high resolution work at relatively high photon energies, even with extended sources (beam
spots).   The Cauchois geometry has been used for detailed studies
of solid state effects on x-ray spectra and for diagnosis of high-
temperature plasmas.   This work is the first use of a transmission
crystal spectrometer for an atomic-collision study.   We illustrate
the utility of the instrument for spectroscopy of ion-excited x
rays by measuring the Ni Kα x-ray spectrum from 61 MeV Ni ions

incident on a Ni target. While the measurement was planned mainly
to establish this utility, it yielded two interesting results.
They are: (1) a systematic departure of experiment and theory for
Kα satellite spacings and (2) new structure in a Kα satellite
spectrum.

The Cauchois Spectrometer

The principles of operation of focusing, curved-crystal x-ray
spectrometers are illustrated in Figure 2 [7]. Reflection instruments use the crystal planes parallel to the large crystal surfaces,
while in transmission spectrometers planes intersecting these surfaces are used. Because of the crystal curvature, the atomic planes
in a transmission crystal diffract radiation from different areas
on a broad source to a line focus. Different x-ray energies arrive
at different points along the focusing (Rowland) circle according
to the Bragg equation

$$n\lambda = 2d \sin \theta = 2d \sin \frac{S}{2R} \tag{1}$$

Here, wavelength λ is diffracted in n^{th} order by a crystal of inter-
planar spacing d. The distance S along the Rowland circle (measured
from the point opposite the crystal center) is $2\theta R$, where 2θ is the

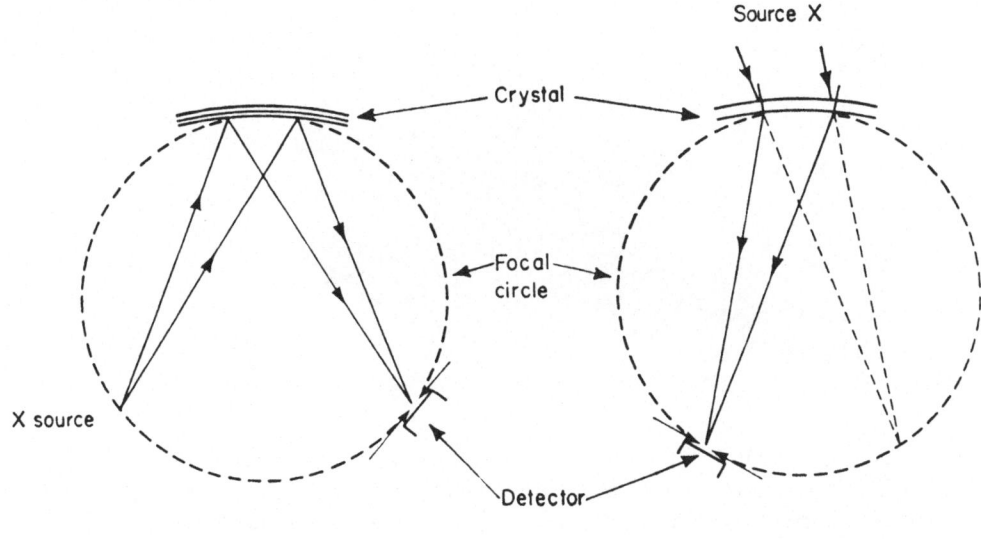

Fig. 2. Principles of curved-crystal x-ray spectrometers from
Bonnelle [7]. The radius of the focal circle is one-half that of
the crystal.

total diffraction angle and R the Rowland circle radius. The
dispersion is obtained by differentiating equation (1),

$$\frac{d\lambda}{\lambda} = \frac{dE}{E} = dS \frac{\cot \theta}{2R} \tag{2}$$

where the photon energy E (in keV) = 12.4/λ (in A). The energy
range resolution and efficiency of a Cauchois spectrometer depend
on the size of the instrument and the size, type and quality of
the crystal. A general description of this type of spectrometer
is given by Blokhin [8]. We discuss the mechanical characteristics,
energy range, resolution and efficiency of the spectrometer used
in the present work in the remainder of this section.

A photograph of our instrument is given in Figure 3. The
beam spot was 8 mm from the crystal center. The (200) planes of
a LiF crystal cleaved to a thickness of 165 μm and elastically
bent to a 20 cm radius were used. An area of the crystal 3 mm
high and 9 mm long was available to diffract x rays. The 50 μm-
wide detector slit was constructed to permit x-ray passage nearly
parallel to the faces of the slit jaws. A P-10 flow proportional
counter with a 4 μm Mylar window was used. For ease of mechanical
fabrication, the slit-detector combination was moved along a straight
line tangent to the 10 cm radius focusing circle at a position

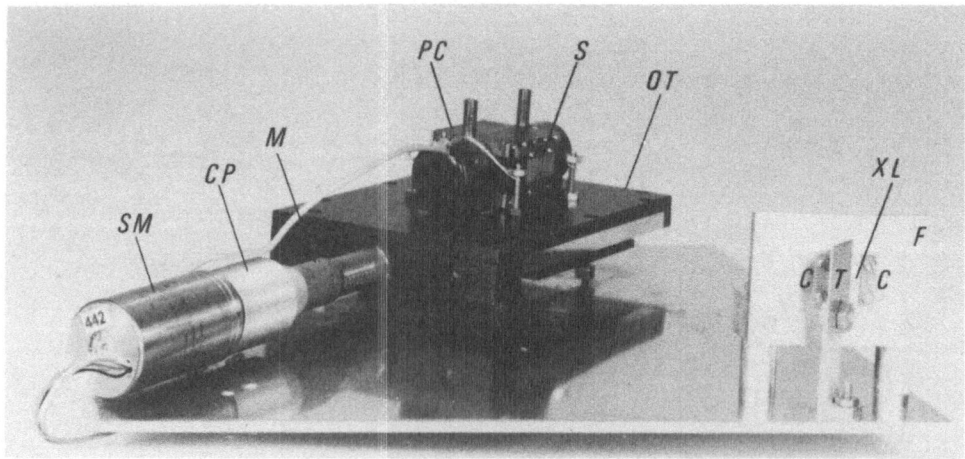

Fig. 3. Photograph of the Cauchois spectrometer used in this work.
The following components are labelled: target T, crystal XL held
on curved form F by clips C, slit S and flow proportional counter
PC on optical table OT. Stepping motor SM (with an integral gear
train) drives OT via the coupling CP on a micrometer M. The motor
bracket has been removed. The projectiles moved from left to right,
parallel to the front edge of the base plate (which is 25 cm long).

corresponding to the Ni Kα energy (7.46 keV). Although the present
instrument was fixed to scan the region near Ni Kα, it is not difficult to build a spectrometer with an adjustable tangency point.
The linear motion was obtained by use of a precision optical table
driven by a computer-controlled stepping motor. The relation between the number of motor steps N and the Bragg angle θ is

$$N = CX = CR \frac{\sin \Delta\theta \sin 2\theta}{\sin[\frac{\pi}{2} - (\theta_o + \Delta\theta)] \sin \theta} \tag{3}$$

where C is a constant, X is the linear distance and $\theta = \theta_o - \Delta\theta$.
N, X and $\Delta\theta$ are all zero at the point of tangency corresponding to
Bragg angle θ_o. Equation (3) can be used to compute the N-E
relation needed to put an energy scale on spectra taken as a function of N. For the present spectrometer, the armature motion of
90°/step, the reduction gears in the motor (494 to 1) and the
micrometer on the optical table yielded a slit displacement of
0.32 μm per step. The N-E curve, which is nearly linear for small
X values, had a slope of 0.026 eV/step at Ni $K\alpha_{1,2}$. Hence the
slit used here corresponded to about 160 steps or 4.2 eV.

A wide range of x-ray energies can be measured with a 10 cm
radius Cauchois spectrometer having a 165 μm thick LiF (200) crystal.
Below about 5 keV, absorption in the crystal limits the ability to
measure all but very intense spectra. The high energy limit of
usefulness is determined by several factors. The small Bragg angles
for x-rays beyond 20 keV require use of a small beam spot and stopping down the crystal length. Otherwise, radiation will enter the
counter directly from the source without being diffracted. The
high energy limit is also influenced by decreasing dispersion and
resolution, and thus is also dependent on available intensity
(which determines the smallest slit that can be tolerated). Dispersion over a 3 keV energy range centered at 7.5 keV is illustrated
in the top of Figure 4. The points were obtained on film by using
characteristic lines from two x-ray tubes as shown in the bottom
of Figure 4. The film recording was done prior to installation
of the counter traverse mechanism by placing a dental x-ray film
curved to a 10 cm radius on the Rowland circle. The spectral
lines in Figure 4 are not well resolved because the x rays struck
the thick emulsion of the dental film obliquely while the densitometer reading was made with the light beam at normal incidence.

The resolution of the present spectrograph is illustrated in
Figure 5. A nickel-target x-ray tube was not available so Cu $K\alpha_{1,2}$
lines were obtained from a Cu target x-ray tube and recorded on a
thin emulsion film. The resolution for the crystal-film combination,
which is less than 10 eV at Cu Kα (8.03 keV), is determined by
the three geometrical factors [8], namely the crystal height, the
length of crystal used and the emulsion thickness, as well as the
crystal rocking curve [9]. Since the slit width used ahead of the

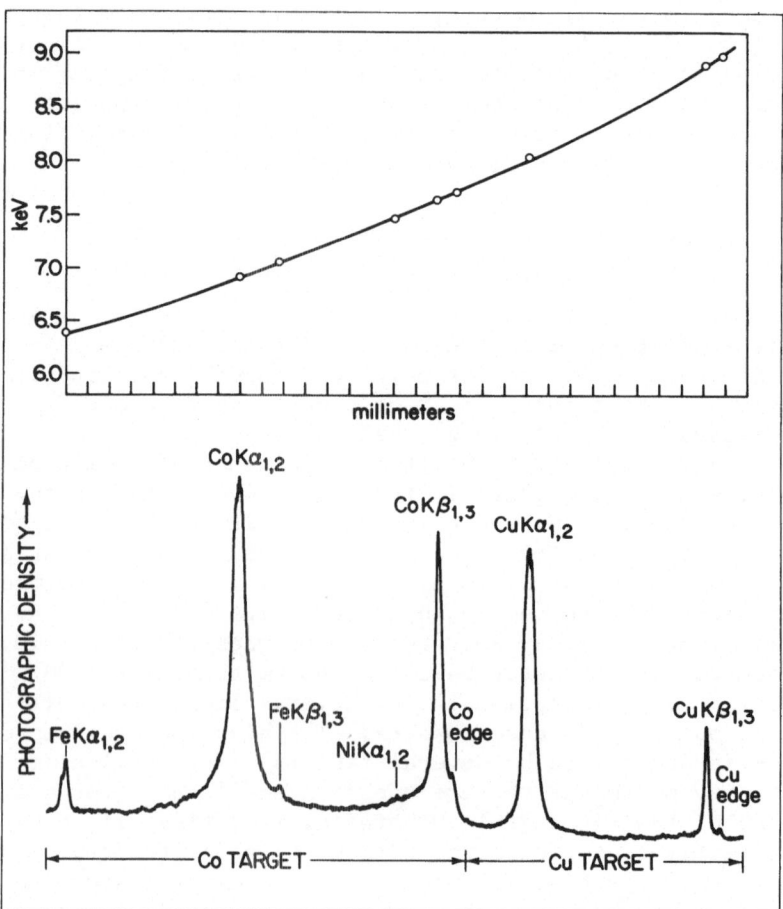

Fig. 4. Partial dispersion curve (top), obtained by exposing a film on the focusing circle to radiation from Co and Cu-target x-ray tubes (bottom). The Co tube, which contained Fe and Ni as impurities, yielded a higher Bremsstrahlung background.

detector in the ion excitation experiment is only 1/4 of the 0.2 mm distance shown in Figure 5, measurement of Cu $K\alpha_{1,2}$ with the detector would have yielded similar resolution to that obtained with film. As usual for dispersive instruments, the dispersion and resolution decrease with increasing photon energy (egn. 2). The change in resolution due to the tangential slit-detector motion over the range in this work is negligible.

The efficiency of the present spectrograph was not measured. It depends on the same factors as the resolution, namely the geometry of the instrument and the characteristics of the crystal.

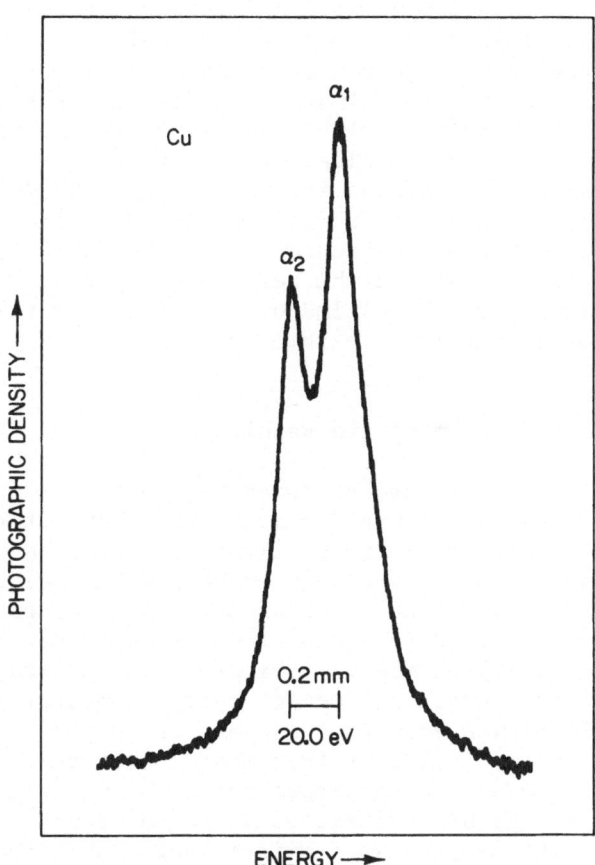

Fig. 5. Densitometer trace of Cu Kα$_1$ and α$_2$ lines from a Cu-target x-ray tube. The slit width used in the ion-excitation measurement was 0.05 mm.

Despite the close target-to-crystal coupling, the efficiency of a transmission crystal spectrometer is not necessarily better than that of a flat crystal instrument. In both cases, for comparable slit and collimator characteristics, it is the integral of the rocking curve of the crystal which primarily determines the spectrograph efficiency. The Cauchois instrument has no collimator blades to obscure radiation from part of the target. But, absorption in the crystal can more than offset this factor. In the present work, the LiF crystal was about 50% transmissive at Ni Kα$_{1,2}$. The best compromise between absorption and diffraction efficiency occurs for a crystal thickness which gives $1/e$ = 37% transmission. Improvements in the detector provide a more straightforward way to improve measured intensity. At Ni Kα$_{1,2}$ use of an Xe-filled counter would improve the count rate by over 50% compared to the Ar-methane

(90-10) gas we used. A longer slit and larger detector window
would also yield a proportionally larger signal.

The usual tradeoff between resolution and intensity applies
to the Cauchois spectrograph. That is, decreasing the slit size
improves the resolution (eqn. 2) but decreases the count rate.
Both trends are approximately linear with slit width. As always,
the resolution should be matched to the scale of the detail in a
spectral measurement in order to minimize the run time. A convenient feature of the Cauchois spectrometer is the ability to continuously and conveniently adjust the resolution by resetting the
detector slit width. Blade collimators for flat crystal instruments do not provide similar flexibility.

The Ion-Excitation Experiment and Results

The 61 MeV Ni^{7+} ions were produced by the Van de Graaff MP
Tandem of the Nuclear Structure Research Laboratory at the University of Rochester. The momentum-analyzed beam of 5 na was focused
to a spot 2 mm in diameter on a 1 mm thick pure nickel target.
The target was mounted to the spectrometer base (Figure 3) and the
entire unit was positioned to locate the target at center of a
21" diameter scattering chamber. A biased beam-integration system
was connected to the target and to an on-line computer in order to
minimize beam fluctuation effects. A computer program permitted
recording of the x-ray intensity from the proportional counter for
a given integrated beam on the target for each spectrometer setting.
After automatic background subtraction, the net count was stored
in a 1024 channel buffer and also printed out. Then the program
moved the spectrometer 200 steps to the next position.

The Ni Kα spectrum obtained in a two hour run is shown in
Figure 6. The first peak (labelled K) is from ions with only a
K inner-electron vacancy initially. It was positioned at 7473 eV,
the weighted Ni $K\alpha_{1,2}$ energy. This may be in error by roughly
8 eV (one-half the $\alpha_1-\alpha_2$ spacing) since peak K might be the $K\alpha_1$
peak. Also projectile stripping affects the absolute energy. In
Al Kα spectra from Al→Al, it was found that Al ions picked up
L-shell electrons between the time of K-shell excitation and decay
[10]. That is, the lower energy Al peaks in a spectrum containing
contributions from both projectile and target atoms were due largely to projectile emission while the high energy Al peaks were
mainly due to target emission. The latter had low intensity in Al
spectra from Al→Mg,Si. If a similar situation exists in the present
Ni→Ni experiment, then peak K arises from projectiles primarily
and has an absolute energy higher than the normal $K\alpha_{1,2}$. The
average charge state of 61 MeV Ni ions is 14 [11]. A shift of
12 eV due to this stripping was calculated using the Herman-Skillman
program [12]. Hence, the absolute energy scale in Figure 6 may
be too low by 15-20 eV. This uncertainty can be avoided if a

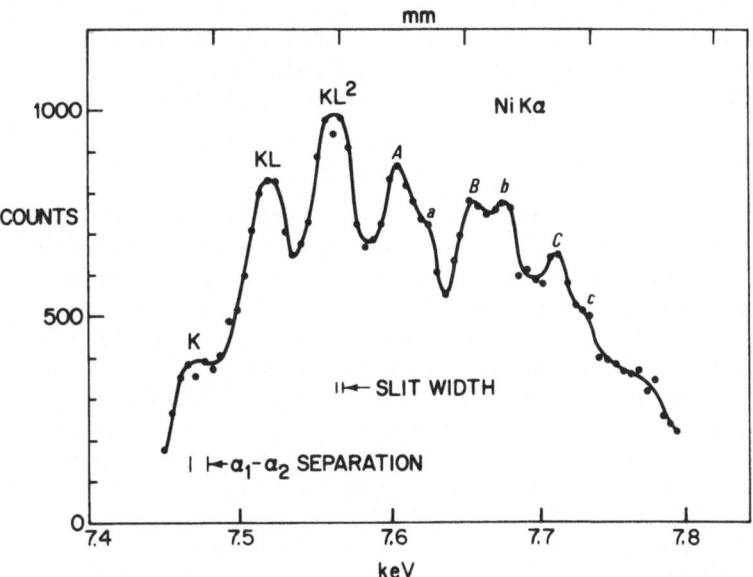

Fig. 6. Ni Kα x-ray specturm produced by 61 MeV Ni ions hitting a thick Ni target (200 steps between points). The labels on the peaks are discussed in the text. Compare the resolution with that in the Ni Kα spectrum in the center of Fig. 1.

light ion beam is available to produce unshifted $K\alpha_{1,2}$ peaks at the time a satellite spectrum is measured. Once a tie point (peak K in this case) is determined for a spectrum, equation (3) can be used to compute the rest of the energy scale, and this was done for Figure 6.

Peaks K and KL^2 in Figure 6 appear to be split, with a component spacing similar to the α_1–α_2 separation. But, the relative intensities of the components are not in the ratio 2:1 as for α_1 and α_2. Except for this possible splitting, the three low energy peaks in Figure 6 are similar to those found in the spectra of lower atomic number elements. However, the high energy structure is unfamiliar.

Discussion

Interpretation of the present results is complicated because the spectrum is a superposition of emission from projectiles and target atoms. Also, Doppler broadening due to projectile motion and target atom recoil introduces some smearing. Two approaches have been used in order to understand the spectrum in Figure 6: (a) comparison of peak spacings for Ni with results for lower atomic number elements, and (b) adiabatic calculations of x-ray transition energies using Hartree-Fock-Slater (HFS) theory [12]. The low-energy structure and the high-energy structure in the Ni $K\alpha$ spectrum will be discussed separately in this section before additional experimental and theoretical work is indicated.

Low energy structure

The dependence of relative satellite energies on atomic number is shown in Figure 7 for the three low energy peaks. Experimental satellite energies, mostly from the literature, were plotted relative to the energy of the parent line ($K\alpha_{1,2}$ for all elements except Fe where the $K\alpha_1$ is the reference line). Several of the energy spacing values were obtained graphically rather than from tabulated energies. Scatter in the data is due in part to errors in picking off peak energies. Also, the various spectra were not equally well-resolved, so the presence of multiplet structure contributes to the scatter. And, although most spectra from which Figure 7 was plotted are from target atoms, projectile data were used in some cases. Missing outer electrons in projectiles cause small shifts in satellite spacings which produce some of the scatter in the empirical data. For Ni, the uncertainty in energy shift values which is shown in Figure 7 (8 eV) is due to the uncertainty in the location of the Ni parent $K\alpha_{1,2}$ lines relative to the spectrum.

Despite scatter in measured satellite spacings, the data indicate a smooth increase which tends to become linear with increasing atomic number. A similar trend is found for the HFS calculations, except the theoretical values are significantly smaller than the experimental curve (22% less for the Ni K-KL spacing). The calculations were done using the optimized exchange parameter [24], although the computed satellite spacings are essentially the same when full-Slater exchange ($\alpha = 1$) is used. Where comparisons could be made, the computed shifts we obtained were within ± 2 eV of HFS shifts reported in the literature. The Herman-Skillman shift calculations were also done for Ni with 14 missing outer electrons. The shifts for such stripped (projectile) ions are only 2-3 eV greater than for unstripped (target) atoms. Hence, there appears to be a significant discrepancy between measured shifts and those computed from HFS theory. The calculated values are consistently low. But, it is nevertheless clear that the K, KL, and KL^2 peaks in Figure 6 for Ni are similar to those found in lighter elements, i.e., they are due to 0, 1, and 2 L-shell vacancies in addition to a K vacancy.

High energy structure

The high energy structure in Figure 6 is more difficult to interpret. It is experimentally reproducible and cannot be due to imperfections in the crystal (since essentially the same region of the crystal was involved across the entire spectrum). The first peak above KL^2 (A) appears from Figure 7 to be KL^3. There is not

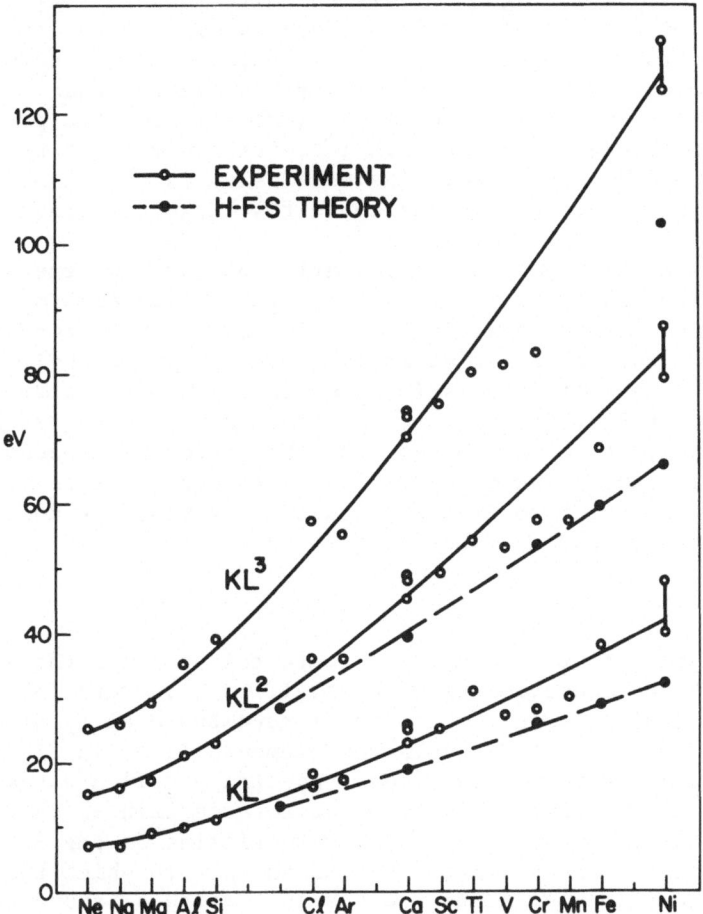

Fig. 7. Plot of KL^N (N = 1,2,3) satellite energies relative to $K\alpha_{1,2}$ ($K\alpha_1$ for Fe). The data come from the following references: Ne (13), Na (14), Mg (15), Al (16), Si (17), Cl (18), Cl-Ar (19), Ca (20,21), Ca-Mn (22), Fe (23), Ni (this work). The Hartree-Fock-Slater calculations were made for 2p spectator vacancies.

sufficient data available to do a reliable extrapolation for KL^4 and KL^5. But strong peaks B and C are found above A at about equal intervals. Hence, it seems that KL^3, KL^4 and KL^5 all appear in the spectrum as peaks A, B, and C in Figure 6, with KL^3 and KL^5 having high energy shoulders (a and c), and KL^4 a twin peak (b). The spacing and relative intensities of these associated structures are not consistent with structure due to α_1 distinct from that due to α_2.

It is tempting to interpret the overall spectrum as the super-position of a more intense target spectrum found at lower energies with a somewhat less intense projectile spectrum shifted to higher energies due to stripping. Calculated shifts due to stripping vary from 12 eV for K to 28 eV for KL^5. Such values are not inconsistent with the separations A-a, B-b and C-c. But we already noted that the projectile ions probably have fewer L vacancies than target atoms. If K, KL and KL^2, a, b, and c were all from the projectile, a very uneven change in shift with increasing number of L-shell vacancies would be indicated. The conclusion is that the additional high energy structure is not due to shifts caused by projectile stripping.

There is another possible explanation of the high energy structure. Larkins has recently calculated L_1-L_{23}M Coster-Kronig energies [25]. He states that for 3d metals, such transitions are energetically forbidden for atoms with one or more L-shell vacancies. Hence, we calculated shifts due to 2s as well as 2p spectator vacancies. For KL^3, the 2p→1s transition energy decreases by 8 eV for each spectator vacancy moved from the 2p to the 2s subshell. This direction and value do not allow us to ascribe the high energy structure in Figure 6 to distinct peaks for various arrangements of vacancies within the L shell.

Future work

The present results clearly indicate the need for the additional high-resolution measurements of K satellite structure in order to interpret the Ni Kα spectrum and to improve the quality of a plot like that in Figure 7. The data for elements Ti-Mn in Figure 7 tend to fall below the solid lines. Spectra for Ti and heavier elements should be remeasured in order to ascertain if this trend is significant (possibly associated with filling of the 3d subshell) or whether it is merely scatter due to the way in which the plot was made. The needed experiments are possible using a Cauchois spectrometer and should also be possible with a reflection crystal instrument. The same projectile and target atoms ($Z_1 = Z_2$) were chosen in the present work to enhance the intensity available to test the suitability of a transmission crystal instrument for spectroscopy of ion-excited x rays. The results show that measurements of x-ray spectra from targets made of Ni and neighboring elements excited by heavy ions with high energies are possible. The factor of two intensity loss for $Z_1 \neq Z_2$ can be recouped by improvements in the detector, as discussed earlier. If intensity permits, lighter projectiles should be used to reduce Doppler effects due to target atom recoil.

Systematic theoretical calculations should be made with Hartree-Fock and relativistic atomic structure programs. There is also a need for multiplet calculations for KL^N (N>2) for vacancies distributed over the 2s and 2p subshells.

Conclusion

The present work shows that a simply-constructed Cauchois spectrometer is well-suited to high resolution measurement of relatively hard x rays produced in very energetic atomic collisions. The discrepancy between empirical and computed Kα shifts due to L-shell vacancies which was found for heavier elements should be investigated by additional experiments and systematic theoretical work. The test Ni Kα spectrum produced by Ni→Ni at 61 MeV contains new high-energy structure, the interpretation of which will also require further measurements and calculations.

Acknowledgments

M. Fatemi deftly cleaved the crystals used in this work. This assistance; discussions with D. B. Brown, B. M. Klein, T. N. Lee and A. R. Knudson; receipt of a preprint from F. P. Larkins; and comments on the manuscript by L. S. Birks and P. G. Burkhalter are greatly appreciated.

References

*On leave from L.P.N.I.N., Université Louis Pasteur, Strausbourg, France.

[1] J. D. Garcia, R. J. Fortner, T. M. Kavanagh, Rev. Mod. Phys. 45, 111 (1973).

[2] R. W. Fink, S. T. Manson, J. M. Palms and P. V. Rao, eds., Proc. of Int. Conf. on Inner Shell Ionization Phenomena and Future Applications (Atlanta, 1972), AEC CONF 720404, V. 2 and 3.

[3] R. C. Jopson, H. Mark and C. D. Swift, Phys. Rev. 127, 1612 (1962).

[4] P. Richard, I. L. Morgan, T. Furuta and D. Burch, Phys. Rev. Ltrs. 23, 1009 (1969).

[5] A. R. Knudson, P. G. Burkhalter and D. J. Nagel in Reference 2, p. 1675.

[6] Y. Cauchois, J. Phys. et. Rad. 3, 320 (1932) and 4, 61 (1933).

[7] C. Bonnelle in "Physical Methods in Advanced Inorganic Chemistry," Ed. by H. A. O. Hill and P. Day, Interscience, New York (1968), p. 45.

[8] M. A. Blokhin, "Methods of X-Ray Spectroscopic Research," Pergamon, Oxford (1965), p. 178.

[9] L. S. Birks, "X-Ray Spectrochemical Analysis" (2nd Ed.), Wiley, New York (1969), p. 36.

[10] A. R. Knudson, P. G. Burkhalter and D. J. Nagel in the proceedings of this conference.

[11] H. R. Betz, Rev. Mod. Phys. 44, 465 (1972).

[12] F. Herman and S. Skillman, "Atomic Structure Calculations," Prentice-Hall, Englewood Cliffs (1963).

[13] R. L. Kauffman, F. Hopkins, C. W. Woods and P. Richard, Phys. Rev. Ltrs. 31, 621 (1973).

[14] C. F. Moore, D. K. Olsen, B. Hodge and P. Richard, Z. Phys. 257, 288 (1972).

[15] D. G. McCrary, M. Senglaub and P. Richard, Phys. Rev. 6A, 263 (1972).

[16] A. R. Knudson, D. J. Nagel, P. G. Burkhalter and K. L. Dunning, Phys. Rev. Ltrs. 26, 1149 (1971).

[17] D. G. McCrary and P. Richard, Phys. Rev. 5A, 1249 (1972).

[18] C. F. Moore, H. H. Wolter, R. L. Kauffman, J. McWherter, J. E. Bolger and C. P. Browne, J. Phys. B. 5, L262 (1972).

[19] A. R. Knudson, P. G. Burkhalter and D. J. Nagel, unpublished.

[20] P. Richard, W. Hodge and C. F. Moore, Phys. Rev. Ltrs. 29, 393 (1972).

[21] J. McWherter, J. Bolger, C. F. Moore and P. Richard, Z. Phys. 263, 283 (1973).

[22] R. L. Kauffman, J. H. McGuire, P. Richard and C. F. Moore, Phys. Rev. 8A, 1233 (1973).

[23] D. Burch, P. Richard and R. L. Blake, Phys. Rev. Ltrs. 26, 1355 (1971).

[24] K. Schwarz, Phys. Rev. 5B, 2466 (1972).

[25] F. P. Larkins, Australian J. Phys., to be published.

CHARGE STATE DEPENDENCE OF Si K X-RAY PRODUCTION IN SOLID AND GASEOUS TARGETS BY 40 MeV OXYGEN ION IMPACT*

J. R. MOWAT,[†] B. R. APPLETON, J. A. BIGGERSTAFF,
S. DATZ, C. D. MOAK, and I. A. SELLIN[†]
Oak Ridge National Laboratory
Oak Ridge, Tennessee 37830, U. S. A.

ABSTRACT

We have studied the yield of silicon K x rays induced
by 2 1/2 MeV/amu oxygen ions in gaseous (SiH_4) and amorphous
solid targets. For a gas density 7.4×10^{14} cm^{-2} the yield
increases approximately exponentially with the charge state
q of the projectile in the region q = 6, 7, 8. A similar,
but smaller, trend is found for a 7 μg/cm^2 solid target.
At this beam energy the most probable oxygen charge state
emerging from a solid Si target is 0^{7+}. The x-ray yield
from the solid is in accord with the emergent charge state.

Introduction

It is well known [1] that when heavy ions pass through matter,
they capture or lose electrons during collisions made with the
atoms in the medium. When the emergent beam is analyzed, a variety
of charge states is found, with some charge states being more
prominent than others. If the medium is thick enough, i.e., if
enough charge exchange collisions are made, the charge of the
exiting ion will have lost its correlation with the charge of the
incident ion. Thus, for thick targets, the charge state distribution in the energing beam should be independent of the charge state
of the incident beam. The distribution is found to depend strongly
upon the energy and nuclear charge of the incident ion, but only
weakly upon the target material. However, it is also known that,
for projectiles of large Z, the average charge emerging from a
solid target is higher than that from a gas target. The actual
average charge of a projectile inside a solid is not known. A
suggestion has been made that the average charge is the same inside
the solid as inside the gas, but that the ion autoionizes on exit
thereby undergoing charge conversion [2]. An alternate hypothesis

claims that the average charge state observed in the emerging beam
is achieved while the ions are still inside the solid [3]. Betz
[1] has recently commented that "...it is still being disputed
whether the large increase is produced inside or outside the solid."

About two years ago Der, Fortner, Kavanagh and Garcia [4]
reported argon L x-ray spectra emitted by 90 keV Ar ions as they
pass through solid (graphite) and gaseous (methane) carbon targets.
The x-ray energies were measured using a crystal spectrometer with
sufficient resolution to separate L x rays from different argon
charge states. The x rays observed in the gas target were identified as 3s-2p or 3d-2p transitions in singly ionized Ar. The
energies of a significant number of x rays emitted by argon ions
in the solid corresponded to 3s-sp transitions in ions with 5 or 6
outer shell vacancies. This result was interesting because the
average charge state of a 90-keV argon beam emerging from a solid
is expected [1] to be 0 or 1, not 5 or 6. The authors suggested
that high resolution x-ray analysis of projectile x rays together
with accurate calculations relating x-ray energy and defect configuration might provide a method for charge-state determinations
for ions moving in solids.

In this paper we report some preliminary results obtained by
a new method for measuring effective charge states in solids.
Recently a group of experimenters from the University of Tennessee
[5] working at Oak Ridge National Laboratory and an independent
group at Kansas State University [6] have reported measurements
of K x-ray yields induced by heavy ions in gas targets. They find
that the yield of both projectile and target x rays is very sensitive to the charge state of the incident ion. The idea occurred
to us that a comparison of relative yields in gas and solid targets
containing the same atomic species might reveal the effective
charge of the ion while it is inside the solid. To this end we
began our study by bombarding solid and gaseous silicon targets
with 40 MeV oxygen ions in charge states 5, 6, 7, 8.

Target Conditions

The targets for such a comparison must meet the following
conditions: Gas targets must be thin enough that there is a low
probability that an incident ion will change charge before producing
a K shell vacancy. In addition the target density and purity must
be well known and stable.

Because of the large cross sections expected and the relatively
large solid target density, we must assume that a single ion might
make as many as three or four collisions as it passes through the
solid. Since target K-shell vacancy production cross sections are
expected to be slightly smaller than projectile charge exchange
cross-sections, there is a good chance that an incoming projectile
will change charge before it produces a target vacancy. A single
collision vacancy production cross section can be extracted from

the data if it can be assumed that that cross section is the same at all points within the solid. Since the cross section depends upon both the energy and the charge state of the projectile, the assumption will be valid only if the target is thick enough to insure that charge state equilibrium is achieved soon after entering the solid and if the target is thin enough so that projectile energy loss can be neglected.

Since the basic idea is to compare single collision cross sections in solid vs gas, factors such as detector efficiency, solid angle and atomic fluorescence yield need not be accurately known.

Apparatus

A beam of 40-MeV O^{6+} ions obtained from the ORNL Tandem Van de Graaff accelerator passes through a thin carbon foil and is upcharged to an average charge state of 7+. Beams of charge state 5+ through 8+ are magnetically steered one at a time into the target region. For these sub-nanoampere beams an average time of about one nanosecond separates successive ions which pass through the solid target in $\sim 10^{-15}$ seconds. Such low beam intensities should leave sufficient time for target atom relaxation between collisions.

The differentially pumped gas cell with effective length 0.42 cm contains SiH_4 at a pressure of 50 millitorr. This amounts to a target density of 7.4×10^{14} atoms/cm^2. At this density single collisions dominate and the probability of a charge changing collision is less than 0.01 [7]. Gas target pressure is monitored by a capacitance manometer. An output signal from the manometer drives a servo-controlled needle valve which establishes the gas flow rate and thereby stabilizes the pressure. Setting accuracy is believed to be ±2 millitorr, and the stability is better than ±1 millitorr. In order to assure target purity the gas feed lines are evacuated back to the storage bottle before a run. With no gas flow the background pressure is less than 0.1 millitorr. About 30 minutes is allowed for target pressure equilibration before taking data.

The beam line upstream of the gas cell is maintained at a vacuum of better than 10^{-5} torr, which guarantees that the charge state purity of the selected beam is maintained up to the entrance to the cell. (<0.2% of the beam charge changes in the beam line.)

The gas feed line can be removed and replaced by a solid target holder mounted inside the cell. The solid target is placed at the center of the field of view of the x-ray detector and is oriented at 36° to the beam.

An amorphous silicon foil, vacuum deposited on a plastic backing serves as the solid target. The target thickness is determined in a separate experiment by counting the fraction of 1 MeV protons that are Rutherford scattered at 90° and comparing that with the fraction scattered from a foil of known thickness. When mounted at 36° to the beam the target thickness is 2.0×10^{17} atoms/cm^2.

This target satisfied the criteria mentioned above with the possible exception of equilibrium thickness. Since incident charge states 6 and 8 need undergo only one charge changing collision in order to reach the average charge (+7) measured emerging from solid silicon at this energy, the foil need not be very thick. (The probability of a single electron capture event by a 40 MeV O^{8+} ion in this target is $\sim$0.8, and the probability of making two collisions is appreciable.)

Silicon K x rays (1.7 keV) are detected in a lithium drifted silicon diode which views the collision region at 90° to the beam direction. Pulses representing detected x rays are amplified and stored according to pulse height in a M.C.A. The jump in detector sensitivity at the silicon dead layer absorption edge does not concern us because, whether the target is gas or solid, the highly ionized target atoms radiate K x rays whose energies are probably all above the absorption edge.

The beam emerging from the target is detected by Rutherford scattering from a 270 μg/cm^2 gold foil into a solid state particle detector. Pulses from the particle detector do not undergo pulse height analysis but are simply counted. Detected x rays are accumulated for a preselected (but arbitrary) number of scattered oxygen ions.

Experimental Results

We have extracted from the measured solid and gas yields the cross section for production of a K x ray in a single collision. These cross sections are plotted in Figure 1 as a function of the charge state of the incident oxygen ion. The gas target data displays the same exponential rise with increasing charge state that has been observed in our laboratory for a number of projectile-target combinations [5]. The origin of this strong charge state effect is still not known. Plotted also are the solid target data which shows a less pronounced, but probably real, projectile charge state dependence. A similar effect has been recently reported for oxygen-ion induced K x rays from solid aluminum targets by Brandt, Laubert, Maurino, and Schwarzschild [7]. They find that the effect diminishes as the foil is made thicker and attribute this charge state effect to dynamic screening; i.e., the time required for the plasma to respond to the disturbance introduced by the penetrating ion. An alternative explanation could lie in the nonequilibrium of the ion charge state. Charge exchange cross sections [8] for 40 MeV O ions in Ar, for example, are on the order of 2 x 10^{-17}cm^2. Thus the Si target thickness ($\sim$ 2 x 10^{17}) should be insufficient to attain equilibrium.

The observed charge state dependence is expected to evolve gradually with increasing target thickness from the rising straight line observed for thin (gaseous) targets to a horizontal straight line at equilibrium thickness. The two lines representing the

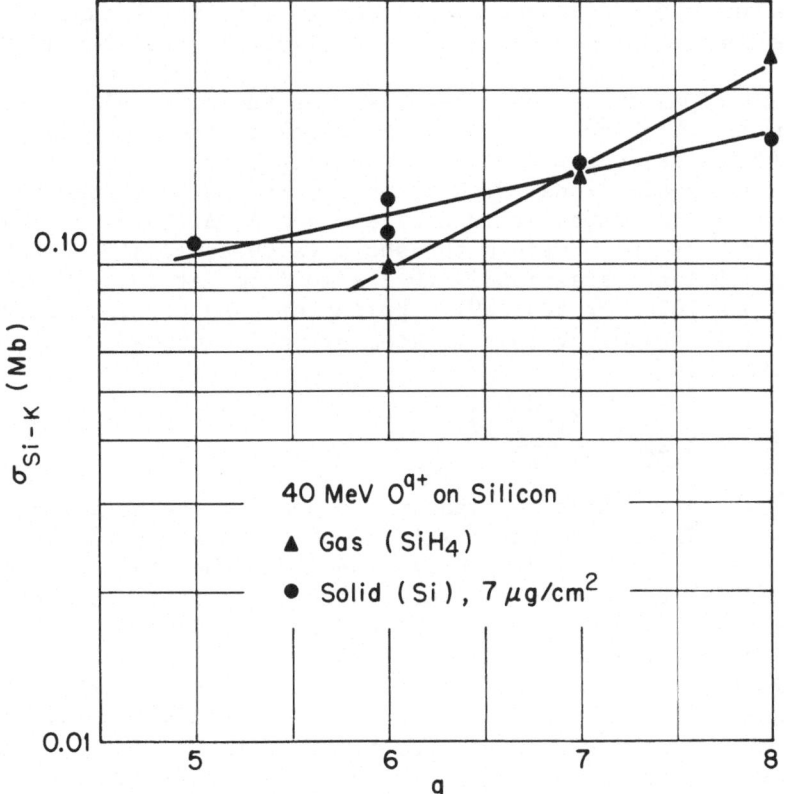

Fig. 1. Silicon K x-ray production cross section in megabarns versus input charge state q for 40 MeV O^{8+} bombardment of SiH_4 gas and solid Si.

extremes in density should intersect at the projectile charge corresponding to the effective equilibrium charge in the solid. The equilibrium charge state distribution of 40 MeV O emerging from Si is strongly peaked at 7+ (20% 6+, 70% 7+, and 10% 8+) For ions as light as oxygen the equilibrium charge distribution emerging from gases is almost the same as for solids. At equilibrium, the input charge dependence of the x-ray cross section would be a horizontal line intersecting the SiH_4 line at charge 7+ if the effective charge in the solid were equal to the most probable emergent charge.

Moreover, since the cross sections for leaving charge 7+ ($\sigma_{7,6}$ and $\sigma_{7,8}$) should be lower at this energy than those for attaining charge 7+ (i.e., $\sigma_{8,7}$ and $\sigma_{6,7}$) one would expect coincidence with the 7+ value of the gas target even before attaining equilibrium. Arguments based on dynamic screening in which the number of bound electrons are not considered also leads to almost the same expected effective charge state. Here $q_{eff} \simeq Z (1-v_o/v) = 7.2$ for 40 MeV O ions.

In principal there could be several effects, unrelated to ion charge state, which could cause differences in the observed yields from solid and gaseous targets. The possibility of energetic Si recoils causing additional Si-K vacancies in the solid can be assessed using the information available on Al-Al(Z=13) collision induced K x rays [9]. Here a threshold is observed at $\sim$50 keV; the cross section rises rapidly to $2 \times 10^{-21} cm^2$ at 200 keV and reaches $\sim 5 \times 10^{-21}$ at 500 keV. The impact parameter, b, required to impart a $\Delta E = 200$ keV to a Si atom by an E = 40 MeV O ion can be estimated from

$$b = Z_1 Z_2 e^2 \ (\frac{M_1/M_2}{E \cdot \Delta E})^{1/2}$$

to be 4×10^{-12} cm which is clearly too small to be noticeable when compared with direct processes with cross sections in the order of $10^{-18} cm^2$. Since the K shell ionization is in the diabatic region (i.e., $hv/I > r_K$) the impact parameter dependence of the probability of ionization is peaked [10] near the K shell radius ($r_K = 0.06$ Å). At this distance the energy transfer is only $\sim$10 eV so that even interactions of the L shell of the recoiling Si with lattice atoms (which could affect the K fluorescence yield) are not possible. Finally, the effect of binding electrons (in the M shell) is not expected to be significant; especially since the Si valence electrons in both SiH_4 and solid Si are in the sp^3 configuration.

In conclusion, the effective charge state for x-ray production by 40 MeV ions in solid Si is in agreement with that anticipated either on the basis of emergent charge state distributions or dynamic screening. Solid state effects due to multiple collisions do not affect the yield. For ions of low Z where no significant difference in emergent charge state distributions exists between gases and solids the effective charge for x-ray production in solids can be obtained from the known emergent charge state distributions.

References

*Research sponsored jointly by Office of Naval Research, NASA, and the U. S. Atomic Energy Commission under contract with Union Carbide Corporation.

†Department of Physics, The University of Tennessee, Knoxville, Tennessee.

[1] H. D. Betz, Rev. Mod. Phys. 44, 465 (1972).

[2] H. D. Betz and L. Grodzins, Phys. Rev. Lett. 25, 211 (1970).

[3] N. Bohr and J. Lindhard, Kgl. Danske Videnskab, Selskab Mat.- Fys.Medd. 28, No. 7 (1954).

[4] R. C. Der, R. J. Fortner, T. M. Kavanagh and J. D. Garcia,
     Phys. Rev. Lett. $\underline{27}$, 1631 (1971).
[5] J. R. Mowat, I. A. Sellin, D. J. Pegg, R. S. Peterson,
     Matt D. Brown and J. R. MacDonald, Phys. Rev. Lett. $\underline{30}$, 1289
     (1973) and Phys. Rev. A $\underline{9}$ (1974).
[6] J. R. MacDonald, L. Winters, M. D. Brown, T. Chiao, and
     L. D. Ellsworth, Phys. Rev. Lett. $\underline{29}$, 1291 (1972).
[7] W. Brandt, R. Laubert, M. Maurino and A. Schwarzschild, Phys.
     Rev. Lett. $\underline{30}$, 358 (1973).
[8] J. R. MacDonald and F. W. Martin, Phys. Rev. A $\underline{4}$, 1965 (1971).
[9] K. Taulbjerg, B. Fastrup, and E. Laegsgaard, Phys. Rev. A $\underline{8}$,
     1814 (1973).
[10] J. M. Hansteen and O. P. Mosebekk, Nucl. Phys. A $\underline{201}$, 541 (1973).

EXCITATION STATES OF PROJECTILES MOVING THROUGH SOLIDS[*]

R. J. FORTNER
Lawrence Livermore Laboratory
Livermore, California 94550, U.S.A.
and
J. D. GARCIA[†]
University of Arizona
Tucson, Arizona 85721, U.S.A.

ABSTRACT

X-ray spectral measurements of S, Cl, and Ar atoms moving in solid carbon targets are used to extract the equilibrium distribution of vacancies in the valence and L-shells of the projectiles. It is found that the state of excitation is much higher than the mean charge measured after the projectile has left the solid. A simple model is used to deduce the expected final charge state distribution, and these are compared to thin foil measurements.

In this paper x-ray spectral measurements of projectiles traversing a solid are used to determine equilibrium distribution of vacancies in the inner and valence shells of the projectile. Disparity between our measurements and the measured emergent charge states [1] is observed. A simple picture is found to bring the two measurements into agreement.

The relevant data used in this paper are presented in Figs. 1 and 2. Figure 1 shows the spectral data [2,3] for S, Cl, Ar → C collisions using a solid graphite target and an incident projectile energy of 90 keV. The arrows in the figure indicate the positions of the normal (unshifted) x-ray transitions. The data indicate that the majority of the x rays are L x rays from the projectile; however, some carbon K x rays are observed. Identification of the x-ray transition is discussed below. Figure 2 shows the x-ray production cross sections for both target and projectile in the Ar → C collision system as determined from the x-ray spectral measurements [4].

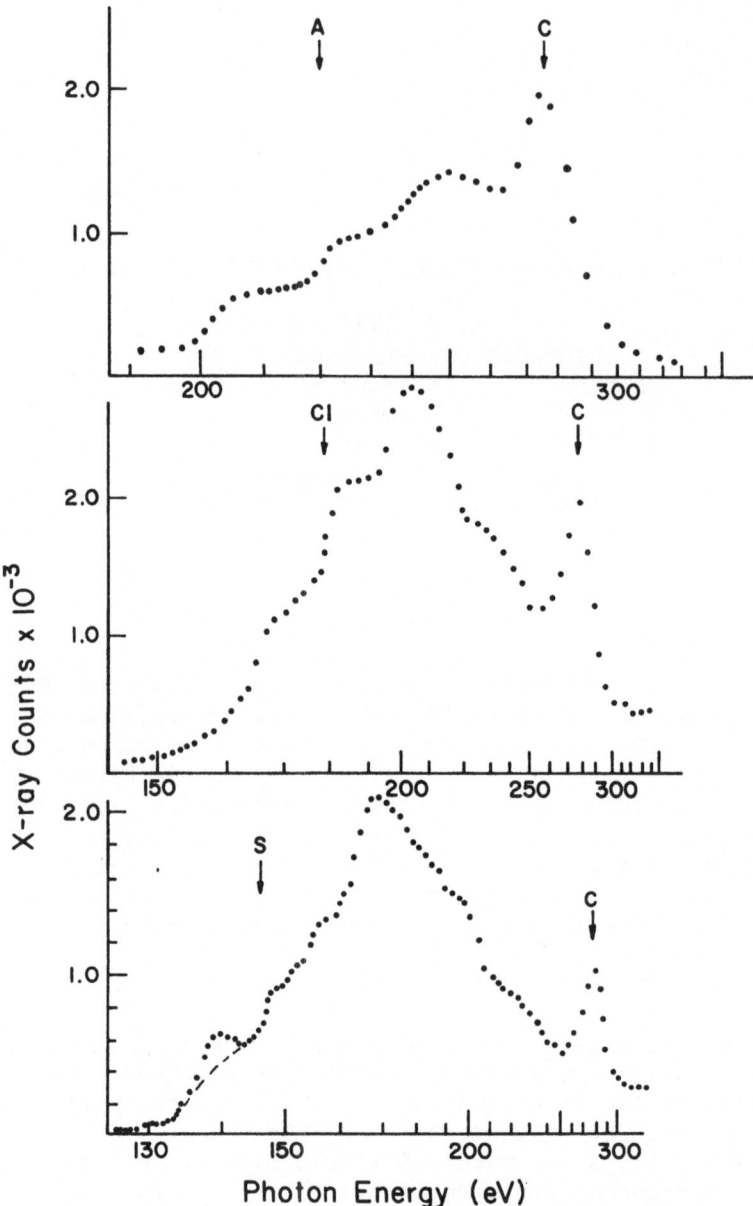

Fig. 1. X-ray spectra for Ar → C (upper diagram), Cl → C (middle diagram), and S → C (lower diagram) collisions in solid (graphite) targets. The position of the "normal" (unshifted) L x rays are indicated by labelled arrows as is the position of the C K x-ray line. The bombarding energy was 90 keV. In the lower diagram, the carbon K x rays produce a second order peak at about 140 keV. The dashed line shows the resultant spectrum when the second order carbon line is subtracted, by using the first order profile.

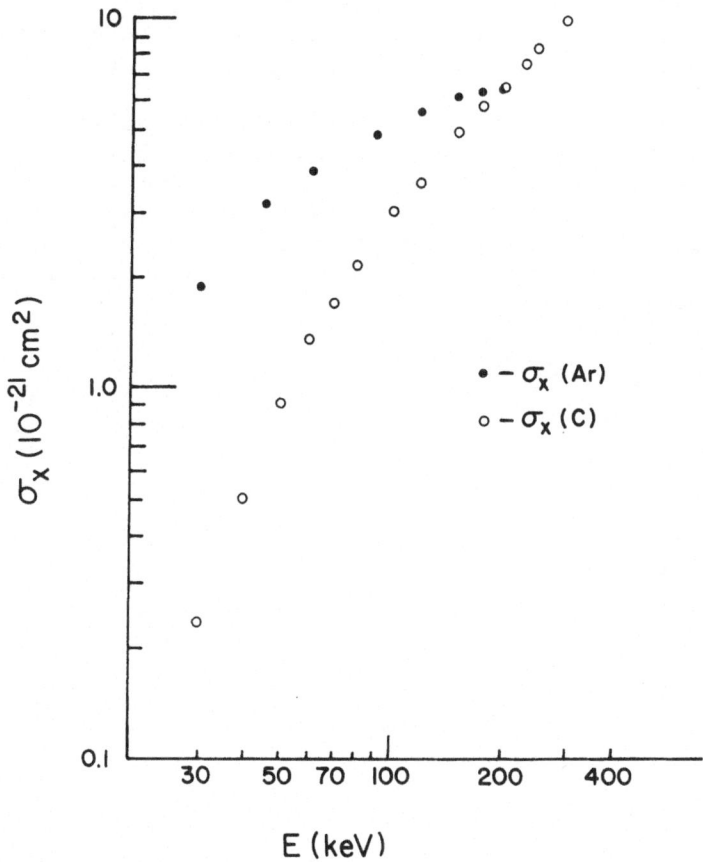

Fig. 2. X-ray production cross sections for Ar_L and C_K resulting from Ar $\rightarrow$ C collisions in solid (graphite) targets.

Multiple L-shell vacancies can be produced in the projectile due to multiple collisions in the solid within the lifetime of an inner shell hole. The lifetime of a ground state sulphur L-shell vacancy is 5×10^{-14} sec [5]. This can be compared with the mean free time for producing an L-shell vacancy, $1/N\sigma v$, where N is the number of carbon atoms per cm^3, σ is the L-shell vacancy production cross section (in cm^2) and v is the velocity of the projectile in cm/sec. The cross section can be estimated [6] to be $\sim 10^{-17}$ cm^2. This leads to a mean free time for producing an L-shell vacancy in 90 keV S^+ - C collisions of about 10^{-14} sec. Thus, each ion having an L-shell vacancy undergoes several collisions of this type during the lifetime of the vacancy. L-shell vacancies are not expected to be produced in the projectile when the 2p binding energy of the projectile exceeds the 1s binding energy of carbon. This is due to changes in the molecular orbital diagrams [7] which are normally used to described these collisions. The changes result in the inner

shell vacancies being produced in the carbon K shell rather than
the projectile. This results in a limitation in the number of
L-shell vacancies of 4, 2, and 1 for S, Cl, and Ar, respectively
[8].

    The general theoretical formulation of this problem is straight-
forward. Let f_i be the fractional probability of beam having i
inner shell vacancies and let q be the maximum number of vacancies
determined by the discussion above. Assuming thin targets and
neglecting double electron processes, the production and decay of
inner shell vacancies in any thickness dx of the solid is determined
by the following equations.

$$f_o + f_1 + \text{---} + f_q = 1 \tag{1}$$

$$\frac{df_o}{dx} = f_1/\nu\tau_1 - N\sigma_o f_o$$

$$\frac{df_1}{dx} = N\sigma_o f_o - f_1/\nu\tau_1 + f_2/\nu\tau_2 - N\sigma_1 f_1$$

$$\text{-----}$$

$$\frac{df_{q-1}}{dx} = N\sigma_{q-2} f_{q-2} - f_{q-1}/\nu\tau_{q-1} + f_q/\nu\tau_q - N\sigma_{q-1} f_{q-1}$$

$$\frac{df_q}{dx} = N\sigma_{q-1} f_{q-1} - f_q/\nu\tau_q$$

where N is the number of target atoms per cubic centimeter, ν is
the velocity of the projectile, τ_i is the lifetime of the projectile
having i inner shell vacancies and σ_i is the cross section for i
to $i + 1$ inner shell vacancies. For q = 1 the solution of Eq. (1)
yield

$$f_1 = \frac{N\sigma_o \nu\tau_1}{1 + N\sigma_o \nu\tau_1} (1 - \exp - [(N\sigma_o + \frac{1}{\nu\tau_1}) x]). \tag{2}$$

For q > 1 solutions of Eq. (1) can be obtained but, the more relevant
quantities, the equilibrium fractions can be determined simply by
setting the derivaties equal to zero. When this is done Eq. (1)
yields

$$f_i = \prod_{j=1}^{i} (N\sigma_{j-1}\nu\tau_j) / [1 + \sum_{k=1}^{q} \prod_{j=1}^{k} (N\sigma_{j-1}\nu\tau_j)] \tag{3}$$

or

$$f_i = N\sigma_{i-1} \nu\tau_i f_{i-1}$$

Equations (2) and (3) can be used to estimate equilibrium fractions
of projectile inner shell vacancies. Before this is done we must
determine the total ionization cross section and the inner shell
lifetimes for the appropriate states. Therefore, we must consider
the effects of M-shell vacancies using the observed x-ray spectra
in Fig. 1.

The L x-ray energy shifts observed can be understood from the
following: the projectile moving in the graphite target suffers
a small-impact-parameter collision which produces the L-shell vacancy, considerable M-shell excitation might also be expected.
However, as the projectile with the inner shell vacancy continues
to move in the solid, other larger-impact-parameter collisions will
change the state of the M-shell by a variety of processes (e.g.,
charge exchange, electron pick-up, etc.). Using a nominal cross
section of 10^{-16} cm^2 for these processes [1] the mean free time for
such events is 10^{-15} sec. Thus, on the average, 50 such events
will take place for the sulphur projectile before the inner shell
vacancy is filled. We conclude that the M-shell of the moving ion
will return to an equilibrium distribution prior to the filling of
the L-shell vacancy. Thus, the observed shifts in the projectile
x-ray spectra are characteristic of an equilibrium distribution of
M-shell vacancies. Note that the particular transitions observed
are only weakly affected by electrons in N or higher shells.

In Table 1 the results of the Hartree Fock (adiabatic) calculations [8] of the energy shift of the $3s \rightarrow 2p$ L x-ray transition
as a function of M-shell excitation are presented. Also included
in the table are theoretical values of the fluorescence yields
[9,10] for each configuration. The chlorine values for energy
shifts and fluorescence yield were obtained by extrapolation between the values for sulphur [8,10] and argon [9-11]. The direct
calculations of sulphur fluorescence yields were compared with
those obtained via statistical corrections [11,12] to ground state
fluorescence yields and only small differences were found (less
than 30%). Additional calculations were done to determine the
effects of multiple inner shell vacancies on x-ray transition energies for the case of sulphur. The results indicated that a positive
shift of ∿25 eV occurs for each additional 2p vacancy. Calculations
also indicated that for sulphur, the position of a $3d \rightarrow 2p$ transition is ∿25 eV higher than a $3s \rightarrow 2p$ transition with the same
initial configuration.

Using the above calculations the analysis of spectra in Fig. 1
is straightforward. In the S$^+$ - C data, for example, the peak in
the spectrum (at 168 eV) corresponds to a $3s \rightarrow 2p$ transition in a
sulphur atom with one 2p vacancy and only one 3s electron in the
M-shell. This peak is enhanced in the spectrum because of a large

Table 1

Theoretical data for multiply ionized S, Cl, and Ar for transitions from the configuration $1s^2\ 2s^2\ 2p^5\ 3s^n\ 3p^m$.

n	m	X-Ray Energy Shift (eV)[a]			Fluorescence Yield x 10^4		
		Sulphur	Chlorine	Argon	Sulphur	Chlorine	Argon
2	6	–	–	N[b]	–	–	1.9
2	5	–	N[b]	2.9	–	1.9	2.7
2	4	N[b]	2.9	7.0	1.9	2.8	4.1
2	3	3.0	7.1	11.8	2.9	4.7	6.9
2	2	7.2	12.0	17.4	5.5	8.8	14.0
2	1	12.1	17.5	23.7	15.6	25.5	40.0
2	0	17.7	24.0	30.8	77.4	105.0	151.
1	6	–	–	–	–	–	1.1
1	5	–	–	–	–	1.05	1.6
1	4	–	–	–	1.0	1.65	2.5
1	3	–	–	–	1.7	2.8	4.3
1	2	11.2	–	–	3.9	6.6	10.9
1	1	15.4	–	–	18.2	32.0	53.9
1	0	21.1	29.2	36.0	10^4	10^4	10^4

a) Hartree–Fock adiabatic calculations [8]. The energy shifts depend primarily on n + m. In that part of the table for n = 1, only significant deviations from this rule are noted.
b) Ground state value; this is the reference point for the shifts.

fluorescence yield (1.0). The other x rays between 147 eV and 168 eV represent the same transition but with differing numbers of M-shell electrons. The L x rays above 168 eV are due to multiple L-shell vacancies (and possibly some $3d \rightarrow 2p$ transitions). In fact, in the sulphur data, x rays corresponding to atoms with as many as 4 L-shell vacancies are observed. The photons at 277 eV are carbon K x rays; the mechanisms responsible for production of these x rays was discussed above. Similar observations can be made for the Cl, Ar – C data. In the chlorine and argon case, x rays due to a maximum of 2 and 1 L-shell vacancies, respectively, can be seen.

The relative probability of the projectile having a given number of M-shell vacancies can be calculated from the x-ray spectra using the data in Table 1. Some assumption must be made concerning the relative probabilities of 3s and 3p vacancies (our calculations indicate that Coster-Kronig transitions in the M-shell are energetically not allowed in an atom with an L-shell vacancy). In our analysis we assumed sequential stripping, but the assumption was tested as indicated below. The x-ray spectra are characteristic of thick targets and thus standard thick target yield analysis is required; however, this made little difference in the results.

In Figure 3 the results of our analysis are present. In the analysis of these spectra corrections for window absorption [13] and crystal reflectivity [14] were made. Assumptions other than sequential stripping (such as statistical distribution of the M-shell vacancies among the M-subshells) resulted in enhancement of the higher states of M-shell excitation. Thus, the sequential stripping assumption results in a <u>lower bound</u> determination of the mean number of M-shell vacancies, which is already larger than the mean emergent charge state (see below).

Using the results presented in Fig. 3, we can determine the mean value of fluorescence yield and the mean life time for the inner shell.

This, in conjunction with Fig. 2 and Eq. (2), can be used to determine equilibrium fraction of argon ions with an inner shell vacancy. The results of this analysis are presented in Table 2. We have included in the table the parameter $r = N\sigma_0 v\tau_1 = \lambda decay/\lambda production$ where the λ's are the appropriate mean free paths for the inner shell vacancies. In Table 3, the relative fraction of L shell vacancies for a 90-keV sulphur projectile determined from

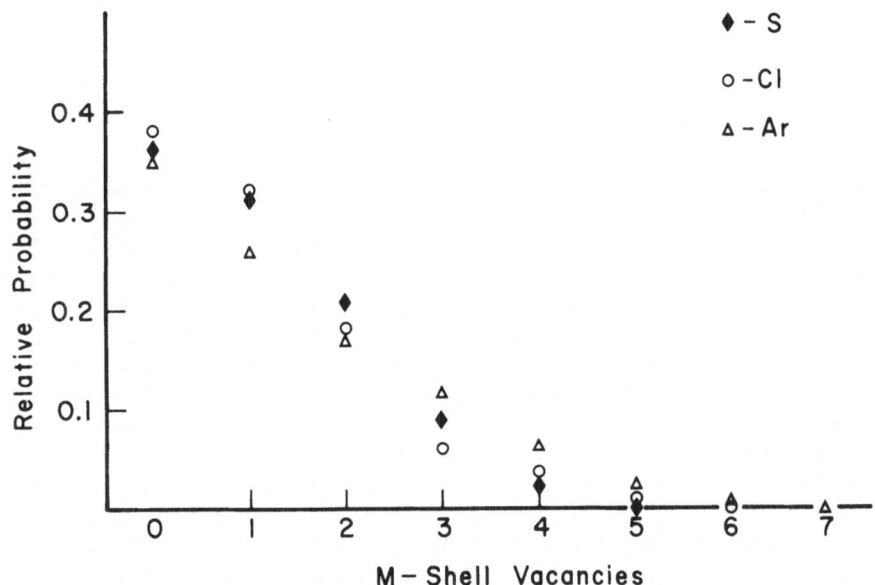

Fig. 3. Equilibrium M-shell vacancy distribution for 90 keV S (diamonds), Cl (open circles) and Ar (triangles) ions traversing a carbon solid target.

Table 2

Data on equilibrium distribution of L-shell vacancies in an argon
beam traversing a solid carbon target.

E (keV)	r	f_1(Ar)	$(N\sigma_o + 1/\nu\tau_1)^{-1}$
30	.20	.17	73 Å
45	.39	.28	75 Å
60	.58	.37	78 Å
90	.88	.47	80 Å
130	1.24	.55	81 Å
180	1.60	.62	82 Å

Table 3

Estimated data on equilibrium distribution of L-shell vacancies in
a 90-keV sulphur beam traversing a solid carbon target.

f_o	— .086
f_1	— .156
f_2	— .232
f_3	— .276
f_4	— .250

Eq. (3), using a cross section $\sigma_i = \frac{6-i}{6} \sigma_o$ where σ_o was estimated
to be 10^{-17} cm^2. We also assumed the lifetime for multiple L-shell
holes was the same as that for a single L-shell hole. As can be
seen from the table, large numbers of L-shell vacancies are expected.

Relevant measurements of the charge state populations of ions
emerging from carbon foils are available only for Ar [15]. In order
to compare our results for projectiles within the solid with these
results, some statement concerning the state of the projectile as
it leaves the solid must be made. Best agreement with our equilibrium distribution for L- and M-shell vacancies in the projectile was
found when we assumed that 1) the missing M-shell electrons are
carried with the ion in higher shells; 2) when the ion emerges from
the solid it loses (a) one electron for each inner shell vacancy
due to Auger processes, and (b) one electron for each electron pair
in shells higher than the M-shell due to autoionizing processes.
The resultant distribution labelled A is shown in Fig. 4 together
with the (interpolated) data of Hvelplund et al. [15].

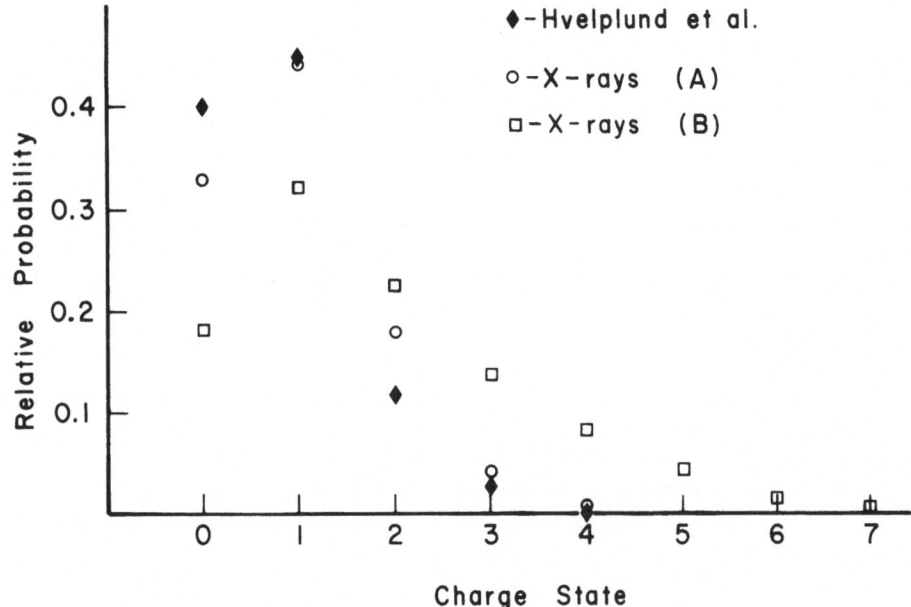

Fig. 4. Charge state distributions for 90 keV Ar projectiles
emerging from a carbon foil. Diamonds denote the measurements of
Hvelplund et al.; circles and squares are the results of the present analysis (see text).

These results indicate that analysis of x-ray spectra can be
used as a practical tool for investigating equilibrium excitation
states of ions traversing a solid. They also provide strong support for the model introduced by Brandt [16]. He proposes that
ions moving in solids with a velocity of less than two atomic units
are completely screened and thus leave the foil essentially neutral
but in a high state of excitation. This is to be contrasted with
the picture that the ion moves through the solid essentially with
its emergent mean charge state. In the present work, if the projectiles emerged without accompanying electrons and with the L-
and M-shell vacancy distributions as observed, the emergent mean
charge states would be the distribution labelled B in Fig. 4. This
predicts higher charge states than those actually observed. It
should be kept in mind that the electrons accompanying the ion in
shells higher than the M-shell will have negligible effects on the
x-ray energies or rates.

References

*Work performed under the auspices of the U. S. Atomic Energy
  Commission
†Supported in part by grants from the NASA, ONR, and NSF.

[1] H. D. Betz, Rev. of Mod. Phys. $\underline{44}$, 465 (1972).

[2] R. C. Der, R. J. Fortner, T. M. Kavanagh, J. D. Garcia, Phys.
     Rev. Letters $\underline{27}$, 1631 (1971).

[3] R. J. Fortner, R. C. Der, T. M. Kavanagh, and J. D. Garcia
     (to be published).

[4] R. J. Fortner, R. C. Der, and J. D. Garcia (to be published).

[5] D. L. Walters and C. P. Bhalla, Phys. Rev. $\underline{A4}$, 2164 (1971).

[6] F. W. Saris and D. Onderdelinden, Physica $\underline{49}$, 441 (1970).

[7] M. Barat and W. Lichten, Phys. Rev. $\underline{A6}$, 211 (1972).

[8] J. D. Garcia, R. J. Fortner, and J. Schofield (to be published).

[9] F. P. Larkins, J. Phys. $\underline{B4}$, L29 (1971).

[10] C. Bhalla, this conference.

[11] F. P. Larkins, J. Phys. $\underline{B4}$, 14 (1971)

[12] R. J. Fortner, R. C. Der, T. M. Kavanagh, and J. D. Garcia,
     J. Phys. B5, L73 (1972).

[13] M. A. Spivack, Rev. Sci. Instr. $\underline{41}$, 1614 (1970).

[14] B. L. Henke, Advances in X-ray Analysis $\underline{7}$, 460 (1963).

[15] P. Hvelplund, E. Laegsgard, J. O. Olsen, and E. H. Pedersen,
     Nuc. Instr. & Methods $\underline{90}$, 315 (1970).

[16] W. Brandt, this conference.

AUTHOR INDEX

Pages 1–478 are found in Volume 1, pages 481–934 in Volume 2

SUBJECT INDEX

Pages 1-478 are found in Volume 1, pages 481-934 in Volume 2